CEREBELLAR DEGENERATIONS: CLINICAL NEUROBIOLOGY

FOUNDATIONS OF NEUROLOGY

SERIES EDITORS:

Louis R. Caplan
Jack Antel
David Dawson

Porter R.J., Schoenberg B.S. (eds): *Controlled Clinical Trials in Neurological Disease* 1990.
ISBN: 0-7923-0613-9.
Plaitakis, A. (ed): Cerebellar Degenerations: Clinical Neurobiology. 1992.

CEREBELLAR DEGENERATIONS: CLINICAL NEUROBIOLOGY

Edited by
ANDREAS PLAITAKIS
Department of Neurology
Mount Sinai School of Medicine
of the City of New York

KLUWER ACADEMIC PUBLISHERS
BOSTON DORDRECHT LONDON

Distributors for North America:
Kluwer Academic Publishers
101 Philip Drive
Assinippi Park
Norwell, Massachusetts 02061 USA

Distributors for all other countries:
Kluwer Academic Publishers Group
Distribution Centre
Post Office Box 322
3300 AH Dordrecht, THE NETHERLANDS

Library of Congress Cataloging-in-Publication Data

Cerebellar degenerations : clinical neurobiology / edited by Andreas
  Plaitakis.
      p. cm. — (Foundations of neurobiology)
   Includes index.
   ISBN 0-7923-1490-5 (hardback)
     1. Cerebellum—Degeneration. 2. Excitatory aminoacids-
  -Pathophysiology. 3. Ataxia—Pathophysiology. 4. Ataxia
  telangiectasia—Pathophysiology. 5. Olivopontocerebellar atrophies-
  -Pathophysiology. I. Plaitakis, Andreas. II. Series.
     [DNLM: 1. Cerebellar Ataxia—etiology. 2. Cerebellar Ataxia-
  -physiopathology.   WI F099I v. 2 / WL 320 C4113]
  RC394.C47C47   1992
  616.8—dc20
  DNLM/DLC
  for Library of Congress   91-35348
                                                                 CIP

Copyright © 1992 by Kluwer Academic Publishers

All rights reserved. No part of this publication may be reproduced, stored in a retrieval system
or transmitted in any form or by any means, mechanical, photocopying, recording, or
otherwise, without the prior written permission of the publisher, Kluwer Academic Publishers,
101 Philip Drive, Assinippi Park, Norwell, Massachusetts 02061.

Printed in the United States of America.

DEDICATION

To my wife, Ingrid, whose love, understanding, and patience have contributed to a family environment supportive of creative work; my daughters, Ariadne and Iris, whose artistic creativity and belief in the power of knowledge and imagination has provided intellectual stimulation; and my parents, Joannis and Aristea, who first suggested the dream of medicine to me.

CONTENTS

Contributing authors	ix
Preface	xv

INTRODUCTION

1. The cerebellum and its disorders in the dawn of the molecular age   1
   A. PLAITAKIS

I. BASIC NEUROSCIENCES OF THE CEREBELLUM

2. Anatomy and neurochemical anatomy of the cerebellum   11
   J. HAMORI

3. Physiology of the cerebellum   59
   M. ITO

4. Amino acid transmitters in the adult and developing cerebellum   89
   R. BALÁZS

5. Glutamate receptors in mammalian cerebellum: Alterations in human ataxic disorders and cerebellar mutant mice   123
   E.D. KOUVELAS, A. MITSACOS, F. ANGELATOU, A. HATZIEFTHIMIOU, P. TSIOTOS, AND G. VOUKELATOU

6. Regional and cellular distribution of glutamate dehydrogenase and pyruvate dehydrogenase complex in brain: Implications for neurodegenerative disorders   139
   C. AOKI, T.A. MILNER, AND V.M. PICKEL

7. The Purkinje cell degeneration mutant: a model to study the consequences of neuronal degeneration 159
   B. GHETTI AND L. TRIARHOU

II. CLINICAL NEUROSCIENCES OF THE CEREBELLUM

8. Classification and epidemiology of cerebellar degenerations 185
   A. PLAITAKIS

9. The cerebellar cortex and the dentate nucleus in hereditary ataxia 205
   A.H. KOEPPEN AND D.I. TUROK

10. Clinical neurophysiology in olivopontocerebellar atrophy 237
    S. CHOKROVERTY

11. Pathophysiology of ataxia in humans 261
    H.C. DIENER AND D. DICHGANS

12. Oculomotor abnormalities in cerebellar degeneration 281
    M. FETTER AND J. DICHGANS

13. Clinical and radiologic features of cerebellar degeneration 305
    A. PLAITAKIS, S. KATOH, AND Y.P. HUANG

III. ETIOPATHOGENESIS OF CEREBELLAR DISORDERS

14. Glutamate dehydrogenase deficiency in cerebellar degenerations 369
    A. PLAITAKIS AND P. SHASHIDHARAN

15. Mitochondrial abnormalities in hereditary ataxias 391
    S. SORBI AND J.P. BLASS

16. Cerebellar disorder in the hexosaminidase deficiencies 403
    W.G. JOHNSON

17. Dominant olivopontocerebellar atrophy mapping to human chromosome 6p 425
    M.A. NANCE AND L.J. SCHUT

18. Positron emission tomography studies of cerebellar degeneration 443
    S. GILMAN

19. Ataxia telengiectasia: A human model of neuroimmune degeneration 461
    M. FIORILLI, M. CARBONARI, M. CHERCHI, AND C. GAETANO

20. Paraneoplastic cerebellar degeneration 475
    J.E. HAMMACK AND J.B. POSNER

Index 499

CONTRIBUTING AUTHORS

C. Aoki, Ph.D.
Laboratory of Neurobiology
Department of Neurology
Cornell University Medical College
New York, New York 10021

F. Angelatou, Ph.D.
Department of Physiology
University Patras, School of Medicine
Patras
GREECE

R. Balázs, Ph.D.
Netherlands Institute for Brain Research
1105 AZ Amsterdam
THE NETHERLANDS

J.P. Blass, M.D., Ph.D.
Departments of Neurology, Neuroscience and Medicine
Cornell University Medical College at
 Burke Medical Research Institute
White Plains, New York

M. Carbonari, M.D.
Department of Clinical Immunology
University of Rome "La Sapienza"
00185 Rome
ITALY

M. Cherchi, M.D.
Department of Clinical Immunology,
University of Rome "La Sapienza"
00185 Rome
ITALY

S. Chokroverty, M.D.
Department of Neurology
UMDNJ Robert Wood Johnson Medical School
New Brunswick, New Jersey 07939

J. Dichgans, M.D.
Department of Neurology
University of Tuebingen
D-7400 Tuebingen 1
GERMANY

H.-C. Diener, M.D.
Department of Neurology
University of Essen
Hufelandstr. 55
4300 Essen 1
GERMANY

M. Fetter, M.D.
Department of Neurology
University of Tuebingen
D-7400 Tuebingen 1
GERMANY

M. Fiorilli, M.D.
Department of Clinical Immunology
University of Rome "La Sapienza"
00185 Rome
ITALY

C. Gaetano, M.D.
Department of Clinical Immunology
University of Rome "La Sapienza"

00185 Rome
ITALY

S. Gilman, M.D.
Department of Neurology
University of Michigan
Ann Arbor, Michigan

B. Ghetti, M.D.
Department of Pathology
Indiana University School of Medicine
Indianapolis, Indiana 46202-5120

J.E. Hammack, M.D.
Department of Neurology
Mem Sloan-Kettering Cancer Center
New York, New York 10021

J. Hámori, Ph.D.
First Department of Anatomy
Semmelweiss University,
1094 Budapest
HUNGARY

A. Hatziefthimiou, M.D.
Department of Physiology
University of Patras, School of Medicine
Patras
GREECE

Y.P. Huang, M.D.
Department of Radiology
Mount Sinai School of Medicine
New York, New York 10029

M. Ito, Ph.D.
Frontier Research Program RIKEN
Wako, Saitama 351-01
JAPAN

W.G. Johnson, M.D.
Department of Neurology
Columbia University
College of Physicians and Surgeons
New York, New York 10032

A.H. Koeppen, M.D.
Department of Neurology
Albany V.A. Medical Center
Albany Medical College
Albany, New York 12208

S. Katoh, M.D.
Department of Radiology
Mount Sinai School of Medicine
New York, New York 10029

E.D. Kouvelas, M.D.
Department of Physiology
University of Patras, School of Medicine
Patras
GREECE

T.A. Millner, Ph.D.
Laboratory of Neurobiology
Department of Neurology
Cornell University Medical College
New York, New York 10021

A. Mitsakos, Ph.D.
Department of Physiology
University of Patras, School of Medicine
Patras
GREECE

M. Nance, M.D.
Department of Neurology
Minneapolis VA Medical Center
Minneapolis, Minnesota 55417

V.M. Pickel, Ph.D.
Laboratory of Neurobiology
Department of Neurology
Cornell University Medical College
New York, New York 10021

A.J. Plaitakis, M.D.
Department of Neurology
Mount Sinai School of Medicine
New York, New York 10029

J.B. Posner, M.D.
Department of Neurology
Memorial Sloan-Kettering Cancer Center
New York, New York 10021

P. Shashidharan, Ph.D.
Department of Neurology
Mount Sinai School of Medicine
New York, New York 10029

L.J. Schut, M.D.
Department of Neurology
Minneapolis V.A. Medical Center
Minneapolis, Minnesota 55417

S. Sorbi, M.D.
Departments of Neurology and Psychiatry
University of Florence
Florence
ITALY

L.C. Triarchou, M.D. Ph.D.
Department of Pathology
Indiana University School of Medicine
Indianapolis, Indiana 46202-5120

P. Tsiotos, M.D.
Department of Physiology
University of Patras, School of Medicine
Patras
GREECE

D.I. Turok, M.D.
Department of Neurology
Albany V.A. Medical Center
Albany Medical College
Albany, New York 12208

G. Voukelatou, Ph.D.
Department of Physiology
University of Patras, School of Medicine
Patras
GREECE

PREFACE

This book encompasses basic and clinical reports on the cerebellum and its primary atrophic disorders, the cerebellar degenerations. Rapid progress has been made in undestading the organization and function of the cerebellum at the neuronal, synaptic, and molecular level. Of particular importance has been the identification of the chemical transmitters utilized by the cerebellar cellular systems. More than any other brain region, the cerebellum utilizes amino acids as its main excitatory and inhibitory neurotransmitters. Excitatory amino acid transmitters, in addition to serving neuronal communication, may also mediate trophic and toxic effects, and as such, they may play a role in neurodegenerative processes.

The cerebellar degenerations were among the first human disorders with primary system atrophy to be studied clinically and pathologically. This field of clinical cerebellar sciences, no longer confined to the previously known descriptive level, is now advancing rapidly, propelled by rapid advances in neuroimaging, immunology, and molecular biology. The advent of CT, MRI, and PET has in recent years permitted the study of central nervous system alterations in living patients, thus contributing substantially to the accuracy of the diagnosis and the classification of these disorders. The nosology of cerebellar degenerations, which has been the subject of much debate for over a century, is presently a dynamic field, with new entities being recognized and old "classic ataxias" being redefined in the light of new genetic evidence.

Many of the cerebellar degenerations are genetically transmitted and, as such, they are the results of gene mutations. The genetic study of some cerebellar degenerations by means of DNA analysis has already advanced our understanding of these disorders. These approaches make possible the prenatal and presymptomatic diagnosis of these diseases, and they may ultimately lead to the identification of the mutant genes. The use of animal models for cerebellar degenerations has contributed substantially to our understanding of these afflictions, particularly with respect to the mechanisms involved in the neurodegenerative processes. Defects in mitochondrial and glutamatergic function and their involvement in premature nerve cell death are now being intensively investigated. Clues provided from the study of animal models have already led to the elucidation of the primary defects for some rare human degenerations, and additional progress is expected to take place in the near future.

One of the main goals of this book is to bridge the basic with the clinical sciences for understanding the cerebellum in health and disease. This volume will be useful to clinicians who care for patients with cerebellar disorders, as well as to the basic and the clinical investigators whose work concerns the cerebellum. Knowledge of advances occurring in basic sciences could help the clinical investigator to develop new approaches and therapeutic strategies for these presently intractable human disorders. Also, the basic investigator could be benefitted from the study of human disorders by identifying new research leads provided by these intractable, yet fascinating, afflictions.

Andreas Plaitakis

1. INTRODUCTION: THE CEREBELLUM AND ITS DISORDERS IN THE DAWN OF THE MOLECULAR AGE

ANDREAS PLAITAKIS

1. HISTORICAL BACKGROUND

The study of the cerebellum has a long and fascinating history, spanning many centuries. As Dow describes in his detailed historical review of cerebellar investigation [1], the Greek physician Herophilus (335–200 BC), known as the "father of anatomy," is generally credited for recognizing the human cerebellum as a distinct brain division. About two millienna later, Sir Thomas Willis (1621–1675) made comparative anatomical observations drawing attention to the characteristic morphologic appearance of the cerebellum in vertebrates. It seems that these observations stimulated interest in understanding the functional role of the cerebellum, which, until then, had remained obscure. It was, however, Luigi Rolando (1773–1831) who began a new era in cerebellar research by developing ablation experiments in an effort to understand the function of this brain area. Based on this work, he correctly suggested that the cerebellum is involved in motor control. Such ablation techniques, along with the subsequently developed stimulation methods, have served as basic experimental tools for understanding the cerebellar physiology for almost two centuries.

While this work was unfolding, clinical observations began to appear over a century ago, when it was realized that disorders exist in the human that cause a selective degeneration and atrophy of the cerebellum and profound

A. Plaitakis (ed.), CEREBELLAR DEGENERATIONS: CLINICAL NEUROBIOLOGY. Copyright © 1992 Kluwer Academic Publishers, Boston. All rights reserved.

disturbances in motor coordination or ataxia. These disorders have provided a fertile ground for studying clinico-pathologic correlations in the human and have also generated further interest in basic cerebellar research. Because many of these disorders are genetically transmitted, they became known as *hereditary cerebellar degenerations* or *ataxias*. Due to the variety of their particular characteristics, these afflictions have attracted a considerable interest, particularly from the nosological point of view. It is, therefore, not surprising that the bulk of the bibliography accumulated over the past 120 years deals primarily with the clinical and pathologic descriptions and the much debated classification problem.

At the basic science level, rapid progress was achieved a few decades ago toward understanding the anatomy and physiology of the cerebellum following the description of the cerebellar cellular systems by Raymond y Cajal. The relatively simple architectural design of the cerebellum proved particularly suitable for understanding physiological processes at the neuronal and synaptic level. More recently, the introduction of new tools and experimental procedures, such as animal models, tissue-culture techniques, brain slices and subcellular fractions, immunocytochemical methods, chemical microanalysis techniques, molecular biology approaches, and neuroimaging applications have furthered our understanding of the cerebellum. Progress has also been made in recent years in the clinical sciences of the cerebellum. This field, no longer confined to the previously known descriptive level, is now progressing quickly, propelled by rapid advances in molecular biology, immunology, and neuroimaging.

2. THE CENTRAL THEME IN THE MODERN ERA OF CLINICAL CEREBELLAR SCIENCES

The central issue faced by the modern clinical neurosciences is the elucidation of the mechanisms involved in the basic phenomenon of neuronal degeneration, which is common to all human disorders with system atrophy. Identification of the factors that cause brain cells to die prematurely is a necessary prerequisite for understanding these disorders and for developing rational therapies. The basic neurosciences of the cerebellum are essential for understanding the fundamental properties of cerebellar systems and for providing insights into the factors that make these systems selectively susceptible to degeneration.

In line with these considerations, the main goal of this book is to bridge the basic with the clinical sciences for a comprehensive understanding of the cerebellum in health and disease. Bringing together the clinical with the experimental sciences could be of particular value for both the basic and clinical investigator whose work concerns the cerebellum. Thus, knowledge of advances taking place in basic sciences could help the clinical investigator to better understand the cerebellar mechanisms in order to develop new approaches and therapeutic strategies for these presently intractable human diseases. Also, the study of human disorders can provide basic investigators with new leads that can help them to further and better target their research goals.

3. BASIC NEUROBIOLOGY

3.1. Anatomy and physiology

It cannot be overemphasized that understanding of the cerebellar mechanisms requires the knowledge of its structure, organization, and function. Accordingly, the first part of this book is devoted to the basic neurosciences. Hamori reviews the present state of knowledge on the anatomy of the cerebellum, including its complex longitudinal organization. Current concepts of neurochemical anatomy, particularly with respect to identification of the cerebellar transmitter systems, are also described in this chapter. Ito, after presenting a short historical review of the fascinating field of cerebellar research over the centuries, provides an up-to-date review of cerebellar physiology, including current views of cellular and signal processing. The role of the cerebellum in various body functions is discussed in this chapter, along with an overall scheme of cerebellar control.

3.2. Amino acids as chemical transmitters in the cerebellum: Toxic and trophic effects

More than any other brain area, the cerebellum utilizes amino acids as its main excitatory and inhibitory transmitters. These chemical messengers and their receptors, in addition to serving normal cellular communication, may also, under pathological conditions, become neurotoxic, mediating neurodegenerative processes. Balázs, after reviewing current knowledge about cerebellar development, discusses the growing body of evidence implicating certain amino acids as cerebellar transmitters. He also reviews present evidence suggesting possible trophic effects of amino acid receptors on cerebellar neurons, observations that may have implications for the pathogenesis of cerebellar disorders. The cellular distribution of the various excitatory amino acid receptors in the mammalian cerebellum and their specific alterations in human ataxic disorders and cerebellar mutant mice is another important subject related to neurotransmitter function of the amino acids, and this is discussed in Chapter 5 by Kouvelas and coworkers.

Amino acid function in the nervous tissue seems to be quite complex, and it may be the key to understanding neurodegenerative processes. The mechanisms controlling the synthesis, catabolism, transport, storage, and release of amino acids have not been as well understood as for other chemical transmitters. It is, therefore, not surprising to note that it has proven difficult to modify chemical transmission in the cerebellum, as a means of therapeutic intervention in cerebellar disorders, by pharmacologic manipulation.

The enzyme systems involved in the synthesis or breakdown of amino acid transmitters are known to play a wider role in cellular metabolism, and as such, they are expected to be present as housekeeping enzymes in all cellular systems. In contrast to these expectations, Aoki and her associates describe the results of recent immunocytochemical investigations that reveal that glutamate dehydrogenase (GDH) and pyruvate dehydrogenase, two important enzymes involved in amino acid and energy metabolism, show a marked regional variation in the mammalian brain, with the distribution of GDH correlating with glutamatergic pathways. As such, these findings seem to have implications in understanding mechanisms that protect nerve cells from the well-established neuroexcitotoxic potentials of glutamate, as well as the factors that determine the topographic distribution of lesions in certain forms of cerebellar degeneration.

The phenomenon of transynaptic degeneration, thought to reflect the loss of trophic effects that neurons may exert on each other at the synaptic level, is certainly in line with the trophic actions of excitatory amino acids described in Chapter 4 by Balázs. Getti and Triarchou have studied cerebellar mutant mice in an effort to understand the transynaptic effects of genetically programmed nerve-cell death in these animals. These observations may be directly relevant to the different patterns of neurodegeneration seen in the various cerebellar disorders in the human.

4. CLINICAL AND MORPHOLOGIC CHARACTERISTICS, AND NOSOLOGY OF CEREBELLAR DEGENERATIONS

In the clinical neurosciences part of this book, the old problem of classification and epidemiology of cerebellar degenerations is addressed in Chapter 8 by Plaitakis. Although these issues have been the subject of several excellent monographs, this is still a dynamic field, with new clinical entities being described each year. Also, classic ataxias, some of which have been known for over 120 years, are now being reexamined with new powerful tools, such as DNA analysis, and these approaches have provided a new ground for a reconsideration of these entities.

The neuropathologic changes occurring in the cerebellar cortex and dentate nucleus are thoroughly reviewed in Chapter 9 by Koeppen and Turok. The reader will find in this chapter that the use of new immunocytochemical techniques have increased our understanding of these disorders by revealing abnormalities, such as dendritic changes, that had not been previously appreciated.

As described in Chapter 3 by Ito, the cerebellum contributes to many aspects of body function and also exerts a substantial control over the autonomic system. Disorders of the cerebellum are known to affect these systems and are often associated with the involvement of the peripheral nerves, giving rise to electrophysiologic abnormalities. This important field and its diagnostic implications are reviewed in Chapter 10 by Chokroverty. The pathophysiologic mechanisms by which cerebellar disorders cause abnormalities of movement and posture as studied by innovative physiologic methods, are presented in Chapter 11 by Diener and Dichgans. In addition, the complex control that the cerebellum exerts on ocular movements and its disturbances in cerebellar disorders are thoroughly discussed in Chapter 12 by Fetter and Dichans.

The clinical, pathologic, and radiologic features of the major forms of cerebellar degeneration are comprehensively reviewed in Chapter 13 by Plaitakis, Katoh, and Huang. Also, the nosologic aspects of these disorders are considered in light of known genetic and biochemical markers. The advent of CT and MRI has contributed substantially to the accuracy of the diagnosis of cerebellar disorders. As described in this chapter, CT and MRI findings often correlate with the basic pathologic changes and the corresponding clinical deficits of cerebellar diseases and, as such, these techniques, by allowing the study of the living neuropathology of the cerebellum, have been proven useful in the classification of these disorders.

5. ETIOPATHOGENESIS OF CEREBELLAR DISORDERS

5.1. The "reverse genetics" approach

Since many of the cerebellar degenerations are genetically transmitted, they are expected to be the results of gene mutations, and these remain to be elucidated. In recent years the use of the so-called "reverse genetics" approach, direct examination of the genes by means of DNA analysis, has proved to be successful in identifying the mutant genes in disorders that lacked any known biochemical defect. The first step of this approach involves mapping the disease locus to specific chromosomal regions on the basis of linkage to specific DNA markers of known chromosomal localization. The availability of a large number of DNA markers spanning the entire genome has markedly facilitated this task in recent years.

Two major cerebellar degenerations, a dominantly inherited form of olivopontocerebellar atrophy (OPCA) and Friedreich's ataxia (FA), have already been mapped to human chromosomes 6 and 9, respectively. In fact, dominant OPCA was one of the first human disorders to be linked to the HLA locus on the sixth chromosome by Yakura and his associates back in 1974, when very few DNA markers were known. As described in Chapter 17 by Nance and Schut, the use of new DNA markers closely linked to the dominant OPCA locus has recently permitted the mapping of this locus to a narrow area of human chromosome 6. The same is also true for the FA gene on human chromosome 9, as described in Chapter 13. In addition to these two classic cerebellar degenerations, ataxia telengiectesia was recently mapped to human chromosome 11.

These advances have now set the stage for the second most difficult step in the "reverse genetics" approach, which involves the identification of the gene itself and of the mutation responsible for the disease. Characterization of the disease gene will, in turn, permit the identification of the protein encoded by this gene and elucidation of its role in cell biology in health and disease.

The recent development of new molecular biology methods, such as the PCR technique, is expected to facilitate these efforts substantially.

Although the genes for the above disorders have not as yet been identified, the linkage approach has already provided the means for the prenatal diagnosis of the above disorders. In addition, linkage analysis of a large number of FA families from around the world has provided evidence for a remarkable genetic homogeneity, suggesting that a single mutation may be responsible for this disorder. The situation may, however, be different for the dominantly inherited forms of cerebellar degeneration, which have been shown by linkage studies to be genetically heterogenous. The availability of large kindreds with disorders not linked to the sixth chromosome, such as the recently studied Cuban pedigree, should allow the identification of the chromosomal locus of these disorders in the near future.

If research efforts succeed in identifying the primary genetic defect for some of the inherited forms of cerebellar degeneration, how will this help in understanding the many sporadic forms of the disease? A similar question has also been posed for other primary neurodegenerations, such as Alzheimer's disease, amyotrophic lateral sclerosis, and Parkinson's disease. It is generally hoped that leads provided from the study of the genetic forms of these disorders could help to elucidate the etiopathogeneis of the more prevelant sporadic forms, but this remains to be seen.

6. ANIMAL MODELS AND UNDERSTANDING OF CEREBELLAR DEGENERATION

In recent years progress has been made toward understanding some genetic and aquired human neurodegenerations, not because of information obtained from the genetic analysis of their inherited forms, but from leads provided by animal models, particularly those produced with the use of trasmissible agents or specific neurotoxins. Thus the study of scrapie, an animal model for transmissible encephalopathy, has led to the elucidation of the genetic and acquired forms of prion diseases, including the Gerstmann-Straüssler syndrome, which often affects the cerebellum.

6.1. Glutamatergic and mitochondrial dysfunction

With respect to neurotoxin-induced neurodegeneration, two pyridine analogues, 3-acetylpyridine and 1-methyl-4-phenyl-1,2,3,6-tetrahydropyridine (MPTP), have been used extensively to selectively damage the inferior olives and substantia nigra, thus producing animal models for OPCA and Parkinson's disease, respectively. Over a decade ago, we followed a biochemical lead provided by the 3-acetyl-pyridine model, namely, inhibition of NADP-dependent oxidoreductases, to search for abnormalities of these enzymes in OPCA patients. As described in Chapter 14 by Plaitakis and Shashidharan, these investigations revealed that the activity of GDH, a predominantly mitochondrial enzyme, was reduced in some forms of this disease and other multisystem atrophies. Given the function of GDH in glutamatergic tarnsmission, these observations have linked a genetic molecular defect with glutamatergic neuroexcitotoxic mechanisms. More recently, investigators used a similar strategy based on the MPTP model to detect another mitochondrial defect in patients with Parkinson's disease.

Sorbi and Blass, in Chapter 15, present further evidence for mitochondrial abnormalities in the cerebellar degenerations by showing that the activities of several mitochondrial enzymes are reduced in such disorders. Abnormalities of pyruvate metabolism had been detected by Blass and others in ataxic encephalopathies over two decades ago. Moreover, recent studies have shown that mitochondrial cytopathies are associated with ataxic disorders, some of which present as late-onset multisystem atrophies. Gilman, in Chapter 18, describes the use of PET technology in the study of cerebellar degenerations. His studies revealed that glucose metabolism is differentially altered in patients with OPCA and Freidreich's ataxia, and these obervations may be relevant to the above evidence implicating mitochondrial abnormalities in these disorders.

Surely these observations raise the important question as to whether this mitochondrial dysfunction reflects a primary abnormality of mitochondrial proteins, as the GDH and the mitochondrial DNA deletions suggest, or results from other defects affecting energy metabolism secondarily. As described in Chapter 14 by Plaitakis and Shashidharan, sequencing of the GDH-specific genes in patients with reduced enzyme activity is expected to provide unequivocal answers to whether these GDH abnormalities are primary to the disease process. Also, identification of the gene mutation causing Friedreich's ataxia and chromosome 6-linked OPCA will certainly provide clues as to whether mitochondrial dysfunction is responsible for these afflictions.

7. LYSOSOMAL DISORDERS

Prior to the advent of recombinant DNA techniques, several disorders characterized by the accumulation of lysosomal material had been elucidated at the enzyme level using conventional biochemical approaches. As Johnson describes in Chapter 16, there are forms of these disorders that can affect the cerebellum predominantly, producing clinical syndromes similar to those of the classic cerebellar degenerations. The molecular basis of these disorders is now under intense investigation following the cloning and characterization of the genes encoding for these enzymes.

8. IMMUNOLOGICAL MECHANISMS

It has been known for years that cerebellar degeneration may develop in the course of cancer and, conversely, patients afflicted by a genetic form of cerebellar degeneration known as ataxia telengiectasia often develop systemic malignancies. Fiorilli and his associates describe in Chapter 19 that the primary genetic defect in ataxia telengiectasia causes altered immunologic function, in addition to inducing cerebellar degeneration. These interesting data suggest that a functional relationship exists between the nervous and the immune systems. On the other hand, Hammack and Posner describe in Chapter 20 that in patients with paraneoplastic cerebellar degeneration, specific autoantibodies reacting with cerebellar Purkinje cell antigens have been detected. These discoveries have not only advanced our understanding of the pathogenetic mechanisms involved, but they have also provided specific diagnostic tests for these disorders.

As the centuries-long cerebellar research enters the dawn of the molecular age, rapid advances are already taking place, while the prospects for major leaps forward are now greater than ever.

REFERENCES
1. Dow R.S., Moruzzi G. (1958). The Physiology and Pathology of the Cerebellum. Minneapolis, MN: University of Minnesota Press.

I. BASIC NEUROSCIENCES OF THE CEREBELLUM

2. ANATOMY AND NEUROCHEMICAL ANATOMY OF THE CEREBELLUM

JÓZSEF HÁMORI

1. ANATOMY OF CEREBELLUM

The cerebellum, or "small brain" is phylogenetically one of the most ancient parts of the brain. Morphologically, perhaps the most characteristic feature of the cerebellum from fishes up to humans is the stereotyped histology of its cortex, which during the last 450 million years (i.e., during the evolution of vertebrates) has changed very little. Gross morphological changes to the cerebellum in the course of its phylogeny have been more quantitative than qualitative, an evolutionary increase in cerebellar cortical volume being accompanied by a similar increase in the development of cerebellar nuclear mass. This has resulted in the emergence of relatively large cerebella in birds and, particularly, in mammals. Sherrington once called the cerebellum the "head ganglion of the proprioceptive system." Although its role in the coordination of motor functions in all vertebrates (though remarkable differences exist between, e.g., amphibia and mammals) is beyond any question, studies over the last 40 years make it clear that the mammalian cerebellum receives (and sends) fibers to several regions of the brain and thus is able to influence many regions and systems within the central nervous system. This holds particularly true for humans, as we know from available clinical observations and studies. The human cerebellum, though not essential to life, is the *sine qua non* for all human activities, from basic motor functions

A. Plaitakis (ed.), CEREBELLAR DEGENERATIONS: CLINICAL NEUROBIOLOGY. Copyright © 1992.
Kluwer Academic Publishers, Boston. All rights reserved.

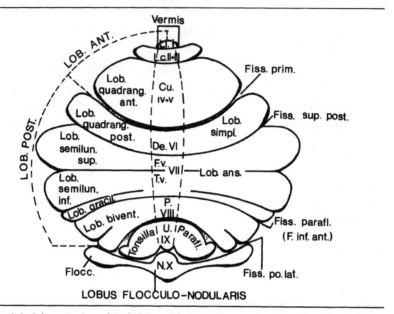

Figure 2-1. Schematic view of cerebellum, with the median vermis, the two large hemisphers, and the lobus flocculo-nodularis. Roman numbers represent Larrel's classification. In the left half of the diagram the classical names of the lobules are shown; in the right half more recent nomenclature is presented. Abbreviations: Li = lingula; Lc = lobulus centralis; Cu = culmen; De = declive; F.v., T.v. = folium and tuber vermis; P = pyramis; U = uvula; N = nodulus; Flocc = flocculus.

through more skilled motor activities, such as speech, which are coordinated with intellectual abilities. In the following chapter, specific morphological characteristics of the human cerebellum will be dealt with, mainly at the gross anatomical level, and will be compared with those of laboratory mammals. Our knowledge of the afferent and efferent connections of the cerebellum derives mostly from experiments on nonhuman species, particularly rodents, cats, and monkeys. However, neuropathological and clinical observation seem to encourage the application of most of these experimental data to human cerebellar function. The intrinsic organization, particularly the synaptic relationships of the cerebellar cortex and the deep nuclei, as studied by the classical light microscopic, as well as ultrastructural and immunocytochemical means, has been investigated primarily in experimental animals. The sporadic data on the structural organization of the human cerebellum [1–10] seem to show that not only the light microscopic, but also the ultrastructural organization [11,12] of the human cerebellar cortex, exhibits basically the same features as do nonhuman cerebella.

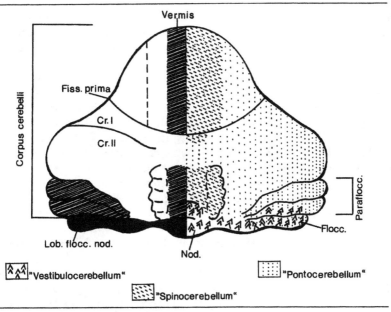

Figure 2-2. Schematic diagram of the cerebellum, showing the three main divisions (anterior and posterior lobe, separated by the deep "fissura prima") and the ancient flocculonodular lobe. Right half: The main regions are coded in accordance with their afferent inputs (spinocerebellum, vestibulocerebellum, and pontocerebellum). Left half: The archicerebellum (black), paleocerebellum (hatching), and neocerebellum (white) are shown. CrI-CrII = "crura" of the posterior lobe.

1.1. Gross anatomy

The cerebellum, located in the posterior fossa of the skull behind the pons and medulla, is separated from the overlying occipital lobes of the cerebral hemispheres by an extension of the dura mater, the tentorium cerebelli. Its base forms the roof of the rostral or metencephalic portion of the fourth ventricle. The most visible parts of the cerebellum are the median vermis and the two large hemispheres lateral to it. Transverse fissures of varying depth divide the cerebellum into lobes, lobules, and folia.

The primary fissure (Figures 2-1 and 2-2) separates the anterior and posterior lobes, which together comprise the corpus cerebelli. The posterolateral fissure separates the large posterior lobe from the small and phylogenetically ancient flocculonodular lobe (= flocculus and nodulus, Figures 2-1 and 2-2). In humans, and in the Cetacea, where the mammalian cerebellum reaches its greatest development in size, it is the middle part of the corpus cerebelli that markedly increases in size, including both the vermal portions and particularly the lateral hemispheric parts. In the Cetacea the most caudal parts of the posterior lobe (paramedian lobe, ventral paraflocculus) exhibit the greatest development [13], contributing to more than the half of the hemispheres. However, in humans these lobes are only of moderate size. The size increase of the human cerebellum is caused by the huge enlargement of the ansiform lobes plus the medial and simplex lobes particularly, although more anterior parts, including the lobus anterior, are also enlarged. Since these "enlarged" portions of the cerebellar hemispheres are phylogenetically the youngest, they are sometimes termed the *neocerebellum*, while other, more ancient portions (including the vermal part of the anterior lobe, the pyramis, uvula, and the paraflocculus) comprise the paleocerebellum. The flocculonodular lobe is phylogenetically the most ancient part of the cerebellum [14,15], hence its name, the *archicerebellum*. A more functional nomenclature utilizes the observations that the archicerebellum receives and sends back afferent and efferent projections from and to the vestibular nuclei: It is perhaps better designated as the *vestibulocerebellum*. The *paleocerebellum* receives afferents mostly from the spinal cord, hence its name *spinocerebellum*. Finally, the large *neocerebellum* is supplied by afferents mostly from the pontine nuclei, and therefore it is called the *pontocerebellum* (Figures 2-2 and 2-3).

The cerebellum is connected to the brain stem by three paired peduncles. The superior peduncle (brachium conjunctivum) proceeds from the upper medial white matter of the cerebellar hemisphere to enter the lateral wall of the fourth ventricle. Most of its fibers go deep into the tegmentum, and all its fibers decussate completely in the midbrain, at the level of the inferior colliculi. In humans, each contains 0.8 million fibers [16], which are in part efferent fibers from the cerebellar nuclei to the red nucleus and thalamus. In addition, the superior peduncle contains afferent fibers. Most are derived from the ventral spinocerebellar tract, but some also come from the red nucleus (rubrocerebellar fibers), and there are also noradrenaline-containing fibers from the locus coeruleus. The middle cerebellar peduncle (brachium pontis) is the largest of the three, and each contains about 20 million contralateral ponto-cerebellar fibers [17].

The inferior cerebellar peduncle (restiform body) ascends laterally from the lateral wall of the fourth ventricle and enters the cerebellum between the superior and middle cerebellar peduncles. The restiform body contains about 0.5 million fibers. Most of these belong to the olivocerebellar pathway, which takes its origin in the inferior olive. The restiform body is also the route of entry for the dorsal spinocerebellar tract, and some fibers, mostly reciprocal, nucleo-olivary axons, leave and enter the cerebellum through this peduncle.

1.2. Longitudinal zonal organization of the cerebellum

Macroscopically, the cerebellum seems to be subdivided by the transverse fissures and sulci, which define the classical lobes and lobules (Figures 2-1 and 2-3). Because of these numerous sulci, and also because of the more or

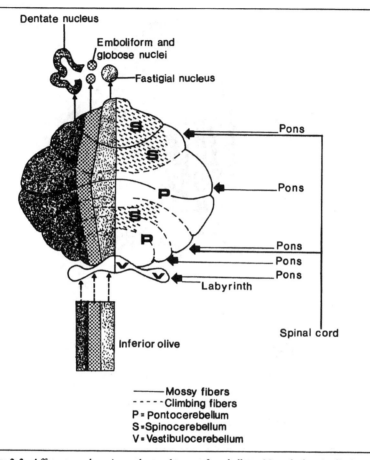

Figure 2-3. Afferents, and corticonuclear pathways of cerebellum. Note the longitudinal organization of olivocerebellar afferents. Corticonuclear connections (left half of the diagram) are also strictly arranged in longitudinal zones.

less transversally oriented folia,[1] the cerebellar cortex not only covers a very large surface area [4], but its longitudinal extent is seven times larger than its lateral extent. In fact, discounting the foldings and considering the cerebellar cortex as an unfolded sheet, its longitudinal dimension would be greater than 1 meter! Nonetheless, the basic structure of cerebellar cortex is uniform, whether derived from vermis or the hemisphere. The cerebellar cortex is basically a transversally subdivided population of lobes, lobules, and folia, all made up of nearly identically organized cell populations. This arrangement is responsible for the highly regular latticelike architecture of the cerebellar

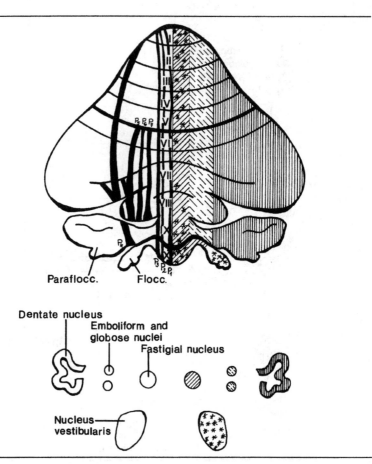

Figure 2-4. Longitudinal zonation of cerebellum based on corticonuclear projection (right half). On the left half are shown mab Q113-positive and -negative bands, representing groups of immunopositive or immunonegative Purkinje neurons (P1–P7) (Modified from Gravel et al. [147], with permission.). These are clearly organized in a longitudinal pattern.

cortex [4], which is clearly seen in the geometrically repetitive, transversely oriented, dendritic arbors of the Purkinje cells and their relationships with the long parallel axons of the granule cells. The axons of the intrinsic granule cells, termed *parallel fibers*, are longitudinally oriented within a folium and intersect at right angles the fanlike dendritic arbor of the Purkinje cells. Two major afferent pathways, the mossy fibers [18,19,20,205], the olivo-cerebellar fibers [21–25], as well as the cortico-nuclear axons of the Purkinje cells, are arranged parasagittally within the folium (Figure 2-4). Earlier studies of the development of the cerebellum [13,26] have clearly shown this longitudinal organization of the cerebellar cortex, as well as the parasagittal zonation of the Purkinje cells' efferent projections to the deep cerebellar nuclei and lateral vestibular nuclei [27-29].

Subsequently, a compartmentalization of afferent and efferent fibers was reported in the myeloarchitectural organization of the cerebellar white matter [18,21], producing marked longitudinal discontinuities in the connectivity of the cortex. In transverse, myelin-stained sections, longitudinal strips of relatively thick Purkinje axons are seen, separated from neighboring Purkinje axon bundles by narrow zones containing thin (presumably afferent) myelinated fibres. The strips containing the thick (Purkinje) fibers are called *compartments*. Each longitudinal compartment also contains all the afferent and efferent fibers associated with the Purkinje cells in that compartment. The entire complex of Purkinje cell compartment, its deep cerebellar nuclear target, and the portion of the inferior olivary nucleus that supplies it is termed a *cerebellar module*.

In most parts of the vermis and the intermediate portion of the cerebellar cortex, this myeloarchitectonic longitudinal organization of the cerebellum is almost strictly sagittal. However, in the more lateral parts, especially caudal to the primary fissure (due to the lateral bending of the central lobules), the zones are no longer positioned strictly parasagittally. However, the orientation of a compartment always remains perpendicular to the long axes of its folia. Voogd [18], using myeloarchitectonic criteria, divided the cerebellar cortex into seven longitudinal zones. A and B are contained within the vermis, C1, C2, and C3 belong mainly in the intermediate zone, while zones D1 and D2 are found laterally. These were later shown to correlate well with other experimental findings.

For example, an electrophysiological study [36] that mapped the termination areas of afferent systems within the cerebellar cortex also confirmed the existence of longitudinal zonation. However, this and more recent anatomical studies indicate that longitudinal organization might be more complex than shown by myeloarchitectonics alone. An afferent-fiber pathway, for instance, may not necessarily supply all of a particular zone, and—as is clearly shown by the somatotopic representation within some cerebellar areas (Figure 2-5)—it might jointly supply neighboring zone(s) as well.

Histochemical evidence also shows a heterogeneity of Purkinje cells within each zone, resulting in further subdivisions of Voogd's original seven zones. Such molecular correlates of parasagittal organization were presented by Scott [37,38], and later by Marani [39,40], who localized the enzyme 5'-nucleotidase within parasagittal bands in the molecular layer of the mouse and rat cerebellar cortices. (Specifically the enzyme was localized to parallel fibers and to Purkinje cell dendrites.) It was found that this enzyme was present or absent in alternating positive and negative longitudinal zones. In fact, each zone of Voogd's A-B-C and D classes could be further subdivided into 5'-nucleotidase-positive and -negative regions. Parasagittal heterogeneity of another enzyme, acetylcholinesterase (whose possible function in the cerebellum is still obscure), has also been demonstrated histochemically in

18 I. Basic neurosciences of the cerebellum

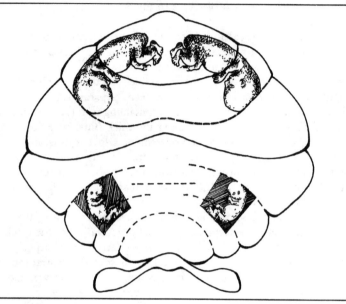

Figure 2-5. Simplified diagram of somatotopic localization in the anterior and posterior lobes of human cerebellum.

the cat cerebellar cortex [39,41,42]. Here the reaction product was found in the molecular layer and appeared as negative and positive parasagittal zones (again, one Voogd zone contains one negative and one positive AChE subzone). Although not identical to that of the cat, a similar AChE staining pattern has been demonstrated in monkey cerebellar cortex [43,44].

Since a similar zonal distribution of AChE staining was found in the inferior olivary complex [40,45], it was suggested that the AChE subdivisions in the cerebellar cortex could be related to olivocerebellar climbing fibers, arranged in longitudinal afferent zones. This assumption was confirmed by combined autoradiographic and AChE histochemical studies [46], which showed a direct correspondence between the ^{3}H-leucine-labeled climbing fibers and AChE-stained strips. This observation gained further support from the studies of Brown [47], who found that chronic lesions of the inferior olive led to an increase of AChE activity in the longitudinal bands.

Immunohistochemical studies have provided further proof of the heterogeneity of cerebellar neurons. Using polyclonal antibodies against cysteine sulfonic acid decarboxylase, Chan-Palay et al. [48] demonstrated that taurine-containing neurons in cerebellar cortex (Purkinje, stellate, basket, and Golgi neurons) are arranged in sagittal microbands, separated by nonimmunoreactive interzones. Ingram et al. [43], using monoclonal antibody B1, observed broad bands of immunopositive and immunonegative Purkinje cells in the monkey cerebellum. However, the B1 positive and negative bands did not seem to correspond to either the taurine-positive or AChE-stained bands of previous studies. Another monoclonal antibody [49], mab Q 113, which binds to a polypeptide antigen confined to a subset of rat cerebellar Purkinje cell bands, was observed to stain parasagittally organized Purkinje cell bands (Figure 2-4) both in the vermis and the hemispheres [50]. A congruence between olivocerebellar and mab Q 113 antigenic zonations has been also demonstrated, suggesting that climbing fiber projections and mab Q 113 positive or negative Purkinje cell phenotypes share a common compartmental organization [51].

The neurochemical heterogeneity of Purkinje cells may be inherent to the heterogeneous development of different Purkinje cell populations and it may be the determining factor in the development of cortical zonal organization [51]. Differences in the time of birth between populations of Purkinje cells with different zonal distributions were described in the rat [52] cerebellar cortex. In rat embryos, Wassef and Sotelo [53] found a differential distribution of Purkinje cells, as shown by the presence or absence of GMP-activated protein kinase activity. They observed that clusters of protein-kinase-positive Purkinje cells originated from a single region of the cerebellar anlage in the caudomedial mantle layer. It is possible that the interdigitation of such chemically different populations of Purkinje neurons is a determining factor in the development of cortical zonal organization. Later, it was, however, found that Zebrin I (mab Q 113) bands and the distribution pattern of spinocerebellar mossy fiber afferents may be determined by different mechanisms inherited by each system. Clearly, the exact mechanism by which differences in efferent (corticonuclear) and afferent connectivity of the zones are established is still obscure and needs further study.

The cortical mapping of responses to natural stimuli also shows a remarkable patchiness of representation. Though at first sight this patchiness may seem random, a detailed analysis of the representation of the trigeminal receptive area, for example, shows a fractured, but nevertheless very systematic and sophisticated, distribution over large areas of the cerebellar cortex. There is a clear correspondence between the zonal organization of perioral skin receptors in the trigeminal sensory nuclei [54] and their representation in the cerebellar cortex. This type of fractured somatotopy was observed and described much earlier by Adrian [55] and by Snider and Stonell [56] (Figure 2-5). It was found [57] that potentials arriving from the hindlimb can be detected in the ipsilateral anterior lobe and also (bilaterally) in the caudalmost part of the paramedian lobule. Similarly, the forelimb is represented both in the caudal part of the anterior lobe and in the paramedian lobe, while the facial area was found to be continuous with the forelimb areas (Figure 2-5). Following tactile stimulation and voluntary movements, functional maps of the human cerebellum have been constructed [58] using positron emission tomography. The observed gross topography was in

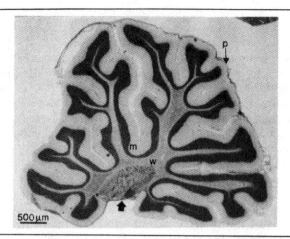

Figure 2-6. Sagittal section of rat cerebellum ("arbor vitae"), including the medial nucleus (arrows). m = molecular layer, which, like the darkly stained granular layer, is continuous throughout the sagittally cut cerebellum; w = white matter; P = pia mater. (Azur II methylene blue staining.)

accordance with previous studies on experimental mammals. More recently, physiological investigations (for a summary, see Welker's excellent review [59]) indicate that the spatial relations of peripheral, cerebral, and tectal projections to the cerebellar cortex are highly organized spacially. It was shown that mossy fiber circuits within the cerebellum are organized in very precise detail, resulting in patchy mosaics of somatosensory projections [60,61]. Somatotopic organization of olivocerebellar climbing fibers to the cerebellar cortex was also demonstrated by Rosina and Provini [62]. It has become clear that focal, mosaiclike representation of many body parts that are peripherally distinct are brought adjacent to each other in cerebellar cortex. The resulting multiple representation of certain body parts clearly increases the choices of cerebellar output, the fine tuning of which will be shaped by the cerebellar cortex, which contains these highly differentiated representational mosaics.

2. MICROSCOPIC ANATOMY OF THE CEREBELLUM
The internal structure of the cerebellum is characterized by a layer of cortex and an internal mass of white matter, within which are located the cerebellar nuclei.

2.1. Cerebellar cortex
The cerebellar cortex is an intricately folded trilaminar sheet of gray matter covering a thin, white lamina in each folium. The folia, in turn, fan out from

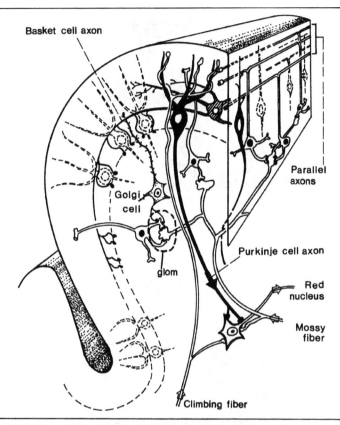

Figure 2-7. Schematic drawing of cerebellar cortex and its main synaptic connections, including the two main afferent pathways (climbing and mossy fibers), the only efferent pathway (Purkinje axons), and the most important intracerebellar neuronal and synaptic types.

the central core of white matter. In midsagittal section, the appearance is that of a leaf of *Thuja occidentalis* and is responsible for the picturesque name given to this view of the cerebellum by the early anatomists, the *arbor vitae* (Figure 2-6). In spite of some regional differences [63], the cerebellar cortex exhibits a nearly identical structure throughout [64–67]. It has three distinct layers [1]: the granular layer, which is superficial to the white matter [2], the surface molecular layer, and [3] sandwiched between the granular layer and molecular layer, the monolayer of Purkinje cell perikarya (Figures 2-6 to 2-9).

2.1.1. The granular layer

The granular layer, in which the mossy fibers, one of the main afferent systems to the cerebellum, terminate, appears as an enormous quantity of

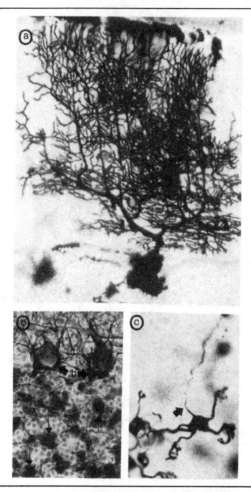

Figure 2-8. A,C: Golgi staining of Purkinje cell (A) and granule cells (C). Note clawlike terminal arborization of granule cell dendrites, and the origin of granule cell axon (arrow). **B:** Silver impregnation demonstrates pericellular baskets (b) around two Purkinje cells and several darkly stained glomeruli (arrow) among tightly packed, moderately stained granule cells.

small, tightly packed granule cells. These cells, round or slightly oval in shape in Nissl stain, range in diameter from 5 to 8 µm and are among the smallest in the brain. An average 133-g human cerebellum contains approximately 41 billion (4.13 × 10) granule cells [68], i.e., half or more of all the brain's nerve cells! The perikaryon of the granule cells consists of a narrow rim of cytoplasm around the nucleus (Figure 2-10). Most have four relatively short dendrites (Figure 2-8) that terminate in small, sometimes longish "claw-like" digits. The thin axon is usually unmyelinated, although a few in the deepest part of the molecular layer may become myelinated. Granule cell axons ascend into the molecular layer (Figures 2-7 and 2-8), where they divide to form a T, after which they are termed *parallel fibers* (see below).

In addition to the small granule cells, larger cell bodies (Figures 2-7 and 2-9), belonging to the so-called Golgi neurons (and to glial cells), are present in far fewer number than are the granule cells. These can be found lying between clusters of granule cells. Golgi neurons comprise two classes [67]: small Golgi cells, 6–11 µm in diameter, which have their dendritic arbors contained within the granular layer, and the larger 9–16 µm in diameter neurons, which send dendritic branches up into the molecular layer. The fine and rich axonal arborization that characterizes both types remains within the granular layer, where delicate, varicose axonal branches preferentially form a network at the periphery of the cerebellar glomeruli [69–70]. Altogether, their number is comparable with Purkinje cells (1 Golgi neuron per 1.5 Purkinje cells in humans) [10].

The cerebellar glomeruli [71] or cerebellar islands (Figure 2-8b) are cell-free areas bounded by clustering granule cells (Figures 2-7 and 2-9), which are partly encapsulated by a thin glial sheath. Their structural organization has been described in several species [72–80] and would appear to be much the same in all species. Within a glomerulus are found (1) granule cell dendrites with their clawlike terminals, (2) terminals of mossy fibers, and (3) Golgi axons, all forming a complex synaptic arrangement (Figures 2-7, 2-9, and 2-10). The usually centrally located mossy terminal contains numerous mitochondria and large spheroidal synaptic vesicles. It forms asymmetric synaptic contacts with the clawlike digits of granule cells. The small varicose axons of the Golgi neurons have ovoid of flat synaptic vesicles and form mostly symmetric synaptic junctions with either the clawlike terminal digits, or with the proximal dendrites of granule cells in the peripheral part of the glomerulus. The number of granule cell dendrites in one glomerulus has been estimated to be from 15 to 20 [65] to 112 [81]. A recent three-dimensional electron microscopic analysis of cerebellar glomeruli in the rat has revealed, that a glomerulus of the simplest type[2] receives 50–54 distinct dendrites, all from different granule cells [79]. Through its three to five terminal digits, each granule cell dendrite might receive three to five synaptic contacts from the excitatory [83] mossy terminal, from the inhibitory Golgi axon [84], or from both. In fact, each granule cell dendrite receives an average of three synaptic contacts from each mossy fiber, and about 60% of all granule cell dendrites receive multiple synaptic contacts from Golgi cell axons. This means there is a very high convergence onto granule cells within a glomerulus (53:1). There is also a rich inhibitory input to more than half of the participating granule nerve cells. The mossy fiber-granule cell relation-

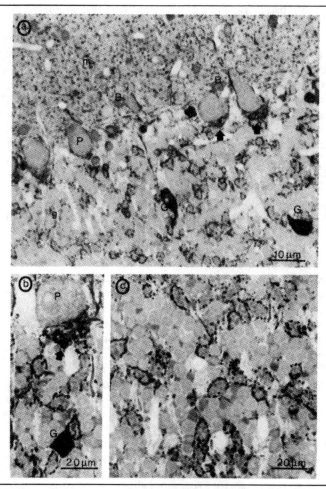

Figure 2-9. A: granular layer (G), Purkinje cells bodies (P), and molecular layer (m) from rat cerebellar cortex immunostained for GABA. G = GABA-positive Golgi neurons. B = moderately GABA stained Basket cells. Arrows show immunopositive pericellular baskets. **B**: Higher magnification view showing Purkinje cell (P), GABA immunostained pericellular "basket" arrow). A GABA-positive Golgi neuron (G) is seen in the lower parts of the figure. Ringlike GABA-positive structures in the granular layer are shown at higher magnification. **C:** Axonal varicosities of Golgi cells, localized at the periphery of the otherwise immunonegative glomeruli. Tightly packed granule cells surrounding the ringlike glomeruli are unstained. (scale = 1 μm.)

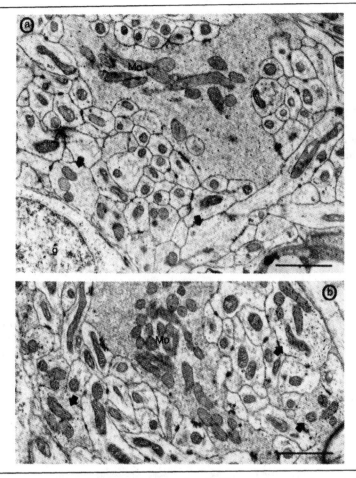

Figure 2-10. Part of cerebellar synaptic glomerulus. **A:** Electron microscopic demonstration of glutamate by immunogold reaction shows strong strong positivity (gold particles = black speckles) within the large mossy terminal (Mo), with moderate activity in granule-cell soma (g) and granule-cell dendrites. Peripheral, Golgi cell axon varicosities (arrows) are immunonegative. **B:** Immunogold GABA stain is localized in small Golgi axons (arrows), while mossy terminals (Mo) and surrounding small granule-cell dendrites are immunonegative.

ship allows for considerable divergence, if one takes into account that a single mossy fiber can send side branches to two neighboring folia [64,85] and supplies at least 44 different glomeruli. This means that a single mossy fiber can excite within neighboring folia more than 2200 granule cells, and this estimate does not even consider other branchings of the same mossy fiber in the course of its ascent through the white matter. This could multiply quite considerably the estimate given above.

About one tenth (or less) of cerebellar glomeruli receive a second type of dendrite [69,86]. These are much larger than the dendrites of granule cells and are derived from both types of Golgi neurons. They are postsynaptic to both mossy fibers and Golgi axons.

The main excitatory input to Golgi cells, as with the granule cells, is provided by mossy fibers [87]. In the case of the small Golgi cells, mossy fibers, terminating either upon the dendrites [88] or the perikarya [67], appear to be the sole excitatory input. The large Golgi neurons, via their ascending dendrites [85] to the molecular layer, also receive a major excitatory input from the parallel fibers. It has been shown experimentally [88] that both types of Golgi cells receive synaptic input from the ascending axons of granule cells. This input can be axodendritic or axosomatic, and can even be to the axon hillocks.

The somata and proximal dendrites of Golgi cells also receive boutons of recurrent Purkinje axon collaterals. These are, in fact, the main inhibitory input to these neurons [67].

2.1.2. Neurotransmitters in the granular layer

Biochemical [89,90] as well immunocytochemical evidence [91,92], unequivocally suggests that the transmitter liberated from granule cells, i.e., from the parallel fibers, is glutamic acid. Accordingly, the cell bodies of granule cells are moderately immunopositive when treated with antiglutamate (Figure 2-10A).

Until recently, the identity of the transmitter in the mossy fiber had not been clearly established. Acetylcholine was considered a possibility, at least in those areas where mossy fibers were observed to stain heavily for acetylcholinesterase (particularly in the archicerebellum, nodulus, and uvula [93–96]. However, most biochemical, physiological [97], and immunocytochemical data failed to detect cholinergic mechanisms, or cholinacetyltransferase activity bound to mossy fibers, in *any* area of the cerebellar cortex.

In contrast to these negative findings, Ojima et al. [98], using monoclonal antibody against ChAT, described cholinacetyltransferase positive mossy fibers predominantly in the archicerebellar cortex. In addition, they also found ChAT-positive fibers in the molecular layers, as well as in some cerebellar nuclei. Obviously, more controlled study is needed to clarify the problem of cholinergic afferentation to the cerebellum.

Substance P [99] and somatostatin [100] have also been mentioned as possible transmitters in some mossy fibers. More recently, by utilizing direct immunocytochemical demonstration of glutamic acid in the cerebellum, it was shown [91] that mossy terminals in the rat cerebellum are rich in this excitatory amino acid. Using a very specific glutamate antiserum [92,101], we were able to demonstrate,[3] in addition to a moderate positivity in granule

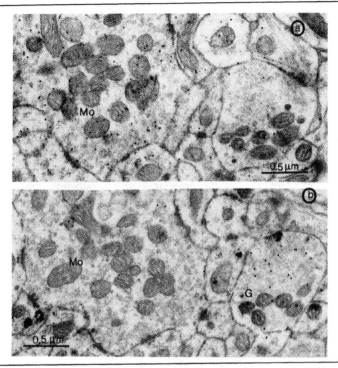

Figure 2-11. Immunocytochemical demonstration of glutamate (**A**) and GABA (**B**) with gold particles as the reaction product in two EM ultrathin sections (five sections apart), utilizing immunogold reaction. **A**: Glutamate is localized in the mossy terminal (Mo) but not in the Golgi axon terminal (G). **B**: In contrast, GABA is found in the Golgi axon but not in the mossy terminal.

cell bodies plus a few granule cell dendrites, a selective and very specific staining of almost all mossy terminals (Figures 2-10A and 2-11A). Recently, a small minority of mossy endings, though containing large, spheroidal synaptic vesicles, were shown to be rich not in glutamate, but in GABA [102], the main inhibitory transmitter in the cerebellum. These rare GABAergic mossy terminals were shown to be the endings of nucleocortical fibers. Although all nucleocortical fibers terminate in the cerebellar cortex as mossy endings [103–105], and Chan-Palay [106] has previously provided indirect evidence that some of the nucleocortical fibers may be GABAergic, arising probably from GABA-positive neurons in the nuclei [107,108], only by the use of the immunogold method for GABA [109,110] has it been possible to obtain direct evidence for the existence of this inhibitory feedback system. At present, the best supposition is that all mossy terminals of extracerebellar origin (and possibly many nucleocortical fibers as well) are excitatory [92], while at least some of the nucleocortical mossy endings are, in contrast to the glutamatergic, excitatory mossy terminals, GABAergic and inhibitory [111].

GABA also appears to be the transmitter in Golgi neurons (Figures 2-9A and 2-9B) [78,112–118]. The somata of small and large Golgi neurons, as well as their dendrites, stain heavily for GABA, as do the axonal varicosities assembled in a "ringlike" fashion [119] at the periphery of the otherwise unstained glomeruli (Figure 2-9C). Using adjacent sections, incubated either for GABA (Figure 2-11B) or for glutamate (Figure 2-11A) antibodies, the distribution of these transmitters in the two types of axon in typical cerebellar glomerulus is clearly seen. GABA is found in the small axonal varicosities at the periphery of the glomerulus. These GABA-positive varicosities contain small, ovoid synaptic vesicles, whereas the glutamate-positive mossy terminals exhibit large, spherical synaptic vesicles.

Concerning GABA receptors, a very intense staining of the granular layer was observed with the aid of monoclonal antibodies against GABA receptors [120]. In human cerebellar cortex, however, the benzodiazepine receptor concentration in the granular layer was found to be only moderately visible using (^{3}H) muscinol autoradiography (muscinol is a well-known GABA-receptor agonist). Nonetheless, a high density of GABA receptors was observed in the human granular layer [121].

To make a simple story even more complicated, it was recently observed [122,123] that in rat cerebellum a subpopulation of Golgi neurons was selectively labeled by glycine antibody. While some Golgi neurons exhibited only glycine positivity, others were GABA positive, and in a few the two inhibitory transmitters coexisted! These results confirmed earlier observations [124] reporting specific uptake by Golgi cells and their processes of (^{3}H) glycine, and agree with the beautiful immunocytochemical demonstration of glycine receptors on granule cell dendrites [125]. It also appears that some Golgi axonal varicosities, which are GABA immunonegative, might be glycinergic and that, as mentioned above, in some neurons the two inhibitory transmitters do indeed coexist. In fact, immunocytochemical staining of cerebellar glomeruli for both GABA and glycine offer a nearly identical picture. Since both substances, glycine as well as GABA, are inhibitory, the morphological framework for inhibitory activity within the glomerulus does not change with the introduction of glycine as a transmitter in some Golgi axon terminals. The physiological mechanism, though, may be different for the two inhibitory transmitters; the discussion of this problem, however, is beyond the scope of the present chapter.

Since Purkinje cells are inhibitory [126–128] and also use GABA as a transmitter [119,129,130], the rich collateral branchings and their intracortical endings onto Golgi cells are also rich in GABA [78]. Considering the longitudinal, zonal organization of climbing afferents, as well as the similarly, longitudinally organized Purkinje axon collaterals from the same zone, it is plausible that disinhibition of glomeruli with the aid of Purkinje axon collaterals to Golgi cell synapses might effectively contribute to the functional "zonation" of the mossy fiber, granule cell system.

Immunocytochemical studies [131–133] have described serotonin-containing fibers in the granular layer. These are morphologically different from mossy or Golgi axons, and are always found in an extraglomerular position. Indeed, it was shown that the granule cells' spontaneous activity is altered by microiontophoretic injections of serotonin to the cerebellar cortex [134]. However, this effect is not related to mossy fiber activity. The corresponding serotonin receptors are located on granule cells, the dendrites of which are located well outside the glomeruli.

2.1.3. Purkinje cells and the molecular layer

Purkinje cells are arguably the most remarkable nerve cell type in the whole central nervous system. The large, ovoid cell bodies, (16–30 μm along minor axes, 21–40 μm along major axes [66,135]) constitute a cellular monolayer, separating the granular and the outermost molecular layers. The Purkinje cell's main dendritic process (Figure 2-8A) ascends through the molecular layer towards the surface of the cortex, while its axon leaves the perikaryon at the opposite (inferior) pole of the cell and, after crossing the granular layer where it becomes myelinated, enters the white matter.

2.1.3.1. THE PURKINJE CELL AXON. The initial segment (IS), i.e., the unmyelinated portion of the axon (see Figure 2-15) is about 17 μm long in the cat and rat [136]. The rather thick myelinated axon is directed either to the cerebellar or vestibular nuclei, where they represent the main (inhibitory) [126,127] synaptic input. Purkinje cell axons also give off recurrent collaterals [85] to the cortex as they cross the granular layer and as they course through the white matter. These collaterals, after some branching, form a myelinated plexus beneath the Purkinje cells (infraganglionic plexus [137]), where they synapse mainly on Golgi nerve cells [67,138], and presumably, on Purkinje cell somata [139]. However, a study that employed the Purkinje-cell specific immunocytochemical marker, guanosine 3',5'-phosphate-guanosine dependent protein kinase [140], failed to verify direct axosomatic contact between recurrent collaterals and Purkinje cells. The recurrent Purkinje axon collaterals that ascend to the molecular layer form the supraganglionic plexus. This is well developed in the cat [141], where the collaterals terminate on basket neurons [142,143] and on Purkinje cell dendrites [144,145]. It is characteristic that the recurrent collaterals, and the collateral plexuses, are organized strictly in the parasagittal plane [67,139,141,146,147], as defined by the topography of the afferent axon terminals [19,21,23,27,148–151]. With the first demonstration that Purkinje cells were inhibitory [126,127], it was also proven that the inhibitory transmitter in Purkinje cells' recurrent collateral endings was GABA. This was shown both by GAD immunoreactivity [112,115,118,119], as well as by direct GABA immunoreactivity [78,114,

116,117] in the Purkinje cell, and particularly in its axon. Since recurrent collaterals are organized in the parasagittal plane and terminate around Golgi cells or basket neurons, their inhibitory action could result in disinhibition of the mossy fiber-parallel fiber chain, or—through inhibition of basket cells— the Purkinje cells themselves, all within the same parasagittal band. However, this is an assumption that needs further physiological verification.

Recently taurine (2-aminoethanesulfonic acid), a free amino acid, was also shown to be localized specifically in Purkinje cells [106,152,153], probably coexisting with GABA. Although it is not certain that taurine has a transmitter role in Purkinje cells, its presence can be used as a morphological identification of the axonal arborization of Purkinje cells. Similarly, the selective presence of the 22-amino-acid polypeptide, motilin, in Purkinje cells and their processes [154] appears to be a good marker for morphological identification of these neuronal elements, either in the cerebellar cortex or in the deep nuclei. Since motilin appears to depress the activity of Deiters neurons [155], this peptide might also contribute to the inhibitory actions of the GABAergic Purkinje cells.

More specific markers for Purkinje cells and their processes are c-GMP-dependent protein kinase [140], and the neuropeptide, cerebellin family [156–158], which selectively stain different regions of Purkinje neurons. These markers are particularly useful in the study of Purkinje cell connectivity, both within the cortex and within the cerebellar nuclei [159].

2.1.3.2. MOLECULAR LAYER. The elegant and beautiful geometry of the cerebellar cortex is due mainly to the two main constituents of this layer, the fanlike dendritic arborizations of the Purkinje neurons, organized strictly in the sagittal or parasagittal plane, and the hundreds of thousands of parallel fibers, which, in contrast, run parallel with the long axis of the folia (Figure 2-7). Two kinds of inhibitory interneurons (basket and stellate cells), plus the afferent excitatory climbing fibers, are also major components of the layer.

The dendritic arbor of the Purkinje cell is two dimensional and is oriented perpendicular to the major axis of the folium. The thickness of this "two-dimensional" dendritic tree is about $8.9\,\mu m$ [65], although, since the dendritic plates of the two principal dendrites of one Purkinje cell may become intermingled in position along the longitudinal axis of the folia, the thickness of the dendritic plate of a single Purkinje cell can reach $18\,\mu m$ or more. In any case, the total plane area encompassed by one single dendritic field was found to be more than $26,000\,\mu m$ in the rat [160]. The dendritic arbor consists of primary and secondary dendrites, which bear relatively few dendritic spines, plus the spiny tertiary branchlets (Figures 2-8A and 2-13).

The main dendrites (particularly the secondary ones) receive at least three types of axonal endings. Two of these contain ovoid synaptic vesicles and are the axon terminals of stellate and basket neurons. The third is the climbing fibers, which contain densely packed spherical synaptic vesicles, a few dense-core vesicles, and bundles of microtubules. These latter fibers, which take their origin in the inferior olive [31,160,161], lose their myelin sheath at the level of Purkinje cells. Their unmyelinated, terminal portion joins the cell body and emerging dendrites of the Purkinje cell and establishes long contacts, particularly with primary and secondary dendrites. The primary branches of climbing fibers give off fine secondary tendrils. Each tendril exhibits several bulbous or beaded enlargements, which contact one to six short spines protruding from the main Purkinje dendrites, (Figure 2-12), particularly at the dendritic branch points. The stubby spines of the secondary dendrites, which are much shorter than the spines of tertiary dendrites (Figures 2-12 and 2-13), are the main postsynaptic receptor sites for the climbing fibers' beaded tendrils. These synaptic contacts are confined to the inner two thirds of the molecular layer and are usually not found in contact with the more superficial, subpial dendrites. Even so, the number of synaptic contacts between one climbing fiber and its target Purkinje cell was estimated in the frog to be 300 [163], and even more in the rat and cat [Hamori, unpublished observation]. The relationship between afferent climbing terminals and Purkinje neurons is typically 1:1, although in immature cerebellar cortex two or more climbing fibers terminate temporarily on one Purkinje cell [164]. This multiple innervation is soon followed by regression of the "supernumerary" fibers, resulting in the adult pattern of mono-innervation [see also 85]. Thus, the powerful excitatory action of climbing fiber impulses on Purkinje cells [165] has, as its morphological parallel, the 1:1 structural arrangement of the climbing fiber-Purkinje cell synaptic complex.

As for the possible excitatory transmitter, the likeliest candidate appears to be aspartate. In patients with an inherited form of olivo-pontocerebellar atrophy [166], a reduction of aspartate in the cerebellar cortex has been found. Similarly, experimental destruction of olivocerebellar fibers by 3-acetylpyridine treatment reduced the level of aspartate in rat cerebellum by 15–26% [167]. Another piece of indirect evidence is the finding by Wiklund et al. [168,169] that olivocerebellar climbing fibers are selectively and retrogradely labeled after the injection of ^{3}H-D-aspartate into the rat's cerebellar cortex or deep nuclei, while (glutamatergic) mossy fiber systems originating in the brain stem or spinal cord remained unlabeled. In opossum cerebellum, enkephalin has also been localized in selected climbing fiber populations within restricted areas of vermal lobules II–VIII and in X [170], providing evidence for chemical heterogeneity within this major afferent system. However, it is not clear whether enkephalin could play a transmitter role, particularly when one considers the suppressive effect of this neuropeptide in other areas of the brain [171]. Recently, another oligopeptide, corticotropin-releasing factor (CRF), has been selectively localized in the olivocerebellar system, first in climbing fiber terminals [172] but subsequently also in mossy afferents [173,174]. However, the possible role of CRF in synaptic transmission mechanisms remains obscure.

Climbing fibers do not synapse on the spiny (tertiary) Purkinje dendrites,

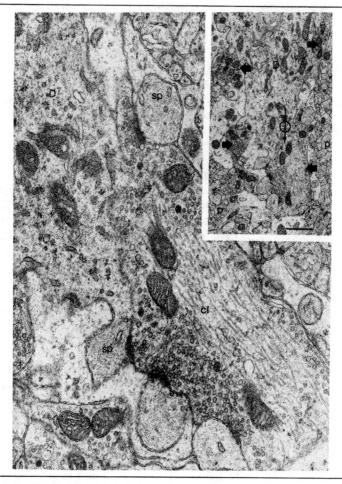

Figure 2-12. Climbing fiber terminals (arrows) surrounding a secondary dendrite (D) of the Purkinje cell. p = parallel fibers. The area at the ringed arrow is also shown under a higher magnification; the large climbing fiber terminal (cf) contains spheroid synaptic vesicles, a few dense core vesicles, and numerous microtubules. sp = spines of secondary dendrite, which are in synaptic contact with the climbing fiber.

which make up, in length and volume, more than half of the total dendritic arborization: The total length of spiny branchlets of one single Purkinje cell is 10 mm in rat [67] and 40 mm in monkey [175]. The spines on the terminal branches are relatively large (1.5–2.0 μm long) [67,176], and the packing density along the dendritic branchlets is high. Along a 10-μm spiny branchlet there are 15–42 spines. As a result, the number of tertiary

2. Anatomy and neurochemical anatomy of the cerebellum   33

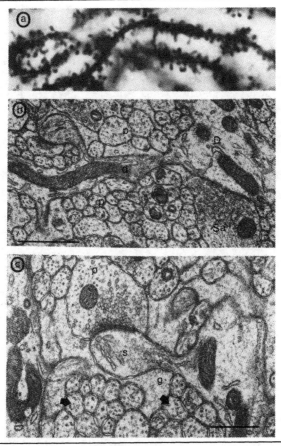

Figure 2-13. Tertiary spiny dendrites of Purkinje cell. **A**: Golgi-impregnated spiny dendrites. **B**: EM of a tertiary, spiny dendrite (d) emerging from a thick dendrite (D). The thick dendrite is postsynaptic to a stellate axon (Sa); parallel fibers (p) are cut transversally. **C**: Higher power view of a synapse between a dendritic spine (s) and a parallel axonal varicosity (p). The latter contains spheroidal synaptic vesicles. Parallel axons (arrows) are cut transversally. g = glial process.

dendritic spines is enormous: In humans one Purkinje cell bears about 50,000 spines [7,175].

The tertiary dendritic spines synapse with parallel fibers. Using histological techniques, the length of the thin (0.09–0.35 µm in diameter) parallel fibers was calculated to be 2.0 mm in cats [176], 3.0 mm in monkeys [175], and 2.6 mm in humans [7]. Electrophysiological recordings [178] have put the probable length of parallel fibers at about 3 mm. However, a degeneration study [179] reported that the average length of parallel fibers in cats was

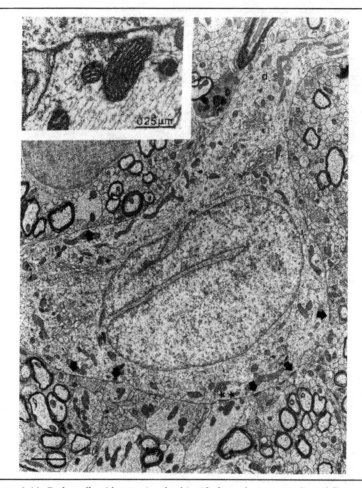

Figure 2-14. Basket cell, with emerging dendrite (d), located among myelinated fibers of the supraganglionic plexus and receiving numerous axosomatic synapses (arrows). One profile (asterisks) has been identified as a basket axon terminal and is shown at higher magnification in the inset. Note small, irregular synaptic vesicles, in addition to numerous neurofilaments. These are characteristic markers of basket axon endings.

6 mm. Later, Mugnaini [180] presented similar values, i.e., 6 mm for parallel fibers in the chicken and in the monkey, suggesting that parallel fibers exhibit an insignificant variation in their maximal and average length, both among species and in relation to cerebellar size.

The parallel fibers, running in bundles (Figures 2-12 to 2-14), synapse with any dendritic spine they may encounter on the way (i.e., not only Purkinje spines, but also those of basket cells and large Golgi neurons). Since all these dendrites run at right angles to the longitudinal axes of the folia, they also cross the longitudinally arranged parallel fibers at right angles. As a result, the vast majority of synapses in the molecular layer are synapses of "overcrossing," [181] where the presynaptic element does not terminate and is only an local, *en passant* spindle-shaped varicosity of the parallel fiber that contacts the dendritic spines.

The synaptic vesicles in the parallel fiber varicosities are spherical, the synapse is asymmetric, and the transmitter is glutamate [91,92,153]. Physiologically, this synapse is excitatory, either on Purkinje cell spines (94% of all synapses formed by parallel fibers) or on non-Purkinje elements (= interneurons, comprising 6% of all parallel fiber synapses).

Of the 400,000 parallel fibers traversing the dendritic arbor of a cat's Purkinje cell [177], about 80,000 establish synaptic contacts with the 60,000–80,000 dendritic spines. That suggests that a parallel fiber would, on average, synapse with every fifth or sixth Purkinje cell dendrite it encounters. This calculation contrasts with the suggestion of Brand et al. [179], who suggested that parallel fibers would synapse with each Purkinje cell dendrite they encountered. In either case, the huge excitatory parallel fiber input is decisive in activating hundreds of Purkinje cells arranged along the longitudinal axis of the folium. In fact, taking an average 6 mm length for parallel fibers, the width of simultaneously activated Purkinje cell column (in the longitudinal axis of the folium) exceeds by several times the width of a single afferent (parasagittal) cortical strip. Obviously, the interplay of the two main excitatory afferent systems, the strictly parasagittal climbing fibers and the transversely oriented parallel fibers, must be decisive in the de facto shaping of functional, parasagittal zones and the resulting corticonuclear units [182].

2.1.3.3. NERVE CELLS IN THE MOLECULAR LAYER. Two types of interneurons, both inhibitory [183,184] and GABAergic [78], are found in the molecular layer. The more superficial stellate cells, with their perikarya, are localized in the superficial half of the molecular layer, and the "deep" stellate cells, called *basket cells*, are found in the deeper parts of this layer (Figure 2-14). It was calculated [185] that the Purkinje cell-stellate ratio is 1:16 to 1:17.5; while in the case of basket cells, this ratio is lower, being only 1:6. Superficial stellate cells usually have longer descending dendrites, while basket neurons, which are present only in birds and mammals, exhibit longer ascending dendrites [85]. The dendrites of both stellate and basket cells, like those of the Purkinje cells, radiate in the transverse direction of the folium. For both neurons, the majority of their dendritic branches are, in fact, confined within the spaces between neighboring Purkinje cell dendritic arbors [65]. The major, excitatory synaptic input, therefore, is the parallel fibers, crossing the dendritic arbors at right angles. As a result, a beam of parallel fibers may excite not only a longitudinal series of Purkinje cells, but also a series of similarly disposed stellate and basket cells. The total number of parallel fibers, synapsing on a single basket cell dendritic tree, is more than 2400 [143], and is somewhat less on the dendrites of (superficial) stellate cells. In addition, both cell types receive a number of inhibitory, GABAergic boutons, mostly on their perikarya [143]. Stellate perikarya receive axonal boutons from other stellate cells, whereas basket-cell perikarya are targeted by Purkinje axon collaterals and basket-cell axon endings from other basket cells (Figure 2-14) [143]. Parallel fibers also form numerous axosomatic contacts on the perikarya of both nerve cell types. In addition, the ascending portions of granule cell axons establish synaptic contacts with the axon hillocks of basket neurons [88]. The ascending granule cell axons also provide a substantial input to Purkinje cell dendrites [186]. In contrast to previous light-microscopic findings [187] and expectations, climbing fibers, even when in contact with stellate or basket-cell somata or dendrites [187,188], fail to establish synaptic contacts with these neurons.

The main difference between stellate and basket cells lies in the arborization pattern of their axons. Stellate cell axons either arborize immediately or run in the transverse plane of the folia before arborizing. Their secondary and tertiary terminal branches contact the dendrites of Purkinje cells, their main postsynaptic target (Figure 2-13B). The main postsynaptic receptors of stellate axons (GABA-B) are not organized, however randomly, but are distributed in parasagittal zones of high and low binding [190].

The axons of basket cells are most characteristic: After a thin initial segment they become rather thick and run at right angles to the parallel fibers in the transverse plane of the folium. These axons span distances of 500–600 µm (i.e., about the distance of 10 neighboring Purkinje cells) and also display secondary side branches, which run along the longitudinal axis of the folium for about 300–320 µm [170], during which time they may contact several successive rows of Purkinje cells. Some of these preterminal branches "ascend" to contact primary and secondary dendrites of the Purkinje cells. Others descend towards the layer of the Purkinje cell bodies, becoming considerably enlarged and varicose in their terminal portions, which then establish multiple axosomatic synaptic contacts with the lower halves only of Purkinje cell somata. About 50 descending basket axon collaterals [65,191] enter the space around the base of the Purkinje cell and its emerging initial axon segment, to form a very complicated, basketlike structure (hence the name, *parent cell*).

The organization of the pericellular basket (Figure 2-8b) which are also called *pinceau* [67], and which comprise one of the most efficient inhibitory synaptic systems in the CNS [183,192], has been described in some detail [66,136,193–196]. These GABAergic basket axons (Figures 2-9A and 2-9B), in addition to several axosomatic synapses, may also come into contact along the Purkinje cell's unmyelinated initial axon segment [136] (Figure 2-15). About 20–25% of the initial segment surface is covered by basket axon terminals, which form three to four axo-axonic synapses (Figure 2-15) per

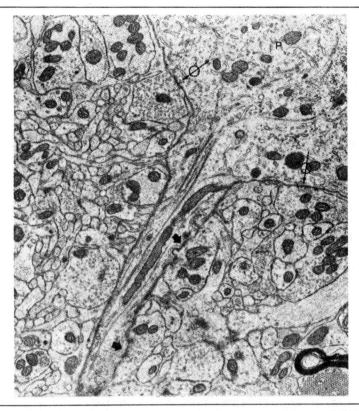

Figure 2-15. Initial axonal segment of Purkinje cell (P) embedded within the pericellular basket, composed of small- and medium-sized basket axon endings and filopodia. The axon receives two synaptic contacts (dark arrows); two axosomatic contacts (circled arrows) are also visible.

Purkinje cell axon, thus representing a significant component of the powerful inhibition of the Purkinje cell by basket axons. Although they are filled with synaptic vesicles and are GABA positive, most basket axons in the pericellular basket terminate "blindly," i.e., nonsynaptically, although they are connected to each other by septate desmosomes [195]. The spatial arrangement of basket cells, their axonal arborization, and the resulting inhibitory pericellular baskets led Szentagothai [65,191–197] to propose a functional model of the cortex in which parallel fiber excitation (spreading in the longitudinal axis of folium) would be shaped and focused to narrow stripes (along the longitudinal axis of folia) by the action of inhibitory, primarily basket, neurons.

2.2. Glial cells of the cerebellar cortex

Two basic types of glial cells are found in the cerebellar cortex. (1) Oligodendrocytes, which, although they occur throughout the cortex, are localized preferentially in areas containing myelinated axons (granular layer and the deepest strata of molecular layer) and (2) astrocytes, which can be further divided into three subclasses [67]. These are the Golgi epithelial cell or Bergmann-glial cell found in association with the Purkinje cells; the so-called velate astrocytes, which are confined to the granular layer; and the rare "smooth" astrocytes, which occur throughout the cortex, particularly in the granular layer.

Bergmann glia are the most characteristic cerebellar glial cell type. These cells give rise to two or more ascending processes, which are sandwiched between two sheets of Purkinje cell dendrites. These ascending glial processes, bear irregular leaflike appendages, which extend horizontally within the molecular layer.

It appears that the Bergmann glial fibers, which ensheath the Purkinje cell surface and its dendritic arbor, isolate longitudinally running bundles of parallel fibers from each other, forming sagittally organized compartments composed of dendritic and axonal elements. Bergmann processes terminate immediately beneath the pia mater with rounded or conical expansions [85].

The processes of velate astrocytes radiate profusely within the granular layer, where they may partially isolate cerebellar glomeruli from cell bodies of granule cells or from other glomeruli. In fact, glomeruli, with the exception of the entering axonal and dendritic profiles, are generally completely ensheathed by the thin lamellar processes of these astrocytes. The smooth astrocytes may also send processes around glomeruli. A more detailed description of these glial types can be found in the book of Palay and Chan-Palay [67].

As to the number of glial cells in the cerebellar cortex, highly differing values have been published: from 9 to 15 per Purkinje cell [198], up to 60 per Purkinje cell [199]. Utilizing immunostaining of oligodendrocytes by the specific marker carboniconhydrase antiserum, and labeling all astrocytes by glial fibrillary acidic-protein antiserum, Ghandour et al. [200] estimated 40–60 glial cells per Purkinje neuron, half of which are astrocytes and half of which are oligodendrocytes.

2.3.1. Cerebellar nuclei

In the human cerebellum there are four distinct paired nuclei embedded deep within the white matter (Figure 2-6). Most laterally and deep in the hemispheric white matter is the largest cell mass, the dentate nucleus. In humans this nucleus is enormous, containing about 284,000 nerve cells [16]. By comparison, in the cat the lateral nucleus, which is the homologue of the dentate nucleus in humans, contains about 6000 nerve cells [201]. In the rhesus monkey, the same nucleus contains 66,000 neurons [202]. Most medial of the four nuclei is the fastigial nucleus (Figures 2-3 and 2-4), called the *nucleus medialis* in other mammals. In the cat, this nucleus was shown to contain about 7000 nerve cells [201], whereas there are 19,000 neurons in the homologous nucleus of the monkey [202]. These comparative values demonstrate that the lateral (dentate) nucleus exhibits a 10X increase from cat to monkey, while the difference in the case of the fastigial nucleus is only 2.5X. This difference agrees with the observed enlargement of the hemispheres and the corresponding increase in corticonuclear connections in primates (and in humans). Between the fastigial and dentate nuclei are the smaller nuclei globosus and emboliformis (nucleus interpositus anterior and posterior in nonprimate mammals), with combined cell numbers of roughly 10,000 in the cat and 60,000 in the monkey [201–202].

Nerve cells in the four nuclei are of different sizes and types [104,201,203]. Most of them [203] are thought to be projection neurons, with only a small minority being local or interneurons [137,204,205].

Recent immunocytochemical studies offer a classification of nuclear cells in relation to their possible transmitter content. The transmitters in most nuclear neurons are probably glutamate and aspartate [206], whereas a minority were shown to contain GABA [208,209].

Electron microscopic studies show that the intrinsic and synaptic organization of the four nuclei is basically similar [104,201,204,207,210]. An average deep nuclear cell receives 20,000 synaptic contacts; 6–17% of all synapses are axosomatic and the rest are axodendritic. Four major types of synaptic boutons can be distinguished.

Type A contain small, pleomorphic synaptic vesicles and cisterns with tubular profiles. This terminal forms intermediate synaptic contacts, very often with dendritic spines. Such synapses are encountered on cell bodies and on dendrites (Figure 2-16a). Since they disappear following separation of the nuclei from the cortex [211], they are identified as axons of Purkinje cells that constitute the main afferent supply to the deep nuclei and comprise some 59–69% of all synaptic endings [201]. On average, the number of Purkinje cell boutons is 11,600/nuclear cell. It has also been calculated that one Purkinje axon supplies 474 individual synapses to 35 nuclear cells, or about 13.5 synapses per nuclear cell (divergence). Thus, on average, one nuclear cell receives input from 860 Purkinje cells. Although these numbers vary among individual nuclei [201] they clearly demonstrate the importance of cortical afferentation to the nuclei. In accordance with the inhibitory GABAergic nature of Purkinje cells, the destruction of these axons through decortication leads to a 70% decrease of glutamate acid decarboxylase activity [212,213]. It has also been shown, both in experimental [201] and in immunocytochemical [50,147] studies, that the Purkinje cell's axonal arbor keeps its parasagittal organizational pattern when projecting to the nuclei.

Type B (Figure 2-16) contain large spheroidal synaptic vesicles and occa-

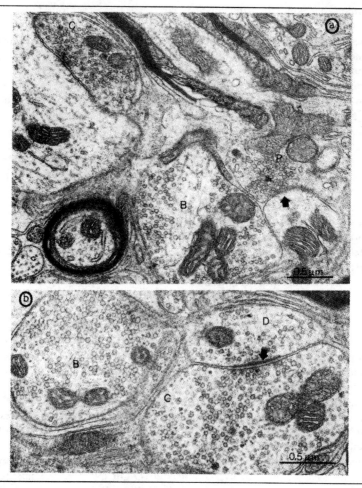

Figure 2-16. Types of axon terminals in nucleus interpositus in the cat. **A**: A Purkinje axon (P) that is partly myelinated forms axodendritic contact (arrow) and contains ovoid-pleomorphic synaptic vesicles. The B-type axon terminal exhibits large, C-type smaller spheroidal synaptic vesicles. **B**: In addition to a B-type terminal with large round vesicles, a C-type terminal with small, round vesicles is seen; the latter forms axo-axonic synapse (arrows) with a D-type axon terminal, which contains small and pleomorphic synaptic vesicles.

sionally a few dense core vesicles constituting 18–22% of all synaptic terminals in the nuclei [201]. Boutons are mostly axodendritic. These terminals were shown by Van der Want et al. [214] to arise from precerebellar mossy fiber sources, or alternatively, to be collaterals of olivocerebellar climbing fibers [215]. These latter, though they make strong, multiple synaptic contacts with the proximal dendrites of Purkinje cells in the cerebellar cortex, synapse only with small (distal) dendrites of nuclear cells. It was also demonstrated that the topographical organization of the climbing fiber collaterals within cerebellar nuclei correlates with the zonal organization of olivocerebellar afferents within the cortex [215].

Type C synaptic boutons are mostly axodendritic and contain spheroidal synaptic vesicles (Figure 2-16), which are considerably smaller in size than those found in type B boutons. They comprise 6–10% of all synaptic endings within the nuclei. It has been suggested that these endings might be recurrent collaterals [204] of the nuclear projection neurons [211,216]. However, their identification as such needs more direct evidence.

Type D boutons are small- to middle-size terminals, localized on dendrites and somata, and contain small, ovoid synaptic vesicles. Since they survive isolation of the nuclei from the cortex, they are most probably of intranuclear origin [201,217]. These boutons, which constitute 6–8% of all synapses in the deep nuclei [201], may be the endings of GABAergic neurons, as demonstrated by Wassef et al. [208] in mouse cerebellum. They could be either recurrent collaterals of GABAergic cerebello-olivary projection neurons [218–220], or simply terminals of local GABAergic interneurons.

In addition to the four types described above, a few, probably aminergic, terminals [221] can be occasionally be found within the deep nuclei. These contain small or large dense-core vesicles and represent only 0.4–3% of all vesicle-containing boutons.

Although most synaptic junctions are axosomatic or axodendritic, complex synaptic arrangements (serial and triadic synapses) between vesicle-bearing profiles have been reported in the cerebellar nuclei of cats [211] and rats [104]. Though the number of such synapses is low (about 1.5% of all observed synapses in the cat's nucleus interpositus), their presence suggests that the neuronal circuitry and local mechanisms in information processing may be more complex than heretofore assumed.

2.3.2. Vestibular nuclei

Although by their location it is extracerebellar, the vestibular nuclear complex can be also considered as a simple form of the cerebellar corticonuclear complex [80]. This is due to the fact that all major (superior, lateral, medial, and descending) vestibular nuclear cell groups receive direct Purkinje cell projections from the vestibulo-cerebellum [222–223] or from the lateral portions of the vermis (Figure 2-4). Although the synaptic organization seems to be similar to that seen in the cerebellar nuclei (particularly in the lateral or Deiters nucleus [224, 227]), the neuronal connections of vestibular nuclei appear to be simpler than those of the "real" intracerebellar nuclei [80]. For example, vestibular nuclear projection neurons do not have recurrent collaterals, as do the projection neurons of the cerebellar nuclei. Nerve cells in the medial, descending, and superior vestibular nuclei are medium to small in size, while those in the lateral (Deiters) nucleus are the characteristic giant (30–45 μm diameter) Deiters neurons. The main afferents to vestibular nuclei are provided by the primary vestibular afferents [228]. Afferents also arise from the spinal cord [229], particularly from the perihypoglossal nuclei [230], from olivocerebellar fibers [126], from the medullary and pontine reticular formations [127], and, of course, from the anterior and posterior parts of the vermis and the vestibulocerebellum. In addition, the cerebellar vermis might also influence the vestibular nuclei via a relay in the fastigial (medial cerebellar nucleus [231,232].

3. CONNECTIVITY OF CEREBELLUM: INTRACEREBELLAR AND EXTRACEREBELLAR

3.1. Corticonuclear and nucleocortical projections

The corticonuclear projection arises solely from Purkinje cell axons. As described previously, more than 60% of all synaptic boutons found in cerebellar nuclei are Purkinje cell axons. As has been well documented [27,28], these are topographically organized within the nuclei in a very precise manner. Accordingly, the vermis, or rather the medial zone of vermis, projects to the fastigial nucleus, while the lateral part of the vermis sends afferents to the lateral vestibular or Deiters nucleus [18,231]. The paramedian lobule and the medial regions of the hemispheres project to the nuclei interpositi (globose and emboliform nuclei), while the larger (lateral) part of the hemispheres, together with the paraflocculus, supply the dentate nucleus. The general tendency is that the more lateral cortical zones project to the lateral cerebellar nuclei [18,233], while most medial (fastigial) nuclei receive inputs from the medial cortical zones. Recently, utilizing Purkinje-cell-specific antibody staining, it has been shown [264] that the parasagittal bands established by Purkinje cells and their axons in the cortex are also maintained in the deep target nuclei.

This close anatomical relationship between cortical and nuclear regions was further documented by introducing the concept of corticonuclear microcomoplex [234,235]. Accordingly, a corticonuclear microcomplex consists of a cerebellar cortical microzone and associated small groups of nuclear (or vestibular) neurons. The existence of such a microcomplex was demonstrated to operate, e.g., in the vestibulo-ocular reflex [236].

The existence of the nucleocortical projection had been predicted by Ramon y Cajal [85] and was later verified by Carrea et al. [237] and Cohen et al. [238]. The first physiological verification of this pathway in the cat was made by Ito and coworkers [127]. Direct morphological proof for the existence of nucleocortical fibers was provided much later, only in the late 1970s [103,239]. A similar pathway in the monkey [240,241] has been also demonstrated. Nucleocortical axons arise from all cerebellar nuclei, and they are, at least partly, collaterals of the nucleofugal axons. Reciprocal [103, 242–244], as well as nonreciprocal, relationships between the cortex and deep nuclei have been described. The original idea of Carrea et al. [237], i.e., that nucleocortical fibers are climbing fibers, was refuted in subsequent studies that presented evidence that these fibers terminate in the granule layer as mossy terminals [105,245]. More recently it has been shown [102] that some of the nucleocortical mossy terminals are GABAergic [111], although the majority of these endings probably contain glutamate [92].

3.2. Afferent and efferent projections of cerebellum

The cerebellum receives information from all kinds of receptors, cutaneous, proprioceptive, vestibular, visual, etc. The afferents carrying this information may enter either directly to the cerebellum or may be relayed through precerebellar relay stations (indirect route). At the same time, the cerebellum sends fibers to several brain centers, mostly through nucleofugal pathways, but partly by means of Purkinje cell axons to the vestibular nuclei. The afferent-efferent connections may be both reciprocal or nonreciprocal [246]. A very important feature of these extracerebellar connections is that the total number of afferents is much higher than the number of efferents (40:1) [16]. This illustrates the importance and operational potentialities in information processing by the cerebellar cortex.

The connections of the cerebellum to extracerebellar regions is one of the most thoroughly studies areas of the CNS. Excellent detailed reviews of these investigations can be found in Ito's [80] or Brodal's [248] books, as well as in the review of Bloedel and Courville [249]. Only a brief summary of the harvested data will be given here.

3.2.1. Afferent pathways

Direct afferent pathways, which reach the cerebellum without being relayed in one of the precerebellar nuclei, are the spinocerebellar pathways (the dorsal, ventral, and rostral spinocerebellar and cuneocerebellar tracts). The dorsal spinocerebellar tract takes its origin in Clarke's column at levels T4 to L2 in humans. The fibers enter the cerebellum via the inferior peduncle (restiform body) and terminate mostly uncrossed in the anterior lobe. Grant [250,251] has shown that in the cat this pathway projects only to cerebellar hindlimb regions. This morphological finding was later corroborated electrophysiologically by Oscarsson [36]. The cervical cord equivalent of the dorsal spinocerebellar tract originates from the lateral cuneate nucleus, which collects afferents from C1 to T4–T5 cord segments. Uncrossed cuneocerebellar fibers also enter the cerebellum via the restiform body and terminate in the posterior part of the anterior lobe [252]. The cuneocerebellar projection carries the head, neck, forelimb, and upper trunk projections.

The ventral spinocerebellar tract, after crossing in the midline of the spinal cord, ascends through the medulla and pons to enter the cerebellum by way of the superior peduncle. This tract is related to the hindlimb and posterior trunk, while the rostral spinocerebellar tract [253] represents the forelimb areas. Rostral tract fibers are uncrossed, and they also enter the cerebellum through the superior peduncle. The termination area of both the ventral and rostral tracts in cerebellar cortex overlaps that of the dorsal spinocerebellar and cerebellocerebellar tract. All four tracts terminate in the cortex as mossy fibers [250,253,254,255], while sending collaterals to the cerebellar nuclei [18,65,256,257].

Vestibulocerebellar pathways consist partly of direct (and in this case, primary) afferent fibers, while others pass indirectly via relay stations. The primary fibers are axons of the vestibular ganglion cells and project primarily to the nodulus and uvula, although a modest projection to the flocculus [258], as well as to other vermal areas, has been described. Secondary, or indirect, vestibulocerebellar fibers are relayed through vestibular nuclei and terminate in the cortex, together with the primary fiber projections. Both vestibulocerebellar tracts end as mossy fibers in the cortex. Their projection to the deep nuclei appears to be very modest: Van der Want et al. [214] demonstrated some projection of secondary fibers to the fastigial nucleus.

The most important relay stations (precerebellar nuclei) for conveying indirect afferent impulses to the cerebellum are the pontine nuclei, the inferior olive, and the nuclei of the reticular formation (tegmental and pontine reticular nuclei, the lateral reticular nucleus, and the paramedian reticular nucleus). With the exception of olivocerebellar fibers, which terminate in the cortex as climbing fibers [161], all precerebellar afferents end intracortically as mossy fibers.

3.2.1.1. PONTOCEREBELLAR TRACT. This massive projection originates from the 20 million (in humans) pontine neurons on each side [17] and enters the cerebel'um by the contralateral brachium pontis or middle peduncle. The main projection targets of pontocerebellar fibers are the hemispheres and the paraflocculus, as well as some regions of the vermis. This tract projects impulses arriving mostly from the cerebral (predominantly frontal) cortex [247], the superior and inferior colliculi, and from the cerebellum, etc. It is interesting that there are no afferents to the deep nuclei from this huge tract [246,259].

3.2.1.2. OLIVOCEREBELLAR TRACT. The axons of olivary cells cross the midline and enter the cerebellum through the contralateral inferior peduncle. As mentioned, the fibers terminate in the cortex as climbing fibers. It is characteristic that each longitudinal cerebellar zone is topographically related to specific regions within the inferior olivary complex. The olivocerebellar tract mediates spinal and cortical impulses, as well as impulses from the red nucleus, mesencephalic reticular formations, superior colliculus, pretectum, and also from the cerebellar nuclei. The tract also sends fibers to the cerebellar nuclei [30,65,215,260-262]. Each cerebellar nucleus receives afferents (probably collaterals of olivocerebellar fibers to the cortex) from a particular subdivision of the olive.

3.2.1.3. RETICULOCEREBELLAR INPUTS. The reticular tegmental nucleus (Bechterew nucleus) supplies primarily the vermis and adjacent regions, particularly Voogd's lobules VI and VII. Afferents to this nucleus [215,263] arrive from the cerebral cortex and, most importantly, from cerebellar nuclei (lateral and anterior interpositus nuclei) [266]. The lateral reticular nucleus sends fibers to the cerebellum through the inferior peduncle, where most fibers terminate in the anterior lobe and vermis, with some fibers terminating to the paramedian lobe. Collaterals enter the dentate [104] and the interpositus nuclei [265].

The paramedian reticular nucleus (PRN) is the smallest of the three reticular nuclei projecting to the cerebellum. Its fibers terminate predominantly in the flocculus and in the vermis of anterior and posterior lobes. Input to PRN is from the spinal cord, the frontoparietal cortex, the vestibular nuclei, and the fastigial nucleus. Collaterals of PRN fibers to cerebellar nuclei are very few [246].

The cerebellum also receives noradrenergic and serotoninergic afferents from the locus coeruleus and from raphe nuclei, respectively. Their target is the entire cerebellar cortex. They also send a limited number of axons to the nuclei [267].

3.2.2. Efferent connections of the cerebellum

Most of these fibers leave the cerebellum via the superior cerebellar peduncle, and a minor portion leave through the inferior peduncle.

The main targets of efferents from the fastigial nucleus are the three reticular nuclei and also the vestibular nuclei. The nuclei interpositi and the dentate nuclei are the sources of cerebellar efferents to the thalamus (ventrolateral nucleus), to the red nucleus, and to other brain-stem nuclei, as well as the oculomotor nuclei, and, of course, the pontine nuclei (cerebellopontine fibers [21]). Cerebellar efferents to vestibular nuclei can be direct (Purkinje axons) from the anterior and posterior vermis, and from the vestibulocerebellum. Indirect vestibulocerebellar fibers take their origin in the fastigial nucleus.

All four cerebellar nuclei participate in the organization of cerebellarolivary "feedback" projection, which is, to a large extent, reciprocal to that of the olivonuclear projection. This pathway leaves the cerebellum through the restiform body.

ACKNOWLEDGMENTS

The author is indebted to Prof. J.R. Haight for reading the manuscript. The invaluable help of Dr. J. Takács during the preparation and composition of the text, as well as the skillful technical assistence of Ms. Zs. Szabó and Ms. E. Borók, are also gratefully acknowledged.

NOTES

1. The cortical tissue corresponding to one elementary folding is called a *folium*, and is about 1 mm wide at the base and has a surface circumference of 3–4 mm. At the base of each folium, the cortex continues without interruption into the next neighboring folium. The

length of folia varies not only between species, but also between different regions of the same cerebellum. Vermal folia are generally the longest, and hemispheric folia are the shortest.
2. "Simple" type glomeruli receive only one mossy "terminal" or *en passant* mossy ending, in contrast to more complex types supplied by two to four mossy terminals.
3. The glutamate antiserum was kindly provided by P. Petrusz.

REFERENCES

1. Ellis R.S. (1919). A preliminary quantitative study of Purkinje cells in normal, subnormal, and senescent human cerebella, with some notes on functional localization. J. Comp. Neurol. 30:229–252.
2. Dow R.S. (1942). The evolution and anatomy of the cerebellum. Biol. Rev. 17:179–220.
3. Jansen J., Brodal A. (1958). Das Kleinhirn. Mollendorff's Handbuch der Mikroskopischen Anatomie des Menschen IV/8. Berlin: Springer-Verlag.
4. Braitenberg V., Atwood R.P. (1958). Morphological observations on the cerebellar cortex. J. Comp. Neurol. 109:1–27.
5. Angevine J.B., Locke S., Yakovlev P.I. (1962). Limbic nuclei of thalamus and connections of limbic cortex. IV. Thalamocortical projection of the ventral anterior nucleus in man. Arch. Neurol. 7:518–528.
6. Rakic P., Sidman R.L. (1970). Histogenesis of cortical layers in human cerebellum, particularly the lamina dissecans. J. Comp. Neurol. 139:473–500.
7. Smoljaninov V.V. (1971). Some special features of organization of the cerebellar cortex. In: I.M. Gelfand, V.S. Gurfinkel, S.V. Fomin, M.L. Tsetlin (eds): Models of the Structural-Functional Organization of Certain Biological Systems. Cambridge, MA: MIT Press, pp. 250–423.
8. Larsell O., Jansen J. (1972). The Comparative Anatomy and Histology of the Cerebellum. The Human Cerebellum, Cerebellar Connections, and Cerebellar Cortex. Minneapolis, MN: University of Minnesota Press.
9. Lange W. (1974). Regional differences in the distribution of Golgi cells in the cerebellar cortex of man and some other mammals. Cell Tiss. Res. 153:219–226.
10. Lange W. (1975). Cell number and cell density in the cerebellar cortex of man and some other mammals. Cell Tissue Res. 157:115–124.
11. Benke B., Hamori J. (1965). Elektronenmikroskopische untersuchung der cerebellaren rindenatrophie. Acta Neuropathol. 5:275–287.
12. Zecevic N., Rakic P. (1976). Differentiation of Purkinje cells and their relationship to other components of developing cerebellar cortex in man. J. Comp. Neurol. 167:27–48.
13. Korneliussen H.K. (1967). Cerebellar corticogenesis in Cetacea, with special reference to regional variations. J. Hirnforsch. 9:151–185.
14. Larsell O. (1934). Morphogenesis and evolution of the cerebellum. Arch. Neurol. Psychiat. 31:373–395.
15. Larsell O. (1937). The cerebellum. A review and interpretation. Arch. Neurol. Psychiat. 38:580–607.
16. Heidary H., Tomasch J. (1969). Neuron numbers and perikaryon areas in the human cerebellar nuclei. Acta Anat. (Basel) 74:290–296.
17. Tomasch J. (1969). The numerical capacity of the human cortico-ponto-cerebellar system. Brain Res. 13:476–484.
18. Voogd J. (1969). The importance of fiber connections in the comparative anatomy of the mammalian cerebellum. In: R. Llinas (ed): Neurobiology of Cerebellar Evolution and Development. Chicago: American Medical Association, pp. 493–514.
19. Scheibel A. (1977). Sagittal organization of mossy fiber terminal systems in the cerebellum of the rat: A Golgi study. Exp. Neurol. 57:1067–1070.
20. Yaginuma H., Matsushita M. (1986). Spinocerebellar projection fields in the horizontal plane of lobules of the cerebellar anterior lobe in the cat: An anterograde wheat germ agglutinin-horseradish peroxidase study. Brain Res. 365:345–349.
21. Voogd J. (1964). The Cerebellum of the Cat. Assen: Van Gorcum.
22. Voogd J. (1967). Comparative aspects of the structure and fibre connexions of the mammalian cerebellum. Prog. Brain Res. 25:94–135.
23. Oscarsson O. (1969). The sagittal organization of the cerebellar anterior lobe as revealed by

the projection patterns of the climbing fiber system. In: R. Llinas (ed): Neurobiology of Cerebellar Organization and Development. Chicago: American Medical Association, pp. 525–532.
24. Chan-Palay V., Palay S.L., Brown J.T., Van Itallie C. (1977). Sagittal organization of olivocerebellar and reticulocerebellar projections: Autoradiographic studies with 35-S-methionine. Exp. Brain Res. 30:561–576.
25. Oscarsson O. (1980). Functional organization of olivary projection to the cerebellar anterior lobe. In: J. Courville, C. De Montigny, Y. Lamarre (eds): The Inferior Olivary Nucleus: Anatomy and Physiology. New York: Raven Press, pp. 279–289.
26. Korneliussen H.K. (1968). On the ontogenetic development of the cerebellum (nuclei, fissures, and cortex) of the rat, with special reference to regional variations in corticogenesis. J. Hirnforsch. 10:379–412.
27. Jansen J., Brodal A. (1940). Experimental studies on the intrinsic fibers of the cerebellum. II. The corticonuclear projection. J. Comp. Neurol. 73:267–321.
28. Jansen J., Brodal A. (1942). Experimental studies on the intrinsic fibers of the cerebellum. III. The corticonuclear projection in the rabbit and the monkey. Skr. Norske Vidensk.-Akad., I. Mat.-nat. Kl. No. 3., pp. 1–50.
29. Chambers W.W., Sprague J.M. (1955). Functional localization in the cerebellum. I. Organization in longitudinal cortico-nuclear zones and their contribution to the control of posture, both extrapyramidal and pyramidal. J. Comp. Neurol. 103:105–130.
30. Groenewegen H.J., Voogd J. (1977). The parasagittal zonation within the olivocerebellar projection. I. Climbing fiber distribution in the vermis of the cat cerebellum. J. Comp. Neurol. 174:417–488.
31. Courville J., Faraco-Cantin F. (1978). On the origin of the climbing fibers of the cerebellum. An experimental study in the cat with an autoradiographic tracer method. Neuroscience 3:797–809.
32. Kawamura K., Hashikawa T. (1979). Olivocerebellar projections in the cat studied by means of anterograde axonal transport of labeled amino acids as tracers. Neuroscience 4:1615–1633.
33. Beyerl B.D., Borges L.F., Swearingen B., Sidman R.L. (1982). Parasagittal organization of the olivocerebellar projection in the mouse. J. Comp. Neurol. 209:339–346.
34. Campbell N.C., Armstrong D.M. (1983). The olivocerebellar projection in the rat: An autoradiographic study. Brain Res. 275:215–233.
35. Eisenman L.M. (1984). Organization of the olivocerebellar projection to the uvula in the rat. Brain Behav. Evol. 24:1–12.
36. Oscarsson O. (1973). Functional organization of spinocerebellar paths. In: A. Iggo (ed): Handbook of Sensory Physiology. Vol. II. Somatosensory System. Berlin: Springer-Verlag, pp. 339–380.
37. Scott T.G. (1963). A unique pattern of localization within the cerebellum. Nature 200:793.
38. Scott T.G. (1964). A unique pattern of localization within the cerebellum of the mouse. J. Comp. Neurol. 122:1–7.
39. Marani E. (1982). The ultrastructural localization of 5′-nucleotidase in the molecular layer of the mouse cerebellum. In: H.F. Bradford (ed): Neurotransmitter Interaction and Compartmentation. New York: Plenum Press, pp. 558–572.
40. Marani E. (1986). Topographic histochemistry of the cerebellum. 5′-nucleotidase, acetylcholinesterase, immunology of FAL. Progr. Histochem. Cytochem. 16:1–144.
41. Ramon Molinar E. (1972). Acetylcholinesterase distribution in the brain stem of the cat. Ergebn. Anat. 46:1–52.
42. Marani E., Voogd J. (1977). An acetylcholinesterase band pattern in the molecular layer of the cat cerebellum. J. Anat. 124:335–345.
43. Ingram V.L., Ogren M.P., Chatot C.L., Gossels J.M., Owens. B.B. (1985). Diversity among Purkinje cells in the monkey cerebellum. Proc. Natl. Acad. Sci. USA 82:7131–7135.
44. Hess D.T., Voogd J. (1986). Chemoarchitectonic zonation of the monkey cerebellum. Brain Res. 369:383–387.
45. Marani E. (1981). Enzyme Histochemistry. In: R. Lahue (ed): Methods in Neurobiology. New York: Plenum Press, pp. 481–581.
46. Voogd J., Hess D.T., Marani E. (1987). The parasagittal zonation of the cerebellar cortex

in cat and monkey: Topography, distribution of acetylcholinesterase, and development. In: J.S. King (ed): New Concepts in Cerebellar Neurobiology. New York: Alan R. Liss, pp. 183-220.
47. Brown B.L. (1985). Changes in acetylcholinesterase staining in the molecular layer of the cat cerebellar cortex following climbing fiber destruction. Anat. Rec. 211:27A.
48. Chan-Palay V., Palay S.L., Wu J.Y. (1982). Sagittal cerebellar microbands of taurine neurons: Immunocytochemical demonstration by using antibodies against the taurine synthesizing enzyme cysteine sulfonic acid decarboxylase. Proc. Natl. Acad. Sci. USA 79:4221-4225.
49. Hawkes R., Colonnier M., Leclerc N. (1985). Monoclonal antibodies reveal sagittal banding in the rodent cerebellar cortex. Brain Res. 333:359-365.
50. Hawkes R., Leclerc N. (1987). Antigenic map of the rat cerebellar cortex: The distribution of parasagittal bands as revealed by monoclonal anti-Purkinje cell antibdoy mabQ113. J. Comp. Neurol. 256:29-41.
51. Sotelo C., Wassef M. (1991). Cerebellar development: Afferent organization and Purkinje cell heterogeneity. Phil. Trans. R. Soc. Lond. B. 331:307-313.
52. Altman J., Bayer S.A. (1985). Embryonic development of the rat cerebellum: III. Regional differences in the time of origin, migration, and settling of Purkinje cells. J. Comp. Neurol. 231:42-65.
53. Wassef M., Sotelo C. (1984). Asynchrony in the expression of cyclic GMP dependent protein kinase by clusters of Purkinje cells during the perinatal development of rat cerebellum. Neuroscience 13:1217-1241.
54. Campbell S.K., Parker T.D., Welker W. (1974). Somatotopic organization of the external cuneate nucleus in albino rats. Brain Res. 77:1-23.
55. Adrian E.D. (1943). Afferent areas in the cerebellum connected with the limbs. Brain Res. 66:289-315.
56. Snider R.S., Stowell A. (1944). Receiving areas of the tactile, auditory and visual systems in the cerebellum. J. Neurophysiol. 7:331-357.
57. Snider R.S. (1950). Recent contributions to the anatomy and physiology of the cerebellum. Arch. Neurol. Psychiat. 64:196-219.
58. Fox P.T., Raichle M.E., Thach W.T. (1985). Functional mapping of the human cerebellum with positron emission tomography. Proc. Natl. Acad. Sci. USA 82:7462-7466.
59. Welker W. (1987). Spatial organization of somatosensory projections to granule cell cerebellar cortex: Functional and connectional implications of fractured somatotopy (Summary of Wisconsin Studies). In: New Concepts in Cerebellar Neurobiology. New York: Alan R. Liss, pp. 239-280.
60. Rushmer D.S., Woolacott M.H., Robertson L.T., Laxer K.D. (1980). Somatotopic organization of climbing fiber projections from low threshold cutaneous afferents to pars intermedia of cerebellar cortex in the cat. Brain Res. 181:17-30.
61. Robertson L.T. (1984). Topographic features of climbing fiber input in the rostral vermal cortex of the cat cerebellum. Exp. Brain Res. 55:445-454.
62. Rosina A., Provini L. (1983). Somatotopy of climbing fiber branching to the cerebellar cortex in cat. Brain Res. 289:45-63.
63. Heinsen Y.L., Heinsen H. (1983). Regionale unterschiede der numerischen Purkinjezelldichte in kleinhirnen von albinoratten zweier stamme. Acta Anat. 116:276-284.
64. Fox C.A., Hillman D.E., Siegesmund K.A., Dutta C.R. (1967). The primate cerebellar cortex: A Golgi and electron microscopic study. Progr. Brain Res. 25:174-225.
65. Eccles J., Ito M., Szentagothai J. (1967). The Cerebellum as a Neuronal Machine. Berlin: Springer.
66. Mugnaini E. (1972). The histology and cytology of the cerebellar cortex. In: O. Larsell, J. Jansen (eds): The Comparative Anatomy and Histology of the Cerebellum: The Human Cerebellum, Cerebellar Connections and Cerebellar Cortex. Minneapolis, MN: University of Minnesota Press, pp. 201-265.
67. Palay S.L., Chan-Palay V. (1974). Cerebellar Cortex. Cytology and Organization. Berlin: Springer-Verlag.
68. Zagon I.S., McLaughlin P.J., Smith S. (1977). Neural populations in the human cerebellum: Estimations from isolated cell nuclei. Brain Res. 127:279-282.
69. Hamori J. (1964). Identification in the cerebellar isles of Golgi II axon endings by aid of

experimental degeneration. In: Proceedings of the Third European Regional Conference, Prague, pp. B291–B292.
70. Hamori J., Szentagothai J. (1966). Participation of Golgi neuron processes in the cerebellar glomeruli: An electron microscopic study. Exp. Brain Res. 2:35–48.
71. Held H. (1897). Beitrage zur struktur der nervenzellen und ihrer fortsatze. Arch. Anat. Physiol. 21:204–292.
72. Gray E.G. (1961). The granule cells, mossy synapses and Purkinje spine synapses of the cerebellum: Light and electron microscope observations. J. Anat. (London) 95:345–356.
73. Palay S.L. (1961). The electron microscopy of glomeruli cerebellosi. Cytology of nervous tissue. In: Proceedings of the Anatomical Society of Great Britain and Ireland. London: Taylor and Francis, pp. 82–84.
74. Larramendi L.M.H. (1969). Morphological characteristics of extrinsic and intrinsic nerve terminals and their synapses in the cerebellar cortex of the mouse. In: W.S. Fields, W.D. Willis (eds): The Cerebellum in Health and Disease. St. Louis: W.H.M. Green, pp. 63–110.
75. Larramendi L.M.H. (1969). Analysis of synaptogenesis in the cerebellum of the mouse. In: R. Llinas (ed): Neurobiology of Cerebellar Evolution and Development. Chicago: AMA-ERF Institute for Biomedical Research, pp. 803–843.
76. Llinas R., Hillman D.E. (1969). Physiological and morphological organization of the cerebellar circuits in various vertebrates. In: R. Llinas (ed): Neurobiology of Cerebellar Evolution and Development. Chicago: AMA-ERF Institute for Biomedical Research, pp. 43–73.
77. Hamori J., Somogyi J. (1983). Differentiation of cerebellar mossy fiber synapses in the rat: A quantitative electron microscope study. J. Comp. Neurol. 220:365–377.
78. Gabbott P.L.A., Somogyi J., Stewart M.G., Hamori J. (1986). GABA-immunoreactive neurons in the rat cerebellum: A light and electron microscope study. J. Comp. Neurol. 251:474–490.
79. Jakab R.L., Hamori J. (1988). Quantitative morphology and synaptology of cerebellar glomeruli in the rat. Anat. Embryol. 179:81–88.
80. Ito M. (1984). The Cerebllum and Neural Control. New York: Raven Press.
81. Palkovits M., Magyar P., Szentagothai J. (1972). Quantitative histochemical analysis of the cerebellar cortex in the cat. IV. Mossy fiber-Purkinje cell numerical transfer. Brain Res. 45:15–29.
82. Mugnaini E., Atluri R.L., Houk J.C. (1974). Fine structure of the granular layer in turtle cerebellum with emphasis on large glomeruli. J. Neurophys. 37:1–29.
83. Eccles J.C., Llinas R., Sasaki K. (1966). The mossy fiber-granule cell relay in the cerebellum and its inhibition by Golgi cell. Exp. Brain Res. 1:82–101.
84. Sasaki K., Strata P. (1967). Responses evoked in the cerebellar cortex by stimulating mossy fiber pathways to the cerebellum. Exp. Brain Res. 3:95–110.
85. Cajal Ramon y (1911). Histologie du Systeme Nerveux de l'Homme et des Vertebres. Paris. Maloine.
86. Hamori J., Somogyi J. (1982). Presynaptic dendrites and perikarya in deafferented cerebellar cortex. Proc. Natl. Acad. Sci. USA 79:5093–5096.
87. Eccles J.C., Sasaki K., Strata P. (1967). A comparison of the inhibitory actions of the Golgi cells and basket cells. Exp. Brain Res. 3:81–94.
88. Hamori J. (1981). Synaptic input to the axon hillock and initial segment of inhibitory interneurons in the cerebellar cortex of the rat. Cell Tissue Res. 217:553–562.
89. Young A.B., Oster-Granite M.L., Herndon R.M., Snyder S.H. (1974). Glutamic acid: Selective depletion by viral induced granule cell loss in hamster cerebellum. Brain Res. 73:1–13.
90. Rohde B.H., Rea M.A., Simon J.R., McBride W.J. (1979). Effects of X-irradiation induced loss of cerebellar granule cells on the synaptosomal levels and the high affinity uptake of amino acids. J. Neurochem. 32:1431–1435.
91. Somogyi P., Halasy K., Somogyi J., Storm-Mathisen J., Ottersen O.P. (1986). Quantification of immunogold labelling reveals enrichment of glutamate in mossy and parallel fibre terminals in cat cerebellum. Neuroscience 19:1045–1050.
92. Hamori J., Takacs J., Pertusz P. (1990). Immunogold electron microscopic demonstration of glutamate and GABA in normal and deafferented cerebellar cortex: Correlation between

transmitter content and synaptic vesicle size. J. Histochem. Cytochem. 38:1767–1777.
93. Friede R.L., Flemming L.M. (1964). A comparison of cholinesterase distribution in the cerebellum of several species. J. Neurochem. 11:1–7.
94. Kasa P., Joo F., Csillik B. (1965). Histochemical localization of acetylcholinesterase in the cat cerebellar cortex. J. Neurochem. 12:31–35.
95. Phillis J.W. (1968). Acetylcholinesterase in the feline cerebellum. J. Neurochem. 15:691–698.
96. Brown W.J., Palay S.L. (1972). Acetylcholinesterase activity in certain glomeruli and Golgi cells of the granular layer of the rat cerebellar cortex. Z. Anat. Entwickl. 137:317–334.
97. Crepel F. (1982). Regression of functional synapses in the immature mammalian cerebellum. Trends Neurosci. 5:266–269.
98. Ojima H., Kawajiri S.I., Yamasaki T. (1989). Cholinergic innervation of the rat cerebellum: Qualitative and quantitative analyses of elements immunoreactive to a monoclonal antibody against choline acetyltransferase. J. Comp. Neurol. 290:41–52.
99. Korte G.E., Reiner A., Karten H.J. (1980). Substance P-like immunoreactivity in cerebellar mossy fibers and terminals in the red-eared turtle, Chrysemys scripta elegans. Neuroscience. 5:903–914. 96.
100. Inagaki S., Shiosaka S., Takatsuki K., Iida H., Sakanaka M., Senba E., Hara Y., Matsuzaki T., Kawai Y., Tohyama M. (1982). Ontogeny of somatostatin-containing neuron system of the rat cerebellum including its fiber connections: An experimental and immunohistochemical analysis. Dev. Brain Res. 3:509–527.
101. Helper J.R., Toomim C.S., McCarthy K.D., Conti F., Battaglia G., Rustioni A., Petrusz P. (1988). Characterization of antisera to glutamate and aspartate. J. Histochem. Cytochem. 36:13–22.
102. Hamori J., Takacs J. (1988). Two types of GABA-containing axon terminals in cerebellar glomeruli of cat: An immunogold-EM study. Exp. Brain Res. 74:471–479.
103. Tolbert D.L., Bantli H., Bloedel J.R. (1976). Anatomical and physiological evidence for a cerebellar nucleo-cortical projection in the cat. Neuroscience 1:205–217.
104. Chan-Palay V. (1977). Cerebellar Dentate Nucleus. Organization, Cytology and Transmitters. Berlin: Springer.
105. Hamori J., Mezey E., Szentagothai J. (1981). Electron microscopic identification of cerebellar nucleocortical mossy terminals in the rat. Exp. Brain Res. 44:97–100.
106. Chan-Palay V. (1982). Neurotransmitters and receptors in the cerebellum: Immunocytochemical localization of glutamic acid decarboxylase. GABA-transaminase and cyclic GMP and autoradiography with H-Muscimol. In: S.L. Palay, V. Chan-Palay (eds): The Cerebellum, New Vistas. Exp. Brain Res Suppl. 6. pp. 552–584.
107. Barber R., Saito K. (1976). Light microscopic visualization of GAD and GABA-T in immunocytochemical preparations of rodent CNS. In: Roberts E., Chase T.N., Tower D.B. (eds): GABA in Nervous System Function. New York: Raven Press, p. 113.
108. Chan-Palay V., Wu J.Y., Palay S.L. (1979). Immunocytochemical localization of GABA-transaminase at cellular and ultrastructural levels. Proc. Natl. Acad. Sci. USA 76:2067–2071.
109. Somogyi P., Hodgson A.J. (1985) Antiserum to γ-aminobutyric acid: III. Demonstration of GABA in Golgi-impregnated neurons and in conventional electron microscopic sections of cat striate cortex. J. Histochem. Cytochem. 33:249–257.
110. Hodgson A.J., Penke B., Erdei A., Chubb I.W., Somogyi P. (1985). Antiserum to γ-aminobutyric acid. I. Production and characterisation using a new model system. J. Histochem. Cytochem. 33:229–239.
111. Hamori J., Takacs J. (1989). Two types of GABA-containing axon terminals in cerebellar glomeruli of cat: An immunogold-EM study. Exp. Brain Res. 74:471–479.
112. Oertel W.H., Mugnaini E., Schmechel D.E., Tappaz M.L., Kopin I.J. (1982). The immunocytochemical demostration of gamma-aminobutyric acid-ergicneurons. In: Chan-Palay V., Palay S.L. (eds): Cytochemical Methods in Neuroanatomy. New York: Alan R. Liss, pp. 297–329.
113. Wu J.Y., Lin C.T., Brandon C., Chan T.S., Mohler H., Richards J.G. (1982). Regulation and immunocytochemical characterisation of glutamic acid decarboxylase. In: V. Chan-Palay, S.L. Palay (eds): Cytochemical Methods in Neuroanatomy. New York: Alan R. Liss, pp. 279–296.

114. Seguela P., Gamrani H., Geffard M., Calas A., Le Moal M. (1985). Ultrastructural immunocytochemistry of γ-aminobutyrate in the cerebral and cerebellar cortex of the rat. Neuroscience 16:865–874.
115. Mugnaini E., Oertel W.H. (1984). An atlas of the distribution of GABAergic neurons and terminals in the rat CNS as revealed by GAD immunocytochemistry. In: A. Bjorklund, T. Hokfelt, M.J. Kuhar (eds): Classical Transmitters and Transmitter Receptors in the CNS, Part II. Handbook of Chemical Neuroanatomy. Vol. 3. Amsterdam: Elsevier, pp. 247–272.
116. Ottersen O.P., Storm-Mathisen J. (1984). Glutamate- and GABA-containing neurons in the mouse and rat brain, as demonstrated with a new immunocytochemical technique. J. Comp. Neurol. 229:374–392.
117. Somogyi P., Hodgson A.J., Chubb I.W., Penke B., Erdei A. (1985). Antiserum to γ-aminobutyric acid: II. Immunocytochemical application to the central nervous system. J. Histochem. Cytochem. 33:240–248.
118. McLaughlin B.J., Wood J.G., Saito K., Barber R., Vaughn J.E., Roberts E., Wu J.Y. (1974). The fine structural localisation of glutamate decarboxylase in synaptic terminals of rodent cerebellum. Brain Res. 76:377–391.
119. Saito K., Barber R., Wu J.Y., Matsuda T., Roberts E., Vaughn J.E. (1974). Immunohistochemical localisation of glutamic acid decarboxylase in rat cerebellum. Proc. Natl. Acad. Sci. USA 71:269–273.
120. Richards J.G., Schoch P., Haring P., Takacs B., Mohler H. (1987). Resolving GABA A/benzodiazepine receptors: Cellular and subcellular localization in the CNS with monoclonal antibodies. J. Neurosci. 7:1866–1886.
121. Faull R.L.M., Villiger J.W., Holford N.H.G. (1987). Benzodiazepine receptors in the human cerebellar cortex: A quantitative autoradiographic and pharmacological study demonstrating the predominance of type I receptors. Brain Res. 411:379–385.
122. Ottersen O.P., Davanger S., Storm-Mathisen J. (1987). Glycine-like immunoreactivity in the cerebellum of rat and Senegalese baboon, *Papio papio*: A comparison with the distribution of GABA-like immunoreactivity and with (^{3}H)glycine and (^{3}H)GABA uptake. Exp. Brain Res. 66:211–221.
123. Ottersen O.P., Storm-Mathisen J. (1987). Localization of amino acid neurotransmitters by immunocytochemistry. Topics Neurosci. 10:250–255.
124. Wilkin G.P., Csillag A., Balazs R., Kingsbury A.E., Wilson J.E., Johnson A.L. (1981). Localization of high affinity (^{3}H)glycine transport sites in the cerebellar cortex. Brain Res. 216:11–33.
125. Triller A., Cluzeaud F., Korn H. (1987). Gamma-aminobutyric acid-containing terminals can be apposed to glycine receptors at central synapses. J. Cell Biol. 104:947–956.
126. Ito M., Yoshida M. (1964). The cerebellar-evoked monosynaptic inhibition of Deiters' neurones. Experientia 20:515–516.
127. Ito M., Yoshida M., Obata K. (1964). Monosynaptic inhibition of the intracerebellar nuclei induced from the cerebellar cortex. Experientia 20:575–576.
128. Ito M., Udo M., Mano M., Kawai N. (1970). Synaptic action of the fastigiobulbar impulses upon neurones in the medullary reticular formation and vestibular nuclei. Exp. Brain Res. 11:29–47.
129. Ito M., Highstein S.M., Fukuda J. (1970). Cerebellar inhibition of the vestibulo-ocular reflex in rabbit and cat and its blockage by picrotoxin. Brain Res. 17:524–526.
130. Fonnum F., Storm-Mathisen J., Walberg F. (1970). Glutamate decarboxylase in inhibitory neurons. A study of the enzyme in Purkinje cell axons and boutons in the cat. Brain Res. 20:259–275.
131. Bishop G.A., Ho R.H. (1985). The distribution and origin of serotonin immunoreactivity in the rat cerebellum. Brain Res. 331:195–207.
132. Bishop G.A., Ho R.H., King J.S. (1985). Localization of serotonin immunoreactivity in the opossum cerebellum. J. Comp. Neurol. 235:301–321.
133. Takeuchi Y., Kimura H., Sano Y. (1982). Immunohistochemical demonstration of serotonin-containing nerve fibers in the cerebellum. Cell Tissue Res. 226:1–12.
134. Armstrong D.L., Hay M., Terrian D.M. (1987). Modulation of cerebellar granule cell activity by iontophoretic application of serotonergic agents. Brain Res. Bull. 19:699–704.
135. Inukai T. (1928). On the loss of Purkinje cells, with advancing age, from the cerebellar

cortex of the albino rat. J. Comp. Neurol. 45:1–31.
136. Somogyi P., Hamori J. (1976). A quantitative electron microscopic study of the Purkinje cell axon initial segment. Neuroscience 1:361–365.
137. Jakob A. (1928). Das Kleinhirn. In: W. Mollendorff (ed): Handbuch der Mikroskopischen Anatomie des Menschen. Berlin: Springer, pp. 674–916.
138. Hamori J., Szentagothai J. (1968). Identification of synapses formed in the cerebellar cortex by Purkinje axon collaterals: An electron microscope study. Exp. Brain Res. 5:118–128.
139. Chan-Palay V. (1971). The recurrent collaterals of Purkinje cell axons: A correlated study of the rat's cerebellar cortex with electron microscopy and tha Golgi method. Z. Anat. Entwickl.-Gesch. 134:200–234.
140. De Camilli P., Miller P.E., Levitt P., Walter U., Greengard P. (1984). Anatomy of cerebellar Purkinje cells in the rat determined by a specific immunohistochemical marker. Neuroscience 11:761–817.
141. Bishop G.A. (1982). The pattern of distribution of the local axonal collaterals of Purkinje cells in the intermediate cortex of the anterior lobe and paramedian lobule of the cat cerebellum. J. Comp. Neurol. 210:1–9.
142. Lemkey-Johnston N., Larramendi L.M.H. (1968). Types and distribution of synapses upon basket and stellate cells of the mouse cerebellum. J. Comp. Neurol. 134:73–112.
143. Leranth C., Hamori J. (1981). Quantitative electron microscope study of synaptic terminals to basket neurons in cerebellar cortex of rat. Z. Mikrosk.-Anat. Forsch. 95:1–14.
144. Larramendi L.M.H., Lemkey-Johnston N. (1970). The distribution of recurrent Purkinje collateral synapses in the mouse cerebellar cortex: An electron microscopic study. J. Comp. Neurol. 138:451–482.
145. King J.S., Bishop G.A. (1982). The synaptic features of horseradish peroxidase labelled recurrent collaterals in the ganglionic plexus of the cat cerebellar cortex. J. Neurocytol. 11:867–880.
146. O'Leary J.L., Petty J., Smith J.M., O'Leary M., Inukai S. (1968). Cerebellar cortex of rat and other animals. A structural and ultrastructural study. J. Comp. Neurol. 134:401–432.
147. Gravel C., Eisenman L.M., Sasseville R., Hawkes R. (1987). Parasagittal organization of the rat cerebellar cortex: Direct correlation between antigenic Purkinje cell bands revealed by mabQ113 and the organization of the olivocerebellar projection. J. Comp Neurol. 265:294–310.
148. Bishop G.A., McCrea R.A., Lighthall J.W., Kitai S.T. (1979). An autoradiographic and HRP study of the projection from the cerebellar cortex to the nucleus interpositus anterior and nucleus interpositus posterior of the cat. J. Comp. Neurol. 185:735–756.
149. Eager R.P. (1966). Patterns and modes of termination of cerebellar cortico-nuclear pathways in the monkey (*Macaca mulatta*). J. Comp. Neurol. 126:551–566.
150. Haines D.E. (1976). Cerebellar cortico-nuclear and corticovestibular fibers of the anterior lobe vermis in a prosimian primate (*Galago senegalensis*). J. Comp. Neurol. 170:67–96.
151. Walberg F., Jansen J. (1964). Cerebellar corticonuclear projection studied experimentally with silver impregnation methods. J. Hirnforschung. 6:338–354.
152. Campistron G., Geffard G., Buijs R.M. (1986). Immunological approach to the detection of taurine and immunocytochemical results. J. Neurochem. 46:862–868.
153. Ottersen O.P. (1987). Postembedding light- and electron microscopic immunocytochemistry of amino acids: Description of a new model system allowing identical conditions for specificity testing and tissue processing. Exp. Brain Res. 69:167–174.
154. Nilaver G., Defendini R., Zimmerman E.A., Beinfeld M.C., O'Donohue T.L. (1982). Motilin in the Purkinje cell of the cerebellum. Nature 295:597–598.
155. Chan-Palay V., Ito M., Tongroach P., Sakurai M., Palay S. (1982). Inhibitory effects of motilin, somatostatin, (leu)enkephalin, (met)enkephalin and taurine on neurons of the lateral vestibular nucleus: Interactions with γ-aminobutyric acid. Proc. Natl. Acad. Sci. USA 79:3355–3359.
156. Slemmon J.R., Danho W., Hempstead J.L., Morgan J.I. (1985). Cerebellin: A quantifiable marker for Purkinje cell maturation. Proc. Natl. Acad. Sci. USA 82:7145–7148.
157. Mugnaini E., Morgan J.I. (1987). The neuropeptide cerebellin is a marker for two similar neuronal circuits in rat brain. Proc. Natl. Acad. Sci. 84:8692–8696.
158. Mugnnini E., Dahl A.L., Morgan J.I. (1988). Cerebellin is a postsynaptic neuropeptide. Synapse 2:125–138.

159. Sotelo C., Alvarado-Mallart R.M. (1987). Reconstruction of the defective cerebellar circuitry in adult Purkinje cell degeneration mutant mice by Purkinje cell replacement through transplantation of solid embryonic implants. Neuroscience 20:1-22.
160. Hollingworth T., Berry M. (1975). Network analysis of dendritic fields of pyramidal cells in neocortex and Purkinje cells in the cerebellum of the rat. Proc. R. Soc. Lond. (Biol.) 270:227-264.
161. Szentagothai J., Rajkovits K. (1959). Über den ursprung der kletterfasern des kleinhirns. Z. Anat. Entwickl. Gesch. 121:130-141.
162. Desclin J.C. (1974). Histological evidence supporting the inferior olive as the major source of cerebellar climbing fibers in rat. Brain Res. 77:365-384.
163. Llinas R., Bloedel J.R., Hillman D.E. (1969). Functional characterization of neuronal circuitry of frog cerebellar cortex. J. Neurophysiol. 32:847-870.
164. Crepel F., Mariani J., Delhaye-Bouchaud N. (1976). Evidence for a multiple innervation of Purkinje cells by climbing fibers in the immature rat cerebellum. J. Neurobiol. 7:567-578.
165. Eccles J.C., Llinas R., Sasaki K. (1966). The excitatory synaptic action of climbing fibres on the Purkinje cells of the cerebellum. J. Physiol. 182:268-296.
166. Perry T.L., Currier R.D., Hansen S., McLean J. (1977). Aspartate-taurine imbalance in dominantly inherited olivocerebellar atrophy. Neurology 27:257-261.
167. Rea M.A., McBride W.J., Rohde R.H. (1980). Regional and synaptosomal levels of amino acid neurotransmitters in the 3-acetylpyridine deafferented rat cerebellum. J. Neurochem. 34:1106-1108.
168. Wiklund L., Toggenburger G., Cuenod M. (1982). Aspartate: Possible neurotransmitter in cerebellar climbing fibers. Science 216:78-80.
169. Wiklund L., Toggenburger G., Cuenod M. (1984). Selective retrograde labelling of the rat olivocerebellar climbing fiber system with (^{3}H)-D-aspartate. Neuroscience 13:441-468.
170. King J.S., Ho R.H., Bishop G.A. (1986). Anatomical evidence for enkephalin immunoreactive climbing fibres in the cerebellar cortex of the opossum. J. Neurocytol. 15:545-559.
171. Schulman J.A. (1981). Anatomical distribution and physiological effects of enkephalin in rat inferior olive. Regul. Pept. 2:125-137.
172. Palkovits M., Leranth C., Gorcs T., Scott Young W. (1987). Corticotropin-releasing factor in the olivocerebellar tract of rats: Demonstration by light- and electron-microscopic immunohistochemistry and in situ hybridization histochemistry. Proc. Natl. Acad. Sci. USA 84:3911-3915.
173. Cummings S., Sharp B., Elde R. (1988). Corticotropin-releasing factor in cerebellar afferent systems: A combined immunohistochemistry and retrograde transport study J. Neurosci. 8:543-554.
174. Van den Dungen H.M., Groenewegen H.J., Tilders F.J.H., Schoemaker J. (1988). Immunoreactive corticotropin releasing factor in adult and developing rat cerebellum: Its presence in climbing and mossy fibres. J. Chem. Neuroanat. 1:339-349.
175. Fox C.A., Bernard J.W. (1957). A quantitative study of the Purkinje cell dendritic branchlets and their relationship to afferent fibres. J. Anat. 91:299-313.
176. Hama K., Kosaka T. (1979). Purkinje cell and related neurons and glia cells under high-voltage electron microscopy. In: H.M. Zimmerman (ed): Progress in Neuropathology. New York: Raven Press, pp. 61-77.
177. Palkovits M., Magyar P., Szentagothai J. (1971). Quantitative histological analysis of the cerebellar cortex in the cat. III. Structural organization of the molecular layer. Brain Res. 34:1-18.
178. Eccles J.C., Llinas R., Sasaki K. (1966). Parallel fibre stimulation and the responses induced thereby in the Purkinje cells of the cerebellum. Exp. Brain. Res. 1:17-39.
179. Brand S., Dahl A.L., Mugnaini E. (1976). The length of parallel fibers in the cat cerebellar cortex. An experimental light and electron microscopic study. Exp. Brain Res. 26:39-58.
180. Mugnaini E. (1983). The length of cerebellar parallel fibers in chicken and rhesus monkey. J. Comp. Neurol. 220:7-15.
181. Hamori J., Szentagothai J. (1964). The "crossing over" synapse. An electron microscope study of the molecular layer in the cerebellar cortex. Acta Biol. Hung. 15:95-117.
182. Ebner T.J., Bloedel J.R. (1987). Climbing fiber afferent system: Intrinsic properties and role in cerebellar information processing. In: New Concepts in Cerebellar Neurobiology. pp. 371-386.

183. Andersen P., Eccles J.C., Voorhoeve P.E. (1964). Postsynaptic inhibition of cerebellar Purkinje cells. J. Neurophysiol. 27:1139–1153.
184. Eccles J.C., Llinas R., Sasaki K. (1966). The inhibitory interneurones within the cerebellar cortex. Exp. Brain Res. 1:1–16.
185. Palkovits M., Magyar P., Szentagothai J. (1971). Quantitative histological analysis of the cerebellar cortex in the cat. II. Cell numbers and densities in the granular layer. Brain Res. 32:15–30.
186. Llinas R. (1982). General discussion: Radial connectivity in the cerebellar cortex. A novel view regarding the functional organization of the molecular layer. Exp. Brain Res. (Suppl.) 6:189–194.
187. Scheibel M.E., Scheibel A.B. (1954). Observations on the intracortical relations of the climbing fibers of the cerebellum. A Golgi study. J. Comp. Neurol. 101:733–763.
188. Desclin J.C. (1976). Early terminal degeneration of cerebellar climbing fibers after destruction of the inferior olive in the rat. Synaptic relationships in the molecular layer. Anat. Embryol. 149:87–112.
189. Hamori J., Szentagothai J. (1980). Lack of evidence of synaptic contacts by climbing fibre collaterals to basket and stellate cells in developing rat cerebellar cortex. Brain Res. 186:454–457.
190. Albin R.L., Gilman S. (1989). Parasagittal zonation of GABA-B receptors in molecular layer of rat cerebellum. Eur. J. Pharmacol. 173:113–114.
191. Szentagothai J. (1965). The use of degeneration methods in the investigation of short neuronal connections. Brain Res. 14:1–32.
192. Andersen P., Eccles J.C., Voorhove P.E. (1963). Inhibitory synapses on somas of Purkinje cells in the cerebellum. Nature (London) 199:655–656.
193. Palay S.L. (1964). Fine structure of cerebellar cortex of the rat. Anat. Rec. 148:419.
194. Hamori J., Szentagothai J. (1965). The Purkinje cell basket: Ultrastructure of an inhibitory synapse. Acta Biol. Hung. 15:465–479.
195. Sotelo C., Llinas R. (1972). Specialized membrane junctions between neurons in the vertebrate cerebellar cortex. J. Cell. Biol. 53:271–289.
196. Palay S.L., Sotelo C., Peters A., Orkand P.M. (1968). The axon hillock and the initial segment. J. Cell. Biol. 38:193–201.
197. Szentagothai J. (1963). New data on the functional anatomy of synapses. Magy. Tud. Akad., Biol. Orv. Tud. Osztal. Kozl. 6:217–227.
198. Clos J., Legrand J. (1973). Effects of thyroid deficiency on the different cell populations of the cerebellum in the young rat. Brain Res. 63:450–455.
199. Balazs R., Hajos F., Johnson A.L., Tapia R., Wilkin G. (1974). Biochemical dissection of the cerebellum. Biochem. Soc. Trans. 2:682–687.
200. Ghandour M.S., Vincendon G., Gombos G. (1980). Astrocyte and oligodendrocyte distribution in adult rat cerebellum: An Immunohistological study. J. Neurocytol. 9:637–646.
201. Palkovits M., Mezey E., Hamori J., Szentagothai J. (1977). Quantitative histological analysis of the cerebellar nuclei in the cat. I. Numerical data on cells and on synapses. Exp. Brain Res. 28:189–209.
202. Gould B.B., Rakic P. (1981). The total number, time of origin and kinetics of proliferation of neurons comprising the deep cerebellar nuclei in the rhesus monkey. Exp. Brain Res. 44:195–206.
203. McCrea R.A., Bishop G.A., Kitai S.T. (1978). Morphological and electrophysiological characteristics of projection neurons in the nucleus interpositus of the cat cerebellum. J. Comp. Neurol. 181:397–420.
204. Matsushita M., Iwahori N. (1971). Structural organization of the interpositus and the dentate nuclei. Brain Res. 35:17–36.
205. Matsushita M., Iwahori N. (1971). Structural organization of the fastigial nucleus. I. Dendrites and axonal pathways. Brain Res. 25:597–610.
206. Monaghan P.L., Beitz A.J., Larson A.A., Altschuler R.A., Madl J.E., Mullett M.A. (1986). Immunocytochemical localization of glutamate-, glutaminase- and aspartate aminotransferase-like immunoreactivity in the rat deep cerebellar nuclei. Brain Res. 363:364–370.
207. Angaut P., Sotelo C. (1973). The fine structure of the cerebellar central nuclei in the cat. II. Synaptic organization. Exp. Brain Res. 16:431–454.

208. Wassef M., Simons J., Tappaz M.L., Sotelo C. (1986). Non-Purkinje cell GABA-ergic innervation of the deep cerebellar nuclei: A quantitative immunocytochemical study in C57BL and in Purkinje cell degeneration mutant mice. Brain Res. 399:125-135.
209. Balini C., Buisseret-Delmas C., Compoint C., Daniel H. (1989). The GABAergic neurones of the c cerebellar nuclei in the rat: Projections to the cerebellar cortex. Neurosci. Lett. 99:251-256.
210. Sotelo C., Angaut P. (1973). The fine structure of the cerebellar central nuclei in the cat. I. Neurons and neuroglial cells. Exp. Brain Res. 16:410-430.
211. Hamori J., Mezey E. (1977). Serial and triadic synapses in the cerebellar nuclei of the cat. Exp. Brain Res. 30:259-273.
212. Fonnum F., Walberg F. (1973). An estimation of the concentration of γ-aminobutyric acid and glutamate decarboxylase in the inhibitory Purkinje axon terminals in the cat. Brain Res. 54:115-127.
213. Houser R.C., Barber R., Vaughn J.E. (1984). Immunocytochemical localisation of glutamic acid decarboxylase in the dorsal lateral vestibular nucleus: Evidence for an intrinsic and extrinsic GABAergic innervation. Neurosci Lett. 47:213-220.
214. Van der Want J.J.L., Gerrits N.M., Voogd J. (1986). Autoradiography of mossy fiber terminals in the fastigial nucleus of the cat. J. Comp. Neurol. 258:70-80.
215. Van Der Want J.J.L., Voogd J. (1987). Ultrastructural identification and localization of climbing fiber terminals in the fastigial nucleus of the cat. J. Comp. Neurol. 258:81-90.
216. J.A. Heckroth, L.M. Eisenman (1988). Parasagittal organization of mossy fiber collaterals in the cerebellum of the mouse. J. Comp Neurol 270:385-394.
217. Chan-Palay V. (1973). Axon terminals of the intrinsic neurons in the nucleus lateralis of the cerebellum. An electron microscope study. Z. Anat. Entwickl.-Gesch. 142:187-206.
218. Angaut P., Sotelo C. (1987). The dentato-olivary projection in the rat as a presumptive GABAergic link in the olivo-cerebello-olivary loop. An ultrastructural study. Neurosci. Lett. 83:227-231.
219. Nelson B., Barmak N.H., Mugnaini E. (1984). A GABAergic cerebello-olivary projection in the rat. Soc. Neurosci. Abstr. 10:539.
220. Buisseret-Delmas C., Batini C., Compoint C., Daniel H., Menetrey D. (1987). The GABAergic neurons of the cerebellar nuclei: A comparison of the projection to the inferior olive, the cerebellar cortex and the bulbar reticular formation. In: The Olivocerebellar System in Motor Control, Satellite Symposium of the 2nd IBRO World Congress of Neuroscience, Turin, 9-12 August, p. 42.
221. Chan-Palay V. (1973). On certain fluorescent axon terminals containing granular synaptic vesicles in the cerebellar nucleus lateralis. Z. Anat. Entwickl.-Gesch. 142:239-258.
222. Dow R.S. (1938). Efferent connections of the flocculonodular lobe in *Macaca mulatta*. J. Comp. Neurol. 68:297-305.
223. Angaut P., Brodal A. (1967). The projection the "vestibulocerebellum" onto the vestibular nuclei in the cat. Arch. Ital. Biol. 105:441-479.
224. Mugnaini E., Walberg F., Hauglie-Hanssen E. (1967). Observations on the fine structure of the lateral vestibular nucleus (Deiters' nucleus) in the cat. Exp. Brain Res. 4:146-186.
225. Mugnaini E., Walberg F., Brodal A. (1967). Mode of termination of primary vestibular fibres in the lateral vestibular nucleus: An experimental electron microscopical study in the cat. Exp. Brain Res. 4:187-211.
226. Sotelo C., Palay S.L. (1968). The fine structure of the lateral vestibular nucleus in the rat. I. Neurons and neuroglial cells. J. Cell. Biol. 36:151-179.
227. Sotelo C., Palay S.L. (1970). The fine structure of the lateral vestibular nucleus in the rat. II. Synaptic organization. Brain Res. 18:93-115.
228. Walberg F., Bowsher D., Brodal A. (1958). The termination of primary vestibular fibers in the vestibular nuclei in the cat. An experimental study with silver methods. J. Comp. Neurol. 110:391-419.
229. Pompeiano O., Brodal A. (1957). Spino-vestibular fibers in the cat: An experimental study. J. Comp. Neurol. 108:353-381.
230. Pompeiano O., Mergner T., Corvaja N. (1978). Commissural, perihypoglossal and reticular afferent projections to the vestibular nuclei in the cat. An experimental anatomical study with the method of the retrograde transport of horseradish peroxidase. Arch. Ital. Biol. 116:130-172.
231. Brodal A., Pompeiano O., Walberg F. (1962). The Vestibular Nuclei and their

Connections. Anatomy and Functional Correlations. Edinburgh: Oliver and Boyd.
232. Van Rossum J. (1969). Corticonuclear and Corticovestibular Projections of the Cerebellum: An Experimental Investigation of the Anterior Lobe, Simple Lobule and the Caudal Vermis in the Rabbit. Thesis. Van Gorcum, Assen.
233. Brodal A. (1974). Anatomy of the vestibular nuclei and their connections. In: H.H. Kornhuber (ed): Handbook of Sensory Physiology. Berlin: Springer-Verlag, pp. 240–352.
234. Haines D.E., Patrick G.W., Satrulee P. (1982). Organization of cerebellar corticonuclear fiber systems. In: S.L. Palay, V. Chan-Palay (ed): The Cerebellum. Berlin: New Vistas, Springer-Verlag, pp. 320–367.
235. Ito M. (1984). The Cerebellum and Neural Control. New York: Raven Press.
236. Ito M. (1991). Structural-functional relationships in cerebellar and vestibular systems. Arch. Ital. Biol. 129:53–61.
237. Carrea R.M.E., Reissig M., Mettler F.A. (1947). The climbing fibers of the simian and feline cerebellum. Experimental inquiry into their origin by lesions of the inferior olives and deep cerebellar nuclei. J. Comp. Neurol. 87:321–365.
238. Cohen D., Chambers W.W., Sprague J.M. (1958). Experimental study of efferent projection from the cerebellar nuclei to the brain stem of the cat. J. Comp. Neurol. 109:233–259.
239. Gould B.B., Graybiel A. (1976). Afferents to the cerebellar cortex in the cat. Evidence for an intrinsic pathway leading from the deep nuclei to the cortex. Brain Res. 110:601–611.
240. Tolbert D.L., Bantli H., Bloedel J.R. (1977). The intracerebellar nucleocortical projection in a primate. Exp. Brain Res. 30:425–434.
241. Tolbert F.L., Bantli H., Bloedel J.R. (1978). Organizational features of the cat and monkey cerebellar nucleocortical projection. J. Comp. Neurol. 182:39–56.
242. Gould B.B. (1979). The organization of afferents to the cerebellar cortex in the cat: Projections from the deep cerebellar nuclei. J. Comp. Neurol. 184:27–42.
243. Dietrichs E. (1981). The cerebellar corticonuclear and nucleocortical projections in the cat as studied with anterograde and retrograde transport of horseradish peroxidase. IV. The paraflocculus. Exp. Brain Res. 44:235–242.
244. Buisseret-Delmas C., Angaut P. (1988). The cerebellar nucleocortical projections in the rat. A retrograde labelling study using horseradish peroxidase combined to a lectin. Neirosci. Lett. 84:255–260.
245. Dietrichs E., Walberg F. (1979). The cerebellar corticonuclear and nucleocortical projections in the cat as studied with anterograde and retrograde transport of horseradish peroxidase. I. The paramedian lobule. Anat. Embryol. 158:13–39.
246. Dietrichs E., Walberg F. (1987). Cerebellar nuclear afferents—where do they origniate? A reevaluation of the projections from some lower brain stem nuclei. Anat. Embryol. 177:165–172.
247. Junck L., Gilman S., Rothley J.R., Betley A.T., Koeppe R.A., Hichwa R.D. (1988). A relationship between metabolism in frontal lobes and cerebellum in normal subjects studied with PET. J. Cereb. Blood Flow Metab. 8:774–782.
248. Brodal A. (1981). Neurological Anatomy, 3rd ed. London: Oxford University Press.
249. Bloedel J.R., Courville J. (1981). Cerebellar afferent systems. In: J.M. Brookhart, V.B. Mountcastle, V.B. Brooks (eds): Handbook of Physiology. Bethesda, MD: pp. 735–829.
250. Grant G. (1962). Spinal course and somatotopically localized termination of the spinocerebellar tracts. An experimental study in the cat. Acta Physiol. Scand. 193:1–45.
251. Grant G. (1962). Projection of the external cuneate nucleus onto the cerebellum in the cat: An experimental study using silver methods. Exp. Neurol. 5:179–195.
252. Rinvik E., Walberg F. (1975). Studies on the cerebellar projections from the main and external cuneate nuclei in the cat by means of retrograde axonal transport of horseradish peroxidase. Brain Res. 95:371–381.
253. Oscarsson O., Uddenberg N. (1964). Identification of a spinocerebellar tract activated from forelimb afferents in the cat. Acta Physiol. Scand. 62:125–136.
254. Miskolczy D. (1931). Ueber die endigungsweise der spinocerebellaren bahnen. Z. Anat. Entwickl.-Gesch. 96:537–542.
255. Brodal A., Grant G. (1962). Morphology and temporal course of degeneration in cerebellar mossy fibers following transection of spinocerebellar tracts in the cat. An experimental study with silver methods. Exp. Neurol. 5:67–87.

256. Ikeda M., Matsushita M. (1973). Electron microscopic observations on the spinal projections to the cerebellar nuclei in the cat. Experientia 29:1280–1282.
257. Robertson B., Grant G., Bjorklund M. (1983). Demonstration of spinocerebellar projections in cat using anterograde transport of WGA-HRP, with some observations on spinomesencephalic and spinothalamic projections. Exp. Brain Res. 52:99–104.
258. Korte G.E., Mugnaini E. (1979). The cerebellar projection of the vestibular nerve in the cat. J. Comp. Neurol. 184:265–278.
259. Brodal P., Dietrichs E., Walberg F. (1986). Do pontocerebellar mossy fibres give off collaterals to the cerebellar nuclei? An experimental study in the cat with implantation of crystalline HRP-WGA. Neurosci Res. 4:12–24.
260. Matsushita M., Ikeda M. (1970). Olivary projections to the cerebellar nuclei in the cat. Exp. Brain Res. 10:488–500.
261. Groenewegen H.J., Voogd J., Freedman S.L. (1979). The parasagittal zonation within the olivocerebellar projection. II. Climbing fiber distribution in the intermediate and hemispheric parts of cat cerebellum. J. Comp. Neurol. 183:551–602.
262. Courville J., Faraco-Cantin F. (1980). Topography of the olivocerebellar projection: An experimental study in the cat with an autoradiographic tracing method. In: J. Courville, D. Montigny, Y. Lamarre (eds): The Inferior Olivary Nucleus: Anatomy and Physiology. New York: Raven Press, pp. 235–277.
263. Hoddevik G.H. (1978). The projection from nucleus reticularis tegmenti pontis onto the cerebellum in the cat. Anat. Embryol. 153:227–242.
264. Hawkes R., Leclerc N. (1986). Immunocytochemical demonstration of topographic ordering of Purkinje cell axon terminals in the fastigial nuclei of the rat. J. Comp. Neurol. 244:481–491.
265. McCrea R.A., Bishop G.A., Kitai S.T. (1977). Electrophysiological and horseradish peroxidase studies of precerebellar afferents to the nucleus interpositus anterior. II. Mossy fiber system. Brain Res. 122:215–228.
266. Brodal A., Lacerda A.M., Destombes J., Angaut P. (1972). The pattern in the projection of the intracerebellar nuclei onto the nucleus reticularis tegmenti pontis in the cat. An experimental anatomical study. Exp. Brain Res. 16:140–160.
267. Hokfelt T., Fuxe K. (1969). Cerebellar monoamine nerve terminals, a new type of afferent fibers to the cortex cerebelli. Exp. Brain Res. 9:63–72.

3. PHYSIOLOGY OF THE CEREBELLUM

MASAO ITO

1. HISTORICAL OVERVIEW

Voluminous data of classical studies performed in the last century through the early years of this century have been compiled in the monograph of Dow and Moruzzi [1]. In brief, Rolando (1809) disclosed the involvement of the cerebellum in motor function, Fourens (1824) pointed out a special role of the cerebellum in the coordination of movements, and Flourens (1842) and Luciani (1891) established functional compensation as a characteristic feature of cerebellar action. Sherrington (1897), Löwenthal and Horseley (1897), Fodera (1923), and Pollock and Davis (1923) recognized inhibition as another characteristic feature of cerebellar action.

In the 1940s to 1950s, Snider and Stowell [2] and Chambers and Sprague [3] demonstrated functional localization in the cerebellum, and Moruzzi [4] revealed the involvement of the cerebellum not only in motor, but also in autonomic, function. Brookhart, Moruzzi, and Snider [5] and Granit and Phillips [6] attempted unit spike recording from cerebellar neurons. In the 1960s, the development of electrophysiological techniques allowed detailed analyses of the neuronal circuitry of the cerebellum, as summarised in the monograph of Eccles et al. [7]. This provoked theoretical modeling of the cerebellum [8,9]. In the 1970s, emphasis shifted to investigation of specific roles played by the cerebellum in various functional subsystems, with record-

A. Plaitakis (ed.), CEREBELLAR DEGENERATIONS: CLINICAL NEUROBIOLOGY. Copyright © 1992.
Kluwer Academic Publishers, Boston. All rights reserved.

ing of signals from cerebellar neurons in an alert animal performing a certain task under cerebellar control. Through the 1980s, another prominent trend was the accumulation of physiological data at cellular levels in close relation to biochemical and immunohistochemical data at molecular levels [10].

This chapter aims at reviewing recent advances in cerebellar studies in terms of both cellular and system physiology. Current views of cellular processes (Section 2) and signal processing (Section 3) in the cerebellum and the contribution of the cerebellum to various forms of bodily function (Sections 4–6) will be introduced, and a general scheme of cerebellar control will be considered (Section 7).

2. CELLULAR PROCESSES IN THE CEREBELLUM

The way in which individual neurons operate as functional units of the cerebellum has been investigated in terms of the electrical properties of resting and active membrane, and the modes of signal transmission at synapses. These have been investigated extensively in in-vivo preprations and in-vitro slice preparations of the cerebellum, and also in acutely isolated or tissue-cultured neurons.

2.1. Electrical properties of resting membrane

Electrical properties of membrane provide a fundamental basis of neuronal function. Resting neuronal membrane exhibits a parallel combination of electrical resistance and electrical capacitance against currents flowing across it. Due to the resistance and capacitance distributed across the membrane and interconnected through intradendritic axial resistance and extracellular fluid resistance, the dendritic shaft of a neuron behaves like a cable along which a potential spreads passively (electrotonically). These cable properties play a key role in determining the transient time course of synaptic potentials and the spatial integration of these potentials in the dendritic arbor of a neuron.

An attempt has been made to build a multicompartmental model of the magnificent dendritic arbor of a Purkinje cell [11]. Dendritic networks containing 86 spiny branchlets have been represented by 1089 coupled compartments (Figure 3-1). In order for this model to behave consistently with experimental observations, an unexpectedly large difference of membrane-specific resistivity between the soma (760 ohm cm^2) and dendrites (45,740 ohm cm^2) had to be assumed. A large value of cytoplasm-specific resistivity (225 ohm cm) and an ordinary value of membrane-specific capacitance (1.16 $\mu F/cm^2$) are suggested. Such a multicompartmental model is useful in visualizing the spatial integration in the dendritic arbor of a Purkinje cell. However, excitation in a dendritic membrane (see Section 2.2) would alter the cable properties of dendrites, and there can also be nonlinear interaction between neighboring synapses activated in a temporal sequence. The spatiotemporal features of dendritic integration would, therefore, be more complicated than is expected in a purely passive model of dendrites.

2.2. Membrane excitation

Neurons generate action potentials, which are all-or-none transient deviations of membrane potential that actively propagate along an axon as impulse signals. An action potential consists of an initial spike potential, which is large in size and fast in its time course, accompanied by a small, slow afterpotential. These potentials are generated by gating of special ion channels in the membrane under the influence of the membrane potential.

Spike potentials generated in the soma of a Purkinje cell and propagated along its axon are ordinary Na spikes, due to opening of the voltage-dependent Na^+ channels and the associated opening of K^+ channels [12]. When spike potentials are evoked at an axon, these propagate up the axon to the soma antidromically but do not further invade the dendrites. In contrast, the dendritic membrane of a Purkinje cell generates Ca spikes due to opening of Ca^{2+} channels and the associated opening of K^+ channels [14] when the membrane is sufficiently depolarized. Ca spikes in Purkinje cell dendrites are slower in their time course than Na spikes in the soma.

When a Ca spike is produced through synaptic transmission from climbing fibers (see Section 2.3), it is usually accompanied by a long-lasting afterpotential, called a *plateau potential* [15]. The plateau potential is also generated by opening of Ca^{2+} channels in the dendritic membrane [13]. The plateau potential is then followed by slow membrane hyperpolarization, which, after repetitive stimulation of a climbing fiber, produces marked membrane hyperpolarization [16]. The composition of Ca^{2+} channels in Purkinje cells differs from other central neurons because of the presence of Ca^{2+} channels called *P channels* [17].

A Purkinje cell in in-vivo conditions spontaneously discharges two types of spike potentials, simple and complex.

A simple spike represents a somatic Na^+ spike triggered by integrated synaptic inputs from granule cells, stellate cells, and basket cells, while a complex spike is a burst train of a few to several spikes triggered by a climbing fiber impulse (see Section 2.3). Component spikes of a complex spike represent Ca spikes generated in dendrites, which then transmit to Na spikes of the soma and axon. Impulses propagating along the axon are largely derived from simple spikes and reflect the integrated sum of passive and active processes occurring in the somadendritic membrane of a Purkinje cell, due to frequency modulation of time analog signals.

The axons of granule cells (parallel fibers) form the molecular layer, which is the most conspicuous assembly of unmyelinated fibers in the central nervous system. These fibers conduct impulses at a velocity of 0.3 m/sec in the cat [18]. The parallel-fiber spike potentials are Na spikes, which are sensitive to tetrodotoxin.

2.3. Synaptic transmission

Synaptic potentials are generated by the action of neurotransmitter substance at a synapse on a soma or dendrite (or even an axon). There are two major

62   I. Basic neurosciences of the cerebellum

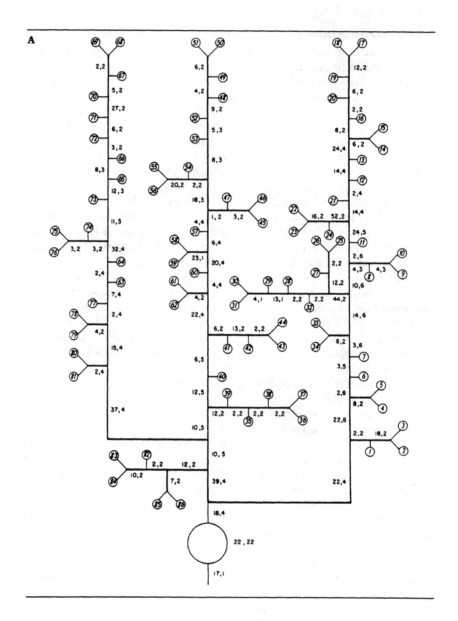

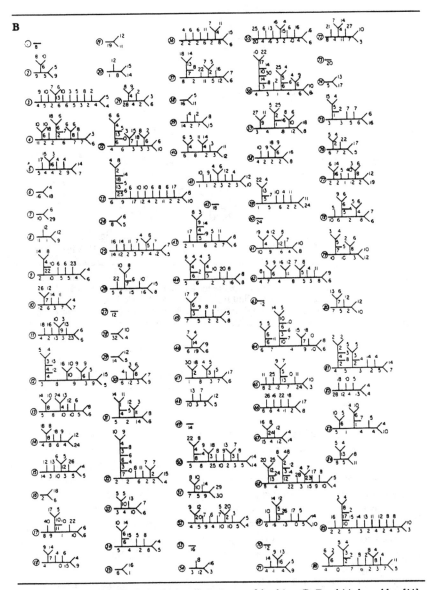

Figure 3-1. Cable model of a Purkinje cell. **A**: Stems of dendrites. **B**: Dendritic branchlets [11].

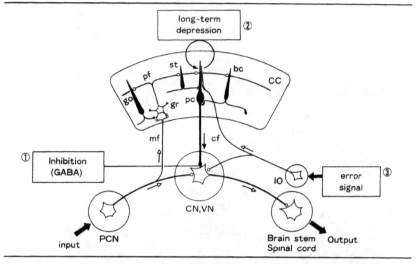

Figure 3-2. Structure of a cerebellar corticonuclear microcomplex. CC = Cerebellar cortical microzone; CN, VN = Cerebellar and vestibular nuclei; PCH = Precerebellar nuclei; IO = Inferior olive; mf = Mossy fiber; cf = Climbing fiber; pf = Purkinje cell; bc = baslet cell; st = stellate cell; gr = granule cell; go = Golgi cell; pf = Parallel fiber. (1–3): Major findings that suggest adaptive operation of the corticonuclear microcomplex.

types of synaptic potentials, EPSPs and IPSPs. EPSPs depolarize and IPSPs hyperpolarize the membrane (i.e., decrease and increase membrane potential), and these exert excitatory and inhibitory action, respectively, upon a postsynaptic cell.

Synaptic transmission is mediated by neurotransmitter molecules released from presynaptic axon terminals and receptor molecules that in the postsynaptic membrane react with the neurotransmitter. In principle, each neuron exerts a unique synaptic action at terminals of its axon through certain neurotransmitter/receptor mechanisms. Granule cells, climbing fibers, and mossy fibers are excitatory, while Purkinje, basket, stellate, and Golgi cells are inhibitory (Figure 3-2). Fusiform neurons in the granular layer (referred to as *Lugaro cells* [19]) project axons to the molecular layer [20], but since their target neurons have not been identified, their synaptic action remains unknown.

Axons of granule cells, which are parallel fibers, release L-glutamate as excitatory neurotransmitter, which in spine synapses of Purkinje-cell dendrites acts on a subtype of glutamate receptors that is selective to AMPA (α-amino-3-hydroxy-5-methyl-4-isoxazole-propionic acid) [21]. The excitatory neurotransmitter of climbing fibers has not yet been identified. L-aspartate has been a likely candidate [22], but L-homocysteate has recently become another likely candidate [23]. The involvement of L-glutamate is not ruled out [23].

IPSPs are generated through synaptic transmission from Purkinje cells, basket cells, and stellate cells, all of which utilize GABA as a neurotransmitter. These IPSPs are mediated by $GABA_A$ receptors and therefore should be due to the gating of Cl^- channels, as has been shown in other types of neurons.

In addition to these ionotropic receptors, coupled with ion channels, there are metabotropic receptors whose activation is linked with inositol-phospholipid metabolism via G-protein, yielding IP_3. IP_3, in turn, acts to release Ca^{2+} ions from intracellular stores. Purkinje cells contain metabotropic glutamate receptors [24,25] and are equipped with well-developed IP_3 receptors and Ca^{2+} stores [26,27]. Ca^{2+} ions so released may be involved in multiple cellular functions, but there is now evidence suggesting their involvement in synaptic plasticity (Section 2.5).

2.4. Synaptic modulation

Synaptic modulation means modification of synaptic transmission by a substance that by itself does not mediate the transmission. A modulator substance would interfere with synaptic transmission at various steps in the production, storage, release, uptake, destruction, or reaction of neurotransmitter substances with postsynaptic receptors. The cerebellum is now known to be equipped with a fairly complex modulatory system for synaptic transmission.

Noradrenaline- and serotonin-containing fibers in the cerebellum make some synaptic contact with Purkinje cells, but they also have a number of beaded swellings from which noradrenaline or serotonin would be released into the nonsynaptic extracellular space, probably exerting diffuse modulatory action upon cerebellar neurons. Noradrenaline is known to enhance [28], and serotonin to depress [29], the sensitivity of Purkinje cells to glutamate, i.e., the putative neurotransmitter of granule cells. Hence, these monoamines would exert modulatory action upon synaptic transmission from granule cells to other cerebellar neurons by regulating the sensitivity of postsynaptic glutamate receptor molecules. Noradrenaline also potentiates the release of glutamate from a slice of cerebellum [30], and it enhances the GABA sensitivity of Purkinje cells [28].

Parallel fibers have $GABA_B$ receptors [31], which are thought to mediate the action of GABA in blocking parallel fiber-to-Purkinje cell transmission [32]. $GABA_B$ receptors are known in the hippocampus to produce IPSPs by gating of the associated K^+ channels [33]. It may be supposed that GABA leaking from inhibitory synaptic sites in the cerebellar cortex acts to depress transmitter release from parallel fibers through $GABA_B$ receptors. Parallel fibers are also equipped with adenosine A1 receptors [34], and accordingly, adenosine acts to block parallel fiber-to-Purkinje cell transmission, presumably by depressing transmitter release from parallel fibers [35].

The effects of deprivation of climbing fibers will now be described due to their relation to synaptic modulation. Lesion of the inferior olive, either surgical or chemical (with 3-acetylpyridine or kainic acid), induces an instantaneous rise in simple spike discharge from Purkinje cells by a factor of 2 [36]. This effect may be due to loss of the sustained hyperpolarization that follows climbing fiber-evoked Ca spikes in Purkinje cell dendrites (Section 2.3) and/or reduced release of adenosine [37], which would normally depress parallel fiber-Purkinje cell transmission (see above). The increased discharge subsides in 2 weeks, returning to normal [38]. Climbing fiber deprivation also causes a reduction of the inhibitory synaptic action of Purkinje cells upon their target neurons in vestibular neurons [39,40]. This effect is accompanied by ultrastructural changes in Purkinje cell axon terminals [41], and there is no recovery. Climbing fibers could have a trophic action upon Purkinje cells by regulating their inhibitory synaptic action.

2.5. Synaptic plasticity

Synaptic plasticity is persistent modulation of synaptic efficacy initiated by the brief period of a triggering event. Long-term depression (LTD) is a type of synaptic plasticity that occurs at the synapse of a parallel fiber on a Purkinje cell when the synapse is activated repeatedly in conjunction with a climbing fiber innervating the same Purkinje cell (Figure 3-3). While the presence of such synaptic plasticity was suggested early on theoretical grounds [8,9], experimental evidence for LTD has become available only recently [42]. LTD includes the four major types of synaptic plasticity described thus far in the nervous system, together with long-term potentiation [43], sensitization [44], and synaptic sprouting [45]. Synaptic plasticity is supposed to play the role of a memory element in the function of a neuronal network. LTD would represent a memory element characteristic of cerebellar neuronal circuitry (Section 3).

It has been shown that the inflow of Ca^{2+} ions into Purkinje cell dendrites under the action of climbing fiber signals plays an essential role in the induction of LTD. Intracellular injection of EGTA, a Ca^{2+} chelator, abolishes LTD [46]. Postsynaptic inhibition in Purkinje cell dendrites by stellate cells also abolishes LTD [47], presumably because of its depressant action on Ca^{2+} inflow into Purkinje cell dendrites that accompanies climbing fiber impulses.

It has also been shown that LTD is due to desensitization of glutamate receptors involved in parallel fiber-Purkinje cell transmission [42]. Iontophoretic application of glutamate or its agonist, quisqualate, to a Purkinje cell dendrite in conjunction with climbing fiber stimulation causes sustained desensitization of quisqualate-specific receptors [48,49]. Application of glutamate paired with direct stimulation of a Purkinje cell, which elicits Ca^{2+} spikes in the

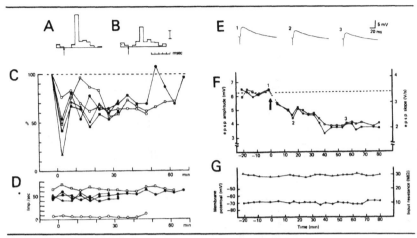

Figure 3-3. Long-term depression in Purkinje cells. **A-D:** Demonstration by an extracellular unit recording from a Purkinje cell in rabbit's flocculus. A vestibular nerve was electrically stimulated to excite the Purkinje cell via a mossy fiber pathway (histogram A). The vestibular nerve was then stimulated with the contralateral inferior olive at 4 Hz for 25 sec. The vestibular nerve responses were depressed thereafter (B). **C:** Rate of simple spike discharge relative to the control value obtained before the conjunction at time 0. **D:** Spontaneous simple spike discharge rate measured simultaenously with C [42]. **E:** Specimen records of EPSPs at the moment indicated in F [1-3]. **F:** Amplitude of EPSPs exemplified by E, plotted against the time before and after conjunction at time 0. **G:** Membrane potential and input resistance simultaneously recorded with F [16].

cell, lead to depression of glutamate sensitivity [50]. Thus, it is apparent that LTD results from the conjoint action of parallel fiber neurotransmitter on quisqualate-specific receptors from outside and Ca^{2+} ions from inside Purkinje cell dendrites.

Long-term desensitization of AMPA-selective glutamate receptors at parallel fiber synapses on Purkinje cells can be induced by activation of AMPA receptors in combination not only with Ca^{2+} spikes, but also with activation of metabotropic glutamate receptors. A brief (1-4 min) exposure of a cerebellar slice to 100 µM quisqualate, which activates both AMPA receptors and metabotropic glutamate receptors, in fact, causes long-term desensitization of the AMPA receptors that lasts over 10 hr [51,52]. Even though it is still unclear how metabotropic glutamate receptors are activated under normal functional conditons, quisqualate-induced desensitization provides a good model system for investigating the molecular mechanisms of LTD [53]. It has been shown that at enhanced Ca^{2+} levels, a chain of messenger reactions is activated to produce NO, to synthesize cGMP, to activate protein kinase G (PKG) (Figure 3-4). PKG may induce desensitization through the phosphorylation of AMPA receptors.

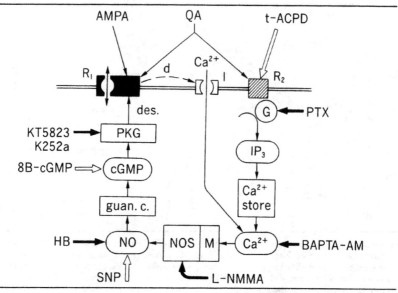

Figure 3-4. Chain of messenger reactions underlying long-term desensitization in cerebellar Purkinje cells. R_1 = AMPA-selective glutamate receptor. R_2 = metabotropic glutamate receptor. I = voltage-sensitive Ca^{2+} ion channel. QA = quisqualate; PTX = pertussis toxin; BAPTA-AM = membrane-soluble Ca^{2+} chelator; L-NMMA = inhibitor of NO synthase; SNP = sodium nitroprusside; HB = hemoglobin; guan. c. = guanylate cyclase; 8B-cGMP = 8-bromo-cyclic GMP; PKG = protein kinase G; KT5823 = PKG-specific inhibitor; K252a = nonselective protein kinase inhibitor; des. = desensitization; d = depolarization [25].

Thus, pertussis toxin, which blocks G-protein; BAPTA-AM, which is a membrane-soluble chelator of Ca^{2+}; L-NMMA (N^G-monomethyl L-arginin), which prevents NO production; hemoglobin, which absorbs NO; and KT5823, which is a specific inhibitor of PKG, effectively block quisqualate-induced long-term desensitization. Accordingly, either L-NMMA or hemoglobin effectively blocks LTD [54,55].

3. SIGNAL PROCESSING IN THE CEREBELLUM

The elaborate neuronal network in the cerebellar cortex is constituted of Purkinje, basket, stellate, Golgi, and granule cells and some other less well-identified neurons (Section 2.3). Impulses from various origins are sent into the cerebellar cortical network through both mossy fibers and climbing fibers, and after signal processing there impulses are sent out through Purkinje cell axons to cerebellar and vestibular neclei. Noradrenergic and serotonergic fibers are also components of the cerebellar cortical network, but in view of the presumed diffuse, modulatory nature of their action

(Section 2.4), these fibers may play a role in regulating functional conditions of the network, rather than directly contributing to signal processing.

Impulse signals have been recorded from the elements of cerebellar neuronal networks. These data, together with the structural data of the networks, provide the bases for assigning roles to cerebellar neurons in the operation of the networks and in the performance of cerebellar systems.

3.1. Mossy fiber-granule cell-Golgi cell transmission

Various modalities of sensory signals are conveyed by mossy fibers. These are of visual, auditory, somesthetic, vestibular (also lateral line organ), or gustatory origin, but no mossy fiber inputs of olfactory origin have been found. Mossy fibers also convey signals related to motor activity, such as the velocity of movement or the position of a limb or the eye. These are not merely of proprioceptive origin, but they imply motor command generated within the central nervous system. Afferent signals from visceral organs are also represented in mossy fiber inputs. It is notable that a large bulk of mossy fibers originating from pontine nuclei, the nucleus reticularis tegmenti pontis, and the lateral reticular nucleus mediate signals from the entire cerebral cortex. These inputs would reflect a variety of cerebral cortical activities.

Mossy fiber signals excite granule cells, and axons of granule cells (parallel fibers), in turn, excite Golgi cells. Golgi cells then inhibit granule cells. In theory, this inhibition has been suggested to prevent granule cells from overactivation by excessive mossy fiber inputs [8], or because of its negative feedback action with a slow decay time course, it may contribute an integrator action to the mossy fiber-to-granule cell transmission, and consequently may act as a phase shifter for mossy fiber signals transmitting to parallel fibers [56]. It has been suggested that a set of signals with varied phase relationships to mossy fiber input are generated at the filter circuit formed by granule cells and Golgi cells (Section 7). Experimental support for this suggestion is provided in that Purkinje cells in the flocculus exhibit modulation of simple spike discharge with a wide variation in phase shift (up to 360°), while sinusoidal head rotation on the horizontal plane activates ipsilateral vestibular mossy fibers in phase, and contralateral mossy fibers 180° out of phase, from ipsilateral head velocity [57].

3.2. Climbing fiber-Purkinje cell pathway

In contrast to mossy fibers arising from a large variety of neural structures, climbing fibers arise only from the inferior olive. Nevertheless, the recording of complex spikes has revealed that climbing fibers convey signals of various modalities [10], including peripheral nociceptive signals [58].

Concerning the functional meaning of climbing fiber signals, numerous interpretations have been proposed. Among earlier suggestions, "comparison between central command and peripheral signals" [59], "instruction of control errors" [60], and "misperformance in pattern recognition" [9] have similar implications in that climbing fibers monitor the performance of a system in which a given cerebellar circuit plays a role, and in that climbing fibers inform Purkinje cells with control errors due to their misperformance. Experimental evidence from studies of various forms of cerebellar control (Sections 4–6) have been accumulating to support this postulate. The fact that climbing fibers convey nociceptive signals [58] also supports this postulate, because nociception would arise from the erroneous performance of motor or autonomic systems.

The postulated control error represented by climbing fiber signals matches with the hypothesis regarding synaptic plasticity in Purkinje cells (Section 2.5). When a control error is produced through climbing fibers, LTD will be induced in parallel fiber-Purkinje cell synapses activated at the same time. This would lead to functional elimination of synapses responsible for the erroneous control, and the cerebellar circuitry would consequently be reorganized. This assumption of the self-organizing capability is the basis for theoretical formulation of the adaptive-learning capability of the cerebellar circuitry (Section 7).

An important recent finding is that cerebellar nuclei contain inhibitory neurons projecting to the inferior olive [61]. High-frequency activation of climbing fibers causes depression of inferior olive neurons, apparently through disinhibition of these inhibitory neurons, since simple spike discharges from Purkinje cells that inhibit nuclear neurons are suppressed following high-frequency climbing fiber stimulation [16]. A functional role of the nuclear inhibitory neurons in cerebellar mechanisms is not clear at the present, but it is possible that these are involved in selective blocking of climbing fiber responses to peripheral stimuli that do not represent control errors, as is typically seen in locomotion or active movement of limbs (Section 5.3).

Climbing fiber signals may also act to produce short-term modification of parallel fiber-Purkinje cell transmission through an as yet unidentified mechanism [62]. No theoretical formulation has been attempted to incorporate these short-term actions of climbing fibers into the capabilities of cerebellar cortical networks.

3.3. Basket cells and stellate cells

Stellate cells supply inhibitory synapses to the dendrites of Purkinje cells, and so would contribute to local computation in dendritic cables (Section 2.1). The simplest form of such a computation is that Purkinje cell dendrites transmit the algebraic sum of excitation and inhibition toward the soma, which may sensitively reflect differential changes in excitatory parallel fiber inputs to Purkinje cells. In addition, since the conditioning of climbing-fiber stimulation with stellate cell inhibition abolishes LTD (Section 2.5), the stellate cell may play a role in regulating the occurrence of LTD [47]. Basket cells supplying inhibitory synapses to somata of Purkinje cells would also contribute to computation in the cerebellar cortical network, even though the precise features of this role are still unclear. Blocking of basket cell inhibition with ionotophoretic application of bicucullin significantly alters mossy fiber responsiveness of floccular Purkinje cells to heard rotation [63]. This observation clearly indicates that basket cells contribute to signal transfer characteristics of a cerebellar cortical network for mossy fiber inputs.

3.4. Corticonuclear microcomplex

An elaborate zonal structure of the cerebellum revealed by recently described cerebellar anatomy has been attached a functional meaning in terms of modules of the cerebellum. A small area of the cerebellar cortex (microzone) receives climbing fibers from a small group of inferior olive neurons and, in turn, sends Purkinje cell axons to a small group of neurons in vestibular or cerebellar nuclei [64]. The climbing fibers projecting to a microzone supply collaterals to the nuclear neurons innervated by Purkinje cells of the microzone. A corticonuclear microcomplex [10] is thus formed through the cerebellar cortex, vestibular or cerebellar nuclei and the inferior olive (Figure 3-2). A corticonuclear microcomplex also receives mossy fibers in a dual manner such that mossy fibers projecting to a microzone supply excitatory synapses to the nuclear cells innervated by Purkinje cells of the microzone. A mossy fiber usually innervates a number of microzones in common, while a climbing fiber shows more limited divergence.

Three major findings of (1) inhibitory action of Purkinje cells on nuclear neurons (Section 2.3), (2) error representation by climbing fiber signals (3.2), and (3) LTD arising from conjunctive activation of climbing fibers and parallel fibers (Section 2.5) conjointly suggest that the mode of operation of a corticonuclear microcomplex is as follows. When climbing fibers inform a microzone about errors in the performance of an associated extracerebellar system, signal transfer characteristics across the microzone will be modified due to induction of LTD. Purkinje cell inhibition on nuclear cells will thus be altered so that the performance of the extracerebellar system associated with these nuclei will be modified. This will eventually lead to improved performance of the extracerebellar system such that error signals in climbing fibers will be minimized.

Elaborate information processing within a cortical microzone has been reproduced with some neuronal network models. In the simple perceptron model [9], the spatial pattern information carried by mossy fibers is distributed in a large number of granule cells from which Purkinje cells pick up appropriate information through selective connections established through learning. In the adaptive filter model of the cerebellum [65], temporal patterns of mossy fiber signals are converted to a set of versions with a varied phase shift, from which Purkinje cells also select through learning.

4. CEREBELLAR CONTROL OF REFLEXES

The human body is equipped with numerous reflexes, both motor and autonomic. Reflexes are essentially classic control systems, each dedicated to a parameter in a control object. Reflexes are represented in phylogentically old parts of the cerebellum, i.e., the flocculo-nodular lobe, vermis, and paravermis.

While most reflex arcs are connected to the cerebellum, a clear exception is the vestibulo-colic reflex, which apparently lacks a cerebellar connection [66,67]. The vestibulocolic reflex is driven by vestibular signals and stabilizes head position through the activity of neck muscles. It relies upon straightforward negative feedback from the output, i.e., head position, to the input, i.e., head movement and positional change, and therefore it could operate effectively by itself without any aid of the cerebellum. However, many other reflexes lack feedback, or even with feedback their operation becomes inefficient at a high frequency range when the feedback is exerted with a relatively long loop time, as is the case with visual feedback. A cerebellar corticonuclear microcomplex inserted into a reflex arc removes this drawback by adaptive correction of misperformance of the reflex [62,63]. This view is supported by experimental data collected in studies on ocular reflexes (Sections 4.1 and 4.2) and classical conditioning of the eye-blink reflex (Section 4.3).

4.1. Ocular reflexes

The vestibulo-ocular reflex (VOR) induces eye movement that compensates for head movement. The major pathway for the VOR is the trineuronal arc, which is composed of the primary vestibular neurons, vestibular relay neurons, and oculomotor neurons (Figure 3-5A).

The trineuronal arc is attached a neural integrator that converts head velocity signals sensed by canals to the eye position signals of oculomotor neurons [69]. Details of the structure and function of the neural integrator are still unknown, but it appears to involve the nucleus prepositus hypoglossi [70].

Studies on the rabbit have disclosed that a microzone in the flocculus is the center for adaptive gain control of the horizontal canal-ocular reflex. This microzone (the H zone) receives vestibular signals as mossy fiber inputs and optokinetic signals as climbing fiber inputs and, in turn, projects Purkinje cell axons to relay cells of the horizontal canalocular reflex in vestibular nuclei. Hence, the H zone constitutes a corticonuclear microcomplex with relay neurons of the horizontal VOR. The H zone can be identified structurally by its specific projection to the medial vestibular nucleus [71] and functionally by local electrical stimulation through a microelectrode, which induces ipsilateral horizontal movement of the eyes [72]. Based on these neuronal connections, it has been hypothesized that the climbing fiber pathway conveys retinal error signals induced by inadequate performance of the VOR, and that these error signals act to modify signal transfer characteristics of the

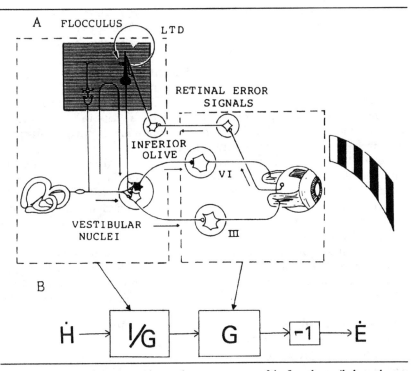

Figure 3-5. Neuronal circuitry and control system structure of the flocculo-vestibulo-ocular system. In A, III, VI, oculomotor and abducens cranial nuclei. Note that inhibitory neurons and their processes are filled with black, excitatory ones are left as hollow. In B, G represents the dynamics of the oculomotor system and 1/G represents the inverse dynamics born in the corticonuclear microcomplex in A (boxed with dotted lines in A). H = head velocity; E = eye velocity.

flocculus for vestibular mossy fiber inputs through induction of LTD at parallel fiber-Purkinje cell synapses [10,73]. This would lead to improvement of the VOR toward minimization of retinal errors. The requirement for such an adaptive mechanism is obvious because the VOR is essentially a feedforward control, lacking direct feedback from the eye to the vestibular organ [68].

The adaptability of the VOR has been tested by subjecting an animal to artificially amplified retinal errors, for example, wearing of dove-prism goggles, which reverse the right-left relationship of the visual field and cause a progressive decrease of gain of the horizontal canal-ocular reflex, and even reversal in polarity [74]. In contrast, wearing of telescopic lenses that magnify the visual field causes a progressive increase of gain of the horizontal canal-ocular reflex [75]. The combination of head rotation and visual field rotation in various phase relationships similarly causes adaptive VOR modification [76]. The adaptability of the VOR so demonstrated is abolished after lesion of the flocculus [77] or the climbing fiber pathway to the flocculus [78]. Recording from H-zone Purkinje cells demonstrated that complex spikes indeed represented retinal errors [79,80], while simple spikes reflected head velocity [57]. Simple spike responsiveness to head rotation was altered in parallel with the VOR adaptation [81]. Furthermore, VOR adaptation has been reproduced in a simulation study with Fujita's adaptive filter model of the cerebellum [82], which is based on the assumption of synaptic plasticity in cerebellar cortical circuitry (Section 3.4).

Observations in monkey flocculus have been at variance with those in rabbits [83]. However, classically defined flocculus in primates has been identified as homologous to the ventral paraflocculus [84], and in fact, H zone Purkinje cells selected by means of local electrical stimulation at the primate flocculus behaved in a manner equivalent to rabbit H-zone Purkinje cells [85,86].

Strong support for the flocculus hypothesis was recently obtained by using hemoglobin, which blocks LTD (Section 2.5). Subdural application of hemoglobin to the flocculus drastically abolished the VOR adaptation in both rabbits and a monkey without affecting the dynamic characteristics of the VOR [87].

In the optokinetic eye-movement response (OKR), the eyes follow a slowly moving visual field. Horizontal OKR is initiated by retinal slip signals, which are forwarded through pretectal nuclei, the nucleus reticularis tegmenti pontis, and relay neurons of the horizontal VOR to abducens and medial rectus motoneurons. Horizontal OKR exhibits adaptability under sustained optokinetic stimulation. Its gain increases toward unity when a striped screen is rotated around the rabbit [80,88]. This OKR adaptability is abolished by floccular lesions, and H-zone Purkinje cells alter their simple spike responsiveness to screen rotation in parallel with the OKR adaptation. These observations indicate that the floccular H zone is also the center of adaptive control of the horizontal OKR. OKR is essentially negative feedback control, but it requires cerebellar aid because of a relatively long loop time in visual feedback. In fact, lesion of the flocculus impairs OKR, especially at a high frequency range.

4.2. Classically conditioned eye-blink reflex

The combination of a corneal air puff with tone stimuli leads to the acquirement of the classically conditioned eye-blink reflex in rabbits [89]. Recent extensive investigations disclosed involvement of the lateral portion of the interpositus nucleus, which projects through the superior cerebellar peduncle to the magnocellular red necleus (Figure 3-4). It has also been demonstrated that air-puff stimuli are conveyed to the cerebellum by climbing fibers, while tone stimuli are mediated by mossy fibers arising from the pontine nuclei.

These neuronal arrangements suggest that tone-induced eye blink is originally mediated by a pathway via pontine nuclei and the interpositus nucleus, but normally it is suppressed by the inhibitory action of Purkinje cells activated by tone-induced mossy fiber inputs, and after conjunction with air puff-induced climbing fiber signals, the tone activation of Purkinje cells is ddpressed due to LTD, thereby releasing the interpositus neurons from Purkinje cell inhibition. Recording from cerebellar cortex in conditioned animals revealed the depression of simple spike discharge from Purkinke cells following tone stimulation [90,91], thus confirming the above prediction. It is noted that climbing fiber responses to air-puff stimulation represent failure of the eye-blink reflex to protect the cornea from being stimulated. This situation is consistent with the general postulate that climbing fibers signal control errors in an associated extracerebellar system. In fact, air-puff-induced climbing-fiber responses in Purkinje cells diminish after they acquire the conditioned reflex [92].

5. CEREBELLAR CONTROL OF COMPOUND REACTIONS

There are a number of compound reactions involving brain-stem structures that may be ranked as higher order in the hierarchy of bodily control mechanisms. Saccadic eye movements, posture, locomotion, and breathing are typical examples of such reactions. These reactions are as mechanistic as reflexes, but differ from reflexes in that they involve elaborate function generators and regulatory mechanisms for the function generators. A compound reaction is often represented in more than one area of the cerebellum. A compound reaction may involve more than one parameter to be controlled, so that one cerebellar area may be devoted to each of the parameters.

5.1. Saccadic eye movement and eye-head coordination

Saccade is quick jerky eye movement to foveate a visual target that is generated by brain-stem saccade generators operating under the influence of the superior colliculus and the cerebral cortex. According to electrical stimulation experiments, saccadic eye movement is represented in three cerebellar areas, i.e., lobules V–VII of the vermis, crus I and II, and the lobulus simplex of the hemisphere [93]. Damage to the vermal cortices of lobules V–VII results in dysmetric saccadic eye movements [94]. Dysmetric saccade induced by partial tenotomy is not recovered after ablation of lobules VI and VII [95]. These effects are indicative of the contribution of the vermal cortices of lobules V–VII to the precision of saccade through its adaptability. Purkinje cells in cat lobules VI and VII exhibit saccade-related simple spikes [96].

In humans and monkeys, visually triggered saccadic eye movement are accompanied by head turning directed toward the visual target. The head turning elicits the VOR, which moves the eyes back, and consequently the gaze is kept stable, regardless of the head movemnt. Even though the exact manner in which the cerebellum intervenes in eye-head coordination is unclear, there may be a corticonuclear microcomplex that adjusts the relative contribution of the visually triggered eye and head movements that is located somewhere in the cerebellar cortex.

5.2. Posture

Posture is held on the basis of the concerted activity of numerous postural reflexes, such as the stretch reflex, crossed-extension reflex, neck reflexes, labyrinthine reflexes, and grasp reflex, and also compound reactions, such as interlimb coordination, placing and hopping reactions, postural adjustment during head and limb movement, and the righting reaction. The exact manner in which these numerous reflexes and reactions are integrated to produce the postural state as a whole is still unclear. A postural center responsible for such integration may exist somewhere in the midbrain, but its structural basis is not yet known. Loss of the righting reflex after intercollicular decerebration may suggest the presence of a neural device for evaluating the gravity center of the body [97] in a rostral region of the midbrain, but no structural correlate has been found.

There is reason to assume that the postural reflexes and reactions operate in close connection with the cerebellum [10], but precise functional localization of these actions have not been demonstrated. In frog Purkinje cells, body inclination evoked stronger complex spike responses in the absence of appropriate compensatory limb movements [98]. This observation is in accordance with the general postulate that climbing fibers represent control errors (Section 3.2). An experiment in rats demonstrated that operant conditioning for balancing on a rotating rod was impaired by lesion of the parvocellular red nucleus, which projects to the inferior olive [99]. This is also consistent with the postulated role of climbing fibers in cerebellar adaptation.

5.3. Locomotion

The generators of locomotion rhythm for hindlimbs are located in the lumbar segments and operate under the control of the midbrain locomotion center [100]. Decerebellate cats still produce locomotion when the midbrain locomotion center is stimulated. Therefore, the cerebellum is not essential for the generation of locomotion rhythm, but it's roles is evident in that locomotion limb movements in decerebellate cats and dogs, or in rats with an X-irradiated cerebellum [101], are poorly coordinated.

Purkinje cells in the vermis of lobule V exhibited a high discharge rate at the stance phase of the ipsilateral limb [102], while the simple spike activity of paravermal Purkinje cells of lobule V was greatest at the time of transition between the stance and swing phases in the ipsilateral limb and was least during midstance [103].

These Purkinje cells did not show a particular response with complex spikes during normal locomotion, but perturbation of locomotion frequently evoked complex spikes [102,104]. Likewise, active movement of a limb, such as spontaneous lifting up and setting down of a limb, did not elicit complex spikes unless the movements were perturbed [105]. These observations are consistent with the postulate of error representation by climbing fibers (Section 3.2).

Golgi cell activity in the paravermis of lobule V has been observed during locomotion to often fluctuate approximately in parallel with the simple spike activity of Purkinje cells [106].

6. CEREBRO-CEREBELLAR INTERACTION

Through the cerebro-cerebellar communication loop, the cerebellum contributes to cerebral functions, as typically represented by voluntary movement control. In the cerebral cortex, motor command seems to be programmed in the premotor area and is executed through the motor area. On the basis of multiple connections to and from cerebral cortical areas, the cerebellum has been assumed to play two major roles in voluntary movement control, i.e., (1) updating of motor programs during execution of movements in the paravermis-interpositus zone and (2) formation of new motor programs in the hemisphere-dentate zones [107].

There are various types of voluntary movements of different complexity, from a simple reaction-time task response to self-initiated complex movement, such as speech, music play, and gymnastic play. In the cat and monkey, various experimental paradigms have been developed for the investigation of neuronal mechanisms of voluntary movement control. However, it is to be noted that these paradigms are mostly for testing the execution of voluntary movement control. The proposed programming function of the cerebellum has largely been disregarded.

Cerebellar studies on human subjects are now given a new tool in the noninvasive measurement of local blood circulation or metabolism of cerebellar tissues. The cerebellar contribution to motor programming, and even to nonmotor cerebral function, is an important subject in cerebellar physiology.

6.1. Smooth-pursuit eye movement

With smooth-pursuit eye movement, gaze is controlled to fixate continuously a small moving target within the fovea. The involvement of the prefrontal and parietal association cortices and the dorsolateral pontine nucleus in smooth eye-movement control has so far been disclosed, but the entity of a smooth-pursuit generator operating under the influence of these structures is still unclear [108,109]. Gaze is also controlled voluntarily in order to fixate a stationary target, presumably through similar neural mechanisms. When gaze is directed to a target fixed to a turntable mounted on the animal, the eyes are fixated to the orbits, causing visual suppression of the VOR.

The role of the cerebellum in smooth-pursuit eye movements is evidenced by the fact that total cerebellectomy abolishes it. Partial lesion and stimulation experiments suggest that smooth-pursuit eye movement is represented in three cerebellar areas, i.e., the flocculus, the vermis of lobules VI and VII, and the lobulus simplex of the cerebellar hemisphere. Simple spike activity related to smooth-pursuit eye movement has been recorded from Purkinje cells in the flocculus [110] and vermis [111]. Complex spikes recorded from floccular Purkinje cells represented a visual response to the foveal slip that results from inaccurate tracking [112], i.e., control errors in smooth pursuit.

6.2. Voluntary limb movement

Various paradigms have been introduced for testing adaptability in voluntary limb movement, e.g., self-initiated limb movement, simple reaction time task, load perturbation, visuo-motor tasks, and finger grip [10].

In a paradigm of load perturbation [113], a monkey was trained to hold a lever in its hand fixedly in a central position, while loads were imposed by a torque motor on the level hand alternately against flexion and extension. The central position was indicated by a light stimulus when the lever was within a prescribed "window." When a shift of load at some unpredictable time knocked the lever and hand out of the window, the monkey had to return to the central position quickly to get a reward.

Just after the load switch, Purkinje cells in the paravermis of lobules III–V exhibited an increase in the occurrence of complex spikes, which persisted for many trials before returning to baseline. Simple spike occurence, by contrast, decreased at the same time after the load switch and remained decreased after the complex spike firing frequency returned to its prior level. These observations are consistent with the view that motor adaptation takes place in the cerebellum through a decrease in the strength of transmission of parallel-fiber synapses on Purkinje cells caused by climbing-fiber signals representing control errors.

In a visually guided, multijoint voluntary arm movement task, a monkey moved a manipulandum over a video screen to place a cursor within displayed start and target boxes. Purkinje cells in lobules IV–VI of the ipsilateral hemiphere and intermediate zone frequently exhibited enhanced complex spike discharges during the initial portion of the movement and/or when movement was redirected to a target box repositioned during arm movement [114]. This is consistent with the postulate of error representation by climbing fibers.

In another visuo-motor task, when a monkey performed visually guided wrist-tracking movements [115], Purkinje cells in the intermediate and lateral parts of cerebellar hemispheres of lobules IV–VI increased or decreased their simple spike discharge rate during task performance. A majority of these Purkinje cells exhibited a phasic increase of complex spike firing at the onset of wrist-tracking movement. Even though the authors of this report related these complex spike responses to the timing of wrist movement, there seems to be no difficulty in interpreting their results as supporting the general postulate of error representation by climbing fibers.

6.3. Speech and mental function

Dysarthria arises in cerebellar disease, most commonly when the superior portion of the left cerebellar hemisphere is damaged [116]. An interesting recent finding is that when a subject is asked to give a verb related to a presented noun, enhanced blood circulation occurs in the right prefrontal cerebral area, and at the same time in the left cerebellar hemisphere [117].

Involvement of the cerebellum in mental activity has been suggested [118] based on comparative anatomy, which indicates that the lateralmost part of the human cerebellum is enlarged dramatically in parallel with the development of cerebral association cortices. The actual involvement of the cerebellum in silent counting and imaginary tennis playing has recently been demonstrated [119].

7. THEORY OF CEREBELLAR CONTROL

It now seems useful to formulate a model view of the cerebellar control function introduced in Sections 4-6. Such a view will explain the experimental observations described thus far and will aid future investigations into more complex aspects of cerebellar function, especially those related to cerebral activity.

7.1. Adaptive control of reflex and reaction

The mode of cerebellar control of reflexes depicted in Sections 4.1 and 4.2 is such that a corticonulcear microcomplex is attached to a reflex arc as a side path whose signal-transfer characteristics are adaptively modifable with reference to error signals conveyed by climbing fibers (Figure 3-6).

An essential feature of this control arrangement is implied in the relationship of feedforward and feedback controls. Feedback is an element of essential importance in classic control systems. A feedforward system, which lacks feedback, is inevitably susceptible to external disturbances, as well as to changes in internal parameters.

Any malfunctioning of such a system would persist or worsen, since there is no feedback to correct it. Nevertheless, a feedforward system has the advantage of being free from a delay in operation due to feedback and also from any complications that the feedback may introduce into the dynamics of the system. In fact, neural control systems seem to prefer the feedforward mode to the feedback one, since in a living body a feedback loop is not always available. Even if a feedback loop is available, it may function only within a certain limited operational range. To be an effective means of neural control, the functional drawbacks in the feedforward operation must be removed by a neural device that replaces a feedback loop. Such a neural device would be composed of a comparator for detecting control errors

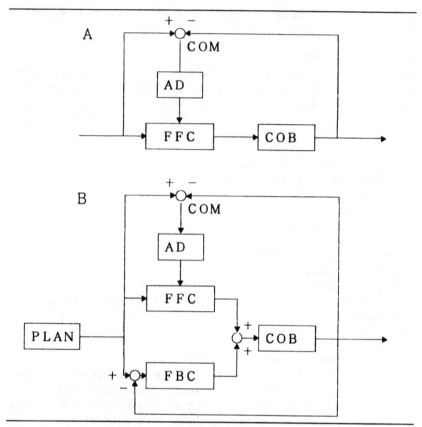

Figure 3-6. Block diagrams for cerebellar control systems. **A**: A feedforward control system with an adaptive mechanism. COB = control object; FFC = feedforward controller; AD = adaptor; COM = comparator; **B**: Scheme for learning voluntary movement control. Learning shifts the Dominant mode of control from feedback to feedforward. FBC = feedback controller.

through a comparison of intended and executed controls, and an adaptor, which, based on control error, acts to correct the performance of a feedforward controller. With such an error-detecting and parameter-adjusting device, the system is converted to an adaptive control system (Figure 3-6A). An essential assumption in the hypothesis is that the cerebellar corticonuclear microcomplex of Figure 3-2 is functional counterpart of the adaptive controller equipped with a comparator and an adaptor of Figure 3-6A.

If the VOR control is perfect, the eye velocity should become equivalent to head velocity, with signs reversed. In that situation, dynamic characteristics of the adaptive controller, including the flocculus, vestibular organ, and VOR relay neurons, should reciprocally be equal to that (G) of the control object, i.e., the oculomotor system, including motoneurons, extraocular muscles, and the eye object (Figure 3-6B). The role of a corticonuclear microcomplex would be defined in terms of control theory as generating dynamics inversely equivalent to the dynamics of a control object. The scheme of adaptive feedforward control of VOR adaptation could be generalized to cerebellar control of compound reactions, such as saccade, locomotion, and posture.

7.2. Learning control of cerebral cortical function

Voluntary movement control is performed by the frontal cortical complex, consisting of the motor, premotor, and supplementary motor cortices. Voluntary movement control is characterized by learning through exercise. At the beginning, it will be executed carefully, relying upon sensory feedback, but after repeated practice with conscious effort, voluntary movement is learned so that it becomes more automatic. This daily experience implies that voluntary movement is learned during repeated practice in a feedback mode, but after learning it is executed in a feedforward mode.

This situation can be represented by a parallel combination of a feedback and an adaptive feedforward system operating upon one and the same control object (Figure 3-6B). Initially, performance of the whole system would rely upon the feedback control, but the feedforward control will take over as soon as it is adapted to the practiced control situation. This scheme would apply to the cerebrocerebellar communication in which the cerebellar hemisphere is connected in parallel to projections from a cerebral cortex to a subcortical nuclei or another area of the cerebral cortex. If the control object is a skeletomuscular system of an arm, the system will function in the following manner. While the cerebral cortex executes arm movement according to a trajectory instructed from an association cortex, relying on visual, proprioceptive, and other types of feedback, the cerebellar corticonuclear microcomplex is adapted to conduct the trajectory formation in a feedforward manner. It is noted that since the whole system aims at forming a trajectory equal to the instructed trajectory, the feedforward system must bear an inverse dynamics of the arm skeletomuscular system. Classic neurological examination, such as the finger-to-nose test, seems to reveal loss of an inverse dynamic model of the finger-arm system in the cerebellum so that trajectory formation cannot be performed accurately without visual feedback.

Recent computer simulation study [43] using this adaptive control scheme demonstrates a remarkable effect of practice in effecting smooth, efficient trajectory formation. It is interesting to note that practice for a trajectory leads to an improved formation of other trajectories as well, as one experiences during the practice of coordinated movements in sports. It is a characteristic feature of motor learning that what to be learned is not an individual movement, but the dynamics of the motor system that operate in a variety of movements.

The control system scheme for voluntary movement can be modified for any type of cerebrocerebellar communication [120]. For example, when one thinks, a cerebral cortical area I would act as a controller on another cortical area II, where an idea or a concept may be handled, such as a limb in Figure 3-6. Such a cortico-cortical system would initially operate in a feedback mode, but after repeated practice the cerebellum would replace the cortical area I, so that thinking would be preformed automatically with less conscious concern [121].

7.3. Comments

The present stage of cerebellar physiology is characterized by the wealth of knowledge at all of the three domains of investigating the cerebellum, i.e., at the cellular and molecular levels, at the levels of complex systems, and at the levels of the human brain. Knowledge at these three levels should be integrated to yield a unified concept of the cerebellum.

The presently adopted concept of the corticonuclear microcomplex and control system models specify the cerebellum as an organ endowing adaptive control abilities to various kinds of control functions, through reflexes to voluntary movement, and even to mental activity. Synaptic plasticity (LTD) plays a key role in the cerebellar cortical network, and a corticonuclear microcomplex is incorporated as a core part of varied forms of neural systems.

It is to be noted here that an adaptive control system refers to memory gained within a ongoing trial. For example, the effect of adaptation in VOR diminishes in a few days, and repeated trials yield equivalent adaptation each time. However, a learning control system utilizes the memory of preceding trials. Hence, as trials are repeated, adaptation becomes faster and faster. This system can "learn to learn." Accordingly, an experience that motor skill once acquired and lost during long disuse may be regained faster on practicing again. In other words, a learning control system is supported by a second memory system. For a moment, the entity of this second memory system is not known, and it will be an important theme of cerebellar physiology to identify it at cellular as well as system levels.

REFERENCES

1. Dow R.S., Moruzzi G. (1958). The Physiology and Pathology of the Cerebellum. Minneapolis, MN: University of Minnesota Press.
2. Snyder R.S., Stowell A. (1944). Receiving areas of the tactile, auditory and visual systems in the cerebellum. J. Neurophysiol. 7:331–357.
3. Chambers W.W., Sprague J.M. (1955). Functional localization in the cerebellum. I. Organization in longitudinal corticonuclear zones and their contribution to the control of posture, both extrapyramidal and pyramidal. J. Comp. Neurol. 103:105–129.
4. Moruzzi G. (1950). Problems in Cerebellar Physiology. Springfield, IL: Charles C. Thomas.
5. Brookhart J.M., Moruzzi G., Snider R.S. (1950). Spike discharges of single units in the cerebellar cortex. J. Neurophysiol. 13:465–486.
6. Granit R., Phillips C.G. (1956). Excitatory and inhibitory processes acting upon individual Purkinje cells of the cerebellum in cats. J. Physiol. (London) 133:520–547.

7. Eccles J.C., Ito M., Szentágothai J. (1967). The Cerebellum as a Neuronal Machine. Heidelberg: Springer–Verlag.
8. Marr D. (1969). A theory of cerebellar cortex. J. Physiol. (London) 202:437–470.
9. Albus J.S. (1971). A theory of cerebellar function. Math. Biosci. 10:25–61.
10. Ito M. (1984). The Cerebellum and Neural Control. New York: Raven Press.
11. Shelton, D.P. (1985). Membrane resistivity estimated for the Purkinje neuron by means of a passive computer model. Neuroscience 14:111–131.
12. Llinás R., Sugimori M. (1980). Electrophysiological properties of in vitro Purkinje cell somata in mammalian cerebellar slices. J. Physiol. (London) 305:171–195.
13. Llinás R., Sugimori M. (1980). Electrophysiological properties of in vitro Purkinje cell dendrites in mammalian cerebellar slices. J. Physiol. (London) 305:197–213.
14. Ross W.N., Lasse-Ross N., Werman, R. (1990). Spatial and temporal analysis of calcium-dependent electrical activity in guinea pig Purkinje cell dendrites. Proc. Roy. Soc. Lond. B, 240:173–185.
15. Ekerot C.-F., Oscarsson O. (1981). Prolonged depolarization elicited in Purkinje cell dendrites by climbing fibre impulses in the cat. J. Physiol. (London) 318:207–221.
16. Sakurai M. (1987). Synaptic modification of parallel fiber-Purkinje cell transmission in the in vitro guinea pig cerebellar slices. J. Physiol. (London) 394:463–480.
17. Llinás R., Sugimori R., Cherksey B. (1989). Voltage-dependent calcium conductances in mammalian neurons. Ann. N. Y. Acad. Sci. 103–111.
18. Eccles J.C., Llinás R., Sasaki K. (1966). Parallel fibre stimulation and the responses induced thereby in the Purkinje cells of the cerebellum. Exp. Brain Res. 1:17–39.
19. Palay S.L., Chan-Palay V. (1974). Cerebellar cortex. New York: Springer-Verlag.
20. Christ H. (1985). Fisiform nerve cells of the granular layer in the cerebellar cortex of the baboon. Neurosci. Lett. 56:195–198.
21. Hirano T., Hagiwara S. (1988). Synaptic transmission between rat cerebellar granule and Purkinje cells in dissociated cell culture. Effects of excitatory-amino acid transmitter agonists. Proc. Natl. Acad. Sci. USA 85:934–938.
22. Wiklund L., Toggenburger G., Cuénod M. (1983). Aspartate. Possible neurotransmitter in cerebellar climbing fibers. Science 216:78–80.
23. Cuénod M., Do K.-Q., Vollenweider F., Zollinger M., Klein A., Streit P. (1989). The puzzle of the transmitters in the climbing fibers, Exp. Brain Res. Suppl. 17:161–176.
24. Blackstone C.D., Supattapone S., Snyder S.H. (1989). Inositol phospholipid-linked glutamate receptors mediate cerebellar parallel-fiber-Purkinje-cell synaptic transmission. Proc. Natl. Acad. Sci. USA 86:4316–4320.
25. Masu M., Tanabe Y., Tsuchida K., Shigeto R., Nakanishi S. (1991). Sequence and expression of a metabotropic glutamate receptor. Nature 349:760–762.
26. Ross C.A., Meldolesi J., Milner T.A., Satoh T., Supattapone S., Snyder S.H. (1990). Inositol 1,4,5-trisphosphate receptor localized to endoplasmic reticulum in cerebellum Purkinje neurons. Nature 339:468–470.
27. Furuichi T., Yoshikawa S., Miyawaki A., Wada K., Maeda N., Mikoshiba K. (1989). Primary structure and functional expression of the inositol 1,4,5-trisphosphate-binding protein P_{400}. Nature 342:32–38.
28. Moise H.C., Woodward D.J., Hoffer B.J., Freedman R. (1979). Interactions of norepinephrine with Purkinje cell responses to putative amino acid neurotransmitters applied by microiontophoresis. Exp. Neurol. 64:493–515.
29. Lee M., Strahlendorf J.C., Strahlendorf H.K. (1986). Modulatory action of serotoninn in glutamate-induced excitation of cerebellar Purkinje cells. Brain Res. 361:107–113.
30. Dolphin A.C. (1982). Noradrenergic modulation of glutamate release in the cerebellum. Brain Res. 252:111–116.
31. Wilkin G.P., Hidson A.L., Hill D.R., Bowery N.G. (1981). Autoradiographic localization of $GABA_B$ receptors in rat cerebellum. Nature 294:584–587.
32. Hacket J.T. (1974). GABA selectively blocks parallel fiber-Purkinje cells synaptic transmission in the frog cerebellum in vitro. Brain Res. 80:527–531.
33. Dutar P., Nicoll R.A. (1988). A physiological role for $GABA_B$ receptors in the central nervous system. Nature 332:156–158.
34. Goodman R.R., Kuhar M.J., Snyder S.H. (1983). Adenosine receptors. Autoradiographic evidence for their location on axon terminals of excitatory neurons. Science 220:967–969.

35. Kocsis J.D., Eng D.L., Bhisikul R.B. (1984). Adenosine selectively blocks parallel fiber-mediated synaptic potentials in rat cerebellar cortex. Proc. Natl. Acad. Sci. USA 81:6531–6534.
36. Colin F., Manil J., Desclin J.C. (1980). The olivo-cerebellar system. I. Delayed and slow inhibitory effects. An overlooked salient feature of cerebellar climbing fibers. Brain Res. 87:3–27.
37. Do K.Q., Vollenweider F.X., Zollinger M., Cuenod M. (1991). Effect of climbing fibre deprivation on the K^+-evoled release of endogenous adenosine from rat cerebellar slices. Eur. J. Neurosci. 3:201–208.
38. Batini C., Billard J.M. (1985). Release of cerebellar inhibition by climbing fiber deafferentation. Exp. Brain Res. 57:370–380.
39. Ito M., Nisimaru N., Shibuki K. (1979). Destruction of inferior olive induces rapid depression in synaptic action of cerebellar Purkinje cells. Nature 227:568–569.
40. Karachot L., Ito M., Kanai Y. (1987). Long-term effects of 3-acetylpyridine-induced destruction of cerebellar climbing fibers on Purkinje cell inhibition of vestibulospinal tract cells of the rat. Exp. Brain Res. 66:229–246.
41. Rossi F., Cantino D., Strata P. (1987). Morphology of Purkinje cell axon terminals in intracerebellar nuclei following inferior olive lesion. Neuroscience 22:99–112.
42. Ito M., Sakurai M., Tongroach P. (1982). Climbing fibre induced depression of both mossy fibre responsiveness and glutamate sensitivity of cerebellar Purkinje cells. J. Physiol. (London) 324:113–134.
43. Bliss T.V.P., Lømo T. (1973). Long-lasting potentiation of synaptic transmission in the dentate area of the anaesthetized rabbit following stimulation of the perforant path. J. Physiol. (London) 232:331–356.
44. Byrne J.H. (1987). Cellular analysis of associative learning. Physiol. Rev. 67:329–439.
45. Tsukahara N. (1981). Synaptic plasticity in the mammalian central nervous system. Annu. Rev. Neurosci. 4:351–379.
46. Sakurai M. (1988). Calcium is an intracellular mediator of the climbing fiber in induction of cerebellar long-term depression. Proc. Natl. Acad. Sci. USA 87:3383–3385.
47. Ekerot C.-F., Kano M. (1985). Long-term depression of parallel fibre synapses following stimulation of climbing fibres. Brain Res. 342:357–360.
48. Kano M., Kato M. (1987). The specific glutamate receptor mechanism involved in cerebellar plasticity. Nature 325:276–279.
49. Kano M., Kato M. (1988). Mode of induction of long-term depression at parallel fibre-Purkinje cell synapses in rabbit cerebellar cortex. Neurosci. Res. 5:544–556.
50. Crepel F., Krupa, M. (1988). Activation of protein kinase C induces a long-term depression of glutamate sensitivity of cerebellar Purkinje cells: An in vitro study. Brain Res. 458:397–401.
51. Ito M., Karachot L. (1989). Long-term desensitization of quisqualate-specific glutamate receptors in Purkinje cells investigated with wedge recording from rat cerebellar slices. Neurosci. Res. 7:168–171.
52. Ito M., Karachot L. (1990). Receptor subtypes involved in, and time course of, the long-term desensitization of glutamate receptors in cerebellar Purkinje cells. Neurosci. Res. 8:303–307.
53. Ito M., Karachot. L. (1990). Messengers mediating long-term desensitization in cerebellar Purkinje cells. NeuroReport 1:129–132.
54. Shibuki K., Okada D. (1991). Endogenous nitric oxide release required for long-ter, synaptic depression in the cerebellum. Nature 349:326–328.
55. Crepel F., Jaillard D. (1990). Protein kinases, nitric oxide and long-term depression of synapses in the cerebellum. NeuroReport 1:133–136.
56. Fujita M. (1982). Adaptive filter model of the cerebellum. Biol. Cybern. 45:195–206.
57. Ghelarducci B., Ito M., Yagi N. (1975). Impulse discharges from flocculus Purkinje cells of alert rabbits during visual stimulation combined with horizontal head rotation. Brain Res. 87:66–72.
58. Ekerot C.-F., Oscarsson O., Schouenborg J. (1987). Stmulation of cat cutaneous nociceptive C-fibres causing tonic and synchronous activity in climbing fibres. J. Physiol. (London) 386:539–546.
59. Miller S., Oscarsson O. (1970). Termination and functional organization of spinoolivocerebellar paths. In: Fields W.S., Willis W.D. (eds): Cerebellum in Health and

Disease. 172–200. St. Louis: W.H. Green, pp. 170–200.
60. Ito M. (1970). Neurophysiological aspects of the cerebellar motor control system. Int. J. Neurol. 7:162–176.
61. Andersson G., Hesslow G. (1987). Inferior olive excitability after high frequency climbing fibre activation in the cat. Exp. Brain Res. 67:523–532.
62. Ebner T.J., Bloedel J.R. (1984). Climbing fiber action on the responsiveness of Purkinje cells to parallel fiber inputs. Brain Res. 309:182–186.
63. Miyashita Y., Nagao S. (1984). Contribution of cerebellar intracortical inhibition to Purkinje cell responses during vestibulo-ocular reflex of alert rabbits. J. Physiol. (London) 351:251–262.
64. Oscarsson O. (1976). Spatial distribution of climbing and mossy fibre inputs into the cerebellar cortex. In: O. Creutzfeldt (ed): Experimental Brain Research, Suppl. 1. Afferent and Intrinsic Organization of Laminated Structures in the Brain, pp. 34–42.
65. Fujita M. (1982). Adaptive filter model of the cerebellum. Biol. Cybern. 45:195–206.
66. Fukuda J., Highstein S.M., Ito M. (1972). Cerebellar inhibitory control of the vestibulo-ocular reflex investigated in rabbit IIIrd nucleus. Exp. Brain Res. 14:511–526.
67. Hirai N., Uchino Y. (1984). Floccular influence on excitatory relay neurons of vestibular reflexes of anterior semicircular canal origin in the cat. Neurosci. Res. 1:
68. Ito M. (1984). Control mechanisms of cerebellar motor system. In: F.O. Schmitt, F.G. Worden (eds): The Neuroscience, Third Study Program, Cambridge, MA: MIT Press, pp. 293–303.
69. Robinson D.A. (1981). Control of eye movements. In: Brooks V.B. (ed): Handbook of Physiology, The Nervous System II, Washington DC: American Physiological Society, pp. 1275–1313.
70. Cheron G., Godaux E. (1987). Disabling of the oculomotor neural integrator by kainic acid injections in the prepositus-vestibular complex of cat. J. Physiol. (London) 394:267–290.
71. Yamamoto M., Shimoyama I. (1977). Differential localization of rabbit's flocculus Purkinje cells projecting to the medial and superior vestibular nuclei, investigated by means of the horseradish peroxidase retrograde axonal transport. Neurosci. Lett. 5:279–283.
72. Nagao S., Ito M., Karachot L. (1985). Eye field in the cerebellar flocculus of pigmented rabbits determined with local electrical stimulation. Neurosci. Res. 3:39–51.
73. Ito M. (1982). Cerebellar control of the vestibulo-ocular reflex around the flocculus hypothesis. Annu. Rev. Neurosci. 5:275–296.
74. Gonshor A., Melvill Jones G. (1976). Extreme vestibuloocular adaptation induced by prolonged optical reversal of vision. J. Physiol. (London) 256:381–414.
75. Miles F.A., Fuller J.H. (1974). Adaptive plasticity in the vestibuloocular responses of the rhesus monkey. Brain Res. 80:512–516.
76. Ito M., Shiida N., Yagi N., Yamamoto M. (1974). The cerebellar modification of rabbit's horizontal vestibulo-ocular reflex induced by sustained head rotation combined with visual stimulation. Proc. Jpn. Acad. 50:85–89.
77. Ito M., Jastreboff J.P., Miyashita Y. (1982). Specific effects of unilateral lesions in the flocculus upon eye movements in albino rabbits. Exp. Brain Res. 45:233–242.
78. Ito M., Miyashita Y. (1975). The effects of chronic destruction of the inferior olive upon visual modification of the horizontal vestibulo-ocular reflex of rabbits. Proc. Jpn. Acad. 51:716–720.
79. Maekawa K., Simpson J.I. (1973). Climbing fiber responses evoked in vestibulocerebellum of rabbit from visual system. J. Neurophysiol. 36:649–666.
80. Nagao S. (1988). Behavior of floccular Purkinje cells correlated with adaptation of horizontal optokinetic eye movement response in pigmented rabbits. Exp. Brain Res., 162, 73:489–497.
81. Dufossé M., Ito M., Jastreboff J.P., Miyashita Y. (1978). A neuronal correlate in rabbit's cerebellum to adpative modification of the vestibulo-ocular reflex. Brain Res. 150:611–616.
82. Fujita M. (1982). Simulation of adaptive modification of the vestibulo-ocular reflex with an adaptive filter model of the cerebellum. Biol. Cybern. 45:207–214.
83. Miles F.A., Lisberger S.G. (1981). Plasticity in the vestibulo-ocular reflex: A new hypothesis. Annu. Rev. Neurosci. 4:273–299.
84. Gerrits N.M., Voogd J. (1989). The topographical organization of climbing and mossy fiber afferents in the flocculus and the ventral paraflocculus in rabbit, cat, and monkey. Exp. Brain Res. 17:26–29.

85. Watanabe E. (1984). Neuronal events correlated with long-term adaptation of the horizontal vestibulo-ocular reflex in the primate flocculus. Brain Res. 297:169–174.
86. Watanabe E. (1985). Role of the primate flocculus in adaptation of the vestibul-ocular reflex. Neurosci. Res. 3:20–38.
87. Nagao S., Ito M. (1991). Subdural application of hemoglobin to the cerebellum blocks vestibuloocular reflex adaptation. NeuroReport 2:193–196.
88. Collewijn H., Grootendorst A.F. (1979). Adaptation of optokinetic and vestibulo-ocular reflexes to modified visual input in the rabbit. In: R. Granit, O. Pompeiano (eds): Progress in Brain Res., Vol. 50, Reflex Control of Posture and Movement, Amsterdam: Elsevier, pp. 771–781.
89. Thompson R.F. (1987). The neurobiology of learning and memory. Science 233:941–947.
90. Donegan N.H., Foy M.R., Thompson R.F. (1985). Neuronal responses of the rabbit cerebellar cortex during performance of the classically conditioned eyelid response. Abstr. Neurosci. Soc. 11:835.
91. Berthier N.E., Moore J.W. (1986). Cerebellar Purkinje cell activity related to the classically conditioned nictitating membrane response. Exp. Brain Res. 63:341–350.
92. Foy M.R., Thompson R.F. (1986). Single unit analysis of Purkinje cell discharge in classically conditioned and untrained rabbits. Abstr. Neurosci. Soc. 12:518.
93. Ron S., Robinson A. (1973). Eye movements evoked by cerebellar stimulation in the alert monkey. J. Neurophysiol. 36:1004–1022.
94. Aschoff J.C., Cohen B. (1971). Changes in saccadic eye movements produced by cerebellar cortical lesions. Exp. Neurol. 32:123–133.
95. Optican L.M., Zee D.S., Miles F.A. (1986). Floccular lesions abolish adaptive control of post-saccadic ocular drift in primates. Exp. Brain Res. 64:596–598.
96. Kase M., Miller D.C., Noda H. (1980). Discharges of Purkinje cells and mossy fibres in the cerebellar vermis of the monkey during saccadic eye movements and fixation. J. Physiol. (London) 300:539–555.
97. Ito M. (1986). Neural systems controlling movements. Trends Neurosci. 9:515–518.
98. Amat J. (1983). Interaction between signals from vestibular and forelimb receptors in Purkinje cells of the frog vestibulocerebellum. Brain Res. 278:287–290.
99. Kennedy P.R., Humphrey, D.R. (1987). The compensatory role of the parvocellular division of the red nucleus in operantly conditioned rats. Neurosci. Res. 5:39–62.
100. Shikk M.L., Orlovsky G.N. (1976). Neurophysiology of locomotor automatism. Physiol. Rev. 56:465–501.
101. Altman J., Sudarshan K. (1975). Postnatal development of locomotion in the laboratory rat. Anim. Behav. 23:896–920.
102. Matsukawa K., Udo M. (1985). Responses of cerebellar Purkinje cells to mechanical perturbations during locomotion of decerebrate cats. Neurosci. Res. 2:393–398.
103. Armstrong D.M., Edgley S.A. (1984). Discharge of Purkinje cells in the paravermal part of the cerebellar anterior lobe during locomotion in the cat. J. Physiol. (London) 352:403–424.
104. Armstrong D.M., Edgley S.A., Lidierth M. (1988). Complex spkes in Purkinje cells of the paravermal part of the anterior lobe of the cat cerebellum during locomotion. J. Physiol. (London) 400:405–414.
105. Gellman R.S., Gibson A.R., Houk J.C. (1985). Inferior olivary neurons in the awake cat: Detection of contact and passive body movement. J. Neurophysiol. 54:40–60.
106. Edgley S.A., Lidierth M. (1987). The discharge of cerebellar Golgi cells during locomotion in the cat. J. Physiol. 392:315–332.
107. Allen G.I., Gilbert P.F.C., Yim T.C.T. (1978). Convergence of cerebral inpus onto dentate neurons in monkey. Exp. Brain Res. 32:151–170.
108. Eckmiller R. (1987). Neural control of pursuit eye movements. Physiol. Rev. 67:797–857.
109. Keller E.L., Heinen S.J. (1991). Generation of smooth-pursuit eye movements: Neuronal mechanisms and pathways. Neurosci. Res. 11:79–107.
110. Miles F.A., Braitman D.J., Dow B.M. (1980). Long-term adaptive changes in primate vestibuloocular reflex. IV. Electrophysiological observations in flocculus of adapted monkeys. J. Neurophysiol. 43:1477–1493.
111. Suzuki D.A., Keller E.L. (1988). The role of the posterior vermis of monkey cerebellum in smooth-pursuit eye movement control. II. Target velocity-related Purkinje cell activity. J.

Neurophysiol. 59:19–40.
112. Stone L.S., Lisberger S.G. (1986). Detection of tracking errors by visual climbing fiber inputs to monkey cerebellar flocculus during pursuit eye movements. Neurosci. Lett. 72: 163–168.
113. Gilbert P.F.C., Thach W.T. (1977). Purkinje cell activity during motor learning. Brain Res. 128:309–328.
114. Wang J.-J., Kım J.H., Ebner T.J. (1987). Climbing fiber afferent modulation during a visually guided, multi-joint arm movement in the monkey. Brain Res. 410:323–329.
115. Mano N., Kanazawa I., Yamamoto K. (1986). Complex-spike activity of cerebellar Purkinje cells related to wrist tracking movement in monkey. J. Neurophysiol. 56:137–158.
116. Lechtenberg R., Gilman S. (1978). Speech disorders in cerebellar disease. Ann. Neurol. 3:285–290.
117. Petersen S.E., Fox P.T., Posner M.I., Mintun, M., Raicle M.E. (1989). Positron emission tomographic studies of the processing of single words. J. Cogn. Neurosci. 1:153–170.
118. Leiner H.C., Leiner A.L., Dow R.S. (1986). Does the cerebellum contribute to mental skills? Behav. Neurosci. 100:443–453.
119. Ingvar P. (1991). On ideation and ideography. In: J.C. Eccles, O. Creutzfeldt (eds): Principles of Design and Operation of the Brain. Exp. Brain Res. Suppl. 21:433–458.
120. Kawato M., Furukawa K., Suzuki R. (1987). A hierarchical neural-network model for control and learning of voluntary movement. Biol. Cybern. 57:169–185.
121. Ito M. (1990). New physiological concepts on cerebellum. Rev. Neurol. (Paris) 146:564–569.

4. AMINO ACID TRANSMITTERS IN THE ADULT AND DEVELOPING CEREBELLUM

R. BALÁZS

1. INTRODUCTION

The cerebellum is a part of the brain in which amino acid transmitters are particularly abundant. Attempts to allocate transmitters to units in the cerebellar circuits have been facilitated by the fact that structural organization in the whole cerebellum is remarkably similar in different vertebrates and involves a limited number of nerve cell types [1,2]. The only efferent cell in the cerebellum is the Purkinje cell, which conveys inhibition onto nerve cells in the deep cerebellar nuclei. Impulse traffice from the cerebellum is determined by two major circuits, the climbing fiber-Purkinje cell and the mossy fiber-granule cell-Purkinje cell circuits, whose activities are modulated by inhibitory interneurones. There is good evidence that the major transmitter of Purkinje cells is gamma-aminobutyric acid (GABA), although the coexistence of this amino acid with certain neuropeptides has also been documented (see Section 5.2.1). The inhibitory interneurones, and basket, stellate, and Golgi cells also seem to operate primarily with GABA. However, it has been proposed that taurine (Tau) may be associated with some of the inhibitory cerebellar neurones (see Section 5.2.2). It seems that the transmitter of the only excitatory nerve cells in the cerebellum, the granule cell, is also an amino acid, most of the evidence favoring glutamate (Glu). In addition to the intrinsic nerve cells, some of the cerebellar afferents are also

A. Plaitakis (ed.), CEREBELLAR DEGENERATIONS: CLINICAL NEUROBIOLOGY. Copyright © 1992.
Kluwer Academic Publishers, Boston. All rights reserved.

aminoacidergic. This applies to the climbing fibers, which project from the inferior olive (IO), and at least in part, to the mossy fibers, whose origin is more heterogeneous (III). The cerebellum also receives noradrenergic and relatively limited serotoninergic and cholinergic innervation, and although concentrations in the adult are very low, certain neuropeptides have also been detected: these systems shall not be considered in detail here, and the reader is referred to the monograph of Ito [2] and chapters in Palay and Chan-Palay [3].

At first, some aspects of the development of the cerebellum that are relevant to the better understanding of the amino acid transmitter systems in the adult will be described briefly. This will be followed by a presentation of the evidence implicating the involvement of amino acid transmitters in the structural units of the major cerebellar circuits. Finally, recent observations implicating amino acids, which play a transmitter role in the adult, as trophic agents during certain stages of the development of nerve cells in the cerebellum will be summarized.

2. CEREBELLAR DEVELOPMENT: A BRIEF OUTLINE

Observations on nerve cell generation in the rat cerebellum have been reviewed recently (for references see Altman, [4]). The cells that are generated earliest are the neurones of the deep cerebellar nuclei (period embryonic day, E13–15, peak E14) and the Purkinje cells (period E13–16, peak E15). Both cell populations are of neuroepithelial origin. Golgi neurones are formed next during the period E19-postnatal day (P) 1 (peak E19). All the other interneurones are postnatal cells that are formed in the secondary germinal matrix, the external granular layer (EGL). The formation of the EGL begins at about E17; a cohort of neuroepithelial cells migrates from the rhombic lip and ultimately invests the whole surface of the cerebellar anlage. External granule cells go through a number of divisions before they opt out of the replication cycle and start differentiating to become either the inhibitory interneurones of the molecular layer (ML) or the excitatory granule cells populating the internal granule cell layer (IGL). The postmitotic cells gather at first in the premigratory zone underneath the replicating zone before migration, and the whole process takes about 2 days. The peak birthdays of the inhibitory interneurones are P6–7 for basket cells and P8–11 for stellate cells. Granule cells are formed during a relatively long period until P21. These are the most numerous single type of cells in the CNS; their total number is similar to that of all the nerve cells in the cerebral cortex. There are very few granule cells formed before P7, and their formation peaks during the latter part of the second postnatal week, when the cell acquisition rate is over 10% per day [5]. Granule cell differentiation starts immediately after the last mitosis. While migrating through the ML they leave behind their axon, the future parallel fibers. Dendrite formation takes place after their arrival in the IGL, where they also receive the first afferent input from the mossy fibers.

There are important differences, in comparison with their cerebellar targets, in the birthdays and the timing of the emission of the axonal projections of the nerve cells that constitute ultimately the various distinct afferent systems of the cerebellum. Noradrenergic cells in the locus coeruleus are formed first (E11–13, peak E12). Their afferents may reach the cerebellum by E14; thus this system might have an influence on the development of the cerebellum before transmission, in the sense of a mature nerve cell, becomes relevant. Although olivary nerve cells, are also formed early (E12-14, peak E13), there is some delay before they reach their final position (E16–19). However, it seems that their axons, the climbing fibers, start to grow towards the cerebellum even before the perikarya reaches the definitive sites in the IO, and the trajectory of their migration corresponds to the future course of the inferior cerebellar peduncle. Electrophysiological observations have shown that the climbing fibers establish synaptic contact with the Purkinje cells, exhibiting certain identifying characteristics similar to the adult as early as P3 [6].

On the other hand, many of the nerve cell groups, such as those in the pontine grey that send mossy fibers to the cerebellum, are formed relatively late (E16-19) and only start to differentiate after their arrival to their final position (E19-22). In general, mossy fibers reach the cerebellum only in the perinatal period and thus are not able to influence the initial phases of cerebellar development. However, their partners, the granule cells, are also late developers, and it seems that the innervation by the mossy fibers has a critical transient influence on the survival and maturation of granule cells (see Section 6.2). Synaptic transmission begins soon after mossy fiber terminals and granule cell dendrites are in physical proximity: by P7 Purkinje cells receive input from the periphery via the mossy fiber-granule cell system [7].

3. AFFERENT SYSTEMS OPERATING WITH EXCITATORY AMINO ACID TRANSMITTERS

3.1. Climbing fibers

There is a general consensus that the main transmitter of the climbing fibers is an excitatory amino acid (EAA) (Table 4-1). Climbing fibers containing corticotropin-releasing factor and enkephalin-like immunoreactivity have been described [8], but these will not be discussed here. It is, however, still controversial which EAA is the principle transmitter at the climbing fiber-Purkinje cell synapse; aspartate (Asp), Glu, and L-homocysteic acid (L-HCA) have been proposed, while a modulatory role for N-acetylaspartylglutamate has also been suggested. The studies aimed at the identification benefitted from the fact that virtually all the climbing fibers originate from the IO; thus in this respect they are relatively homogeneous. Furthermore, there is a simple and effective way to deplete the cerebellum relatively selectively from climbing fibers by destroying with 3-acetylpyridine (3-AP) nerve cells in the IO [9].

Table 4-1. Attempts to identify the transmitters of the climbing fibers

ELECTROPHYSIOLOGY
Purkinje cells are endowed with excitatory amino acid (EAA) receptors: in adults mainly AMPA/quisqualate receptors, but it is claimed by some,[1] but not by others,[2] that during development also NMDA receptors; pharmacology shows certain unconventional features[2-5]
Unique features of transmission at the climbing-fiber (CF) vs. parallel-fiber (PF) Purkinje-cell synapses (differences in reversal potentials)[4]

UPTAKE AND RETROGRADE TRANSPORT OF D-[^{3}H]ASPARTATE
Only CF and neurones in the inferior olive (IO) are labelled[6] (however[7])

ASPARTATE IMMUNOCYTOCHEMISTRY
Neurones in IO heavily labelled, but these are discrepancies concerning cerebellar (CBL) structures[8-10]

EFFECTS OF CLIMBING-FIBER DELETION (3-ACETYLPYRIDINE)
a) *Amino acid levels*
   Asp ↓[11,12]; O[13]
   Glu O[11,12]
   Tau ↓[12]
b) *Uptake of EAA*
   D-[^{3}H]Asp ↓[14]
c) *Signal transduction*
   Elevation of cGMP levels evoked by CF stimulation in vivo[15] or by depolarization in vitro ↓[14]
d) *Stimulus-coupled release of putative transmitters and neuromodulators*
   Asp   ↓ (CBL hemispheres only)[16,17]; 0 (whole CBL)[14]
   Glu   ↓ (rel. little, both vermis and hemispheres)[16,17]; O (whole CBL)[14]
   L-HCA ↓ (severe, both vermis and hemispheres)[17]
   Adenosine ↓ (severe, both vermis and hemispheres)[17]

[1] Dupont et al. [11].
[2] Llano et al. [96].
[3] Mayer and Westbrook [22].
[4] Kimura et al. [12].
[5] Joels et al. [95].
[6] Wiklund et al. [27].
[7] Levi et al. [30].
[8] Campistron et al. [148].
[9] Madl [149].
[10] Aoki [150].
[11] Nadi et al. [100].
[12] Rea et al. [151].
[13] Perry et al. [152].
[14] Foster and Roberts [16].
[15] Biggio and Guidotti [17].
[16] Vollenweider et al. [20].
[17] Cuénod et al. [19].

As expected from developmental studies, climbing fibers reach the cerebellum very early, and functional synapses are detectable from P3 in the rat (see 2). Climbing fibers coursing through the IGL lose their myelin sheath at about the level of the Purkinje cells. They arborize along the main dendrites of the Purkinje cells, usually within the inner two thirds of the ML. During early development multiple climbing fibers also synapse with somatic spines on the Purkinje cells. This is a transient phenomenon that is influenced by factors such as the thyroid state [10] and synaptic contact with parallel fibers

[11], while the 1:1 climbing fiber to Purkinje cell ratio characteristic of the adult is established in the rat by about P13.

The mature climbing fiber response is characterized by Purkinje cell complex spikes. Significantly, after excitation there is a marked depression of Purkinje cell activity. In addition, climbing fiber activity results in a tonic inhibition of the single spikes induced by the stimulation of parallel fibers [2]. In particular, the simultaneous activation of climbing fiber and parallel fiber inputs leads to long-term depression of transmission (LTD) through the parallel fiber-Purkinje cell synapse (see Chapter 3).

The first indication that acidic amino acids may serve as transmitters of the excitatory input to the Purkinje cells was provided by electrophysiological studies. However, it turned out to be very difficult to establish which EAA is the transmitter of the climbing fibers and whether the same transmitter is involved at the parallel fiber Purkinje cell synapses. The proposal of Kimura et al. [12] that Asp is the likely climbing fiber transmitter was based on circumstantial evidence: the pharmacological profile of inhibition of Ca^{2+}-evoked focal Purkinje cell responses to exposure to Asp resembled more the effect of climbing fiber stimulation than that obtained with Glu. However, recent investigations on synaptic transmission between rat IO neurones and cerebellar Purkinje cells in culture have not supported this proposal [13].

The most convincing evidence implicating EAA, and Asp in particular, as the climbing fiber transmitter has been provided by neurochemical investigations on the effect of 3-acetylpyridine (3-AP)-induced lesion of olivary neurones, although the evidence is not free of controversy (Table 4-1). In some of the studies, a decrease in the level of Tau was also noted (see Section 5.2.2). This and some of the other metabolic defects may involve the formation of the ineffective cofactor from 3-AP, 3-AP-adenine nucleotide, which is responsible for decreases in reactions leading to reduced cerebellar contents of the Tau precursors, cysteine sulphinic acid, and cysteic acid; a cerebellum specific low K_m cysteine dioxygenase is implicated in these alterations [14].

Although conclusions from observations on the high-affinity uptake of putative transmitters must be considered with caution, there is evidence indicating that nerve terminals are endowed with high-affinity transport carriers for the transmitters with which they operate [15]. It has been reported, using cerebellar slices from 3-AP treated rats, that there is a marked reduction in the uptake of D-[^{3}H]Asp, a metabolically stable analogue of the physiologically relevant EAA, with which it shares a common transport carrier [16]. Furthermore, there is evidence indicating that signal transduction, in terms of elevation of cGMP levels, elicited by increased climbing-fiber impulse traffic after harmaline administration or cold exposure, is markedly reduced in the cerebella of 3-AP treated rats [17]. Since climbing-fiber deletion did not impair either in vivo or in vitro the Glu-evoked elevation of cerebellar cGMP, it has been argued that in the intact cerebellum

EAA released from climbing-fiber terminals stimulate receptors linked to guanylate cyclase activation in the postsynaptic cell. Both the in vivo and the in vitro studies after climbing-fiber depletion implicated the Purkinje cells, which are not directly affected by the 3-AP treatment, as the primary site of the climbing-fiber-induced cGMP elevation. However, there is recent evidence which indicates that although cGMP possibly plays an important role in Purkinje cells, these cells, in quantitative terms, are not the major sites of the cGMP response to EAA stimulation in the cerebellum [18] (Section 4.2).

Effects of IO destruction on Ca^{2+}-dependent evoked amino acid release from cerebellar preparations offer a good possibility for the identification of the climbing-fiber transmitter(s), but the findings are controversial. In contrast to earlier reports [16], Cuénod and his collaborators reported a pronounced reduction in the evoked release of Asp and a less marked decrease in the fractional release of Glu in cerebellar slices from 3-AP treated rats, provided these were from the hemispheres and not from the vermis [19,20]. In addition, these authors found marked massive reductions in the stimulus-coupled release of a sulphur-containing EAA, L-HCA, and of adenosine, a putative neuromodulator (see Section 5.1.2).

Cuénod et al. [19] argue that studies that failed to implicate climbing fibers in Asp/Glu release were conducted on slices from the whole cerebellum containing persistent climbing fibers in the vermis derived from 3-AP-resistant neurones in the caudal part of the medial accessory nucleus. It is, however, unlikely that this is the whole explanation of the discrepancy, in view of the massive destruction of the olivary neurones by 3-AP treatment and the almost complete suppression of the evoked release of L-HCA and adenosine not only in the hemispheres, but also in the vermis. Nevertheless, these findings provide evidence that at least a significant proportion of the climbing fibers operate with EAA, in particular with Asp, as their transmitter.

Although compared with Asp the concentration of L-HCA in the cerebellum is very low, these observations have suggested that climbing-fiber stimulation involves the release of this amino acid. HCA has been recognized for some time as a mixed agonist acting on ionotropic Glu receptors [21,22] with higher potency on the N-methyl-D-aspartate (NMDA) class than on the non-NMDA receptor types. Recent observations by Grandes et al. [23] have shown, however, that in contrast to evoked release, climbing-fiber destruction does not affect the cerebellar level of L-HCA, which is primarily localized in glial structures. The authors suggest that climbing-fiber activation elicits L-HCA release indirectly, through a factor that could be either the transmitter EAA activating glial receptors [24] or intercellular messengers, such as arachidonic [25] or nitric oxide [18].

An interesting consequence of climbing-fiber deletion was the severe reduction of the K^+-induced release of adenosine from cerebellar slices [19].

This, like the reduction of the evoked release of L-HCA, was not restricted to the hemispheres. In view of the powerful inhibitory effect of adenosine on evoked Glu release, a neuromodulatory role of adenosine released by the stimulation of climbing fibers has been proposed [19,26].

Autoradiographic studies on intact cerebellum have supported the view that an acidic amino acid is the transmitter of the climbing fibers [27]. Microinjection of D-[^{3}H]Asp into the cerebellum resulted in pronounced labelling of glial cells, but uptake into granule cells was not detectable. These findings are consistent with biochemical observations which showed that among the various cell types isolated from the cerebellum, astrocytes have the highest capacity for Glu uptake, the V_{max} being nearly 30-fold higher than the estimate in the granule cell-enriched fractions [28]. The uptake of D-[^{3}H]Asp or [^{3}H]Glu by granule cells is also relatively low in slices from adult cerebellum, although the parallel-fiber terminals are significantly labelled [29], and in more recent studies accumulation of the labelled acidic amino acids in granule cells has been clearly demonstrated [30].

In contrast to granule cells or to mossy fibers, Wiklund et al. [27] found that climbing fibers and cells in the contralateral olive are well labelled after D-[^{3}H]Asp injection into the cerebellum. The distribution of labelled cells in the olivary subnuclei after injections into different cerebellar areas agreed with the olivocerebellar organization previously described [31]. Furthermore, it was found that initially every olivary neurone sends collaterals to deep cerebellar nuclei. Although these results support the view that an acidic amino acid serves as the transmitter of the climbing fibers, they do not prove that Asp may be the transmitter. D-Asp is transported by the same carrier as Glu and Asp. Furthermore, a relatively low uptake of an acidic amino acid does not disqualify a neuronal structure that may, nevertheless, operate with such a substance as the transmitter. Thus D-Asp uptake into granule cells (see above) or mossy-fiber terminals [27] is relatively low, but there is independent evidence in favor of EAA being the major transmitters of these structures (see Section 3.2)

3.2. Mossy fibers

In contrast to climbing fibers, these fibers originate from diverse sources and include spinal, vestibular, reticular, and pontine afferents (for refs. Palay and Chan-Palay [1]). Results concerning the identification of the transmitter(s) of the mossy fibers are forthcoming. In earlier studies, choline acetyltransferase was located immunocytochemically in some mossy fibers and cerebellar glomeruli [32]: neuropeptides, especially substance P and somatostatin, have also been implicated, although it should be noted that the concentration of many peptides show a marked decline in the cerebellum following a peak in the early postnatal period [33]. Even GABAergic mossy fibers have been described, these originating from the deep cerebellar nuclei [34]. However, the predominant transmitters of the mossy fibers seem to be EAA.

Immunocytochemical studies have shown that a significant proportion of these fibers are glutamatergic (e.g., Somogyi et al. [35]). Granule cells, in turn, are endowed with the different classes of EAA receptors [36,37]. In slices the response of granule cells to mossy-fiber stimulation could be completely blocked by EAA receptor agonists, the short latency potentials being mediated predominantly via non-NMDA receptors, while the delayed responses also involve NMDA receptors [38]. It seems that, compared with adults, NMDA receptors play a more important role in the mossy-fiber evoked granule-cell response during development when they are contributing—subject to the presence of glycine (Gly)—to the fast synaptic activity [39].

4. GLUTAMATERGIC AND GABAERGIC NEUROTRANSMISSION: RECEPTORS AND SIGNAL TRANSDUCTION SYSTEMS

As cerebellar nerve cells receive extensive synaptic input mediated by amino acid transmitters, which are also the means of communication of these cells with their targets, it is appropriate to consider briefly here the machinery of the cellular processing of the aminoacidergic signals.

4.1. Excitatory amino acid receptors

Different types of EAA receptors have been detected in granule cells. Currently two major classes of EAA receptors are recognized: ionotropic receptors, which form a ligand-gated ion channel, and metabotropic receptors. The latter receptors are coupled through a GTP binding (G) protein to phospholipase C (PLC), catalyzing the hydrolysis of phosphatidyl inositol biphosphate (PI) to inositol triphosphate (IP_3) and diacylglycerol (DAG), which serve as second messengers.

Pharmacologically, ionotropic receptors have been subdivided and named after their preferred agonists. In the cerebellum the following subtypes are relevant: the NMDA, the AMPA (α-amino-3-hydroxy-5-methyl-4-isoxazole-propionic acid)/QA (quisqualic acid) and the kainate (KA) receptors [21,22,40]. Fast excitatory synaptic activity is mediated, in general, through non-NMDA receptors, whereas NMDA receptors seem to be involved primarily in more subtle functions. This is due to certain unique properties of the NMDA receptors [22]: The permeability of the ion channel is relatively high not only to monovalent cations, as in the case of the non-NMDA receptors, but also to Ca^{2+}. Furthermore, Mg^{2+} inhibits ion conductance at resting membrane potential, but the blockade is relieved by membrane depolarization, thus conferring to the NMDA receptor the properties of a Hebbian synapse: depolarization of the cell by a coincident synaptic stimulation activates the NMDA receptor, leading to the amplification of the signal. Such mechanisms have been implicated in plastic changes in the nervous system underlying experience-dependent fine tuning of synaptic positions during development, as well as certain forms of learning, both during development and in the adult [41–43].

A more rational classification must await further developments in the molecular biological characterization of the Glu receptors. Current observations are consistent with the view that ionotropic Glu receptors comprise two independent families. The AMPA/KA receptors, already cloned [44,45], are composed of different subunits (presently at least nine are known), which constitute an ion channel. It seems that the subunit composition determines critical properties, such as the agonist preference and the rate and degree of desensitization of the receptors. There are developmental and regional differences in the expression of the subunits. For example, the expression of one of the subunits (Glu R4), which shows high sequence similarity with the other three KA/AMPA receptor subunits (Glu R1,2,3), is detectable over the entire brain during development, but postnatally the mRNA levels are usually reduced compared with those in the later embryonic stages [45]. On the other hand, the expression of the Glu R5 gene, which seems to represent a separate type of Glu receptor subunit, becomes more restricted spatially during development and transcript levels are downregulated. Extensive changes are evident in the rat/mouse cerebellum during development: until P12 both subunits are well expressed throughout the cerebellar cortex, but later the mRNA for Glu R5 is virtually confined to the Purkinje cells, whereas that for Glu R4 is detectable in significant amounts in all the different layers, including the IGL. At least 4 of the KA/AMPA receptors are expressed in 2 alternatively spliced variants (designated "flip" and "flop" [46]) showing distinct expression patterns in rat brain and developmental profile characteristics [47].

The NMDA receptor has been purified and thought to comprise 4 subunits [48,49]. Of these, the agonist recognition site has been detected in cultured granule cells and shown to be regulated by environmental cues [49a]. Two different DNA clones specific for the NMDA receptor have been reported. One, encoding for a protein structurally similar to the AMPA/KA receptors (M_r about 100 KD), was expressed in Xenopous oocytes and found to show the physiologic and pharmacologic characteristics of the NMDA receptor [49b]. The other is thought to encode for the glutamate-binding subunit (M_r about 60 KD) of an NMDA complex [49c].

Much of the information on the functional role of the different types of EAA receptors in the cerebellum is the result of pharmacological studies. In contrast to broad-spectrum antagonists, agents specific to classes of the ionotropic Glu receptor were only available for some time for the NMDA receptor, exemplified by competitive antagonists, such as gamma-phosphone derivatives of Glu (e.g., 2-amino-5-phosphonovalerate, APV), by noncompetitive antagonists acting on the ion channel (sigma opioids or dissociative anaesthetics, e.g., dextromethorphan or MK-801; some of these drugs have already been in clinical use, but it has only been realized recently that they are NMDA antagonists) and by substances acting on the allosteric glycine and polyamine sites [40]. Recently more selective AMPA/KA receptor antagonists have also become available, such as quinoxaline diones [50] (although at higher concentrations they also block the NMDA receptors at the Gly site).

Concerning the metabotropic Glu receptor (mGluR), relatively selective agonists, such as trans-1-aminocyclopentane-1,3-dicarboxylic acid (t-ACPD) and antagonists (e.g., L-amino phosphonoproprionate, APP) have only recently been identified [25]. It seems that, in comparison with adults, the metabotropic Glu response is greater in the immature brain, while APP inhibition is smaller.

The functional role of the mGluR is not yet clarified. The two second messengers resulting from the activation of the receptor initiate important intracellular events, namely, the IP_3-induced mobilization of Ca^{2+} from intracellular stores and critical protein phosphorylations through the activation via Ca^{2+} and DAG of protein kinase C(PKC). PKC may exert, in turn, a negative feedback on the metabotropic response. Coincident stimulation of the two major classes of EAA receptors may also lead to the release—catalyzed by the Ca^{2+}-dependent phospholipase A2—of arachidonic acid, which has been recently implicated as a retrograde messenger from postsynaptic to presynaptic structures [25]. There are other indications of the role of the mGluR in synaptic plasticity. The activity is maximal during the period of massive synaptogenesis, and it seems to be involved in strengthening synaptic positions during the establishment of the retinotectal map. With respect to the cerebellum, mGluRs have been implicated recently in LTD in the cerebellum (see Chapter 3). Finally, a role for the mGluR in excitotoxicity has also been proposed (for refereces see Schoepp et al. [25]). β-N-methyl-amino-alanine, implicated in the aetiology of amyotrophic lateral sclerosis-Parkinsonism-dementic complex of Guam (Guam disease) is both a metabotropic and ionotropic Glu receptor agonist. Furthermore, in models of brain injury, such as resulting from cerebral ischemia, the response to mGluR stimulation is markedly increased in the damaged tissue.

The mGluR has been cloned recently [51]. It has no significant sequence similarity to conventional G-protein-coupled receptors and has a unique structure, with large hydrophilic sequences at both sides of seven putative membrane-spanning domains. The mGluR is distributed throughout the brain, the highest levels being detected in the cerebellum (Purkinje cells in particular), the olfactory bulb, and the hippocampus.

4.2. Transduction systems

The transduction of the effect of stimulation of EAA surface receptors into the cell interior may be mediated through cation fluxes, leading to the elevation of cytoplasmic free Ca^{2+} levels either directly (via the NMDA receptor) or indirectly (via the activation of voltage-sensitive Ca^{2+} channels

(VSCC) by KA/AMPA receptor-induced depolarization). In addition, Ca^{2+} mobilization from intracellular stores may occur as a result of the stimulation of the mGluR, whose intracellular effector is a PLC. It has been proposed that under resting conditions PLC is repressed and that the inhibition is relieved by a G protein activated upon ligand occupancy of a metabotropic receptor, such as mGluR [52].

At least five immunologically distinct PLC isoenzymes have been identified. These are separate gene products with low similarities in the amino aicd sequence. According to the nomenclature of Rhee et al. [52], the neuronal forms are the gamma and β isoenzymes, the former ubiquitous and the latter relatively restricted to brain regions, while δ is expressed in astrocytes. In the cerebellum PLCα is highly concentrated in Purkinje cells and PLCβ is expressed in granule cells, while the rank order of mRNAs for the differnt PLCs is PLCα > β > gamma > δ [53]. The different regional and cellular expression suggests that each isoenzyme has a distinct function in processing the physiological response of different cell types to a variety of external stimuli and each is regulated differently (also involving differential coupling to discrete G proteins). Regulation may involve protein phosphorylation, since PKC activating phorbol esters inhibit PLC activity, which is also reduced by cAMP and cGMP.

The IP3-binding protein is highly concentrated in the cerebellum, particularly in Purkinje cells [53]. It has recently been discovered that the IP_3 receptor is identical with P400, a protein missing in cerebellar mutants deficient in Purkinje cells [54]. The P400 cDNA has been cloned and the amino acid sequence established. It is a 250-kDa glycoprotein with multiple membrane-spanning sequences and it is proposed that the N terminal is on the cytoplasmic face of the membrane, whereas the short C terminal is located in the lumen of the endoplasmic reticulum (ER), which is an important Ca^{2+} storage site, or on the outside of the plasma membrane, where it may be instrumental for the activation of an IP_3-sensitive Ca^{2+} channel. The predicted structure resembles the ryanodine receptor, which mediates Ca^{2+} release from the sarcoplasmic reticulum during excitation-contraction coupling. Although P400 mRNA is at a very high level in the cerebellum, predominantly in Purkinje cells, it shows wide tissue distribution.

The other messenger of the metabotropic response, DAG, activates PKC which, in turn, exerts its effect through phosphorylation of specific proteins. There are many putative functions for PKC, including involvement in neurotransmitter release, modulation of ion conductance, regulation of receptor interaction with components of the signal transduction apparatus, gene expression, and cell proliferation [55]. In addition, a role of PKC has also been proposed in mediating both trophic effects, including emission of neurites (for references see Cambray-Deakin and Burgoyne, [56]) and glutamate-induced toxic effects on cultured nerve cells [57].

PKC represents a large family of proteins that have a common structural design with four conserved and a number of variable regions (MW range

68–83 kDa; for review, see Nishizuka, [55]). The CNS contains not only a high level of PKC activity, but also one subtype, gamma, which is brain specific. It is concentrated, in addition to the hippocampus and cerebral cortex, in Purkinje cells, being detectable throughout the cell, including the axon terminals. The β subtype, although derived from a single gene, comprises two species as a result of differential splicing: In the cerebellum, granule cell bodies contain βI, while βII is abundant in nerve terminals and dendrites in the molecular layer. It is, therefore, evident that the machinery of the same transduction system exhibits remarkable differences depending on nerve cell types and even on compartments within the cell (e.g., different PKC and PLC subspecies in Purkinje and granule cells). Kinetic differences among the various subspecies have already been noted in vitro, but the full meaning of such differences will only come to light when the physiological target proteins of the various members of the PKC family have been identified. It has been proposed that PKC activity can lead to positive forward reactions, such as the expression of a particular gene, while synergism between the PKC and Ca^{2+} pathways underlies a variety of cellular responses to external stimuli [55]. In addition, PKC activity can also trigger negative feedback control, e.g., by shutting off PI hydrolysis via phosphorylation of PLC and by reducing free $[Ca^{2+}]_i$ by activating Ca^{2+}-ATPase. PKC itself may also be subject to regulation. Phorbol esters mimic DAG in activating and translocating the cytosolic enzyme to lipid-rich membrane compartments. However, these changes, in turn, seem to promote the degradation of the enzyme. It should be noted that Glu receptor stimulation can lead to PKC activation, which seems to be critical in the cascade, ultimately leading to excitotoxic death of "mature" neurones. This view is based upon the observation that nerve cells can be protected against the toxic effect of Glu, either by blocking PKC activity or by prior downregulation of PKC [57].

Ca^{2+} is one of the most important second messengers in neural cells. Elevation of its concentration as a result of external stimuli (like Glu receptor stimulation) from the resting low cytoplasmic level can activate a wide range of reactions, including some that can influence directly the firing of cells (e.g., opening of K^+ channels), while others may have longer lasting consequences. Here only the Ca^{2+}-induced activation of cGMP formation and of Ca^{2+}-calmodulin (CaM)-dependent protein kinase II (CaM-kinase II) will be considered.

Increased activity in excitatory pathways results in an increase in the production of cGMP, which is considered to be one of the second messengers in signal-transduction pathways. The cGMP elevation is particularly striking in the cerebellar cortex, where a significant part of this response to Glu seems to be mediated through NMDA receptors during development. It has recently been discovered that rises in cGMP do not take place necessarily in the neurones stimulated by Glu, but in other cells via an intervening intercellular factor (for review, see Garthwaite [18]). This factor has the properties of

EDRF (endothelium-derived relaxing factor), which has been identified as nitric oxide (NO). The enzyme catalyzing the formation of NO (and citrulline) from arginine (NO synthase, NOS) has been purified (for review, see Snyder and Bredt [58]). The molecular weight is 150 kDa, and it requires NADPH and Ca^{2+}/CaM for activity. The enzyme in the brain and blood vessels is identical, but it differs from the macrophage enzyme. From neural cells, only nerve cells seem to express NOS: in the cerebellum granule cells and inhibitory interneurones contain the enzyme, which, surprisingly, is not detectable in Purkinje cells, although these cells are particularly rich in guanylate cyclase and cGMP-dependent protein kinase (PKG). As NO, a diffusible molecule with a very short half-life, is a powerful activator of the soluble guanylate cyclase, the following scenario seems to account for the Glu-induced events in the cerebellum [18]: the elevation in $[Ca^{2+}]_i$ activates, in combination with CaM, NOS in granule cells and inhibitory interneurones. NO diffuses from these cells to activate guanylate cyclase in adjacent glia and neuronal structures, such as Purkinje cells or mossy-fiber terminals. It is consistent with this view that immunocytochemically detectable cGMP is heavily concentrated in glial structures in the cerebellum. In addition, NO may also serve as a transmitter, for example, in the nonadrenergic, noncholinergic (NANC) neurones in the periphery.

It seems that an important function of NO in the CNS is to relay information about postsynaptic NMDA receptor activation to neighboring nerve cells and glia. cGMP elevation in these cells is critical in initiating molecular events that lead to changes, for example, in synaptic strength, underlying activity-dependent organization of afferent fibers with respect to their target neurones during development.

cGMP may serve as a direct regulator of a membrane cation channel that underlies the transduction of light signals in retinal photoreceptor cells. In the cerebellum, however, it is likely that the effects of cGMP are mediated through PKG. This enzyme is two- to fourfold more concentrated in cerebellum than in other brain regions, primarily on account of the high levels in Purkinje cells [59].

Interactive effects involving cGMP and other second messenger systems have also been noted, although information on brain preparations is limited. Nevertheless, it is worth mentioning that cGMP-dependent protein phosphorylation leads to the inhibition of agonist-charged receptor-induced activation of G proteins and of the interaction between these G proteins and PLC in certain systems (for references, see Garthwaite [18]). Interactions with the cAMP systems have also been noted via cGMP-stimulated or cGMP-inhibited phosphodiesterases.

Protein phosphorylation is one of the most important processes in the signal transduction cascade involving Ca^{2+}. CaM kinase II is the major multifunctional protein-phosphorylating enzyme in eukaryotic tissues [60]. The brain enzyme is a multimeric complex (about 600 kDa) composed of variable numbers of 50-kDa (α) and 58/60-kDa (β/β′) subunits, which both express activity and are encoded by related but different genes. The total amount and the ratio of the two types of subunits vary in different brain regions. In the forebrain, the kinase constitutes 0.5–2% of the total protein and the α to β/β′ ratio is 3:1. The kinase concentration in the cerebellum is less and the ratio of α to β/β′ subunits is 1:4. However, even within the cerebellum, there is a marked difference in the subunit composition of the enzyme in various cell types: The α subunit is detectable only in the Purkinje cells, while the β/β′ subunit is present in both the Purkinje cells and the interneurones [61]. On the other hand, glial cells do not seem to contain CaM kinase II. The enzyme is relatively concentrated in postsynaptic densities and may play a role in transmitter release and long-term modulation of synaptic transmission, since, while being able to phosphorylate a wide range of proteins, its substrates include components of the cytoskeleton and synapsin-1. When phosphorylated the latter protein dissociates from synaptic vesicles, thus permitting exocitotic transmitter release. In addition, elevation of $[Ca^{2+}]_i$—induced, for example, by Glu receptor stimulation—will result through activation by Ca^{2+}/CaM in the autophosphorylation of the kinase, leading to the release of the enzyme from the membrane-cytoskeleton complex. The autophosphorylated kinase does not require any longer Ca/CaM for activity, thus providing a mechanism for prolonging the activity of the enzyme after the transient elevation of Ca^{2+}_i has subsided (for refs. see Cohen [60]). This mechanism seems to play an important role in the trophic effects exerted by EAA at a critical stage of the maturation of granule cells (see last section).

4.3. GABA receptors

Inhibitory neurotransmission in the supraspinal regions of the mammalian brain involves predominantly the activation of synaptic Cl^- channels by the transmitter GABA. The channel, which is an integral part of the $GABA_A$ receptor, is allosterically modulated by both benzodiazepines (BZ) and barbiturates, which are therapeutically useful drugs. Analysis of the purified receptor protein originally identified two major subunits (α and β, M 48–53 and 55–57 kDa), whose structure, deduced from cloned complementary DNAs, showed about 70% homology [62]. Coexpression of these subunits in heterologous systems generated receptors with pharmacological properties similar to those of their neuronal counterparts, with the exception that no BZ binding/responsiveness and GABA cooperativity could be detected. This deficit was, however, in part correctable: a novel $GABA_A$ receptor subunit, gamma$_2$, when coexpressed with α and β subunits, restored BZ responsiveness to the receptor. Each of these subunits, with the exception of a recently discovered unique δ, are heterogeneous, but the variants within each subunit type show very high sequence homology (for references, see Shievers et al. [63]).

Early studies have established that the density of bicuculline-sensitive

GABA binding sites is high in the cerebellum, particularly in the IGL. More recent investigations showed that the situation is relatively complex, since the distribution of the different subunits, and thus the combination of the subunits to make up the heterooligomeric GABA receptor, show regional variation. Within the α subtype, the rank order of abundance is α1>α3>α2, and the highest level of α1 occurs in the cerebellum. Even within regions differences have emerged in the subunit composition of the receptor, depending on the cell type. For example, it seems that in the cerebellum the predominant subunit composition of the receptor in Purkinje cells is α1β2gamma2, whereas it is α1β1δ in granule cells (the stoichiometry is unknow): this provides a structural basis for the earlier recognized differences in the pharmacological properties of GABA receptors in these two cell types (high BZ binding in the molecular layer, predominantly in Purkinje cell dendrites and lack of BZ binding with extensive, high-affinity muscimol binding in granule cells). Furthermore, the importance of the subunit types in modulating the properties of the receptor is indicated by the observation that the sensitivity to GABA is different, depending on which α subunit is part of the receptor and that the gamma2 and gamma1 subunits are apparently restricted to the pharmacologically different neuronal and glial receptor, respectively. It seems, therefore, that differential expression of multiple GABA-receptor subtype genes generates a diversity of GABA responses. Such differential expression of homologous, but functionally distinct receptor subtypes may be an important mechanism that contributes to synaptic plasticity [63].

The cerebellum also contains bicuculline-insensitive GABA binding sites, which display high affinity for β-p-chlorophenyl GABA (baclofen); these are defined as $GABA_B$ *receptors*. Whereas $GABA_A$ receptors are concentrated in the IGL, $GABA_B$ receptors are confined almost exclusively to the ML [64]. Nevertheless, it seems that the $GABA_B$ receptors are also located primarily on granule cells. Studies on cerebellar mutants have indicated that $GABA_B$ receptors are located primarily on parallel fibers and terminals [65]. These receptors are negatively coupled through a Gi protein to adenylate cyclase. In addition, baclofen can inhibit histamine- and 5HT-stimulated PI hydrolysis and can interfere, in part, with Ca^{2+} influx through VSCC in cerebellar granule cells (the mechanism of these effects is not yet clarified) (Wojcik et al. [66] and references therein).

5. INTRINSIC NEURONAL STRUCTURES

5.1. Excitatory interneurones: granule cells

These are the most numerous nerve cell types in the CNS. In addition, their generation continues for some time after the other cerebellar neurones have already been formed. These circumstances have contributed to the successful isolation of these cells using either methods of cell separation or tissue culture techniques (for review see Garthwaite and Balázs [28]). As a consequence our knowledge about the properties of these cells is relatively advanced, although it must be taken into account that the cell preparations comprise only the perikarya, whereas the cultured granule cells develop in the virtual absence of their synaptic partners. Nevertheless, these preparations have provided important information, especially in comparison with separated and/or cultured glial cells, concerning certain biochemical characteristics of nerve cells in general and unique features of granule neurones in particular. In addition, granule cell cultures have contributed significantly to the better understanding of the cell biological role of glutamatergic and GABAergic transmission (see Section 6).

Granule cell cultures are usually derived from early postnatal (P6–8) rodent cerebellum, and their great asset is that they are relatively homogeneous in terms of both cellular composition (about 90% of the cell being granule neurones) and maturation [67,68]. During development in culture, the cells express certain features of the neuronal phenotype, such as voltage-sensitive transmembrane Na^+ and Ca^{2+} fluxes, as well as depolarization-induced Ca^{2+}-dependent amino acid release (Balázs et al. [69,70] and references therein). Furthermore, in agreement with evidence consistent with glutamatergic transmission at the mossy fiber-granule cell synapse (see Section 3.2), EAA receptors and molecular events associated with receptor stimulation have been detected in granule cells. Hitherto, receptor binding has been examined primarily in the whole cerebellum: Autoradiographic studies showed significant EAA receptor binding in the granule cell layer, including relatively high levels of KA receptors [40]. Although NMDA receptors are detectable in the adult, the linked ion channels, if expressed at all, exhibit unusual binding properties [71] and this may be relevant concerning the apparent changes in the sensitivity of granule cells to NMDA when maturation is completed (see Section 6.4). AMPA receptors, which seem to play an important role in neurotransmission in the adult cerebellum [2] are apparently concentrated in the molecular layer [72].

5.1.1. Parallel fibers may operate with glutamate as the transmitter
The axons of granule cells, the parallel fibers make synaptic contact with the tertiary dendritic spines on the Purkinje cell. In addition, they innervate all the inhibitory interneurones of the cerebellar cortex. It seems that granule cells both receive and transmit glutamatergic stimuli. Electrophysiological observations (reviewed in other chapters) are consistent with this view, namely, that an EAA, presumably Glu, is the transmitter of the parallel fibers: Glu mimics parallel-fiber-induced excitation of Purkinje cells; reversal of Glu-induced potentials in Purkinje cells is closer to the equilibrium potential for parallel fiber- than for climbing fiber-evoked EPSP; and the influence of climbing-fiber impulses on Glu sensitivity of Purkinje cells and on parallel-fiber activation of Purkinje cells is similar. However, much of the relevant evidence has been derived from neurochemical studies that were facilitated by the possibility of relatively selective experimental deletion of

Table 4-2. An excitatory amino acid, most likely glutamate, is the putative transmitter of the parallel fibers

EVIDENCE FROM STUDIES ON CEREBELLA WITH ALTERED CELLULAR COMPOSITION
1. Agranular cerebellum
   Selective effects on Glu levels
      X-irradiation   ↓[1], O [2,3]
      Virus infection   ↓[4]
      Methylaxozymethanol ↓[5]
      Granuloprival mutants ↓[6]
   Uptake of labelled acidic amino acids (AAA)
      X-irradiation   ↓[7]
      Virus infection   ↓[4]
   Stimulus-coupled transmitter release
      X-irradiation   ↓[3]
2. Granule-cell sparing lesions
   Selective effect on Glu levels O[8]
   Uptake of labelled acidic amino acids   ↓[8]
3. Evidence from granule-cell cultures
   Stimulus-coupled and Ca^{2+}-dependent Glu release[9]
   This property, however, is not selective to Glu[10]

[1] Valcana et al. [153].
[2] Patel and Baláns [154].
[3] Sandoval and Cotman [76].
[4] Young et al. [155].
[5] Slevin et al. [156]
[6] McBride et al. [157].
[7] Rhode et al. [158].
[8] Tran and Snyder [159].
[9] Levi and Gallo [74]
[10] Kingsbury et al. [78].

granule cells (e.g., by X-irradiation during the period of postnatal cell proliferation in the cerebellum) and by the availability of granule cell deficient mutant mice.

Neurochemical findings are summarized in Table 4-2. The findings on Glu levels in agranular cerebellum are somewhat ambiguous, probably because the metabolic Glu pool has not been affected under these conditions. Results on high-affinity EAA uptake sites were initially also equivocal, since labelling of granule cells with tritiated acidic amino acids was not detected [27,29], although significant uptake over parallel fiber terminals was noted [29]. In later studies granule cell labelling, especially after longer exposure to D-[^{3}H]Asp, was clearly demonstrated [30]. Moreover, most of the observations on granule-cell-depleted preparations indicated a significant reduction in labelled acidic amino acid uptake (Table 4-2).

One of the most reliable neurochemical criteria for the transmitter/modulator role of a substance is its Ca^{2+}-dependent release upon depolarization. For example, although depolarization can also elicit the release of the Glu analogue, D-[^{3}H]Asp, from astrocytes, this, in contrast to that from granule cells, is not Ca^{2+} dependent [73]. The outcome of such investigations has been consistent with the possibility that Glu is indeed the transmitter of the parallel fibers (for review see Levi and Gallo [74]). Ca^{2+}-dependent evoked Glu release is not detectable in slices from immature cerebellum, in which only few of the adult contingent of parallel fibers are present [73]. The development of this neuronal property could be followed in granule cells in culture: It is undetectable in the first 2 days in vitro (DIV) but increases strikingly from 4 DIV [74]. This time course is similar to the development of other neuronal phenotypic characteristics, such as voltage-sensitive Ca^{2+} channels in granule cell cultures [75]. With this background, it is therefore a persuasive finding as to the nature of the parallel-fiber transmitter that depolarization-induced Ca^{2+}-dependent release of Glu was markedly reduced in preparations from the agranular cerebellum [76].

The interpretation of the results presented in Table 4-2 needs qualification. As mentioned before, the acidic amino acid carrier cannot discriminate among the various EAA. Further, Ca^{2+}-dependent depolarization-induced release in cerebellar preparations is not selective to Glu: Asp and certain neutral amino acids are also released from both cerebellar slices [76] and granule-cell cultures [78]. It is worth noting that one of these neutral amino acids is Gly, which has a critical role in potentiating the effect of NMDA receptor stimulation [79] (see also Section 5.2.2).

Modulation of evoked Glu release has been described in cerebellar preparations. Thus in granule-cell cultures GABA inhibits, via low-affinity $GABA_A$ receptors the K^+-induced release of preloaded D-[^{3}H]Asp or newly formed Glu [80] (Section 6.1). Furthermore, in cerebellar slices noradrenaline was shown to enhance the K^+-evoked release of newly synthesized [^{3}H]Glu through α2-adrenergic receptors and to inhibit release through β-adrenergic receptors [81]. Modulation of evoked Glu release by 5HT agonists has also been described [82]. However, the most extensive studies on the modulation of glutamatergic transmission have explored the effect of adenosine.

5.1.2. Adenosine and the cerebellum

In addition to being a vital intermediary metabolite, adenosine also plays an important role in neurotransmission in both the central and peripheral nervous system (for review, see Snyder [83]). In this respect, it is of interest that the concentration of adenosine in the CNS is the highest in the cerebellum, where it is especially abundant in the perikarya of the Purkinje cells [84]. The influence of adenosine on neuronal activity is mediated through membrane receptors (classified as A1 and A2), which are linked through G proteins to adenylate cyclase, with A1 receptors inhibiting and A2 receptors stimulating cyclase activity. Studies on cerebellar mutants have indicated that in the cerebellum adenosine receptors are concentrated in granule cells, with a preferential localization on parallel fibers and terminals (for references see Snyder [83]).

Adenosine is a potent inhibitor of synaptic activity in the CNS, primarily involving the inhibition of presynaptic release of transmitters. For example, in cultured cerebellar granule cells A1 agonists inhibit depolarization-induced Glu release [85]. The inhibition can be prevented, in fact converted, to stimulation by pertussis toxin (PTX)-induced ADP ribosylation of the Gi protein that couples the adenosine-charged receptor to adenylate cyclase.

Inhibition of Ca^{2+} influx seems to be instrumental in the influence of adenosine on transmitter release [86]. It has been argued that the mechanism involves adenosine effects on ion currents, rather than on cAMP levels. It seems that adenosine receptors are coupled to a K^+ channel via a PTX-sensitive G protein, and can also inhibit the N type of VSCC, which is believed to be responsible for Ca^{2+} influx triggering transmitter release.

In the cerebellum adenosine can block parallel-fiber, but not climbing-fiber elicited synaptic activity [26]. This finding is consistent with the preferential localization of the A1 receptors to parallel fibers and their terminals (see above). Interestingly, climbing-fiber deletion results in a marked depression of depolarization-induced adenosine release [19]. However, the site of adenosine release has not yet been defined: In addition to the climbing-fiber terminals, this could also be localized postsynaptically to the Purkinje cells. Nevertheless, these observations suggest that adenosine may also be involved in the mechanism of LTD (see Chapter 3).

It should be noted here that both the localization to the parallel fibers and the negative coupling of the A1 receptors to adenylate cyclase via a PTX-sensitive Gi protein are remarkably similar to the properties of the $GABA_B$ receptors (see Section 4.3).

5.2. Cells operating with inhibitory transmitters, primarily with GABA

5.2.1. Purkinje cells

These are the only efferent cells of the cerebellar cortex and have the distinction of being the first to be identified by Obata and colleagues in electrophysiological studies as GABAergic cells. This proposal soon received neurochemical support by the demonstration that stimulation of the cerebellar cortex leads to the release of GABA into the fourth ventricle [87] and that glutamate decarboxylase (GAD) and GABA in the deep cerebellar nuclei are mainly derived from the terminals of the Purkinje cells [88]. GAD purification was promptly followed by the immunocytochemical detection of GABAergic (GAD-positive) structures in the cerebellum, including the Purkinje cells (for review see Chan-Palay [34]). Local injections of GAD antibodies led to the recognition that Purkinje cell projections in the deep cerebellar nuclei conform to traditional distributions (e.g., Voogd [31]) and that reciprocal pathways exist between the deep nuclei and the Purkinje cells (Chan-Palay [34]).

It is generally accepted that virtually all Purkinje cells express GAD. However, Tau and certain neuropeptides, such as motilin, somatostatin, and enkephalin, have also been detected in these cells (Chan-Palay [34]). Nevertheless, all these substances depress the activity of Deiters neurones, thus irrespective of the neuroactive substances contained in Purkinje cells, the inhibitory action of these cells upon their targets is not in question. However, these findings, together with observations on a number of chemical markers, which are differentially distributed among Purkinje cells, cast doubt on the assumed homogeneity of Purkinje cells (Palay [89]). One such marker is cysteine sulphinic acid decarboxylase (CSD) (the rate-limiting enzyme of Tau biosynthesis), which is contained in certain Purkinje cells and is arranged in longitudinal stripes. Other differential markers also appear in longitudinal stripes, which may or may not overlap with the CSD stripes.

Probably the first enzyme activity that was shown to display a longitudinal zonal pattern of distribution in the cerebellar cortex was 5′-nucleotidase [90], which may play a role in the formation of adenosine, a powerful modulator of evoked Glu release (see Section 6.1.2). A longitudinal zonal distribution of acetylcholinesterase was also detected in the cerebellar cortex, together with regional differences in the enzyme activity in the inferior olive [91]. More recently, a number of cerebellar antigens (named *zebrins*) have been identified, using monoclonal antibodies that are confined to subsets of Purkinje cells arranged in parasagittal bands [92]. Anterograde transport mapping indicated that both the topography of the olivocerebellar and the mossy-fiber projections correlate with the bands of Purkinje cells stained using anti-zebrin antibodies.

In addition to this persistent heterogeneity, Sotelo and colleagues have shown that Purkinje cells display a transient biochemical heterogeneity during early development [93]. Immunocytochemical detection of cGMP-dependent protein kinase, vitamin D-dependent calcium binding protein, and a Purkinje cell-specific glycoprotein showed for each antigen the appearance in the perinatal period of a different mosaic of positive and negative clusters of Purkinje cells, thus indicating a differential expression of parts of the same genotype by clusters of Purkinje cells during their development. The Purkinje cell compartmentation during development is concomitant with the ingrowth of the cerebellar afferents. Once synaptogenesis starts, this biochemical heterogeneity disappears and all Purkinje cells express the three tested antigens.

The localization of GABA-receptor binding in he cerebellum has been studied using [^{3}H]muscimol, a GABA agonist [34]. In electromicroscopic studies, binding sites were detected in association with somata and primary and secondary dendritic shafts of Purkinje cells, with axons and terminals of basket cells in the pinceau around the initial axons segment of the Purkinje cells, as well as on the dendrites of granule and Golgi cells.

As described earlier, it seems that most of the excitatory afferent input to the Purkinje cells operate with EAA as the transmitter. Accordingly, Purkinje cells are endowed with EAA receptors. However, the Purkinje cell EAA receptors show some unusual characteristics (see Chapter 3). In particular, mature Purkinje cells have no NMDA receptors, although it is believed that these receptors are expressed transiently during development (and persistently in the "staggerer" mouse mutant, which has no parallel-fiber synapses) (for references, see Dupont et al. [94]). Furthermore, the non-NMDA receptors of these cells display certain properties, such as voltage-sensitive rectification, Mg^{2+} blockade, and increased Ca^{2+} conductance upon activation, which resemble those associated with NMDA receptors in other cells [95], although single–channel conductance is of the same low value as for non-NMDA receptors on other nerve cells [96].

Initial studies showing that in the cerebellum excitation is usually associated with an increase, and inhibition with a decrease, in cGMP levels have been interpreted as reflecting primarily changes in the Purkinje cells. This view was strengthened by the observation that a cGMP-dependent protein kinase and a substrate of this enzyme are greatly enriched in Purkinje cells. Nevertheless, recent studies have indicated that Purkinje cells do not play the main role in maximal cGMP accumulation in the cerebellum, which seems to involve an interaction between nerve cells, especially granule cells, and glia [18] (see Section 4.2).

Llinás and Sugimori [97,98] have observed that activity in Purkinje cells is associated with an increase in Ca^{2+} conductance, including the generation of dendritic spikes. The spatial and temporal dynamics of intracellular calcium concentration have recently been followed in spontaneously firing Purkinje cells using microfluorometric imaging of the fluorescent calcium indicator fura-2 [99]. It is expected that EAA receptor stimulation is also involved in the Ca^{2+} response. Purkinje cells seem to be especially rich in proteins binding and processing Ca^{2+} (e.g., calbindin, calcineurine, CaM kinase II, Ca^{2+}- actvated ATPase; see references in Ross et al. [53]), which may be involved in the regulation of free Ca^{2+} levels.

5.2.2. Inhibitory interneurones

Both immunocytochemical studies with antisera to GAD or GABA, and [³H]GABA uptake autoradiography showed that the inhibitory interneurones in the cerebellum, the Golgi, and the basket and stellate neurones are GABAergic cells. Some observations were, however, consistent with the view that some of the stellate cells may use Tau as their transmitter. The Tau content of the cerebellum is substantially reduced after preventing the genesis of stellate cells [100]. Furthermore, Tau can potently inhibit Purkinje-cell firing [101]. Nevertheless, it is still controversial whether Tau is a genuine transmitter or plays a neuromodulatory role [102]. Tau-like immunoreactivity has been found to be very low in stellate cells, while it is enriched in all Purkinje cells [103], which also contain the enzyme involved in Tau biosynthesis (CSD; see Section 5.2.1) in sagittal cortical bands.

The proposal that Tau may serve as a transmitter in the cerebellum is supported by recent findings using cultures containing granule cells and inhibitory interneurones [104]. Elevated K^+ (up to 40 mM) induced a Na^+-

and Ca^{2+}-dependent Tau release, regulated by dihydropyridine-sensitive VSCCs. However, it is not yet known which nerve cell type is the source of this synaptic-like release, and certain properties of the stimulated efflux are not consistent with those usually associated with transmitters (such as long delay and Ca^{2+} independence under certain conditions; see also Schousboe and Pasantes-Morales [105]).

There is less ambiguity concerning the transmitter of the other inhibitory interneurone in the molecular layer, the basket cells. These contain both GAD and GABA transport carriers, and are thought to be GABAergic cells. In contrast to Purkinje, in which the NMDA receptor-ionophore complex seems to be expressed only transiently during development [11], electrophysiological evidence indicates that stellate and basket cells do respond to NMDA, even in the adult [106].

In comparison with the other inhibitory neurones in the cerebellum, the Golgi cells show the unique characteristic of possessing high-affinity transport carriers not only for GABA, but also for Gly. This was realized while studying the properties of a very pure preparation of the cerebellar glomeruli containing the giant mossy-fiber rosette surrounded by amputated granule-cell dendritic digits, which on their periphery are in contact with axon terminals of Golgi cells [107]. Both GAD activity [107] and GABA receptor binding [108] are relatively concentrated in these glomerulus particles, which also exhibit high-affinity [^{3}H]GABA transport selectively by the Golgi axon terminals [15]. These findings are consistent with neurophysiological evidence indicating that GABA is the transmitter of the Golgi neurones. However, glomerulus particles also show high-affinity [^{3}H]Gly uptake, which is predominantly localized over the Golgi axon terminals [109], while GABA and Gly do not share the same carrier [110]. These studies have also demonstrated the selective uptake of [^{3}H]Gly into Golgi cells in cerebellar slices, and have been extended and confirmed recently by the immunocytochemical detection of Gly and GABA in the cerebellum [103]. The latter studies indicated that there are about equally large subpopulations of Golgi neurones that contain either GABA or Gly, while about 40% of these cells contain both amino acids.

With respect to functional consequences, it should be noted that Gly does not seem to serve as a major transmitter in supraspinal regions, including the cerebellum, and that the inhibitory action of Golgi cells is unaffected by the Gly-receptor antagonist strichine [111]. Furthermore, strichine-sensitive Gly binding has not been detected in cerebellar preparations [109]. On the other hand, strichine-insensitive Gly binding has been observed in the cerebellum, and a remarkable correspondence between the distribution throughout the brain of such sites and NMDA receptors has recently been noted [112]. Furthermore, it has been established that Gly markedly potentiates the effect of NMDA receptor stimulation through binding to an allosteric site on the receptor-ionophore complex [40,79]. Thus NMDA receptors may be the sites of Gly binding in the cerebellar glomeruli. Kingsbury et al. [78] have observed that not only Glu, but also Gly is released in a Ca^{2+}-dependent manner upon depolarization of cultured granule cells. Furthermore, Gly is an obligatory coagonist for eliciting responses to NMDA in cultured granule cells [37] and to synaptic stimulation of these cells in slices [39]. It was suggested that the high-affinity transport system localized in Golgi cells provides a means of inactivation in the cerebellar glomeruli of Gly, released either from the mossy-fiber-stimulated granule cells [78] or from the Golgi terminals, and so is contributing to the regulation of granule cell NMDA currents [39].

6. TROPHIC EFFECTS OF AMINO ACID TRANSMITTERS

6.1. Trophic effects of GABA

The first indication that GABA may exert neurotrophic effects was obtained by the demonstration that in the superior cervical ganglion, chronic GABA administration can induce the formation of new dendritic spines, postsynaptic thickenings, and the development of synaptogenetic capacity (for review, see Wolff et al. [113]). Further studies have provided supporting evidence for the trophic role of this amino acid by demonstrating the effects on neuronal ultrastructure and neurite emission, both in primary cultures and in cell lines (see various chapters in Redburn and Schousbou [114]). The possibility that GABA may indeed have a trophic role has been underlined by recent observations showing that the differentiation of the GABAergic system is very early relative to other transmitter-identified neurones in the CNS of rat embryos [115] and that GABAergic cells in the retina may fulfil a pathfinder role [116].

By virtue of the very low number of GABAergic cells, cerebellar granule cell cultures have permitted the study of the influence of GABA on neuronal development. It has been observed that the addition of GABA or GABA agonists, while having no effect on the survival of granule cells, promotes their morphological and functional differentiation (in terms of increases in the cytoplasmic density of rough endoplasmic reticulum, Golgi apparatus, and coated and other types of vesicles, as well as of neurite emission) [117]. Effects on biochemical maturation were indicated by the increase in the amount of total protein and of proteins relatively concentrated in nerve cells and by the induction of low-affinity GABA receptors [80,118]. Pharmacological characterization showed that these effects are mediated through conventional GABA receptors. The effects of GABA are manifested rapidly—ultrastructural changes by 1 hr and the first biochemical effects, the expression of low-affinity GABA receptors and an increase in the amount of the high-affinity GABA receptors, by 3 and 6 hr. Also in vivo elevated GABA usually results in an upregulation of the receptor [119]; but it has not yet been established what the contribution of de novo synthesis and of receptor unmasking is to this effect. However, in granule-cell cultures, low-affinity receptors are not induced by GABA when protein synthesis is blocked, and the induction of the low-affinity receptor is restricted to an early developmental state of the granule cells [80].

Information on mechanisms underlying the trophic effect of GABA is limited. Both in the superior cervical ganglion and in granule-cell cultures, the trophic effect of GABA is mimicked by NaBr [113-120]. These findings, together with the effect of other means of hyperpolarization, have suggested that the hyperpolarizing action of $GABA_A$ receptor stimulation is of critical importance in the trophic influence [120].

It should also be mentioned here that treatment with a nonmetabolizable GABA agonist also resulted in significant effects on maturation of the CNS in vivo [118]. The most impressive effect in the cerebellum was an advancement of the developmental changes affecting GAD-specific activity.

It has recently been reported that under certain conditions, $GABA_B$ receptor stimulation can also influence the maturation of nerve cells [121]. In contrast to the stimulation of neurite growth in serum-containing media (S^+), GABA resulted in an inhibition in chemically defined medium. This effect was mimicked by baclofen (a $GABA_B$ agonist), which had no influence on neurite growth in S^+.

6.2. Trophic effects of excitatory amino acids

It has been discovered recently that cerebellar granule cells in vitro, as nerve cells in vivo, develop characteristic survival requirements. For the cultured granule cells, these could be met by chronic membrane depolarization (with ≥ 20 mM K^+ in the medium; e.g., Gallo et al. [75]) or by exposure to EAA (for review Balázs and Hack [122]). The analysis of these effects led to the formulation of the hypothesis that the in vitro survival requirements mimic the influence of the first innervation received by the immediately postmigratory granule cells from the glutamatergic mossy fibers [123]. The survival of differentiating granule cells is, therefore, an example of the trophic influence exerted by afferent inputs, while the hypothesis also implies that the effect is mediated through neurotransmitters and involves the active responsiveness of the postsynaptic cell.

Detailed studies showed that granulecell survival is promoted by agonists of the different ionotropic Glu receptors, NMDA being the most effective [123-127]. The trophic influence of both K^+-induced membrane depolarization and EAA treatment proved to be mediated through Ca^{2+} influx, either through the L type of VSCC (activated by depolarization evoked by K^+ or KA/AMPA) or through the NMDA receptor-linked ion channel. It seems that, irrespective of the route of Ca^{2+} entry, the next step in the cascade leading ultimately to cell survival involves CaM-mediated reactions and CaM kinase II [75-128].

Recent observations have shown that the cell-survival promoting effect of

EAA is not restricted to the cerebellar granule cells, as similar effects have been reported for neurones in cultures of dissociated spinal cord [129] and of cortical slabs [130] (for a compilation of papers on the trophic effect of Glu and GABA, and the role of these amino acid transmitters in plastic changes in nervous tissues, see Balázs [131]). Furthermore, EAA also exert trophic effects other than promoting cell survival. Thus Pearce et al. [132] have noted that NMDA-receptor stimulation promotes neurite emission from cultured granule cell. This effect seems to involve PKC activation [55]. It is of interest that Glu can also influence the outgrowth of processes of hippocampal pyramidal cells, differentially affecting dendrites and axons in a dose-dependent manner, and thus modulating the development of neuronal shape [133]. However, in contrast to granule cells, hippocampal neurones respond to Glu by a reduction of dendritic growth, and, although Ca^{2+} is apparently the second messenger for both cell types, the receptors involved are the non-NMDA preferring subclass.

It has recently been found that NMDA can also promote the biochemical differentiation of cerebellar granule cells. Employing the conditions that have been shown to promote cell survival. Moran and Patel [134] also noted that NMDA receptor stimulation results in the induction of glutaminase, an enzyme playing an important role in the formation of the transmitter of these cells. The effect was not related to cell survival and could be inhibited by selective NMDA receptor antagonists and by blockers of protein synthesis. Similar mechanisms seem also to operate in differentiating neurones other than granule cells. Patel et al. [135] have found that in medial frontal forebrain cultures, NMDA can induce the expression of the "transmitter" enzymes, GAD and glutaminase, although choline acetyl transferase (CAT) remains unaffected. CAT is, however, induced in these nerve cells by polypeptide factors, such as NGF or NTF3 [136], indicating differential effects of trophic factors in terms of cell types.

In addition to trophic effects, EAA are also involved in plastic changes in both the developing and the adult CNS. It seems that NMDA receptors again play a critical, although presumably not exclusive, role. It has been shown that, during development, these receptors influence the fine tuning of synaptic positions (for review, see Constantine-Paton et al. [43], while in the adult they play a role in long-term potentiation in the hippocampus, in spatial learning, and in long-term depression in the cerebellum [41,42] (see Chapter 3).

6.3. Toxic effect of excitatory amino acids

It is now well established that EAA in excess, can exert neurotoxic effects (see, e.g., Olney [137] who coined the term *excitotoxicity*, and a number of recent reviews [138-140]). It has also been proposed that EAA may play an important role in a wide range of neurodegenerative processes. The evidence is most convincing when the degenerative process is acute, such as after status epilepticus, hypoglycemia, cerebral ischemia, and trauma. A major role of endogenously released EAA is indicated by the potent protection provided by EAA receptor antagonists in experimental models of these disorders. NMDA receptors are believed to play a major role in the selective neuronal loss occurring under these conditions, although a stringent correlation of the distribution of the damage with NMDA-receptor densities is not evident and, protection by new antagonists of non-NMDA receptors [50] highlights the involvement of the latter receptors in causing the injury.

Excitotoxic mechanisms are also believed to contribute to the pathogenesis of various chronic neurodegenerative disorders, including olivopontocerebellar atrophy, (see Chapter 14). Alzheimer's disease, and amyotrophic lateral sclerosis, while neurolathyrism and Guam disease (which display certain features of the diseases mentioned) have been tentatively linked to plant toxins with excitotoxic properties.

In vitro models (brain slices and cultured cells) and Glu analogues (which are less actively metabolized and address specific EAA receptor classes) have been used in order to better understand the excitotoxic mechanisms. Detailed studies of Garthwaite and colleagues on cerebellar slices (reviewed in Meldrum and Garthwaite [141]) showed that different types of neurones display pronounced variations in their vulnerability to a given agonist and different agonists can selectively kill different populations of neurones. Thus KA caused necrosis of Purkinje cells and inhibitory interneurones at a concentration lower than that required to damage granule cells. Selective toxicity was also evident in the immature cerebellum, but the cells were much less sensitive than in the adult. AMPA/QA had only slight effects in the immature cerebellum on Purkinje cells and the inhibitory interneurones, which become somewhat more vulnerable with increasing age to these substances. On the other hand, the sensitivity of granule cells to the neurotoxic action of these Glu analogues is persistently relatively low.

The toxic effect of NMDA in the cerebellum is also age dependent, but it is remarkably different from that elicited by KA or QA. Purkinje cells are not sensitive, either in the immature or in the adult; Golgi cells become more vulnerable with age, whereas the sensitivity of basket and stellate cells does not change much with maturation. On the other hand, granule cells undergo an intriguing change in chemosensitivity with maturation.

6.4. Both the trophic and the toxic effect of EAA is maturational stage dependent

On the basis of information on the trophic and the toxic effect of NMDA on the cerebellum in vivo and in vitro, it would appear that replicating and migrating granule cells are not responsive to NMDA, which is toxic only to differentiating granule cells, although those in the upper part of the IGL are not adversely affected [142]. The latter cells have just completed migration, and according to our hypothesis [124] (Section 6.2) these are the cells that require glutamatergic stimulation for survival and maturation (this being provided by the mossy fibers in vivo or by Glu agonist treatment in vitro). The observations of Rakic and Sidman [143] suggest that mossy fibers may indeed exert such a trophic influence. In the cerebellar mutant "weaver," the genetic defect involves granule cells, which die soon after their generation. A few cells survive and these are innervated in ectopic position by aberrant mossy fibers.

As the maturation of the granule cells progresses, the trophic influence of NMDA-receptor stimulation [124] is superceded by a toxic effect [142]. Recent studies (including our own) indicate that in culture the maturation of granule cells can also reach a stage when EAA become toxic, rather than trophic factors (see also Manev et al. [57]). It seems that both the trophic and the toxic effects of NMDA are mediated through the same mechanism, i.e., increased Ca^{2+} influx [124–127,138,144]. Our knowledge is, however, limited concerning the mechanism(s) underlying the maturational stage-dependent life-or-death response of the cell to the same signal, although it seems that specific protein kinases are involved—the CaM kinase II in the trophic and PKC in the toxic effects [57,128].

In connection with the trophic effects of EAA and their involvement in neuronal plasticity, the observation that NMDA can induce in cultured granule cells the expression of early-response genes [145], such as *c-fos*, is potentially important [146,147]. The products of these genes, in specific combinations, may serve as transcription factors, thus eliciting a critical influence on gene expression. However, we do not yet know whether the induction of early response genes is part of the mechanism underlying the long-term effects of EAA or represents only an epiphenomemon. Thus, although much further work mus be done to elucidate the mechanism(s) underlying the trophic and plastic influence of amino acid transmitters in the CNS, the demonstration of these effects indicate that signalling systems, such as involved in specific neurotransmission, can also fulfil different functions that during development are determined by the maturational state of the cell that receives the signal.

ACKNOWLEDGMENTS

The excellent secretarial assistance of Ms. Olga Pach is gratefully acknowledged. R.B. receives support from ZWO and the Van den Houten Foundation.

REFERENCES

1. Palay S.L., Chan-Palay V. (eds). (1974). Cerebellar Cortex-Cytology and Organization. Berlin: Springer.
2. Ito M. (1984). The Cerebellum and Neural Control. New York: Raven Press.
3. Palay S.L., Chan-Palay V. (eds). (1982). The Cerebellum—New Vistas. Exp. Brain Res. Suppl. 6. Berlin: Springer.
4. Altman J. (1982). Morphological development of the rat cerebellum and some of its mechanisms. In: V. Chan-Palay, S. Palay (eds): The Cerebellum: New Visas. Berlin: Springer Verlag, pp. 8–49.

5. Patel A.J., Balázs R., Johnson A.L. (1973). Effect of undernutrition on cell formation in the rat brain. J. Neurochem. 20:1151–1165.
6. Puro D.G., Woodward D.J. (1977). Maturation of evoked climbing fiber input to rat cerebellar Purkinje cells (I). Exp. Brain Res. 28:85–100.
7. Puro D.G., Woodward D.J. (1977). Maturation of evoked mossy fiber input to rat cerebellar Purkinje cells (II). Exp. Brain Res. 28:427–441.
8. Palkovits M., Léránth C., Görcs T., Young W.S. (1987). Corticotropin-releasing factor in the olivocerebellar tract of rats: Demonstration by light- and electron-microscopic immunohistochemistry and in situ hybridization histochemistry. Proc. Natl. Acad. Sci. USA 84:3911–3915.
9. Balaban C.D. (1985). Central neurotoxic effecs of intraperitoneal administered 3-acetylpyridine, harmaline and nicotinamide in Sprague-Dawley and Long-Evans rats: A critical review of 3-acetyl-pyridine neurotoxicity. Brain Res. Rev. 9:21–42.
10. Hajós F., Patel A.J., Balázs R. (1973). Effect of thyroid deficiency on the synaptic organization of the rat cerebellar cortex. Brain Res. 50:387–401.
11. Dupont J.-L., Gardette R., Crepel F. (1987). Postnatal development of the chemosensitivity of rat cerebellar Purkinje cells to excitatory amino acids. An in vitro study. Dev. Brain Res. 34:59–68.
12. Kimura H., Okamoto K., Sakay Y. (1985). Pharmacological evidence for L-aspartate as the neurotransmitter of cerebellar climbing fibers in the guinea-pig. J. Physiol. 365:103–109.
13. Hirano T. (1990). Synaptic transmission between rat inferior olivary neurons and cerebellar Purkinje cells in culture. J. Neurophysiol. 63:181–189.
14. Ida S., Ohkuma S., Kimori M., Kuriyama K., Morimoto N., Ibata Y. (1985). Regulatory role of cystein dioxygenase in cerebral biosynthesis of taurine. Analysis using cerebellum from 3-acetylpyridine-treated rats. Brain Res. 344:62–69.
15. Wilkin G.P., Wilson J.E., Balázs R., Schon F., Kelly J.S. (1974). How selective is high affinity uptake of GABA into inhibitory nerve terminals? Nature 252:397–399.
16. Foster G.A., Roberts P.J. (1983). Neurochemical and pharmacological correlates of inferior olive destruction in the rat: Attenuattion of the events mediated by an endogenous glutamate-like substance. Neuroscience 8:277–284.
17. Biggio G., Guidotti A. (1976). Climbing fiber activation and 3',5'-cyclic guanosine monophosphate (cGMP) content in cortex and deep nuclei of cerebellum. Brain Res. 107:365–373.
18. Garthwaite J. (1991). Glutamate, nitric oxide and cell-cell signalling in the nervous system. Trends Neurosci. 14:60–67.
19. Cuénod M., Do K.-Q., Vollenweider F., et al. (1988). The puzzle of the transmitters in the climbing fibers. In: P. Strata (ed): The Olivocerebellar System in Motor Control. Exp. Brain Res., Suppl. 17. Berlin: Springer, pp. 161–176.
20. Vollenweider F.X., Cuénod M., Do K.Q. (1990). Effect of climbing fiber deprivation on release of endogenous aspartate, glutamate and homocysteate in slices of rat cerebellar hemispheres and vermis. J. Neurochem. 54:1533–1540.
21. Watkins J.C., Evans R.H. (1981). Excitatory amino acid transmitters. Ann. Rev. Pharmacol. Toxicol. 21:165–204.
22. Mayer M., Westbrook G.L. (1987). The physiology of excitatory amino acids in the vertebrate central nervous system. Prog. Neurobiol. 28:197–276.
23. Grandes P., Dok Q., Morino P., et al. (1991). Homocysteate, an excitatory transmitter candidate localized in glia. Submitted.
24. Barres B.A., Chun L.L.Y., Corey D.P. (1990). Ion channels in vertebrate glia. Ann. Rev. Neurosci. 13:441–474.
25. Schoepp D., Bockaert J., Sladeczek F. (1990). Pharmacological and functional characteristics of metabotropic excitatory amino acid receptors. Trends Pharmacol. Sci. 11:508–515.
26. Kocsis J.D., Eng D.-L., Bhisitkul R.B. (1984). Adenosine selectively blocks parallel-fiber-mediated synaptic potentials in rat cerebellar cortex. Proc. Natl. Acad. Sci. USA 81:6531–6534.
27. Wiklund L., Toggenburger G., Cuénod M. (1984). Selective retrograde labelling of the rat olivocerebellar climbing fiber system with D-[3H]aspartate. Neuroscience 13:441–468.
28. Garthwaite J., Balázs R. (1981). Separation of cell types from the cerebellum and their properties. Adv. Cell Neurobiol. 2:461–489.

29. Wilkin G.P., Garthwaite J., Balázs R. (1982). Putative amino acid transmitters in the cerebellum. II. Electron microscopic localization of transport sites. Brain Res. 244:69–80.
30. Levi G., Aloisi F., Giotti M.T., Gallo V. (1984). Autoradiographic localization and depolarization induced release of acidic amino acids in differentiating granule cell cultures. Brain Res. 290:77–86.
31. Voogd J. (1982). The olivocerebellar projection in the cat. In: S.L. Palay, V. Chan-Palay (eds): The Cerebellum—New Vistas. Exp. Brain Res., Suppl. 6. Berlin: Springer, pp. 134–161.
32. Kan K.-S., Chao L.-P., Forno L.S. (1980). Immunohistochemical localization of choline acetyltransferase in the human cerebellum. Brain Res. 193:165–171.
33. McGregor G.P., Woodhams P.L., O'Shaughnessy D.J., et al. (1982). Developmental changes in bombesin, substance P, somatostatin and vasoactive intestinal polypeptide in the rat brain. Neurosci. Lett. 28:21–27.
34. Chan-Palay V. (1982). Neurotransmitters and receptors in the cerebellum: Immunocytochemical localization of glutamic acid decarboxylase, GABA-transaminase, and cyclic GMP and autoradiography with ^{3}H-muscimol. In: S.L. Palay, V. Chan-Palay (eds): The Cerebellum: New Vistas. Berlin: Springer Verlag, pp. 552–584.
35. Somogyi P., Halasy K., Somogyi J., et al. (1986). Quantification of immunogold labelling reveals enrichment of glutamate in mossy and parallel fiber terminals in cat cerebellum. Neuroscience. 19:1045–1050.
36. Cull-Candy S.G., Mathie A., Symonds C., Wyllie F. (1989). Noise and single channels activated by excitatory amino acids in rat cerebellar granule neurones. J. Physiol. 400: 189–222.
37. Van der Valk J.B.F., Resink A., Balázs R. (1991). Membrane depolarization and the expression of glutamate receptors and cultured cerebellar granule cells. Europ. J. Pharmacol. 201:247–250.
38. Garthwaite J., Brodbelt A.R. (1989). Synaptic activation of N-methyl-D-aspartate and non-N-methyl-D-aspartate receptors in the mossy fiber pathway in adult and immature rat cerebellar slices. Neuroscience 29:401–412.
39. D'Angelo E., Rossi P., Garthwaite J. (1990). Dual-component NMDA receptor currents at a single central synapse. Nature 346:467–470.
40. Monaghan D.T., Bridges R.S., Cotman C.W. (1989). The excitatory amino acid receptors: Their classes, pharmacology, and distinct properties in the function of the central nervous system. Ann. Rev. Pharmacol. Toxicol. 29:365–402.
41. Cotman C.W., Monaghan D.T. (1988). Excitatory amino acid neurotransmission: NMDA receptors and Hebb-type synaptic plasticity. Ann. Rev. Neurosci. 11:61–80.
42. Collingridge G.L., Singer W. (1990). Excitatory amino acid receptors and synaptic plasticity. Trends Pharmacol. Sci. 11:290–296.
43. Constantine-Paton M., Cline H.T., Debski E. (1990). Patterned activity, synaptic convergence, and the NMDA receptor in developing visual pathways. Ann. Rev. Neurosci. 13:129–154.
44. Hollmann M., O'Shea-Greenfeld A., Rogers S., Heinemann S. (1989). Cloning by functional expression of a member of the glutamate receptor family. Nature 342:643–648.
45. Bettler B., Boulter J., Hermans-Borgmeyer I., et al. (1990). Cloning of a novel glutamate receptor subunit, GluR5: Expression in the nervous system during development. Neuron 5:583–595.
46. Sommer B., Keinänen K., Verdoorn T.A., et al. (1990). Flip and Flop: A cell specific functional switch in glutamate-operated channels of the CNS. Science 249:1580–1585.
47. Momjer H., Seeburg P.H., Wisden W. (1991). Glutamate-operated channels: Developmentally early and mature forms arise by alternative splicing. Neuron 6:799–810.
48. Ikin A.F., Kloog Y., Sokolovsky M. (1990). N-methyl-D-aspartate/phencyclidine receptor complex of rat forebrain: Purification and biochemical characterization. Biochemistry 29:2290–2295.
49. Chen J.-W., Cunningham M.D., Galton N., Michaelis E.K. (1988). Immune labelling and purification of a 71-kDa glutamate binding protein from brain synaptic membranes. Possible relationship of this protein to physiological glutamate receptors. J. Biol. Chem. 263:417–426.
49a. Balázs R., Resink A., Hack N., et al. (1991). NMDA treatment and K^+-induced depolarization selectively promote the expression of an NMDA-preferring class of the ionotropic

glutamate receptors in cerebellar granule cells. Neurosci. Lett. (in press).
49b. Moriyoshi K., Masu M., Ishii T., et al. (1991). Molecular cloning and characterization of the rat NMDA receptor. Nature 354:31–37.
49c. Kumar K.N., Tilakaratne N., Johnson P.S., et al. (1991). Cloning of a cDNA for the glutamate-binding subunit of an NMDA receptor complex. Nature 354:70–73.
50. Sheardown M.S., Nielsen E.O., Hansen A.J., et al. (1990). 2,3-dihydroxy-6-nitro-7-sulfamoyl-benzo(F)quinoxaline: A neuroprotectant for cerebral ischaemia. Science 247: 571–574.
51. Masu M., Tanabe Y., Tsuchida K., et al. (1991). Sequence and expression of metabotropic glutamate receptor. Nature 349:760–765.
52. Rhee S.G., Suh P.-G., Ryn S.-H., Lee S.Y. (1989). Studies of inositol phospholipid-specific phspholipase C. Science 244:546–550.
53. Ross C.A., Bredt D., Snyder S.H. (1990). Messenger molecules in the cerebellum. Trends Neurosci. 13:216–222.
54. Furuichi T., Yoshikawa S., Miyawaki A., et al. (1989). Primary structure and functional expression of the inositol 1,4,5-triphosphate-binding protein P_{400}. Nature 342:32–38.
55. Nishizuka Y. (1988). The molecular heterogeneity of protein kinase C and it implications for cellular regulation. Nature 334:661–665.
56. Cambray-Deakin M.A., Ader J., Burgoyne R.D. (1990). Neuritogenesis in cerebellar granule cells in vitro: A role for protein kinase C. Dev. Brain Res. 53:40–46.
57. Manev H., Costa E., Wroblewski J.T., Guidotti A. (1990). Abusive stimulation of excitatory amino acid receptors: A strategy to limit neurotoxicity. FASEB J. 4:2789–2797.
58. Snyder S.H., Bredt D.S. (1991). Nitroc oxide as a neuronal messenger. Trends Pharmacol. Sci. 12:125–128.
59. Lohmann S.M., Walter V., Miller P.E., et al. (1981). Immunocytochemical localization of cyclic GMP-dependent protein kinase in mammalian brain. Proc. Natl. Acad. Sci. USA 78:653–657.
60. Cohen P. (1988). The calmodulin-dependent multiprotein kinase. In: P. Cohen, C.B. Klee (eds): Molecular Aspects of Cellular Regulation, Vol. 5. Amsterdam: Elsevier, pp. 145–193.
61. Walaas S.I., Lai Y., Goelick F.S., et al. (1988). Cell-specific localization of the alpha-subunit of Ca/CaM dependent protein kinase II in Purkinje cells in rodent cerebellum. Mol. Brain Res. 4:233–242.
62. Barnard E.A., Darlison M.G., Seeburg P.H. (1987). Molecular biology of the $GABA_A$ receptor: The receptor/channel superfamily. Trends Neurosci. 10:502–509.
63. Shievers B.D., Killisch I., Sprengel R., et al. (1989). Two novel $GABA_A$ receptor subunits exist in distinct neuronal subpopulations. Neuron 3:327–337.
64. Wilkin G.P., Anderson A.L., Hill D.R., Bowery N.G. (1981). Autoradiographic localization of $GABA_B$ receptors in the cerebellum. Nature 294:584–587.
65. Wojcik W.J., Neff N.H. (1984). Gamma-aminobutyric acid B receptors are negatively coupled to adenylated cyclase in brain, and in the cerebellum these receptors may be associated with granule cells. Mol. Pharmacol. 25:24–28.
66. Wojcik W.J., Travagli R.A., Costa E., Bertolino M. (1990). Baclofen inhibits with high affinity the L-type-like voltage-dependent calcium channel in cerebellar granule cells cultures. Neuropharmacology 29:969–972.
67. Thangnipon W., Kingsbury A., Webb M., Balázs R. (1983). Observations on rat cerebellar cells in vitro: Influence of substratum, potassium concentration and relationship between neurones and astrocytes. Dev. Brain Res. 11:177–189.
68. Kingsbury A., Gallo V., Woodhams P., Balázs R. (1985). Survival, morphology and adhesion properties of cerebellar interneurones cultured in chemically defined and serum supplemented medium. Dev. Brain Res. 17:17–25.
69. Balázs R., Gallo V., Atterwill C.K., et al. (1985). Does thyroid hormone influence the maturation of cerebellar granule neurons? Biomed. Biochem. Acta 44:1469–1482.
70. Balázs R., Gallo V., Kingsbury A.E., et al. (1987). Factors affecting the survival and maturation of nerve cells in culture. In: H.H. Althouse, W. Seifert (eds): Glial-Neuronal Communication in Development and Regeneration. Berlin: Springer, pp. 285–312.
71. Maragos W.F., Penney J.B., Young A.B. (1988). Anatomic correlation of NMDA and ^{3}H-TCP-labeled receptors in rat brain. J. Neurosci. 8:493–501.

72. Greenamyre J.T., Olson J.H.M., Penney J.B., Jr., Young A.B. (1985). Autoradiographic characterization of N-methyl-D-aspartate-, quisqualate-, and kainate-sensitive glutamate binding sites. J. Pharmacol. Exp. Ther. 233:254–263.
73. Levi G., Gordon R.D., Gallo V., et al. (1982). Putative amino acid transmitters in the cerebellum. I. Depolarization-induced release. Brain Res. 239:425–445.
74. Levi G., Gallo V. (1986). Release studies related to the neurotransmitter role of glutamate in the cerebellum: An overview. Neurochem. Res. 11:1027–1042.
75. Gallo V., Kingsbury A., Balázs R., Jørgensen O.S. (1987). The role of depolarization in the survival and differentiation of cerebellar granule cells in culture. J. Neurosci. 7:2203–2213.
76. Sandoval M.E., Cotman C.W. (1978). Evaluation of glutamate as a neurotransmitter of cerebellar parallel fibers. Neuroscience 3:199–206.
77. Bosley T.M., Woodhams P.L., Gordon R.D., Balázs R. (1983). Effects of anoxia on the stimulated release of amino acid neurotransmitters in the cerebellum in vitro. J. Neurochem. 40:189–201.
78. Kingsbury A., Gallo V., Balázs R. (1988). Stimulus-coupled release of amino acids from cerebellar granule cells in culture. Brain Res. 448:46–52.
79. Johnson J.W., Ascher P. (1987). Glycine potentiates the NMDA response in cultured mouse brain neurons. Nature 325:529–531.
80. Meier E., Belhage B., Drejer J., Schousboe A. (1987). The expression of GABA receptors on cultured cerebellar granule cells is influenced by GABA. In: D.A. Redburn, A. Schousboe (eds): Neurotrophic Activity of GABA during Development, New York: Alan R. Liss, pp. 139–159.
81. Dolphin A.C. (1982). Noradrenergic modulation of glutamate release in the cerebellum. Brain Res. 252:111–116.
82. Raiteri M., Maura G., Bonnano G., Pittalunga A. (1986). Differential pharmacology, localization and function of two 5HT1 receptors modulating transmitter release in rat cerebellum. J. Pharmacol. Exp. Ther. 237:644–648.
83. Snyder S.H. (1985). Adenosine as a neuromodulator. Ann. Rev. Neurosci. 8:103–124.
84. Braas K.M., Newly A.C., Wilson V.S., Snyder S.H. (1986). Adenosine-containing neurons in the brain localized by immunocytochemistry. J. Neurosci. 6:1952–1961.
85. Dolphin A.C., Prestwick S.A. (1985). Pertussis toxin reverses adenosine inhibition of neuronal glutamate release. Nature 316:148–150.
86. Fredholm B.B., Dunwiddie T.V. (1988). How does adenosine inhibit transmitter release? Trends Pharmacol. Sci. 9:130–134.
87. Obata K., Takeda K. (1969). Release of gamma-aminobutyric acid in the fourth ventricle induced by stimulation of the cat's cerebellum. J. Neurochem. 16:1043–1047.
88. Fonnum F., Storm-Mathisen J., Walberg F. (1970). Glutamate-decarboxylase in inhibitory neurons. A study of the enzyme in Purkinje cell axons and boutons in the cat. Brain Res. 20:259–275.
89. Palay S.L. (1992). Heterogeneity in Purkinje cells. In: R. Llinás, C. Sotelo (eds): Neurobiology of the Cerebellar Systems. New York: Springer-Verlag, in press.
90. Scott T.G. (1964). A unique pattern of localization within the cerebellum of the mouse. J. Comp. Neurol. 122:1–7.
91. Marani E. (1982). Topographic Enzyme Histochemistry of the Mammalian Cerebellum. 5′ Nucleotidase and Acetylcholinesterase. Thesis, University of Leyden.
92. Hawkes R. (1992). Zebrins: Molecular correlates of compartmentation in the cerebellum. In: R. Llinás C. Sotelo (eds): Neurobiology of the Cerebellar Systems. Oxford: Oxford University Press, in press.
93. Wassef M., Zanetta J.P., Brehier A., et al. (1985). Transient biochemical compartmentalization of Purlinje cells during early cerebellar development. Dev. Biol. 111:129–137.
94. Dupont J.-L., Fournier E., Gardette R., Crepel F. (1984). Effect of excitatory amino acids on Purkinje cell dendrites in cerebellar slices from normal and staggerer mice. Neurscience 12:613–619.
95. Joels M., Yool A.J., Gruol D.L. (1989). Unique properties of non-N-methyl-D-aspartate excitatory responses in cultured Purkinje neurons. Proc. Natl. Acad. Sci. USA 86:3404–3408.

96. Llano I., Marty A., Johnson J.W., et al. (1988). Patch-clamp recording of amino acid-activated responses in "organotypic" slice cultures. Proc. Natl. Acad. Sci. USA 85:3221–3225.
97. Llinás R., Sugimori M. (1980). Electrophysiological properties of in vitro Purkinje cell somata in mammalian cerebellar slices. J. Physiol. 305:171–195.
98. Llinás R., Sugimori M. (1980). Electrophysiological properties of in vitro Purkinje cell dendrites in mammalian cerebellar slices. J. Physiol. 305:197–213.
99. Tank D.W., Sugimori M., Connor J.A., Llinás R.R. (1988). Spatially resolved calcium dynamics of mammalian Purkinje cells in cerebellar slices. Science 242:773–777.
100. Nadi N.S., Kanter D., McBride W.J., Aprison H.H. (1977). Effects of 3-acetyl-pyridine on several putative neurotransmitter amino acids in the cerebellum and medulla of the rat. J. Neurochem. 28:661–662.
101. Okamoto K., Kimura H., Sakai Y. (1983). Evidence for taurine as an inhibitory neurotransmitter in cerebellar stellate interneurons: Selective antagonism by TAG (6-aminomethyl-3-methyl-4H,1,2,4-benzothiadiazine-1,1-dioxide). Brain Res. 265:163–168.
102. Curtis D., Leah J.D., Peet M. (1982). Lack of specificity of a "taurine antagonist". Brain Res. 244:198–199.
103. Ottersen O.P., Storm-Matthisen J. (1987). Localization of amino acid neurotransmitters by immunocytochemistry. Trends Neurosci. 10:250–255.
104. Philibert R.A., Rogers K.L., Dutton G.R. (1989). Stimulus-coupled taurine efflux from cerebellar neuronal cultures: On the roles of Ca^{++} and Na^+. J. Neurosci. Res. 22:167–171.
105. Schousboe A., Pasantes-Morales H. (1989). Potassium-stimulated release of [^{3}H]taurin from cultured GABAergic and glutamatergic neurones. J. Neurochem. 53:1309–1315.
106. Crepel F., Dhangal S.S., Sears T.A. (1982). Effect of glutamate, aspartate and related derivatives on cerebellar Purkinje cell dendrites in the rat: An in viro study. J. Physiol. 329:297–317.
107. Balázs R., Hajós F., Johnson A.L. (1975). Subcellular fractionation of rat cerebellum: An electron microscopic and biochemical investigation. III. Isolation of large fragments of the cerebellar glomeruli. Brain Res. 86:17–30.
108. Kingsbury A., Wilkin G.P., Patel A.J., Balázs R. (1980). Distribution of GABA receptors in the rat cerebellum. J. Neurochem. 35:739–742.
109. Wilkin G.P., Csillag A., Balázs R. (1981). Localization of high affinity [^{3}H]glycine transport sites in the cerebellar cortex. Brain Res. 216:11–33.
110. Wilson J.E., Wilkin G.P., Balázs R. (1976). Metabolic properties of a purified preparation of large fragments of the cerebellar glumeruli: Glucose metabolism and amino acid uptake. J. Neurochem. 26:957–965.
111. Bisti S., Iosif G., Marchesi G.F., Strata P. (1972). Pharmacological properties of inhibitions in the cerebellar cortex. Exp. Brain Res. 14:24–37.
112. Bowery N.G. (1987). Glycine binding sites and NMDA receptors in brain. Nature 326:338.
113. Wolff J.R., Joo F., Kesa P. (1987). Synaptic, metabolic and morphogenetic effects of GABA in the superior cervical ganglion of rats. In: D.A. Redburn and A. Schousboe (eds): Neurotrophic Activity of GABA during Development. New York: Alan R. Liss, pp. 221–252.
114. Redburn D., Schousboe A. (eds). (1987). Neurotrophic Activity of GABA during Development. New York: Alan R. Liss.
115. Lauder J., Han V.K.M., Henderson P., et al. (1986). Prenatal ontogeny of the GABAergic system in the rat brain: An immunocytochemical study. Neuroscience 19:465–493.
116. Messersmith, E.K., Redburn D.A. (1990). Kainic acid lesioning alters development of the outer plexiform layer in neonatal rabbit retina. Int. J. Dev. Neurosci. 8:447–461.
117. Hansen G.H., Meier E., Abraham J., Schousboe A. (1987). Trophic effects of GABA on cerebellar granule cells on culture. In: D.A. Redburn, A. Schousboe (eds): Neurotrophic Activity of GABA During Development, New York: Alan R. Liss, pp. 109–138.
118. Meier E., Jørgensen O.S., Schousboe A. (1987). Effect of repeated treatment with a GABA receptor agonist on postnatal neural development in rats. J. Neurochem. 49:1462–1470.
119. Beart P.M., Scatton B., Lloyd K.G. (1985). Subchronic administration of GABAergic agonists elevates 3H-GABA binding and produces tolerance in striatal dopamine catabolism. Brain Res. 335:169–173.

120. Belhage B., Hansen G.H., Schousboe A. (1990). GABA agonist induced changes in cerebellar granule cells is linked to hyperpolarization of the neurons. Int. J. Dev. Neurosci. 8:473–479.
121. Michler A. (1990). Involvement of GABA receptors in the regulation of neurite growth in cultured embryonic chick tectum. Int. J. Dev. Neurosci. 8:463–472.
122. Balázs R., Hack N. (1990). Trophic effects of excitatory amino acids in the developing nervous system. In: Y. Ben-Ari (ed): Excitatory Amino Acids and Neural Plasticity. New York: Plenum Press, pp. 221–228.
123. Balázs R., Hack N., Jørgensen O.S. (1988). Stimulation of the N-methyl-D-aspartate receptor has a trophic effect on differentiating granule cells. Neurosci. Lett. 87:80–86.
124. Balázs R., Jørgensen O.S., Hack N. (1988). N-methyl-D-aspartate promotes the survival of cerebellar granule cells in culture. Neuroscience 27:437–451.
125. Balázs R., Hack N., Jørgensen O.S., Cotman C.W. (1989). N-methyl-D-aspartate promotes the survival of cerebellar granule cells: Pharmacological characterization. Neurosci. Lett. 101:241–246.
126. Balázs R., Hack N., Jørgensen O.S. (1990). Selective stimulation of excitatory amino acid receptor subtypes and the survival of cerebellar granule cells in culture: Effect of kainic acid. Neuroscience 37:251–258.
127. Balázs R., Hack N., Jørgensen O. (1990). Interactive effects involving different classes of excitatory amino acid receptors and the survival of cerebellar granule cells in culture. Int. J. Dev. Neurosci. 8:347–358.
128. Balázs R., Hack N., Resink A., et al. (1992). Trophic effect of excitatory amino acids on differentiating granules cells: involvement of calcium and other second massengers. Mol. Neuropharmacol, in press.
129. Brenneman D., Yu C., Nelson P.G. (1990). Multi-determinate regulation of neuronal survival: Neuropeptides, excitatory amino acids and bioelectrical activity. Int. J. Dev. Neurosci. 8:371–378.
130. Ruijter J.M., and Baker R.E. (1990). The effects of potassium-induced depolarisation, glutamate receptor antagonists and N-methyl-D-aspartate on neuronal survival in cultured neocortex explants. Int. J. Dev. Neurosci. 8:361–370.
131. Balázs R. (ed) (1990). Plastic and Trophic Effects of Amino Acid Transmitters in the Developing Nervous System. Int. J. Dev. Neurosci. 8(4):345–503.
132. Pearce I.A., Cambray-Deakin M.A., Burgoyne R.D. (1987). Glutamate acting on NMDA receptors stimulate neurite outgrowth from cerebellar granule cells. FEBS Lett. 223:143–147.
133. Mattson M.P. (1988). Neurotransmitters in the regulation of neuronal cytoarchitecture. Brain Res. Rev. 13:179–212.
134. Moran J., Patel A.J. (1989). Stimulation of the N-metyl-D-aspartate receptor promotes the biochemical differentiation of cerebellar granule neurons and not astocytes. Brain Res. 486:15–25.
135. Patel A.J., Hunt A., Sanfeliu C. (1990). Cell-type specific effects of N-methyl-D-aspartate on biochemical differentiation of subcortical neurons in culture. Int. J. Dev. Neurosci. 8:379–389.
136. Thoenen H. (1991). The changing scene of neurotrophic factors. Trends Neurosci. 14:165–170.
137. Olney J.W. (1978). Neurotoxicity of excitatory amino acids. In: E.G. McGeer, J.W. Olney, P.L. McGeer (eds): Kainic Acid as a Tool in Neurobiology. New York: Raven Press, pp. 95–121.
138. Choi D.W. (1987). Glutamate neurotoxicity in cortical cell cultures. J. Neurosci. 7:357–368.
139. Choi D.W., Rothman S.M. (1990). The role of glutamate neurotoxicity in hypoxic-ischemic neuronal death. Ann. Rev. Neurosci. 13:171–182.
140. Balázs R. (1989). Excitotoxic mechanisms and neurological disease. Curr. Opin. Neurol. Neurosurg. 2:929–935.
141. Meldrum B., Garthwaite J. (1990). Excitatory amino acid neurotoxicity and neurodegenerative disease. Trends Pharmacol. Sci. 11:379–387.
142. Garthwaite G., Garthwaite J. (1986). In vitro neurotoxicity of excitatory amino acid analogues during cerebellar development. Neuroscience 17:755–767.

143. Rakic P., Sidman R.L. (1973). Organization of the cerebellar cortex secondary to deficit of granule cells in weaver mutant mice. J. Comp. Neurol. 152:133–162.
144. Garthwaite G., Garthwaite J. (1986). Neurotoxicity of excitatory amino acid receptor agonists in rat cerebellar slice: Dependence in calcium concentration. Neurosci. Lett. 66:193–198.
145. Morgan J.J., Curran T. (1991). Stimulus-transmission coupling in the nervous system: Involvement of inducible proto-oncogenes fos and jun. Ann. Rev. Neurosci. 14:421–452.
146. Székely A.M., Barbaccia M.L., Alho H., Costa E. (1989). In primary cultures of cerebellar granule cells the activation of N-methyl-D-aspartate-sensitive glutamate receptors induces c-fos mRNA expression. Mol. Pharmacol. 35:401–408.
147. Didier M., Roux P., Piechaczky M., Verrier B., Bockaert J., Pin J.-P. (1989). Cerebellar granule cell survival and maturation induced by K^+ and NMDA correlate with c-fos proto-oncogene expression. Neurosci. Lett. 107:55–62.
148. Campistron G., Buijs R.M., Geffard M. (1986). Specific antibodies against aspartate and their immunocytochemical applications in the rat brain. Brain Res. 365:179–184.
149. Madl J.E., Beitz A.J., Johnson R.L., Larson A.A. (1987). Monoclonal antibodies specific for fixative-modified aspartate: Immunocytochemical localization in the rat CNS. J. Neurosci. 7:2639–2650.
150. Aoki E., Semba R., Kato K., Kashiwamata S. (1987). Purification of specific antibodies against aspartate and immunocytochemical localization of aspartergic neurones in the rat brain. J. Neurochem. 21:755–765.
151. Rea M.A., McBride W.J., Rohde B.H. (1980). Regional and synaptosomal levels of amino acid neurotransmitters in the 3-acetylpyridine deafferented rat cerebellum. J. Neurochem. 34:1106–1108.
152. Perry T.L., MacLean J., Perry T.L., Jr., Hansen S. (1976). Effects of 3-acetylpyridine on putative neurotransmitter amino acids in rat cerebellum. Brain Res. 109:632–635.
153. Valcana T., Hudson D., Timiras P.S. (1972). Effects of X-irradiation on the content of amino acids in the developing rat cerebellum. J. NeuroChem.. 19:2229–2232.
154. Patel A.J., Balázs R. (1975). Effect of X-irradiation on the biochemical maturation of rat cerebellum: Metabolism of [^{14}C]glucose and [^{14}C]acetate. Radiat. Res. 62:456–469.
155. Young A.O., Oster-Granit M.L., Herndon R.M., Snyder S.H. (1974). Glutamic acid: Selective depletion by viral induced granule cell loss in hamster cerebellum. Brain Res. 73:1–13.
156. Slevin J.T., Johnston M.V., Biziere K., Coyle J.T. (1982). Methylazoxymethanol acetate ablation of mouse cerebellar granule cells: Effects on synaptic neurochemistry. Dev. Neurosci. 5:3–12.
157. McBride W.J., Aprison M.H., Kusano K. (1976). Contents of several amino acids in the cerebellum, brain stem and cereberum of the "staggerer", "weaver" and "nervous" neurologically mutant mice. J. Neurochem. 26:867–870.
158. Rohde B.H., Rea M.A., Simm J.R., McBride W.J. (1979). Effects of X-irradiation induced loss of cerebellar granule cells on the synaptsomal levels and the high affinity uptake of amino acids. J. Neurochem. 32:1431–1435.
159. Tran V.T., Snyder S.H. (1979). Amino acid neurotransmitter candidates in rat cerebellum: Selective effects of kainic acid lesions. Brain Res. 167:345–353.

5. GLUTAMATE RECEPTORS IN MAMMALIAN CEREBELLUM: ALTERATIONS IN HUMAN ATAXIC DISORDERS AND CEREBELLAR MUTANT MICE

E.D. KOUVELAS, A. MITSACOS, F. ANGELATOU, A. HATZIEFTHIMIOU, P. TSIOTOS, AND G. VOUKELATOU

1. INTRODUCTION

The concept that excitatory amino acids (EAA) function as neurotransmitters mediating neuronal excitation is now well established. Their receptors are among the most abundant in the mammalian central nervous system, and at least five types of EAA receptors exist with significantly distinct functions. For a recent review, see Monaghan et al. [1]. Excitatory amino acid receptors appear not only to mediate normal synaptic transmission along excitatory pathways, but also to participate in mechanisms underlying the phenomena of development and plasticity of the nervous system [1]. Furthermore, overactivation of selected excitatory amino acid receptors can also mediate neuronal degeneration. Evidence is accumulating that brain damage associated with anoxia, stroke, hypoglycemia, epilepsy, and several degenerative diseases (Huntington's chorea, olivopontocerebellar atrophy, OPCA amyotrophic lateral sclerosis) may be at least partially produced by excessive activation of excitatory amino acid receptors [2–4].

Cerebellar tissue is an ideal model for the study of excitatory amino acid receptors because (1) this tissue is very well characterized developmentally and anatomically, (2) almost all the neuronal cells of cerebellar cortex receive EAA input and bear EAA receptors, and (3) the cellular localization of the different types of EAA receptors can be determined by

A. Plaitakis (ed.), CEREBELLAR DEGENERATIONS: CLINICAL NEUROBIOLOGY. Copyright © 1992.
Kluwer Academic Publishers, Boston. All rights reserved.

Table 5-1. Kinetic parameters of L-[^{3}H] glutamate binding to human cerebellar membranes

Age	Tris-HCl			Tris-Acetate		
	B_{max} ($\frac{pmoles}{mg.prot.}$)	K_d (nM)	Hill Co.	B_{max} ($\frac{pmoles}{mg.prot.}$)	K_d (nM)	Hill Co.
1 hr	13.52 ± 0.29	127.05 ± 12.13	0.99	3.93 ± 0.57	167.13 ± 22.34	0.99
1 yr	17.61 ± 3.10	138.54 ± 13.33	0.98	4.53 ± 1.01	127.61 ± 5.67	0.99
7 yrs	11.01 ± 1.81	125.02 ± 14.70	0.99	—	—	—
23 yrs	9.31 ± 2.10	140.6 ± 16.31	0.98	2.08 ± 0.31	131.18 ± 9.86	0.98
45–46 yrs	8.46 ± 1.64	110.66 ± 21.22	0.99	1.94 ± 0.09	115.06 ± 10.74	0.97
75–80 yrs	9.35 ± 0.25	130.5 ± 3.51	0.99	2.14 ± 0.09	123.17 ± 7.57	0.98

The cerebella used were one cerebellum of each of the ages of 1 hr, 1 year, 7 years, and 23 years; two cerebella of the age 45–46 years; and four cerebella of the age 75–80 years. For each cerebellum three independent triplicate determinations were performed. Values shown are the averages ± SEM of these determinations. Data in part from [7].

Table 5-2. Inhibition of L-[^{3}H] glutamate-specific binding to human cerebellar membranes by different amino acid analogues

	IC$_{50}$ (μM)	
Amino acid	Tris-HCl	Tris-acetate
L-glutamate	0.345 ± 0.015	0.52 ± 0.025
L-aspartate	0.35 ± 0.048	0.78 ± 0.038
Ibotenate	5.50 ± 0.25	7.0 ± 0.36
Quisqualate	6.3 ± 0.28 nM	15.85 ± 2.09
	15.85 ± 1.32	
L-homocysteic acid	18.1 ± 3.1	26.55 ± 1.45
D-glutamate	28.1 ± 3.0	31.8 ± 4.3
D-aspartate	20.0 ± 2.1	14.5 ± 2.6
% Inhibition at 100 μM		
NMDA	41.6 ± 7.14	44.58 ± 2.34
DL-APB	21.8 ± 5.3	12.48 ± 4.7
Kainic acid	28.0 ± 3.7	29.2 ± 6.4

Values shown are the averages ± SEM of five triplicate determinations performed in brains of the age of 45–80 years. For each inhibition curve 9–11 concentrations, ranging from 1 nM to 1 mM of each drug, were used. Data in part from [7].

using cerebellar tissues that lack selected types of cells (human OPCA and mouse neurological mutants).

This article is a review of our recent studies of the different types of EAA binding sites in normal and atrophic human and mouse cerebellum [5–9].

2. STUDIES IN HUMAN CEREBELLUM

2.1. Binding, developmental, and pharmacological properties of L-glutamate binding sites in normal human cerebellum

Kinetic studies for specific binding to membranes in human cerebellum at 1 hr after birth, 1 year, 7 years, 23 years, 45–46 years, and 75–80 years of age, and for concentrations of L-[^{3}H] glutamate up to 700 nM, were performed in Tris-HCl and Tris-acetate buffer. Scatchard-type plots reveal what appears to be one type of binding sites for both Tris-HCl and Tris-acetate buffer. From the Scatchard-type plots, K_d and B_{max} values were calculated and the results are shown in Table 5-1. As seen in this table, high levels of specific binding are already found at birth, and even greater levels are found at 1 year. Specific glutamate binding was lower at age 7 and declined even further at ages 23, 46, and 80. Specific binding in Tris-HCl buffer was about 3.5–4.5 times higher than that measured in Tris-acetate buffer. No significant differences were observed in the K_d values measured in Tris-acetate and Tris-HCl buffer, and no significant changes of these values were observed in cerebellar tissues from subjects of different ages. Hill plots of these data were linear (r = 0.93–0.99), with slopes ranging from 0.97 to 0.99, indicating the absence of cooperative interactions in L-[^{3}H] glutamate binding.

The specificity and pharmacological properties of L-glutamate binding sites were investigated in both Tris-HCl and Tris-acetate buffer. The results of these experiments are summarized in Table 5-2.

2.1.2. L-glutamate binding sites in atrophic human cerebellum

Kinetic studies for L-[^{3}H] glutamate-specific binding in the membranes of cerebellar cortical tissues from four OPCA patients were performed in Tris-HCl and Tris-acetate buffer.

Two of these patients (cases 1 and 2) were members of the Schut-Haymaker OPCA kindred [10], which is of Dutch extraction and in which early-onset and autosomal dominant transmission with complete penetrance of the mutant gene has been well documented [11]. A linkage of the gene locus to the HLA complex of the sixth chromosome has been established [11]. The other two patients (cases 3 and 4) were from unrelated families and were affected by forms of OPCA that are associated with distinct GDH abnormalities. One of these patients (case 3) was a 15-year-old boy afflicted by cerebellar ataxia, blindness, opthamoplegia, and dementia since 7 years of age. His 42-year-old father and a 14-year-old brother were similarly affected since 27 and 6 years of age, respectively. In this family no linkage between the gene defect and the HLA locus has been found. The disorder in this family is probably inherited in an autosomal dominant manner [12]. The last patient (case 4) was a 71-year-old man suffering from a late-onset sporadic OPCA associated with deficiency of the membrane-bound GDH in neural and extraneural tissues [13,14].

Scatchard type plots of the results from these studies again reveal one type of binding sites. As seen in Tables 5-3 and 5-4, specific L-[^{3}H] glutamate binding in OPCA cerebella was significantly lower compared to the specific binding observed in the cerebella of normal individuals. As in normal individuals, binding in the cerebella of OPCA patients in Tris-HCl buffer was significantly higher than that measured in the same patients under Cl$^-$-free

Table 5-3. Kinetic parameters of L-[^{3}H] glutamate binding to human OPCA cerebellar membranes

	Tris-HCl			Tris-Acetate		
	B_{max} ($\frac{pmoles}{mg.prot.}$)	K_d (nM)	Hill Co.	B_{max} ($\frac{pmoles}{mg.prot.}$)	K_d (nM)	Hill Co.
OPCA-1	2.09 ± 0.313	131.54 ± 6.75	0.99	0.51 ± 0.0443	148.04 ± 5.61	0.99
OPCA-2	1.60 ± 0.029	147.31 ± 11.93	0.99	0.17 ± 0.122	75.10 ± 0.480	0.99
OPCA-3	2.99 ± 0.344	183.15 ± 15.45	0.99	0.86 ± 0.431	131.44 ± 11.58	0.99
OPCA-4	5.03 ± 0.049	139.93 ± 4.11	0.99	1.08 ± 0.0313	108.49 ± 12.15	0.99

Values shown are the averages ± SEM of four triplicate determinations. Data in part from [7].

Table 5-4. L-[^{3}H] glutamate and [^{3}H] QNB-specific binding in OPCA and control cerebella

Ligand	Control	OPCA	% change	p value
L-[^{3}H] glut. B_{max} (pmoles/mg prot.) Tris-HCl	9.04 ± 0.29	2.77 ± 0.85	−70	<0.001
L-[^{3}H] glut. B_{max} (pmoles/mg prot.) Tris-acetate	2.05 ± 0.05	0.65 ± 0.19	−69	<0.001
[^{3}H] QNB Sp. binding at 1 nM (fmoles/mg prot.)	19.5 ± 0.05	20.3 ± 0.10	+4	<0.5

Values for L-[^{3}H] glutamate binding are averages ± SEM from seven controls (age 23–80 years) and four OPCA brains. Values for [^{3}H] QNB are means ± SEM from four controls (age 75–80) and four OPCA brains. Data in part from [7].

conditions. All K_d values in both Tris-HCl and Tris-acetate buffer were within the range of values observed in normal adult individuals. Hill plots of these data were also linear (r = 0.93–0.98), with the slopes equal to 0.99 (Table 5-3).

QNB binding in OPCA cerebella did not show significant differences when compared to the binding of normal individuals (Table 5-4).

2.1.3. Autoradiographic study of L-glutamate binding sites in normal human cerebellum

For the quantitative autoradiographic study of the distribution of L-glutamate binding sites in normal human cerebellum, frozen coronal sections (15 µm) were incubated with 200 mM L-[^{3}H] glutamate in Tris-HCl or Tris-acetate buffer (50 mM, pH 7.1). As seen in Table 5-5, in the absence of Cl$^-$ ions, L-glutamate-specific binding was more pronounced in the granule cell layer. In the presence of Cl$^-$ ions, an increase of about 2.5 times was observed in the molecular layer, and therefore under this condition specific L-glutamate binding was more pronounced in the molecular layer.

The inhibition of L-[^{3}H] glutamate binding by different excitatory amino

Table 5-5. Specific L-[^{3}H] glutamate binding in the layers of control and OPCA human cerebellum

Tissue	L-[^{3}H] glutamate bound (pmol/mg of protein)	
	Molecular layer	Granule cell layer
Control (Tris-HCl)	0.589 ± 0.041 (n = 4)	0.318 ± 0.035 (n = 4)
Control (Tris-acetate)	0.195 ± 0.028	0.325 ± 0.054
OPCA (Tris-HCl)	0.273 ± 0.042 (n = 7)	0.265 ± 0.047 (n = 7)

Values are expressed as means ± SEM of values from four or seven brains. The ligand concentration was 200 nM. OPCA = olivoponto cerebellar atrophy. Data in part from [8].

Table 5-6. Inhibition of specific L-[^{3}H] glutamate binding in the layers of control and OPCA human cerebellum by various excitatory amino acid agonists and antagonists

	% Inhibition of specific L-[^{3}H] glutamate binding			
	Molecular layer		Granule cell layer	
Compound	Control	OPCA	Control	OPCA
Glutamate	98.34 ± 0.84	95.94 ± 1.57	97.13 ± 1.46	95.00 ± 2.20
NMDA	22.02 ± 2.57	40.96 ± 5.90	51.11 ± 3.89	50.26 ± 3.30
APV	21.91 ± 2.82	40.34 ± 3.88	46.79 ± 3.26	53.83 ± 3.38
CPP	19.79 ± 2.03	34.94 ± 2.82	52.60 ± 5.43	60.56 ± 2.55
Quisqualate	75.07 ± 5.84	59.24 ± 5.17	80.11 ± 6.21	63.54 ± 3.47
Quisqualate*	56.81 ± 3.86	—	32.63 ± 6.62	—
AMPA	39.03 ± 3.92	18.09 ± 2.35	22.15 ± 5.83	14.48 ± 2.94
CNQX	42.48 ± 3.58	17.07 ± 4.00	38.15 ± 6.75	23.49 ± 3.57
Kainate	31.93 ± 2.71	21.77 ± 2.99	25.98 ± 4.41	22.73 ± 3.06

Values are expressed as means ± SEM of values from four to seven brains. The assay was done in Tris-HCl buffer using a ligand concentration of 200 nM. Each compound was at a concentration of 100 µM. OPCA = olivopontocerebellar atrophy. This table from [8] with permission.
*2.5 µM concentration.

acid agonists and antagonists was measured in Tris-HCl buffer and at the concentrations of 100 µM (concentration of maximal inhibition). Quisqualate was used at concentrations of 100 µM and 2.5 µM. The results of these experiments are shown in Table 5-6 and Figure 5-1. As was expected, L-[^{3}H] glutamate binding was effectively inhibited by unlabelled L-glutamate. The NMDA and NMDA antagonists APV and CPP showed significantly higher inhibition in the granule cell than in the molecular layer. Quisqualate at a concentration of 100 µM was a very potent inhibitor of L-[^{3}H] glutamate binding, with almost equal potency in both cerebellar layers. Quisqualate at a concentration of 2.5 µM produced a higher inhibition in the molecular than in the granule cell layer. AMPA, which is a quisqualate receptor agonist; CNQX, which is a quisqualate receptor antagonist; and kainate produced a slightly higher inhibition of L-[^{3}H] glutamate-specific binding in the molecular than in the granule cell layer.

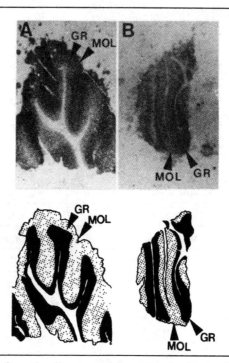

Figure 5-1. Autoradiographs of human cerebellar sections incubated with 200 nM L-[^{3}H]glutamate in 50 mM Tris-HCl buffer (top) and drawings of the corresponding Nissl-stained sections (bottom). Total L-[^{3}H]glutamate binding in (**A**) control human cerebellum; (**B**) human cerebellum with olivopontocerebellar atrophy (case 2 of Table 5-7). MOL = molecular layer; GR = granule cell layer. Magnification ×6.3. This figure from [8] with permission.

2.1.4. Autoradiographic study of L-glutamate binding sites in OPCA human cerebellum

Quantitative autoradiographic study of L-glutamate binding was performed in seven OPCA cerebellar tissues in Tris-HCl (50 mM, pH 7.1). A significant reduction of L-[^{3}H] glutamate binding was observed in all OPCA cerebellar tissues studied (Table 5-5, Figure 5-2). However, as shown in Table 5-7, several differential characteristics in each group of patients were detected. Patients who were members of the Schut-Haymaker kindred (cases 1–4) showed a significant decrease of L-[^{3}H] glutamate-specific binding in only the molecular layer. Patients who suffered from a late-onset sporadic OPCA (cases 5 and 6) showed a significant decrease in both the molecular and granule cell layer. Finally, the patient who suffered from a form of OPCA inherited in a dominant manner and who died at the age of 15 (case 7) also showed a significant decrease in both cerebellar layers.

In the molecular layer of OPCA cerebellar tissues, inhibition of L-[^{3}H]

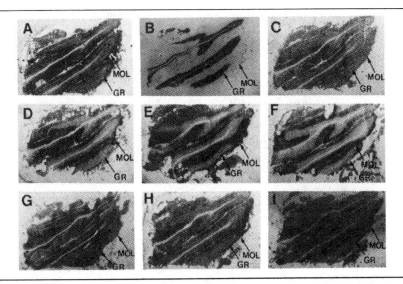

Figure 5-2. Autoradiographs of control human cerebellar sections incubated with 200 nM L-[³H]glutamate in 50 mM Tris-HCl buffer in the presence of various excitatory amino acid agonists and antagonists. **A:** Total L-[³H]glutamate binding. **B:** Corresponding Nissl-stained section of total binding. **C:** 100 μM kainate. **D:** 100 μM NMDA. **E:** 100 μM APV. **F:** 100 μM CPP. **G:** 2.5 μM quisqualate. **H:** 100 μM AMPA. **I:** 100 μM CNQX. MOL = molecular layer; GR = granule cell layer. This figure from [8] with permission.

Table 5-7. Specific L-[³H] glutamate binding in the cerebellar layers of seven patients who died with OPCA

Tissue	L-[³H] glutamate bound (pmol/mg of prot.)		% Decrease compared to control	
	Molecular layer	Granule cell layer	Molecular layer	Granule cell layer
OPCA-1	0.389 ± 0.038	0.439 ± 0.013	33.95[1]	—
OPCA-2	0.290 ± 0.014	0.360 ± 0.060	50.76[1]	—
OPCA-3	0.226 ± 0.011	0.297 ± 0.035	61.63[1]	—
OPCA-4	0.386 ± 0.042	0.334 ± 0.085	34.46[1]	—
OPCA-5	0.072 ± 0.008	0.109 ± 0.011	87.77[1]	65.72[2]
OPCA-6	0.239 ± 0.074	0.165 ± 0.023	59.42[1]	48.11[1]
OPCA-7	0.330 ± 0.065	0.153 ± 0.052	43.97[2]	51.88[2]

Values are expressed as means ± SEM of three separate experiments. The assay was done in Tris-HCl buffer using a ligand concentration of 200 nM. OPCA = olivopontocerebellar atrophy. Comparisons were made with Student's t test. [1]$p < 0.01$; [2]$p < 0.05$. This table from [8] with permission.

glutamate binding by NMDA, APV, and CPP was doubled in comparison to the control tissues, whereas inhibition of binding by AMPA and CNQX was reduced by about 50%. In this layer inhibition of binding by 100 μM quisqualate was also reduced. A slight decrease of inhibition of L-[³H]

Table 5-8. L-[^{3}H] glutamate-specific binding in normal "nervous" and "staggerer" mouse cerebellum

Layer of cerebellum	Normal	"Nervous"	"Staggerer"
K_d (nM)			
Molecular	482.40 ± 1.77	418.74 ± 6.65	414.54 ± 9.44
Granular	444.30 ± 31.51	376.18 ± 3.83	309.96 ± 10.43
B_{max} (pmoles/mg protein)			
Molecular	2.993 ± 0.334	2.690 ± 0.448	1.203 ± 0.295
Granular	1.137 ± 0.225	0.620 ± 0.047	0.538 ± 0.255

Values are means ± SEM from six normal, three "nervous", and three "staggerer" cerebellar tissues.

glutamate-specific binding by kainate was observed. In the granule cell layer of all OPCA cerebellar tissues, inhibition of L-[^{3}H] glutamate-specific binding by the different excitatory amino acid agonists and antagonists was similar to that observed in control tissues (Table 5-6).

3. STUDIES IN MOUSE CEREBELLUM

Frozen cerebellar sections (15 µM) from "nervous" and "staggerer" mice (purchased from Jackson Laboratories) and their normal littermates were incubated with varying concentrations of L-[^{3}H] glutamate (50–800 nM), in 50 mM Tris-citrate buffer, pH 7.1. "Nervous" mutant is an autosomal recessive mutation that results in a selective degeneration of cerebellar Purkinje cells [15]. "Staggerer" mutant mouse is also a recessive mutation that results in the progressive loss of mainly granule cells. Autoradiograms were generated and a densitometric analysis of each cerebellar layer was performed for Scatchard analysis, taking into account the quenching coefficient of the cerebellar grey matter according to Geary and Wooten [16]. The results of these experiments are shown in Table 5-8 and Figure 5-3. As seen in this table, the dissociation constants (K_d) of the molecular and granule cell layer were very similar in all kinds of cerebellar tissues studied. However, significant differences were observed in the maximum number of binding sites (B_{max}). In normal cerebellum, the B_{max} value of the granule cell layer is more than double the B_{max} value of the molecular layer. In "nervous" cerebellum, the B_{max} of the molecular layer was reduced by almost 50% in comparison to controls, whereas in the granule cell layer the B_{max} value was almost unchanged. In "staggerer" cerebellum, both layers showed a 60% reduction of the B_{max} value in comparison to controls.

4. DISCUSSION

This study investigated:
1. The binding, pharmacological, and developmental properties of L-glutamate binding sites in the cerebellum of neurologically normal human

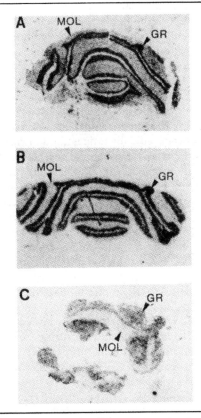

Figure 5-3. Autoradiographs of mouse cerebellar sections incubated with 300 nM L-[^{3}H] glutamate in 50 mM Tris-citrate buffer, pH 7.1. Total L-[^{3}H] glutamate binding in (**A**) normal cerebellum, (**B**) "nervous" cerebellum, and (**C**) "staggerer" cerebellum. The age of "nervous" and "staggerer" mice was 94 and 21 days old, respectively. MOL = molecular layer; GR = granule cell layer.

individuals and in atrophic cerebella from patients with olivopontocerebellar degeneration.
2. The distribution of the different types of L-glutamate binding sites in the above tissues using quantitative autoradiography.
3. The distribution of L-glutamate binding sites in normal, "nervous," and "staggerer" mouse cerebellum with the technique of quantitative autoradiography.

Our results from membrane homogenates indicated the presence of one class of binding sites in normal human cerebellum, the affinity (K_d value) of which was similar to that previously reported for rat cerebellar tissue [17]. A previous study using freeze-thawed membranes isolated from human cerebellum have also shown one class of L-glutamate binding sites [18]. However, the affinity reported for L-glutamate binding was lower than that found in this study, and this may be related to the freezing and thawing of membranes. The addition of chloride ions produced a marked increase in glutamate-specific binding, which was not associated with a change in the affinity. However, the pharmacological properties were found to be different, particularly with respect to the inhibition curve of quisqualate. Thus, this compound produced a biphasic inhibition curve in the presence of chloride ions and a monophasic curve in the absence of chloride ions. These results are similar to those previously described for rat brain [19–21]. Our autoradiographic studies have shown that L-glutamate binding was enchanced in the presence of chloride ions in the molecular layer, and this result is similar to that previously described for rat cerebellum [20]. As our autoradiographic study has shown, this layer of human cerebellum is rich in quisqualate-sensitive binding sites, and the increased L-glutamate binding in the presence of chloride ions in this layer is probably associated with the appearance of a binding site subtype with very high affinity to quisqualate. The physiological function of these effects is not clear, and further investigation is needed in this direction.

Developmental studies in human brain are difficult to perform because suitable human material is not readily available. Although our data are based on a small number of individuals, they clearly suggest that significant developmental changes may occur in glutamate binding sites in the human cerebellum. Taking into consideration the results of previous developmental studies in human cerebellum [22,23], one may hypothesize that the development of L-[^{3}H] glutamate binding sites follows the patterns of cytoarchitectonic differentiation of the neuronal elements of the molecular and internal granular layer of the cerebellum.

The autoradiographic studies of human and mouse cerebellum presented in this article indicate the presence of L-glutamate binding sites in both the molecular and granule cell layers. These results are in agreement with those obtained in autoradiographic studies using rat, mouse, and chick cerebellar tissues [20,24–27]. Considerable evidence supports the concept that L-glutamate serves as the excitatory neurotransmitter of parallel fibers of granule cells in the molecular layer of cerebellum [28]. Recent evidence suggests that efferents to the cerebellar cortex (climbing and mossy fibers) may also use excitatory amino acid neurotransmitters [29–33]. Therefore, one would expect that all cerebellar cortical neurons would have binding sites for excitatory amino acids.

The cellular localization of the different types of excitatory amino acid receptors can be determined by using tissues that lack selected types of cells. Cerebellar tissues from OPCA patients are of this kind, because earlier anatomical studies have established that the degeneration of Purkinje cells is a consistent finding in OPCA. Degeneration of granule cells was also observed in certain forms of OPCA [34,35]. Similarly the "nervous" mutant mouse is an autosomal, recessive mutation that results in selective degeneration of cerebellar Purkinje cells [15].

The "staggerer" mouse is also a single-gene recessive mutation, characterized by a grossly underdeveloped cerebellar cortex [36]. The molecular layer is thin and the granule cell layer has few cells. Purkinje cells are scattered in the granule cell layer. In the inner granular layer, granule cells undergo a degeneration process, and by 28 days virtually all "staggerer" granule cells have degenerated [36–39].

Our autoradiographic studies showed a 55% decrease of L-glutamate-specific binding in the molecular layer of OPCA patients. A decrease of about the same order was also observed in the molecular layer of "nervous" mouse cerebellum. These results probably reflect the degeneration of Purkinje cells and suggest that more than 50% of L-glutamate binding sites in the molecular layer of both human and mouse cerebellum are localized on the dendrites of Purkinje cells. In three OPCA patients (cases 5–7 of Table 5-7), a marked decrease of L-glutamate binding was observed in both the molecular and granule cell layers. This decrease probably reflects a degeneration of both Purkinje and granule cells. Furthermore, these results suggest that different mechanisms, which remain to be elucidated, may be responsible for the changes observed in the distinct forms of OPCA studied.

A 50% decrease of L-[^{3}H] glutamate-specific binding was also observed in the molecular layer of "staggerer" cerebellum. Three factors at least may contribute for this decrease: (1) the alterations of the tertiary spines of the dendrites of Purkinje cells, (2) the decrease in size of the dendritic tree of these cells, and (3) the decreased number of Purkinje cells. A more pronounced decrease of L-[^{3}H] glutamate binding was observed in the granule cell layer of "staggerer" mouse cerebellum. This decrease is probably due to the degeneration of granule cells. Indeed in this layer at the age of 21 days (which was the age of the mice that we used), the number of intact granule cells is very limited. However, the remaining specific binding was rather high (1.2 pmole/mg protein). This result is not surprising, because this layer of "staggerer" mouse contains, in addition to the remaining granule cells, several medium- or large-size neuronal cells, which are probably Purkinje and Golgi cells scattered in this layer [39]. Furthermore, recent studies support the concept that astrocytes also contain excitatory amino acid receptors [40,41]. A study of the pharmacological properties of the L-glutamate binding sites remaining in "staggerer" mouse cerebellum is in progress in our laboratory.

In the present study, the cellular localization of NMDA, -kainate and quisqualate-sensitive, binding sites in human cerebellum was also studied. Because of the small amount of tissue that was available, especially from

OPCA patients, the inhibition of binding of L-[^{3}H] glutamate by the different agonists and antagonists was measured only at a concentration of 100 µM (the concentration of maximal inhibition). Because of reports showing that quisqualate inhibits L-[^{3}H] glutamate binding in a biphasic manner in both human and rat brain [7,19,20], this drug was used in the control tissues in two concentrations (2.5 µM and 100 µM).

In the molecular layer of normal human cerebellum, quisqualate at the concentration of 100 µM inhibited approximately 75% of L-[^{3}H] glutamate specific binding. However, this inhibition is most likely associated with binding to both quisqualate- and kainate-sensitive binding sites. Taking as a criterion for the measurement of quisqualate-sensitive binding sites the inhibition of L-[^{3}H] glutamate specific binding by 100 µM of AMPA, or CNQX, or 2.5 µM quisqualate (39–56%), we can conclude, from our results, that quisqualate-sensitive binding sites are the most abundant type of binding sites in the molecular layer of human cerebellum. Parallel fiber synapses on Purkinje cell dendrites are the most numerous synapses in the molecular layer. It is, therefore, possible that the quisqualate-sensitive binding sites in the molecular layer is associated with this type of synapse. In support of this hypothesis are our results and those from other laboratories [42], which indicate a decrease of quisqualate-sensitive binding sites in the absence of Purkinje cells in the OPCA cerebellar tissue. Thus, quisqualate-sensitive binding sites of both ionotropic and metabotropic type appear to be located on Purkinje cells [43]. The remaining quisqualate-sensitive binding sites in the OPCA cerebellar tissue are probably located in the remaining Purkinje cells and on basket, stellate, and Golgi cells, with which parallel fibers also synapse. The same conclusions were also recently deduced from a similar study on normal and "nervous" mouse [27].

In OPCA cerebellar tissue, an increase in NMDA-sensitive sites was observed in the molecular layer. This result provides strong evidence that few, if any, NMDA binding sites are located on the dendrites of Purkinje cells. This conclusion also agrees with the results of the above-mentioned study in mouse cerebellum and with the results of electrophysiological studies in rat cerebellar Purkinje cells [27,44–46]. Furthermore, the only cells in rat cerebellar cortex that are vulnerable to NMDA toxicity are stellate, Golgi, and basket cells [47]. Thus, in our opinion, the increase of NMDA-sensitive binding sites could be attributed to the decrease of cellular structures that do not bear this kind of receptors (dendrites of Purkinje cells).

In normal human cerebellar cortex, NMDA-sensitive binding sites were more abundant in the granule cell layer than in the molecular layer. The density of these binding sites in the granule cell layer was higher than that of the quisqualate-sensitive binding sites. Similar results have been previously reported for rat and mouse cerebellum [20,24,27]. No significant change of the NMDA- and quisqualate-sensitive binding was observed in the granule cell layer of the OPCA tissues in comparison to the controls. Such a result suggests that both binding sites are located on the same cellular elements, and thus, a degeneration of these elements does not change the percent contribution of NMDA- and quisqualate-sensitive binding.

The kainate-sensitive L-glutamate binding sites were detected in both the molecular and granule cell layer of human cerebellum, with slightly higher levels in the molecular layer.

In the present work using displacement studies, we investigated the localization of the different types of L-glutamate binding sites in human cerebellum. However, in view of recent reports suggesting the existence of multiple NMDA and quisqualate receptor sites and/or states [48–51], further experiments are needed to characterize and locate each of these subtypes in human brain.

REFERENCES

1. Monaghan D.T., Bridges R.J., Cotman C.W. (1989). The excitatory amino acid receptors: Their classes, pharmacology, and distinct properties in the function of the central nervous system. Annu. Rev. Pharmacol. Toxicol. 29:365–402.
2. Rotman S.M., Olney J.W. (1987). Excitotoxicity and the NMDA receptor. Topics Neurol. Sci. 10:299–302.
3. Plaitakis A., Berl S., Yahr M.D. (1982). Abnormal glutamate metabolism in an adult onset degeneration neurological disorder. Science 216:193–196.
4. Plaitakis A., Constantakakis E., Smith J. (1988). The neuroexcitotoxic amino acids glutamate and aspartate are altered in the spinal cord and brain in amyotrophic lateral sclerosis. Ann. Neurol. 24:446–449.
5. Angelatou F., Mitsacos A., Goulas V., Kouvelas E.D. (1987). L-Aspartate and L-glutamate binding sites in developing normal and "nervous" mutant mouse cerebellum. Int. J. Devel. Neurosci. 5:373–381.
6. Mitsacos A., Angelatou F., Kouvelas E.D. (1986). Quantitative autoradiographic characterization of Cl^--independent glutamate binding sites in mouse brain. In: Abstracts of Sixth ESN General Meeting, Prague, p. 198.
7. Tsiotos P., Plaitakis A., Mitsacos A., Voukelatou G., Michalodimitrakis M., Kouvelas E.D. (1989). L-glutamate binding sites of normal and atrophic human cerebellum. Brain Res. 481:87–96.
8. Hatziefthimiou A., Mitsacos A., Mitsaki E., Plaitakis A, Kouvelas E. (1990). Quantitative autoradiographic study of L-glutamate binding sites in normal and atrophic human cerbellum. Neurochem. Intern. 16S1:42.
9. Hatziefthimiou A., Mitsacos A., Mitsaki E., Plaitakis A., Kouvelas E.D. (1990). Quantitative autoradiographic study of L-glutamate binding sites in normal and atrophic human cerebellum. J. Neurosci. Res. 28:361–375.
10. Schut, J.W., Haymaker W. (1951). Hereditary ataxia: Pathological study of 5 cases of common ancestry. J. Neuropathol. Clin. Neurol. 1:183–213.
11. Haines J.L., Schut L.J., Weitkamp L.R., Thayer M., Anderson V.E. (1984). Spinocerebellar ataxia in a large kindred. Age at onset, reproduction and genetic linkage studies. Neurology 34:1542–1548.
12. Hussain M.M., Zannis V.I., Plaitakis A. (1989). Characterization of glutamate dehydrogenase isoproteins purified from the cerebellum of normal subjects and patients with degenerative neruological disorders, and human neoplastic cell lines. J. Biol. Chem. 254:20730–20735.
13. Plaitakis A., Berl S., Yahr M.D. (1984). Neurological disorders associated with deficiency of glutamate dehydrogenase. Ann. Neurol. 15:144–153.
14. Plaitakis A., Smith J. (1986). Biochemical and morphological changes of brain in a patient dying of GDH deficient olivoponto-cerebellar atrophy. Ann. Neurol. 20:182–183.
15. Landis C.S. (1973). Ultrastructural changes in the mitochondria of cerebellar Purkinje cells of nervous mutant mice. J. Cell Biol. 57:782–797.

16. Geary W.A., Wooten G.F. (1985) Regional tritium quenching in quantitative autoradiography of the central nervous system. Brain Res. 336:334–336.
17. Sharif N.A., Roberts P.J. (1981). Regulation of cerebellar L-[^{3}H] glutamate binding: Influence of guanine nucleotides and Na$^+$ ions. Biochem. Pharmacol. 30:3019–3022.
18. Cross A., Skan N., Slater P. (1986). Binding sites for [^{3}H] glutamate and [^{3}H] aspartate in human cerebellum. J. Neurochem. 47:1463–1468.
19. Fagg G.E., Mena E.E., Cotman C.W. (1983). L-glutamate receptor populations in synaptic membranes. Effects of ions and pharmacological characteristics. In: P. Mandel, F.V. De Feudis (eds): CNS Receptors-From Molecular Pharmacology to Behavior. New York: Raven Press, pp. 199–209.
20. Greenamayre J.T., Olson J.M.M., Penney J.B., Young A.B. (1985). Autoradiographic characterization of N-methyl-D-aspartate-, quisqualate- and kainate-sensitive glutamate binding sites. J. Pharmac. Exp. Ther. 233:254–263.
21. Werling L.L., Nadler J.V. (1984). Couplex binding of L-[^{3}H] glutamate to hippocampal synaptic membranes in the absence of sodium. J. Neurochem. 38:1050–1062.
22. Raaf J., Kernohen J.W. (1944). A study of the external granular layer in the cerebellum. Am. J. Anat. 75:151–172.
23. Zecevic N., Rakic P. (1976). Differentiation of Purkinje cells and their relationship to other components of developing ceebellar cortex in man. J. Comp. Neurol. 167:27–48.
24. Halpain S., Wieczorek C.M., Rainbow T.C. (1984). Localization of L-glutamate receptors in rat brain by quantitative autoradiography. J. Neurosci. 4:2247–2258.
25. Mitsacos A., Dermon C.R., Stassi K., Kouvelas E.D. (1990). Localization of L-glutamate binding sites in chick brain by quantitative autoradography. Brain Res. 513:348–352.
26. Monaghan D.T., Yao D., Cotman C.W. (1985). L-[^{3}H] glutamate binds to kainate- NMDA- and AMPA-sensitive binding sites: An autoradiographic analysis. Brain Res. 340:378–383.
27. Olson J.M.M., Greenamyre J.T., Penney J.B., Young A.B. (1987). Autoradiographic localization of cerebellar excitatory amino acid binding sites in the mouse. Neuroscience 22:913–923.
28. Fagg G.E., Foster A.C. (1983). Amino acid neurotransmitters and their pathways in the mammalian central nervous system. Neuroscience 9:701–719.
29. Beitz A.J., Larson A.A., Monaghan P., Alshuler R.A., Mullet M.A., Madl J.E. (1986). Immunohistochemical localization of glutamate, glutaminase and aspartate amino-transferase in neurons of the pontine nuclei of the rat. Neuroscience 17:741–753.
30. Foster G.A., Roberts P.J. (1983). Neurochemical and pharmacological correlates of inferior olive destruction in the rat: Attenuation of the events mediated by an endogenous glutamate-like substance. Neuroscience 8:277–284.
31. Freeman M.E., Lane J.D., Smith J.E. (1983). Turnover rates of amino acid neurotransmitters in regions of rat cerebellum. J. Neurochem. 40:1441–1447.
32. Garthwaite J., Brodbelt A.B. (1989). Synaptic activation of N-methyl-D-aspartate and non-N-methyl-D-aspartate receptors in the mossy fibers pathway in adult and immature rat cerebellar slices. Neuroscience 29:401–412.
33. Rea M.A., McBride W.J., Rohde B.H. (1980). Regional and synaptosomal levels of amino acid neurotransmitters in 3-acetylpyridine deafferented rat cerebellum. J. Neurochem. 34:1106–1108.
34. Kish S.J., P.P. Robitaille Y., Currier J., Gilbert J., Schut L., Warsh J.J. (1989). Cerebellar [^{3}H] inositol 1,4,5,triphosphate binding is markedly decreased in human olivopontocerebellar atrophy. Brain Res. 489:373–376.
35. Koeppen S.J., Barron K.D. (1984). The neuropathology of olivopontocerebellar atrophy. Adv. Neurol. 41:225–243.
36. Sidman R.L., Lane P.W., Dickie M.M. (1962). Staggerer a new mutation in the mouse affecting the cerebellum. Science 137:610–612.
37. Sotelo C., Changeux J.-P. (1974). Transynaptic degeneration "en cascade" in the cerebellar cortex of staggerer mutant mice. Brain Res. 67:519–526.
38. Landis D.M.D., Sidman R.L. (1978). Electron microscope analysis of postnatal histogenesis in the cerebellar cortex of staggerer mutant mice. J. Comp. Neurol. 179:831–864.
39. Herrup K., Mullen R.J. (1979). Regional variation and absence of large neurons in the cerebellum of the staggerer mouse. Brain Res. 172:1–12.

40. Backus K.H., Kettenmann H., Schachner M. (1989). Pharmacological characterization of the glutamate receptor in cultured astrocytes. J. Neurosci. Res. 22:274–282.
41. Usowicz M.M., Gallo V., Cull-Candy S.G. (1989). Multiple conductance channels in type-2-cerebellar astrocytes activated by excitatory amino acids. Nature 339:380–383.
42. Makowiec R.L., Albin R.L., Cha J-H.J., Young A.B., Gilman S. (1990). Two types of quisqualate receptors are decreased in human olivopontocerebellar atrophy cerebellar cortex. Brain Res. 523:309–312.
43. Blackstone C.B., Supattapone S., Snyder S.H. (1989). Inositol phospholipid-linked glutamate receptors mediate cerebellar parallel-fiber-Purkinje-cell synaptic transmission. Proc. Natl. Acad. Sci. USA 86:4316–4320.
44. Crepel F., Dupont J.L., Gradette R. (1983). Voltage clamp analysis of the effects of excitatory amino acids and derivatives on Purkinje cell dendrites in rat cerebellar slices maintained in vitro. Brain Res. 279:311–315.
45. Joels M., Yol A.J., Gruol D.L. (1989). Unique properties of non-N-methyl-D-aspartate excitatory responses in cultured Purkinje neurons. Proc. Natl. Acad. Sci. USA 86:3404–3408.
46. Quinlan J.E., Davies J. (1985). Excitatory and inhibitory responses of Purkinje cells, in the rat cerebellum in vivo, induced by excitatory amino acids. Neurosci. Lett. 60:39–46.
47. Garthwaite G., Garthwaite J. (1984). Differential sensitivity of rat cerebellar cells in vitro to the neurotoxic effects of excitatory amino acid analogues. Neurosci. Lett. 48:361–367.
48. Sladeczek F., Recasens M., Bockaert J. (1988). A new mechanism for glutamate receptor action: Phosphoinositide hydrolysis. Topics Neurol. Sci. 11:545–549.
49. Sugiyama H., Ito I., Hirono C. (1987). A new type of glutamate receptor linked to inositol phospholipid metabolism. Nature 325:531–533.
50. Palmer E., Monaghan D.T., Cotman C.W. (1988). Glutamate receptors and phosphoinositide metabolism: Stimulation via quisqualate receptors is inhibited by N-methyl-D-aspartate receptor activation. Mol. Brain Res. 4:161–165.
51. Baudry M., Evans J., Lynch G. (1986). Excitatory amino acids inhibit stimulation of phoshatidylinositol metabolism by aminergic agonist in hippocampus. Nature 319:329–331.

6. REGIONAL AND CELLULAR DISTRIBUTION OF GLUTAMATE DEHYDROGENASE AND PYRUVATE DEHYDROGENASE COMPLEX IN BRAIN: IMPLICATIONS FOR NEURODEGENERATIVE DISORDERS

CHIYE AOKI, TERESA A. MILNER, AND VIRGINIA M. PICKEL

1. INTRODUCTION

Glutamate dehydrogenase (GDH) [L-glutamate:NAD oxidoreductase; E.C. 1.4.1.2] and pyruvate dehydrogenase complex (PDHC) [pyruvate dehydrogenase (PDH), E.C. 1.2.4.1; dihydrolipoyl transacetylase, E.C. 2.3.1.2; dihydrolipoyl dehydrogenase, E.C. 1.6.4.3; PDHa kinase, E.C. 2.7.1.9; and PDHb phosphatase, E.C. 3.1.3.43] are mitochondrial enzymes involved in the formation of cellular energy through the tricarboxylic acid (TCA) cycle [1]. In contrast to what might be expected from the universal requirements for these metabolically vital enzymes, a number of studies recently have shown heterogeneity in the regional distributions of GDH [2-6] and PDHC [7] with respect to each other and compared to cytochrome oxidase (CyO), another mitochondrial enzyme of the TCA cycle [1]. Additionally, GDH and PDHC are markedly different in their cellular distributions, with GDH enriched principally in glial cells [2-6] and PDHC occurring at high levels in neuronal perikarya [7]. Both regional and cellular differences in the distribution of these mitochondrial enzymes are believed partially to reflect the contrasting involvement of GDH and PDHC in the metabolism of transmitters, glutamate, and acetylcholine (ACh), respectively; whereas GDH catalyzes the interconversion of alpha-ketoglutarate and glutamate [8], PDHC is known to be important for the synthesis of ACh within cholinergic terminals [9-12]. The level of CyO, on the other hand, correlates with

A. Plaitakis (ed.), CEREBELLAR DEGENERATIONS: CLINICAL NEUROBIOLOGY. Copyright © 1992.
Kluwer Academic Publishers, Boston. All rights reserved.

neuronal activity levels [13,14] without specific association with any single transmitter [15].

We focus this review on (1) the prerequisite immunolabeling conditions needed for light microscopic detection of GDH and PDHC in brain; (2) the mitochondrial localization of GDH by immunogold electron microscopy; (3) comparisons of the light microscopic distribution in rat brain of GDH, PDHC, and CyO among each other; and (4) with certain identified transmitters. The implications of these findings are discussed as they may pertain to neurodegenerative disorders in humans.

2. LABELING CONDITIONS

Light microscopic detection of immunoreactivity to mitochondrial enzymes is highly dependent on the procedures for tissue preparation [2,3,7]. Using the peroxidase-antiperoxidase method [16], we have found labeling for both GDH and PDHC to be optimal within Vibratome sections from brains that are fixed by rapid perfusion through the ascending aorta with 4% paraformaldehyde (4°C). The labeling generally is enhanced by the use of detergents, such as Triton X-100, to increase penetration of the antisera. Regardless of the concentration of Triton X-100 in incubation buffers, immunolabeling for PDHC within these tissues is detected principally within neuronal perikarya [7]. In contrast, the labeling for GDH appears to be neuronally associated at lower and glially associated at higher concentrations of detergent [2]. Difficulty in detecting neuronal labeling for GDH in the presence of Triton X-100 is probably attributable to masking by the more intense labeling of glia, or alternatively, to the greater solubility of neuronal GDH. Identification of the GDH-labeled processes as glial has been confirmed [2] by combining immunoperoxidase labeling for GDH with immunoautoradiographic localization of an antiserum against glial fibrillary acidic protein (GFAP), a glial marker [17].

Mild tissue fixation and detergents that are required for optimal immunolabeling probably reflect the necessity for antisera against GDH and PDHC to penetrate through inner compartments of mitochondria to contact antigenic sites. The demonstration of glial vs. neuronal labeling for GDH, depending on the presence of Triton [2], may reflect the existence of at least two forms of GDH, differing either in their molecular structure and/or subcellular localization (however, see below, Ultrastructural Localization of GDH). This possibility has previously been suggested based on the detection of GDH activity in different subcellular fractions (nuclear and mitochondrial) [18,19], as well as by differences in the susceptibilities of GDH to detergent extraction [20].

3. ULTRASTRUCTURAL LOCALIZATION OF GDH

The GDH antiserum used in the light-microscopic study by Aoki et al. [2,3] was produced in rabbits and was biochemically characterized for its

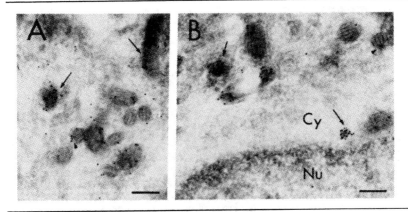

Figure 6-1. Ultrastructural localization of immunogold labeling for GDH within frozen ultrathin sections sampled from the Purkinje cell layer of the cerebellar cortex. Gold particles (15 nm) are associated primarily with mitochondria in the neuropil (A and B) and in the perikaryal cytoplasm (Cy) (lower arrow in B). As shown here, gold particles are rarely seen within the nucleus (Nu). Within mitochondrial profiles sectioned through the middle (in which the mitochondrial cristae are clearly visible), gold particles are found closer to the periphery than directly over the mitoplast (arrowheads in A and B). Larger clusters of gold particles are associated with eccentrically sectioned mitochondria (arrows, in which the cristae are not visible). Bar = 0.5 μm.

specificity by Drs. J.C.K. Lai et al. [19] at the Dementia Research Service, Burke Rehabilitation Center, White Plains, New York. Additional specificity of the antiserum for recognizing mitochondrial sites has been established using Tokuyasu's method of immunogold labeling in ultrathin frozen sections [21]. Within the cerebellar cortex, clusters of immunogold particles occur selectively over subcellular organelles having cristae or other characteristics of mitochondria (Figure 6-1). Heterogeneity in immunolabeling can also be observed. First, not all mitochondria exhibit immunogold labeling. Second, clusters of gold particles over mitochondria vary in number from 3 to over 20. Third, within mitochondrial profiles sectioned through the center, where the cristae (folds of inner mitochondrial membrane) are clearly visible (arrowheads in Figure 6-1), gold particles are found eccentrically, rather than directly over the mitoplast. Larger clusters of gold particles are found over mitochondrial profiles lacking clearly visible cristae (i.e., sectioned off center, close to the outer membrane; arrows in Figure 6-1). The eccentric immunogold labeling of mitochondria is also evident in postembedded tissue [3]. Results using postembedded tissue indicate that mitochondria of both neurons and glia are immunoreactive for GDH [3]. These observations confirm the light-microscopic localization of GDH within neurons and glia. Moreover, they show that GDH in brain, like GDH in other tissues [22], occurs within the mitochondrial matrix, but is prin-

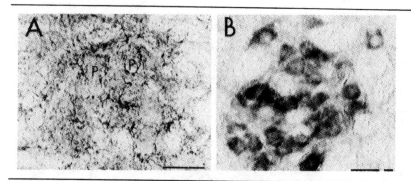

Figure 6-2. Comparative lightmicroscopic localization of GDH and PDHC in the nucleus ambiguus using immunoperoxidase-labeling methods. **A**: Photomicrograph shows processes with GDH immunoreactivity surrounding large unlabeled perikarya (P). **B**: Photomicrograph of a near-adjacent section through the same region illustrated in A shows the localization of PDHC in neuronal perikarya. Differential interference contrast optics. Bar = 40 μm.

cipally seen nearer to the mitochondrial outer membrane. Nuclear GDH [18,19], if present, was not recognized by the antiserum or occurs in levels below detectability.

4. RELATIVE DISTRIBUTIONS OF THREE MITOCHONDRIAL ENZYMES: GDH, PDHC, AND CYTOCHROME OXIDASE

The mitochondrial localization of GDH [3], PDHC [23], and CyO [14] in brain suggests that the distribution of these enzymes may parallel that of mitochondria, and thus, each other. Indeed, in a few select regions, correspondence among the three enzymes is evident. For example, the nucleus ambiguus in the brainstem reticular formation shows discrete immunolabeling for GDH and PDHC (Figure 6-2), as well as high levels of histochemical reactivity for CyO in adjacent sections [3]. Other areas with overlapping distributions of all three enzymes include the caudate-putamen, the entorhinal and primary olfactory cortices, the nuclei basalis of Meynert, the central gray, the lateral reticular and paramedian nuclei, and the dorsal motor nuclei of vagus (Figure 6-3).

In most regions of the brain, however, the topographic distributions of GDH and PDHC immunoreactivity differ significantly from each other (Figure 6-3), as well as from the histochemical reactivity for CyO [3]. The cerebellar cortex is one such region that clearly depicts these differences (Figure 6-4). In this region, intense cytochemical reactivity for all three enzymes is evident within the granular layer (GL in Figure 6-4). However, closer examination of this layer indicates that GDH immunoreactivity and CyO histochemical reactivity are more intense than PDHC immunoreactivity. Additionally, immunoreactivity for GDH is seen within fine ramifying processes, while the labeling for PDHC is more diffuse. In contrast, the histochemical reactivity for CyO appears more cellular. Variations are also seen in the Purkinje cell layer. Intense immunoreactivity for PDHC is detectable within Purkinje cell perikarya (horizontal arrow in Figure 6-4B). This pattern sharply contrasts with the absence of histochemical reactivity for CyO within this layer (asterisks over the halo of perikarya in Figure 6-4D). The distribution of PDHC immunoreactivity also differs markedly from GDH immunoreactivity, which is concentrated within fine processes closely encircling unlabeled Purkinje cell perikarya (asterisks over the halo of an unlabeled perikaryon in Figure 6-4A). Furthermore, the veil-like radiating processes in the molecular layer that are moderately immunoreactive for GDH (ML, arrows in Figure 6-4A) appear to not contain detectable levels of PDHC or of CyO. The close resemblance between these and some of the more superficial GFAP-immunoreactive processes in the molecular layer of the cerebellum (Figure 6-4C) is analogous to that seen in the cerebral cortex and other regions [2]. This localization supports conclusions made by Kaneko et al. [4] and Wenthold et al. [5,6] that the GDH-labeled processes in the cerebellum are glial. Other brain regions that also exhibit divergent distribution of the three mitochondrial enzymes include the neocortices, the hippocampal formation, the septal nuclei, the thalamic nuclei, nuclei of the solitary tracts, and brainstem regions projecting to the cerebellum, including the external cuneate nuclei and the inferior olive (Figure 6-3).

5. RELATION OF GDH AND PDHC TO SPECIFIC TRANSMITTER PATHWAYS

The low-intensity cytoplasmic labeling for GDH and PDHC seen in virtually all neurons probably is attributable to levels of the enzymes associated with metabolic pools of glutamate and acetyl-coenzyme A, respectively. On the other hand, the heterogeneous, intense immunoreactivity detectable in the presence of Triton X-100 may reflect the enzymes' associations with transmitter pools of glutamate and acetyl-coenzyme A.

5.1. The relation between GDH-immunoreactivity and glutamatergic pathways

The regional distribution of GDH within glial cells and processes through rostrocaudal extents of rat brain (Figure 6-3) differs from that of other mitochondrial enzymes but is remarkably similar to the reported glutamatergic pathways. These glutamatergic pathways have been identified by the uptake of ^{3}H-glutamate or aspartate [24], the binding of glutamatergic ligands [25,26], the immunocytochemical reactivity to antibodies directed against protein-conjugated forms of glutamate [27], or the histochemical staining for glutamate dehydrogenase [28–30].

In the cerebellar cortex, the laminar distribution of N-methyl-D-aspartate (NMDA)-specific binding sites occurs in the granular layer at levels four

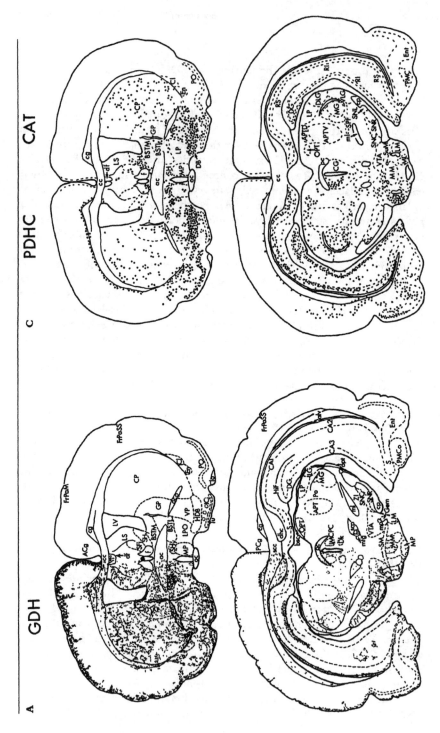

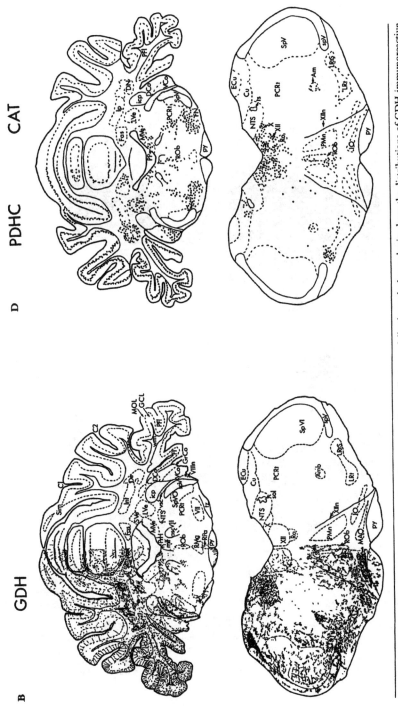

Figure 6-3. Camera-lucida drawings of coronal sections arranged rostrally to caudally through the rat brain show the distributions of GDH-immunoreactive processes (left-hand page) and of PDHC and choline acetyltransferase (CAT)-immunoreactive neuronal perikarya (right-hand page). Bar = 2.28 mm at all except the most caudal level, where it is 1.14 mm. See appendix for abbreviations. Figure 6-3B is from [2] and [3] with permission.

145

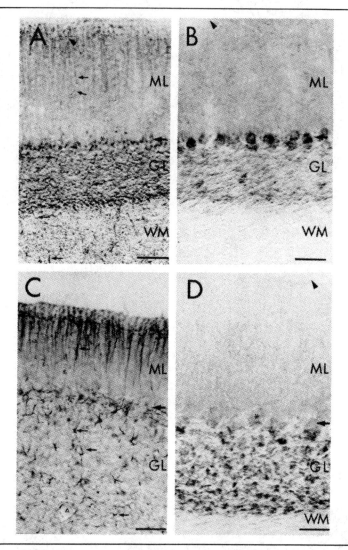

Figure 6-4. Photomicrographs of near-adjacent coronal sections through the cerebellar cortex showing the laminar distributions and cytological characteristics of profiles immunoreactive for GDH (**A**), PDHC (**B**), GFAP (**C**), and histochemically reactive for CyO (**D**). ML = molecular layer; large horizontal arrow in right margins = Purkinje cell layer; GL = granular layer; WM = white matter. Arrowhead = pial surface; asterisks = Purkinje cell perikarya. Small arrows point to GDH-immunoreactive processes in A and GFAP-immunoreactive glial processes in C. Triangle in C = blood vessel lumen. Differential interference contrast optics. Bar = 50 μm. See text for further details.

times greater than in the molecular layer or white matter [25]. This laminar pattern correlates well with the distribution of GDH demonstrated previously using histochemical methods [30] and with the lightmicroscopic immunocytochemical localization of GDH (Figures 6-3C, 6-4, and described above). Since glutamatergic binding sites are also detectable within the molecular layer [26], moderate levels of GDH immunoreactivity observed within the radial fibers of this layer may reflect their involvement with non-NMDA glutamatergic synapses.

Dense distribution of GDH-immunoreactive glial processes spanning radially through layers I through III of the rostral neocortex (Figure 6-3A, left) mirrors the distribution of binding sites for glutamatergic ligands [25,26], as well as robust uptake sites for ^{3}H-aspartate [24]. Similarly, hindbrain regions known to receive cortical afferents [rev. in 31], which most likely are glutamatergic [32], also exhibit intense GDH immunoreactivity. These include (1) the reticular nuclei (reticulotegmental, lateral reticular, and paramedian nuclei), (2) the sensory relay nuclei (gracile, cuneate, external cuneate, sensory trigeminal, and solitary tract nuclei), (3) the inferior olive, and (4) the pontine nuclei (Figure 6-3C). Of these, the pontine, lateral reticular, paramedian, inferior olive, and external cuneate nuclei also contain glutamatergic neurons that project to the cerebellum [rev. in 27]. Immunoreactivity that is detectable within these latter nuclei, in spite of only moderate densities of glutamatergic receptors, suggests that *intensities* of GDH labeling may reflect factors other than the *degree* of glutamatergic innervation. It is interesting that these and all other areas with *intense* GDH immunoreactivities exhibit high levels of CyO [3,33]. Conceivably, glutamatergic transmission may be highly energy consuming, thus requiring high levels of CyO [13,14]. However, comparisons of glutamate binding sites to the CyO distribution indicate otherwise, since not all glutamatergic areas have correspondingly high levels of CyO activities (e.g., stratum radiatum of CA1 of the hippocampus, Figure 6-3A). This and other regions that are glutamate receptive, but that do not have intense labeling for GDH, are also not strongly reactive for CyO [3]. An explanation consistent with these findings is that GDH levels in glia are determined by at least two factors: (1) the presence of glutamatergic afferents, which is necessary but not sufficient for the high levels of GDH, and (2) chronically active synaptic transmission within the area.

The enrichment of glial GDH in areas undergoing active glutamatergic transmission is interesting in light of their recognized role in the compartmentation of metabolic and transmitter pools of glutamate [reviewed in 34]. In addition to their well-documented role in the control of interstitial ion composition and volume [reviewed in 35], astrocytes take up glutamate with a capacity that is even higher than that of neurons [36]. Thus, astrocytes can presumably take up glutamate released from glutamatergic axon terminals, convert glutamate to alpha-ketoglutarate with the aid of GDH, and return the product to the TCA cycle of neurons and glia [8]. Alternatively, glutamine synthetase within astrocytes may catalyze the conversion of glutamate to glutamine for its return to the amino-acid pool of astrocytes and neurons [37]. The importance of this neuron-glia interaction is manifested in clinical cases where neurodegeneration is correlated with a systemic deficiency of GDH (discussed below).

5.2. Relationship between the distribution of PDHC and cholinergic neurons

Immunocytochemical labeling for choline acetyltransferase (CAT), the acetylcholine-synthesizing enzyme, has been detected in vibratome sections that are almost adjacent to the sections immunoreacted for PDHC [7]. As can be seen in Figures 6-3, 6-5A/B, and 6-6A/B, the cytological features and distribution of perikarya labeled with the two antigens are similar in a large number of regions. These include the caudate-putamen, septal nuclei, motor nuclei of the trigeminal nerve, dorsal motor nuclei of the vagus, the hypoglossal nuclei, and nucleus ambiguus. There are, however, notable exceptions. These include neurons immunoreactive for PDHC, but not for CAT, occurring in the deepest layers of neocortices (Figure 6-5D), the primary olfactory cortex (Figure 6-5E), the hippocampal formation, the entorhinal cortex, the lateral vestibular nuclei, and the cerebellar Purkinje cells (Figure 6-4B). Some of the regions that lack CAT immunoreactivity contain neurons that stain for acetylcholinesterase, the cholinergic degradative enzyme [7]. The distribution of PDHC-immunoreactive neurons may, alternatively, or in addition, reflect the occurrence of peptidergic neurons, such as somatostatin, in the caudate-putamen (Figure 6-5B) [38], which, together with some cholinergic neurons, also contain high levels of another enzyme, NADPH-diaphorase [39,40].

Selectivity of PDHC with respect to neurotransmitters in the CNS is supported by their low levels in regions containing catecholaminergic neurons. While adrenergic neurons of the C1 group in the rostral ventrolateral medulla contain a large number of mitochondria per unit cytoplasmic area [41], PDHC-immunoreactive neurons are sparse within this region (Figure 6-6A/C). In contrast, dopaminergic neurons in the lateral substantia nigra (Figure 6-3B) and noradrenergic neurons in the locus coeruleus [7] are appreciably labeled for PDHC. Conceivably, these differences may reflect more extensive cholinergic afferents, since both the substantia nigra and locus coeruleus contain acetylcholinesterase-positive cells [7,42,43].

6. FUNCTIONAL IMPLICATIONS

Regional heterogeneity in the levels of the mitochondrial enzymes—GDH, PDHC, and CyO—in brain could account for at least some of the areal differences [44,45] in response to extrinsic insults or metabolic disorders. For example, the cerebral cortex is one of the areas most easily damaged by ischemic insults [46]. Accordingly, the cerebral cortex exhibits high levels of

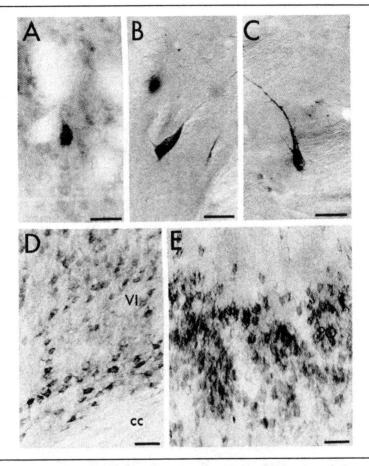

Figure 6-5. Photomicrographs of coronal sections through the caudate-putamen depict neurons of similar sizes and shapes that are immunolabeled for PDHC (**A**), choline acetyltransferase (**B**), and somatostatin (**C**). Photomicrographs of coronal sections through layer VI of the somatosensory cortex (**D**) and the primary olfactory cortex (**E**) show the distributions of neurons containing PDHC immunoreactivity but not CAT. Differential interference contrast optics. Bar = 25 μm in A through C, 50 μm in D and E.

all three enzymes: (1) GDH in glia of the superficial layers, (2) CyO in the granular layer, and (3) PDHC in neuronal perikarya of deeper layers. The involvement of mitochondria and glucose utilization in metabolic insults also is suggested by the organelles' microvacuolation in early stages of ischemia [47] and the exacerbation of neuronal damage due to the administration of glucose prior to ischemia [46].

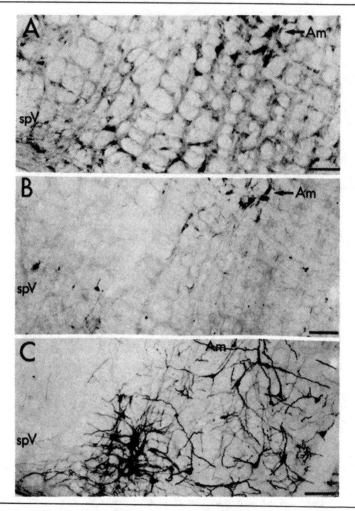

Figure 6-6. Photomicrographs of coronal sections through the rostral ventrolateral medulla and nucleus ambiguus (Am) showing the distribution of PDHC-immunoreactive neurons (**A**), choline acetyltransferase (CAT)-labeled perikarya and processes (**B**), and phenylethanolamine N-methyltransferase (PNMT)-immunoreactive neurons and processes (**C**). PDHC-immunoreactive neurons are distributed similarly to CAT-labeled cells in the nucleus ambiguus, but are not found in regions of the rostral ventrolateral medulla that contain PNMT-labeled neurons. Bright-field optics. Bars = 100 μm.

The etiology of several neurodegenerative disorders may involve the interaction of excitatory transmitters, such as L-glutamate and ACh, that are not appropriately metabolized [48]. L-glutamate is considered to be excitotoxic based on observations of neuronal death following synaptic release of L-glutamate in vitro [49] or exogenous administration of L-glutamate and its analogs in vivo [50–52] and in vitro [53]. The toxicity is manifested under abnormal physiological conditions, such as anoxia, ischemia, epileptic seizures, and exogenous glucocorticoid infusions [50–53]. The implication of PDHC and GDH in neurodegenerative disorders is almost exclusively correlative. For example, reductions in PDHC and ACh are seen in the cortex and hippocampal formation of Alzheimer's patients [54–57]. These areas receive strong afferent inputs from cholinergic neurons in the nucleus basalis of Meynert [58,59], the latter of which are glutamate receptive and thus are susceptible to excitotoxins [60]. Also, in the striatum of patients suffering with Huntington's chorea, neurodegeneration, hypothesized to be caused by glutamate or related endogenous excitotoxins [61], is accompanied by the reduction of both PDHC [55,62,63] and ACh [64,65], among other transmitters [reviewed in 66]. The detection of high levels of PDHC, GDH, and CyO in the nucleus basalis of Meynert and the caudate-putamen (striatum) of rat is consistent with the idea that the loss of cholinergic neurons in these areas follows from their receptivity to L-glutamate or its analogs. Clearly, additional studies are needed to establish the relative significance of these enzymes in complex diseases such as Alzheimer's dementia.

The importance of GDH, specifically in mental health, is suggested by the consequences of its deficiency. Several laboratories have shown correlations between a systemic deficiency of GDH (assumed from measurements in leukocytes and fibroblasts) and extrapyramidal motor disorders involving characteristic atrophies in the olivo-ponto-cerebellar pathway [67–69], which are reviewed in Chapter 20 of this book. These authors suggest that the atrophies may be caused by toxicity from neuronally released L-glutamate that cannot enter metabolic pools. The heterogeneous distribution of GDH and CyO in normal brains may partially account for the particular vulnerability of the olivo-ponto-cerebellar pathway. Presumably, high levels of GDH observable in this pathway are *required* due to the chronically active transmission that causes the frequent release of neuronal glutamate. Thus, glutamate toxicity due to a global deficiency of GDH would be most damaging in these regions. Plaitakis and his colleagues point out that GDH-deficient patients exhibit "multiple-system degeneration," involving alterations not only of the olivo-ponto-cerebellar pathway, but also of the basal ganglia, the hippocampal formation and other limbic structures, the cortex, the oculomotor nerve, the lateral vestibular nucleus, hypothalamus, and thalamus [69]. The detection of high densities of GDH immunoreactivity in glial processes of these areas strongly supports the concept that glial GDH plays an important role of protecting glutamate-receptive neurons by maintaining an adequate turnover of glutamate released from nearby terminals. Future ultrastructural studies to simultaneously localize GDH-containing glial processes and glutamatergic terminals should further our understanding of this important neuron-glia interaction.

ACKNOWLEDGMENTS

We wish to thank Dr. Brian Andrews for teaching CA the immunogold method to label frozen ultrathin sections during the Neurobiology Course at Marine Biological Laboratory (Woods Hole, MA). This work was supported by NIH grants NS07782 (CA), EY08055 (CA), BRSG 2-507-RR-07062026 (CA), HL18974 (TAM & VMP), MH42834 (TAM), and HL18974 (VMP); NIMH grant MH40342 (VMP); and NSF grant BNS87-06790 (VMP).

REFERENCES

1. Tzagoloff A. (1982). Mitochondria. New York: Plenum, 342 pp.
2. Aoki C., Milner T.A., Sheu K.-F.R., Blass J.P., Pickel V.M. (1987). Regional distribution of astrocytes with intense immunoreactivity for glutamate dehydrogenase in rat brain: Implications for neuron-glia interactions in glutamate transmission. J. Neurosci. 7:2214–2231.
3. Aoki C., Milner T.A., Berger S.B., Sheu K.-F.R., Blass J.P., Pickel V.M. (1987). Glial glutamate dehydrogenase: Ultrastructural localization and regional distribution in relation to the mitochondrial enzyme, cytochrome oxidase. J. Neurosci. Res. 18:305–318.
4. Kaneko T., Akiyama H., Mizuno N. (1987). Immunohistochemical demonstration of glutamate dehydrogenase in astrocytes. Neurosci. Lett. 77:171–175.
5. Wenthold R.J., Altschuler R.A., Skaggs K.K., Reeks K.A. (1987). Immunocytochemical characterization of glutamate dehydrogenase in the cerebellum of the rat. J. Neurochem. 48:636–643.
6. Madl J.E., Clements J.R., Beitz A.J., Wenthold R.J., Larson A.A. (1988). Immunocytochemical localization of glutamate dehydrogenase in mitochondria of the cerebellum: An ultrastructural study using a monoclonal antibody. Brain Res. 452:396–402.
7. Milner T.A., Aoki C., Sheu K.-F.R., Blass J.P., Pickel V.M. (1987). Light microscopic immunocytochemical localization of pyruvate dehydrogenase complex in rat brain: Topographical distribution and relation to cholinergic and catecholaminergic nuclei. J. Neurosci. 7:3171–3190.
8. Dennis S.C., Clark J.B. (1977). The pathway of glutamate metabolism in rat brain mitochondria. Biochem. J. 168:521–527.
9. Tucek S., Cheng S.-C. (1974). Prominence of the acetyl group of acetylcholine and compartmentation of acetyl-CoA and Krebs cycle intermediates in the brain in vivo. J. Neurochem. 22:893–914.
10. Gibson G.E., Jope R., Blass J.P. (1975). Decreased synthesis of acetylcholine accompanying impaired oxidation of pyruvic acid in rat brain minces. Biochem. J. 148:17–23.
11. Jope R.S., Jenden D.J. (1980). The utilization of choline and acetyl coenzyme A for the synthesis of acetylcholine. J. Neurochem. 35:318–325.
12. Sterri S.H., Fonnum F. (1980). Acetyl-CoA synthesizing enzymes in cholinergic nerve terminals. J. Neurochem. 35:249–254.
13. Wong-Riley M.T.T., Merzenich M.M., Leake P.A. (1978). Changes in endogenous enzymatic reactivity to DAB induced by neuronal inactivity. Brain Res. 141:185–192.
14. Wong-Riley M. (1979). Changes in the visual system of monocularly sutured or enucleated cats demonstrable with cytochrome oxidase histochemistry. Brain Res. 171:11–28.
15. Carroll E.W., Wong-Riley M. (1985). Correlation between cytochrome oxidase staining and the uptake and laminar distribution of tritiated aspartate, glutamate, GABA and glycine in the striate cortex of squirrel monkey. Neuroscience 15:959–976.
16. Sternberger L.A. (1979). Immunocytochemistry, 2nd ed. Englewood Cliffs, NJ: Prentice Hall.
17. Bignami A., Eng L.F., Dahl D., Uyeda C.T. (1972). Localization of the glial fibrillary acidic protein in astrocytes by immunofluorescence. Brain Res. 43:429–435.
18. DiPrisco G.T., Casola L. (1975). Detection of structural differences between nuclear and mitochondrial glutamate dehydrogenase by the use of immunoadsorbents. Biochemistry 14:4679–4683.
19. Lai J.C., Sheu K.-F.R, Kim Y.T., Clarke D.D., Blass J.P. (1986). The subcellular

localization of glutamate dehydrogenase (GDH): Is GDH a marker for mitochondria in brain? Neurochem. Res. 11:733–744.
20. Colon A.D., Plaitakis A., Perakis A., Berl S., Clarke D.D. (1986). Purification and characterization of soluble and particulate glutamate dehydrogenase from rat brain. J. Neurochem. 46:1811–1819.
21. Griffiths G. (1985). The Tokuyasu frozen section technique for antigen localization. EMBL course in immunocytochemistry.
22. Knecht E., Martinez-Ramon A., Grisolia S. (1986). Electron microscopic localization of glutamate dehydrogenase in rat liver mitochondria by an immunogold procedure and monoclonal and polyclonal antibodies. J. Histochem. Cytochem. 34:913–922.
23. Sheu K.-F.R., Lai J.C.K., Blass J.P. (1983). Pyruvate dehydrogenase phosphate (PDHb) phosphatase in brain activity, properties and subcellular localization. J. Neurochem. 40:1366–1372.
24. Fonnum F., Soreide A., Kvale I., Walker J., Walaas I. (1981). Glutamate in cortical fibers. Adv. Biochem. Psychopharmacol. 27:29–42.
25. Monaghan D.T., Cotman C.W. (1985). Distribution of N-methyl-D-aspartate-sensitive L-^{3}H-glutamate binding sites in rat brain. J. Neurosci. 5:2902–2919.
26. Halpain S., Wieczorek C.M., Rainbow T.C. (1984). Localization of L-glutamate receptors in rat brain by quantitative autoradiography. J. Neurosci. 4:2247–2258.
27. Ottersen O.P., Storm-Mathisen J. (1984). Glutamate and GABA-containing neurons in the mouse and rat brain, as demonstrated with a new immunocytochemical technique. J. Comp. Neurol. 229:374–392.
28. Rothe F., Schmidt W., Wolf G. (1983). Postnatal changes in the activity of glutamate dehydrogenase and aspartate aminotransferase in the rat nervous system with special reference to the glutamate transmitter metabolism. Dev. Brain Res. 11:67–74.
29. Schunzel G., Wolf G., Rothe F., Seidler E. (1986). Histophotometric evaluation of glutamate dehydrogenase activity of the rat hippocampal formation during postnatal development, with special reference to the glutamate transmitter metabolism. Cell. Mol. Neurobiol. 6:31–42.
30. Wolf G., Schunzel G. (1987). Glutamate dehydrogenase in aminoacidergic structures of the postnatally developing rat cerebellum. Neurosci. Lett. 78:7–11.
31. Carpenter M.B. (1978). Core Text of Neuroanatomy. 2nd ed., Baltimore, MD: Williams and Wilkins, 354 pp.
32. Streit P. (1984). Glutamate and aspartate as transmitter candidtates for systems of the cerebral cortex. In: E.G. Jones, A. Peters (eds): Cerebral Cortex, Vol. 2. New York: Plenum, pp. 119–144.
33. Wong-Riley M.T.T. (1976). Endogenous peroxidase activity in brain stem neurons as demonstrated by their staining with diaminobenzidine in normal squirrel monkey. Brain Res. 108:257–277.
34. Hertz L., Kvamme E., McGeer E.G., Schousboe A. (eds). (1983). Glutamine, Glutamate and GABA in the Central Nervous System. New York: Alan R. Liss.
35. Abbott N.J. (1986). The neuronal microenvironment. Trends Neurosci. 9:3–8.
36. Huck S., Grass F., Hortnagl H. (1984). The glutamate analogue alpha-aminoadipic acid is taken up by astrocytes before exerting its gliotoxic effect in vitro. J. Neurosci. 4:2650–2657.
37. Bradford H.F., Ward H.K., Thomas A.J. (1978). Glutamine—a substrate for nerve endings. J. Neurochem. 30:1454–1459.
38. Johansson O., Hokfelt T., Elde R.P. (1984). Immunohistochemical distribution of somatostatin-like immunoreactivity in the central nervous system of the adult rat. Neuroscience 13:265–339.
39. Vincent S.R., Johansson O., Hokfelt T., Skirboll L., Elde R.P., Terenius L., Kimmel J., Goldstein M. (1983). NADPH-diaphorase: A selective histochemical marker for striatal neurons containing both somatostatin and avian pancreatic polypeptide (APP)-like immunoreactivities. J. Comp. Neurol. 217:252–263.
40. Vincent S.R., Satoh K., Armstrong D.M., Fibiger H.C. (1983). NADPH-diaphorase: A selective histochemical marker for the cholinergic neurons of the pontine reticular formation. Neurosci. Lett. 43:31–36.
41. Milner T.A., Pickel V.M., Park D.H., Joh T.H., Reis D.J. (1987). Adrenergic neurons on the rostral ventrolateral medulla of the rat: I. Electron microscopic localization of

phenylethanolamine N-methyltransferase. Brain Res. 411:28-45.
42. Butcher L.L., Talbot K. (1978). Acetylcholinesterase in rat nigroneostriatal neurons: Experimental verification and evidence for cholinergic dopaminergic interactions in the substantia nigra and caudate-putamen complex. In: L.L. Butcher (ed): Cholinergic-Monoaminergic Interactions in the Brain. New York: Academic Press, pp. 25-95.
43. Albanese A., Butcher L.L. (1979). Locus coeruleus somata contain both acetylcholinesterase and norepinephrine: Direct histochemical demonstration of the same tissue section. Neurosci. Lett. 4:101-104.
44. Smith M.-L., Auer R.N., Siesjo B.K. (1984). The density and distribution of ischemic brain injury in the rat following 2-10 min of forebrain ischemia. Acta Neuropathol. (Berlin) 64:319-332.
45. Wieloch T. (1985). Neurochemical correlates to selective neuronal vulnerability. Progress Brain Res. 63:69-85.
46. Pulsinelli W.A., Brierley J.B., Plum F. (1982). Temporal profile of neuronal damage in a model of transient forebrain ischemia. Ann. Neurol. 11:491-498.
47. Brierley J. (1986). Cerebral hypoxia. In: Greenfield's Neuropathology. M. Blackwood, A. Corsellis (eds): London: Arnold, pp. 43-85.
48. Olney J.W. (1985). Excitatory transmitters and epilepsy-related brain damage. Int. Rev. Neurobiol. 27:337-362.
49. Rothman S. (1984). Synaptic release of excitatory amino acid neurotransmitter mediates anoxic neuronal death. J. Neurosci. 7:1884-1891.
50. Sapolsky R.M. (1985). A mechanism for gluticorticoid toxicity in the hippocampus: Increased neuronal vulnerability to metabolic insults. J. Neurosci. 5:1228-1232.
51. Coyle J.T., Bird S.J., Evans R.H., Gulley R.L., Nadler R.L., Nadler J.V., Nicklas W.J., Olney J.W. (1981). Excitatory amino acid neurotoxins: Selectivity, specificity, and mechanisms of action. Neurosci. Res. Prog. Bull. 19:331-427.
52. Lucas D.R., Newhouse J.P. (1957). The toxic effect of sodium L-glutamate on the inner layers of the retina. Arch. Opthalmol. 58:193-201.
53. Choi D.W., Maulucci-Gedde M., Kriegstein A.R. (1987). Glutamate neurotoxicity in cortical cell culture.
54. Perry R.H., Blessed G., Perry E.K., Tomlinson B.E. (1980). Histochemical observations on cholinesterase activities in the brains of elderly, normal and demented (Alzheimer's type) patients. Ageing 9:9-16.
55. Sorbi S., Bird E.D., Blass J.P. (1983). Decreased pyruvate dehydrogenase complex activity in Huntington and Alzheimer brain. Ann. Neurol. 13:72-78.
56. Price D.L. (1984). Neuropathology of Alzheimer's disease. In: Alzheimer's Diseases and Related Disorders. Research and Management. W.E. Kelly (ed): Springfield, IL: Charles C. Thomas, pp. 81-104.
57. Sheu K.-F.R., Kim Y.T., Blass J.P., Weksler M.E. (1985). An immunochemical study of the pyruvate dehydrogenase deficit in Alzheimer's disease brain. Ann. Neurol. 17:444-449.
58. Mesulam M.-M., Mufson E.J., Wainer B.H., Levey A.I. (1983). Central cholinergic pathways in the rat: An overview based on an alternative nomenclature (Ch1-Ch6). Neuroscience 10:1185-1201.
59. Amaral D.G., Kurz J. (1985). An analysis of the origins of the cholinergic and noncholinergic septal projections to the hippocampal formation of the rat. J. Comp. Neurol. 240:37-59.
60. Ueki A., Miyoshi K. (1987). Changes in cholinergic markers following kainic acid lesion of the ventral globus pallidus in rat. Jpn. J. Psychiatry 41:87-96.
61. Coyle J.T., Schwarcz R. (1976). Lesion of striatal neurons with kainic acid provides a model for Huntington's chorea. Nature 263:244-246.
62. Butterworth J., Yates C.M., Reynolds G.P. (1985). Distribution of phosphate-activated glutaminase, succinic dehydrogenase, pyruvate dehydrogenase and gamma-glutamyl transpeptidase in post-mortem brain from Huntington's disease and agonal cases. J. Neurol. Sci. 67:161-171.
63. Graveland G.A., Silliams R.S., DiFiglia M. (1985). Evidence for degenerative and regenerative changes in neostriatal spiny neurons in Huntington's disease. Science 227:770-773.
64. Bird E.D., Iversen L.L. (1974). Huntington's chorea. Postmortem measurement of glutamic

acid decarboxylase, choline acetyltransferase and dopamine in basal ganglia. Brain 97:457–472.
65. McGeer P.L., McGeer E.G., Fibiger H.C. (1973). Choline acetylase and glutamic acid decarboxylase in Huntington's chorea. A preliminary study. Neurology 23:912–917.
66. Kowall N.W., Ferrante R.J., Martin J.B. (1987). Patterns of cell loss in Huntington's disease. Trends in Neurosci. 10:24–29.
67. Yamaguchi T., Hayashi K., Murakami H., Ota K., Maruyama S. (1982). Glutamate dehydrogenase deficiency in spinocerebellar degenerations. Neurochem. Res. 7:627–636.
68. Duvoisin R.C., Chokroberty S., Lepore F., Nicklas W. (1983). Glutamate dehydrogenase deficiency in patients with olivopontocerebellas atrophy. Neurology 33:1322–1326.
69. Plaitakis A., Berl S., Yahr M.D. (1984). Neurological disorders associated with deficiency of glutamate dehydrogenase. Ann. Neurol. 15:144–153.

APPENDIX: ABBREVIATIONS USED IN FIGURE 3

ac	anterior commissure
ACg	anterior cingulate cortex
acp	anterior commissure, posterior
alv	alveus of hippocampus
Am, Amb	nucleus ambiguus
APT	anterior pretectal nucleus
APTD	anterior pretectal nucleus, dorsal
APTV	anterior pretectal nucleus, ventral
ascVII	ascending fibers of facial nerve
bsc	brachium of superior colliculus
BSTL	bed nucleus of stria terminalis, lateral
BSTM	bed nucleus of stria terminalis, medial
C1	crus 1 of ansiform lobule of cerebellum
C2	crus 2 of ansiform lobule of cerebellum
CA1	field CA1 of Ammon's horn of hippocampus
CA2	field CA2 of Ammon's horn of hippocampus
CA3	field CA3 of Ammon's horn of hippocampus
cc	corpus callosum
CG	central grey
cg	cingulum
Cl	claustrum
CP	caudate putamen
cp	cerebral peduncle, basal
Cu	cuneate nucleus
DB	nucleus of diagonal band
DCo	dorsal cochlear nucleus
De, DN	dentate nucleus of cerebellum
df	dorsal fornix
DG	dentate gyrus of hippocampal formation
dhc	dorsal hippocampal commissure
Dk	nucleus of Darkschewitsch
DLG	dorsal lateral geniculate nucleus
ECu	external cuneate nucleus
En	endopiriform nucleus
Ent	entorhinal cortex
f	postcommissural fornix
Fas	fastigial nucleus of cerebellum
Fl	flocculus of cerebellum
FN	facial nucleus
fr	fasciculus retroflexus
FrPaM	frontoparietal cortex, motor area
FrPaSS	frontoparietal cortex, somatosensory area
GCL	granule cell layer of cerebellar cortex

Gem	gemini nucleus
GP	globus pallidus
GrCo	granule cell layer of cochlear nucleus
HDB	horizontal limb of diagonal band
HiF	hippocampal fissure
ICj	islands of Calleja
icp	inferior cerebellar peduncle
IMCPC	interstitial magnocellular nucleus of posterior commissure
Int, Ip	interpositus nucleus of cerebellum
IO	inferior olive
LG	lateral geniculate nucleus of thalamus
LM	lateral mammillary nucleus
lo	lateral olfactory tract
LP	lateral posterior nucleus of thalamus
LPO	lateral preoptic area
LRt	lateral reticular nucleus
LRts	lateral reticular nucleus, subtrigeminal
LS	lateral septal nucleus
LV	lateral ventricle
LVe	lateral vestibular nucleus
MAO	medial accessory olive
MG	medial geniculate nucleus of thalamus
ML	medial mammillary nucleus, lateral
ml	medial lemniscus
mlf	medial longitudinal fasciculus
MM	medial mammillary nuclei, medial and lateral
MOL	molecular layer of cerebellar cortex
MP	medial preoptic area
mp	mammillary peduncle
mtg	mammillotegmental tract
MVe	medial vestibular nucleus
NTS	nucleus of solitary tract
oc	optic chiasm
OPT	nucleus of olivary pretectal nucleus
opt	optic tract
pc	posterior commissure
PCg	posterior cingulate cortex
PCRt	parvocellular reticular nucleus
PFl	paraflocculus
PMCo, PMC	posteromedial cortical nucleus of amygdala
PMn	paramedian reticular nucleus
PO	primary olfactory cortex
Po	posterior complex of thalamus
PrH, PP	prepositus hypoglossal nucleus
py	pyramidal tract
RI	regio inferior of the hippocampus
RPa	raphe pallidus nucleus
RMg	raphe magnus nucleus
Ro	nucleus of Roller
ROb	raphe obscurus
RS	regio superior of the hippocampus
S	subiculum
scc	splenium of corpus callosum
scp	superior cerebellar peduncle
SFi	septofimbrial nucleus
SHy, SH	septohypothalamic nucleus
Sim	simple lobule of cerebellum
SM	supramammillary nucleus

SNC	substantia nigra, pars compacta
SNL	substantia nigra, lateral
SNR	substantia nigra, pars reticulata
sol	solitary tract
SPF	subparafascicular thalamic nucleus
spV	spinal tract trigeminal nerve
SpVI	nucleus of spinal trigeminal nerve, interpositus
SpVO, SpV	nucleus of spinal trigeminal nerve, oral
SuM	supramammillary nucleus
SVe	superior vestibular nucleus
ts	tractus solitarius
Tu	olfactory tubercle
tz	trapezoid body
VCo	ventral cochlear nucleus
VII	facial nucleus
VIIIn	vestibulocochlear nerve
VLG	ventral lateral geniculate nucleus
VP	ventral pallidum
vsc	ventral spinocerebellar tract
VTA	ventral tegmental area
wm	white matter of cerebellar cortex
X	dorsal motor nucleus of vagus
XII	hypoglossal nucleus
XIIn	hypoglossal nerve
ZI	zona incerta of thalamus

7. THE PURKINJE CELL DEGENERATION MUTANT: A MODEL TO STUDY THE CONSEQUENCES OF NEURONAL DEGENERATION

BERNARDINO GHETTI AND LAZAROS C. TRIARHOU

1. INTRODUCTION

A distinct feature of nervous systems is that the selective loss of a neuronal population, regardless of the cause, may initiate a cascade of regressive changes in presynaptic and postsynaptic neurons. This property can be explained by the interdependence of neurons for survival and normal function, which derives from the synaptic mode of neuronal connectivity and the trophic interactions of neurons at the level of the synapse.

Neuronal loss is the ultimate cellular manifestation in degenerative conditions due to genetic causes or in acquired conditions resulting from injury of viral, toxic, ischemic, or metabolic origin.

One way to investigate neuronal loss and its consequences experimentally is by using laboratory mice with neurological mutations, which are characterized by selective neuronal death resulting from discrete, genetically induced lesions, and thus offer experimental paradigms for the study of altered neuronal circuitries. In broad terms, nerve cell death in mutant mice can fall under two general categories: first, loss of cells during postnatal ontogenesis, which is relevant to developmental disorders, and second, neuronal degeneration following maturation of the nervous system, which is relevant to neurodegenerative conditions of the adult. Mutant mice falling into either of these two categories and featuring a genetically determined loss

A. Plaitakis (ed.), CEREBELLAR DEGENERATIONS: CLINICAL NEUROBIOLOGY. Copyright © 1992.
Kluwer Academic Publishers, Boston. All rights reserved.

of specific neuronal populations can provide useful insights into the biology of nerve-cell death and indirectly contribute to the unraveling of abiotrophic mechanisms operating in human neurological disorders. By using such animals, one has the flexibility of predetermining the temporal stages of the various degenerations to study, while in human neurodegenerative diseases, and particularly in the cerebellar heredoataxias, it is usually difficult to establish whether an observed atrophy or loss (or both) of a particular neuronal group is *primary* (i.e., due to a genetic effect) or *secondary* (i.e., due to trans-synaptic degeneration).

In neurological mutant mice, one can recognize two orders of degenerative phenomena: (1) a *primary* loss of neurons that are programmed to degenerate through a more or less direct action of the mutant allele and (2) a *secondary* (or *trans-synaptic*) neuronal degeneration or atrophy of cells that are presynaptic or postsynaptic to the neurons-targets of the mutation. In most cases, trans-synaptic cell death proceeds slower than genetically determined neuronal death; therefore, secondary losses can in general be best detected and measured at advanced stages of the disorder. Trans-synaptic regressive changes may involve neuronal numbers, perikaryonal size, and dendritic arborization. The principal variables that are known to affect the outcome of trans-synaptic degeneration are age, species, and neural site [1].

In this chapter we present an overview of studies in *pcd* mutant mice with particular emphasis on the changes that occur in neurons presynaptic and postsynaptic to Purkinje cells.

2. GENETICALLY DETERMINED DEGENERATION OF PURKINJE CELLS IN THE *pcd* MUTANT

2.1. Brief overview of the normal postnatal development of Purkinje cells

Much of the morphogenetic development of Purkinje cells in the cerebellum of the normal mouse takes place between postnatal days 6 and 15. The size of the perikaryon reaches the dimensions of mature cells by day 10. During the so-called funnel stage, a compilation of free ribosomes in the basal part of the perikaryon is seen, along with an upward displacement of the nucleus. That polyribosomal mass normally disappears after postnatal day 14. Perisomatic processes of Purkinje cells, referred to as somatic thorns or spines, are present in large numbers from postnatal day 6–10, and then decline, such that by day 14 the majority of them are gone. Apical dendrites emerge from the Purkinje cell soma around day 7; by day 10, there is a profound transformation of the soma and dendrites. Finally, the synaptogenesis of incoming afferents on Purkinje cells is mainly made between postnatal days 7 and 14 [2].

2.2. Purkinje cell degeneration in the mutant

The Purkinje cell degeneration (*pcd*) mutant mouse is characterized by a genetically determined loss of virtually all Purkinje cells, taking place after the maturation of the cerebellum, between 17 and 45 days of age [3,4]. In that respect, it provides a useful model for understanding pathogenetic

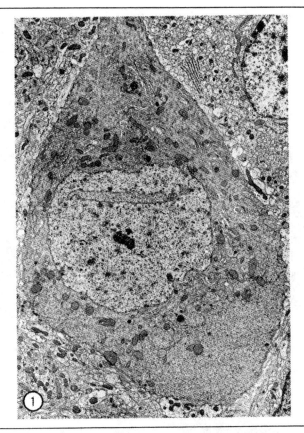

Figure 7-1. Electron microscopic appearance of a Purkinje cell soma from a 17-day-old *pcd* mouse. Note the accumulation of ribosomes in the basal pole of the cytoplasm and the formation thereof of a polysomal mass. Uranyl acetate and lead citrate. ×4,300 (B. Ghetti, unpublished micrograph).

mechanisms underlying some of the human cerebellar ataxias. The *pcd* allele is recessive and has been localized to mouse chromosome 13 [5]. Homozygous females (*pcd/pcd*) are fertile, whereas homozygous males are sterile.

In the cerebellum of 17-day-old *pcd* mutant mice, Purkinje cells appear to have a normal morphology, except that a metachromatic mass, consisting of polyribosomes (Figure 7-1), is present in the basal pole of the perikaryon; some Purkinje cell somata and dendrites undergo dark degeneration and cytoplasmic shrinkage (Figure 7-2).

At 23 days of age, the degenerative process of Purkinje cells has advanced and cell debris is found in the molecular and Purkinje cell layers (Figure 7-2),

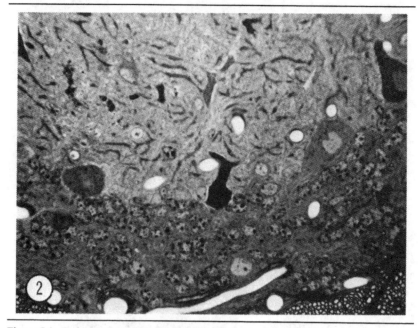

Figure 7-2. Purkinje cells undergoing dark degeneration in the cerebellar cortex of a 23-day-old *pcd* mutant mouse. Note the concomitant degeneration of Purkinje cell dendrites in the molecular layer. One-micrometer-thick Epon section stained with toluidine blue. ×400 (B. Ghetti, unpublished).

as well as in the granule cell layer and in the deep cerebellar nuclei, resulting from the fragmentation of dendrites, perikarya, and Purkinje axon terminals. Nevertheless, surviving Purkinje cells are still found.

As already mentioned, the loss of Purkinje cells in *pcd* mutants progresses rapidly. It is estimated that about 25–50% of Purkinje cells have already degenerated by postnatal days 22–24 [3]. The loss of Purkinje cells takes place in clusters, in such a way that the remaining Purkinje cells are organized in parasagittal bands in coronal sections of cerebellum with a symmetrical disposition relative to the midline [6] (Figures 7-3 and 7-4). Eventually (i.e., after 45–50 days of age), more that 99% of the Purkinje cells disappear and the *pcd* cerebellum becomes devoid of immunoreactivity for specific Purkinje cell protein markers, such as calcium-binding protein (Figure 7-5, see also Wassef et al. [6]), polypeptide PEP-19 [7], and cyclic GMP-dependent protein kinase [8]. In histological preparations, one can see cell debris throughout the molecular, Purkinje, and granule cell layers, as well as in the subcortical white matter and the deep cerebellar nuclei. At later ages (6 months and beyond), a remarkable atrophy is observed of the cerebellar

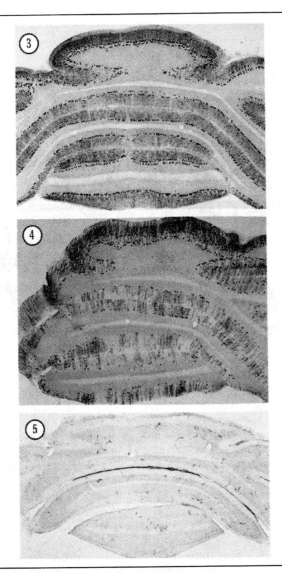

Figures 7-3 to 7-5. Calcium-binding protein (calbindin) immunoreactivity in the cerebellum of normal C57BL/6J (Figure 7-3) and *pcd* mutant mouse at 22 days (Figure 7-4) and 4 months (Figure 7-5) of age. Sections have been cut at the coronal plane. Notice the appearance of parasagittal bands in Figure 7-4, resulting from the loss of Purkinje cells in distinct zones. Following the complete loss of Purkinje cells in Figure 7-5, calbindin immunoreactivity also disappears. ×18 (L.C. Triarhou and B. Ghetti, unpublished).

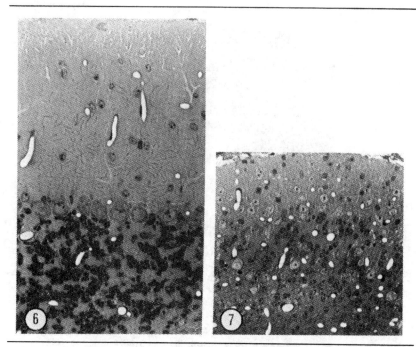

Figures 7-6 and 7-7. One-micrometer-thick plastic sections showing the normal (Figure 7-6) and *pcd* cerebellar cortex (Figure 7-7) after Purkinje cell loss. Notice the cortical atrophy in the mutant and, also, the presence of dark-degenerating osmiophilic granule cells, probably a consequence of Purkinje cell loss. Toluidine blue. ×200 (B. Ghetti, unpublished).

cortex, which essentially consists only of molecular and granule cell layers (Figures 7-6 and 7-7). The overall size of the cerebellum is substantially reduced as well (Figures 7-8 and 7-9); the average change in cerebellar weight is from 41 mg at 23 days to 22 mg at 300 days, in contrast to normal mouse cerebellum, which is about 55–60 mg (weights of fixed tissues, Triarhou et al. [9]).

3. ASTROCYTIC REACTION

A reaction of Bergmann glial radial fibers accompanies the degeneration of Purkinje cells. Reactive astrocytes participate in the process of phagocytosis of Purkinje cell debris. Compared to the cerebellum of age-matched +/+ mice (Figure 7-10), Bergmann glial radial fibers are hypertrophic, as shown by immunocytochemistry for glial fibrillary acidic protein (GFAP) (Figure 7-11) or vimentin [10] or by electron microscopy [11]. The severity of this glial reaction reaches its peak in the cerebellum of 40- to 50-day-old *pcd* mutants.

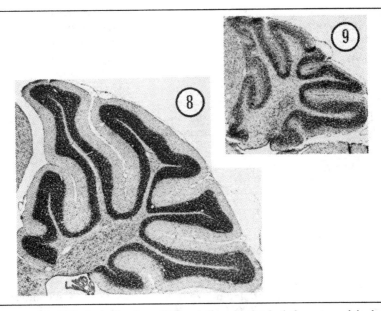

Figures 7-8 and 7-9. Sagittal sections of *pcd* cerebellum showing both the cortex and the deep cerebellar nuclei in 23- and 300-day-old mutants, respectively. Ten-micrometer-thick paraffin sections, stained with gallocyanin. ×25 (From Triarhou et al., [9], with permission.)

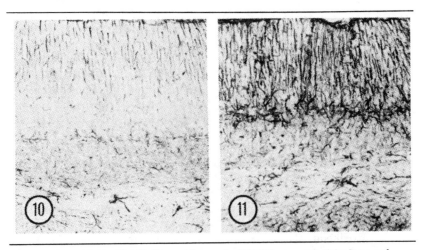

Figures 7-10 and 7-11. Glial fibrillary acidic protein (GFAP) immunoreactivity in normal (Figure 7-10) and *pcd* mutant mouse at 45 days of age (Figure 7-11). Note the hypertrophy of Bergmann fibers in the mutant. ×160 (A.C. Chang and B. Ghetti).

The cerebellum of *pcd* mutant mice, in which neuronal degeneration elicits a glial response, was found to contain increased amounts of immunoreactive nerve growth factor (NGF), which is detectable by radioimmunoassay [12]. Normally, levels of such an NGF are found in several areas of the mouse brain, including the cerebellum, brain stem, and hemispheres (however, this immunoactive NGF is not biologically active in the chick embryo sensory ganglia bioassay). The results with *pcd* mutant mice show a parallel increase in glial activity and NGF expression and suggest that C.N.S. glia may produce a NGF-like protein for central neurons.

4. SEQUENCE AND EXTENT OF DEGENERATIVE EVENTS IN PRESYNAPTIC NEURONAL SYSTEMS

4.1. Granule cells

A slow, progressive degeneration of granule cells follows the rapid, genetically determined degeneration of Purkinje cells in *pcd* mutants [13]. Osmiophilic debris of granule cells are readily seen at 3 months of age and up (Figure 7-7). Parallel fiber terminals are found without their postsynaptic Purkinje cell element (Figure 7-12).

A quantitative study of granule cells in normal and *pcd* mutant mice, ranging in age from 17 to 600 days, showed an evident reduction in granule cell number at 3 months of age, suggesting that numerous cells had already degenerated by that time. The severest losses take place between 3 and 12 months of age. In controls, granule cell numbers in cross sections of the declive and tuber vermis were 5808 ± 295 (mean $\pm$ SEM) before 1 month and 5546 ± 335 after 1 year of age. The corresponding figures in mutants were 5740 ± 154 and 612 ± 26. Granule cell loss amounted to about 90% and followed an exponential decay ($R^2 = 0.947$), which was highly significant ($p < 0.0001$) [14].

At the same time, the molecular layer (which normally contains the dendrites of Purkinje cells and the axons of granule cells) becomes atrophic; in the course of 1 year, its thickness in *pcd* mutants is reduced to about one fourth of normal.

Counts of parallel fiber profiles (the axons of granule cells) in electron micrographs of the molecular layer of cerebellar cortex showed a substantial reduction in their number as well. The density of parallel fibers per square micrometer of molecular layer neuropil is 3.91 in control mice, and amounts to 3.99, 3.39, and 2.44 in 6-, 9-, and 12-month-old mutants, respectively [B. Ghetti et al., in preparation].

In all likelihood, the loss of granule cells is trans-synaptic, due to the primary degeneration of Purkinje cells. Granule cell degeneration is also observed in other mutants with primary Purkinje cell deficits, such as the staggerer (*sg/sg*), nervous (*nr/nr*), and Lurcher (*Lc/+*); it is thought to be trans-synaptic in those situations as well [15-17]. In essence, it appears as if

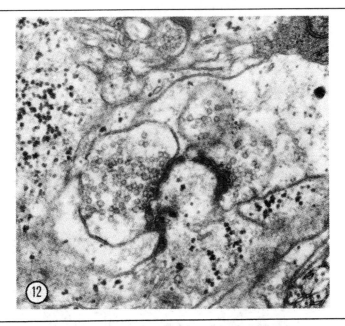

Figure 7-12. Parallel fiber terminals in the molecular layer of a *pcd* mutant mouse lacking a postsynaptic Purkinje cell element. The endings are now apposed to an astroglial process; however, they have maintained their synaptic densities. Uranyl acetate and lead citrate. ×38,000 (B. Ghetti, unpublished micrograph).

Purkinje cells are essential to the continued survival of their presynaptic granule cells in the cerebellar cortex.

4.2. Basket cells

In the normal adult cerebellum, basket-cell axons form a plexus around the somata of Purkinje cells, precisely in the manner of "baskets" (Figure 7-13). When Purkinje cells degenerate in *pcd* mice, these conglomerates of basket-cell axons remain without their "contents" (Figure 7-14). In spite of the loss of their postsynaptic Purkinje cells, many of them survive initially. However, at later stages of the disease (Figure 7-15), basket-cell axons become reorganized in the atrophied cerebellar cortex, and their number appears reduced, implying that several of them undergo regressive changes or cellular death. The loss of baskets can be better estimated if one further takes into account the cerebellar atrophy: the *absolute* number of baskets in *pcd* mutants becomes even smaller when calculated per cerebellum rather than per unit of tissue (i.e., after correcting for the changes in tissue volume). As in the case of granule cells, these changes of basket cells are probably due to the primary loss of Purkinje cells.

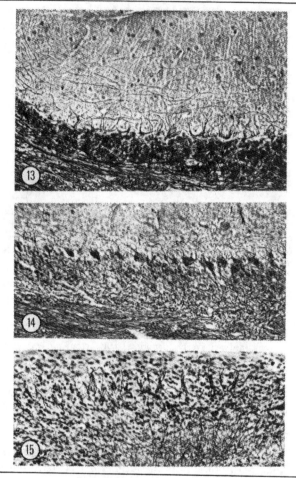

Figures 7-13 to 7-15. Demonstration of basket-cell axons in 6-month-old normal mouse cerebellum (Figure 7-13), in a 6-month-old *pcd* mutant (Figure 7-14) and in a 12-month-old mutant (Figure 7-15). In the normal cortex, basket axons surround Purkinje cell somata; in the 6-month-old *pcd* mutant, Purkinje cells have been lost and basket axons are empty of their contents; in the 12-month-old mutant, basket axons have reorganized their geometry and appear to be regressive. De Myer silver impregnation. ×180 (B. Ghetti and L.C. Triarhou, unpublished).

4.3. Inferior olivary complex

The loss of Purkinje cells in *pcd* mutant mice deprives inferior olivary neurons of their major postsynaptic target. The inferior olivary complex of *pcd* mutant mice at 17 days of age does not differ in cell number from control mice. However, inferior olivary neurons in 23-day-old mutants are 23%

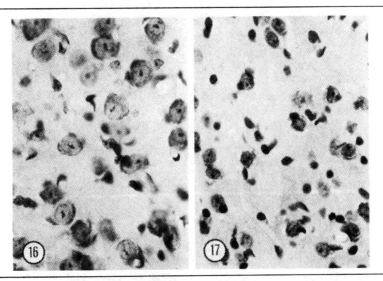

Figures 7-16 and 7-17. Histological appearance of inferior olivary neurons in 23- and 300-day-old *pcd* mutant, respectively. Note the substantial cell atrophy in the elder mutant. Ten-micrometer-thick paraffin sections, stained with gallocyanin. ×500 (From Ghetti et al. [18], with permission.)

fewer than in age-matched controls, and in 300-day-old mutants they are 48% fewer than in controls ($p < 0.001$ in both cases). The decline in the number of inferior olivary neurons in *pcd* mice between days 17 and 300 is 49% ($p < 0.0001$) [18]. Interestingly, the cell number in the inferior olive appears to be stabilized at that point, such that the amount of cell loss at 450 days of age is also about 50% [19]. The medial accessory olive appears to be less affected than the principal and dorsal accessory olives.

The atrophy of the inferior olive as a structure is accompanied by an atrophy of individual inferior olivary neurons as well (Figures 7-16 and 7-17). The mean neuronal diameter in control mice is 11.6 μm at 23 days and 10.8 μm at 300 days of age. The corresponding values in *pcd* mutants are 11.5 μm and 8.7 μm. Diameters in old mutants were significantly smaller than those in both age-matched controls and young mutants ($p < 0.001$). These findings suggested that in the mature olivocerebellar system the stability of inferior olivary neurons depends on the state of their postsynaptic Purkinje cells.

4.4. Noradrenaline system

The origin of the noradrenaline (NA) innervation of the cerebellar cortex lies in neurons of the dorsal part of the nucleus locus coeruleus and also in

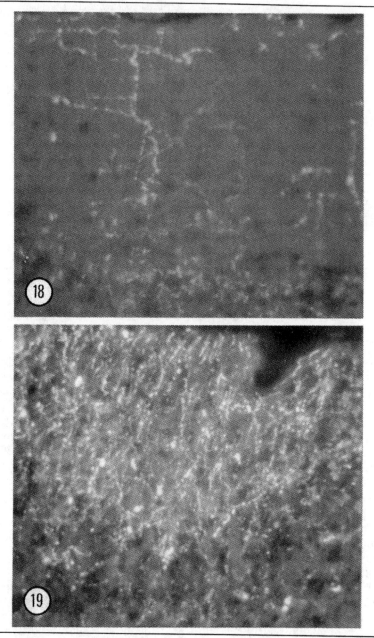

Figures 7-18 and 7-19. Noradrenaline histochemical fluorescence in normal (Figure 7-18) and *pcd* mutant cerebellum (Figure 7-19) at 6 months of age. Note the persistence of noradrenergic fibers in the mutant and the increase in fiber density that follows cerebellar cortical atrophy. Glyoxylic acid fluorescence histochemistry. ×240 (L.Y. Koda and B. Ghetti, unpublished).

neurons of the fields A5 and A7, and those of the nucleus subcoeruleus. Based on data from ultrastructural and physiological studies, it has been shown that Purkinje cells constitute a target for NA fibers. In the mouse, histofluorescent NA fibers form linear and tortuous profiles through the granule cell layer, form pericellular arrays alongside Purkinje cell somata, and branch into radially and longitudinally oriented chains of varicosities (Figure 7-18).

Purkinje cell degeneration mutant mice were examined during the course of Purkinje cell death and at 3, 5, 6, 9, and 12 months of age (L.Y. Koda and B. Ghetti, unpublished and ref. [20]). Glyoxylic acid fluorescence histochemistry for catecholamines was used to determine the modality of response of NA fibers to the profound structural alterations of the cerebellar cortex. In the mutants, a progressive increase in the density of NA varicosities accompanies the progressive shrinkage of the molecular layer (Figure 7-19); this is most conspicuous at 6–12 months of age, when the molecular layer is depleted both of Purkinje cell dendrites and parallel fibers. NA fibers in these zones form dense parallel bundles of varicose profiles; their density reached about 600% of normal at 9–12 months of age. The progressive increase in the density of NA varicosities in the molecular layer of mutant mice is most likely due to the altered geometry of the cerebellar cortex and not to newly sprouted fibers. It appears that the health of the environment surrounding the NA fibers in the cerebellar cortex has little influence on their anatomical integrity.

4.5. Serotonin system

Purkinje cells constitute one of the cerebellar cell types that are innervated by raphé serotonin (5-HT) neurons. Granule cells also constitute a major cell target. In normal mice, 5-HT immunoreactive fibers are distributed to all cerebellar lobules, with an anterior to posterior preference gradient. While all cortical layers receive 5-HT immunoreactive fibers, their density appears higher in the granule cell layer. Following the degeneration of Purkinje cells in *pcd* mutant mice, 5-HT-immunoreactive fibers survive and become compressed in a smaller volume of tissue, which results from cerebellar atrophy, thus giving the appearance of a higher fiber density than normal (Figures 7-20 and 7-21). Like noradrenergic fibers, it seems that the survival of 5-HT axons in the cerebellum is not influenced by the death of two major targets and the events causing Purkinje cell death (primarily) and granule cell death (secondarily).

4.6. Ultrastructure of monoamine-containing nerve terminals

Fixation of tissue with potassium permanganate is used to demonstrate small granular vesicles in monoamine (MA) nerve terminals, including both catecholamines and indoleamines (Figure 7-22). In control mice, MA nerve terminals are found mainly in apposition with Purkinje cell dendrites. One

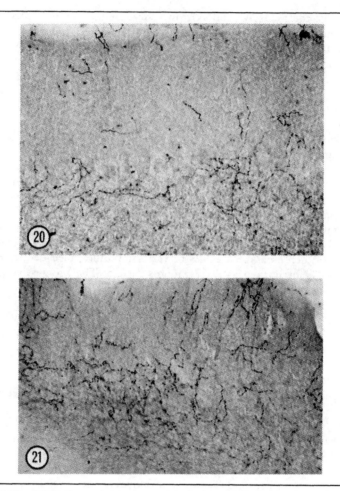

Figures 7-20 and 7-21. Serotonin immunoreactivity in the cerebellar cortex of normal (Figure 7-20) and *pcd* mutant mouse (Figure 7-21) at 6 months of age. The survival of serotoninergic axons does not appear to be affected by the death of Purkinje and granule cells. Anti-serotonin antiserum was a gift of Dr. H.W.M. Steinbusch. ×200 (L.C. Triarhou and B. Ghetti, unpublished).

MA nerve terminal is found in an area of molecular layer of approximately 4000 μm² or larger [21].

After the degeneration of Purkinje cells in *pcd* mutant mice, MA terminals can still be seen in the cerebellar cortex by electron microscopy, and with a higher incidence: at 45 days of age, the average incidence of MA nerve terminals in the molecular layer of the declive and tuber vermis is 1 per

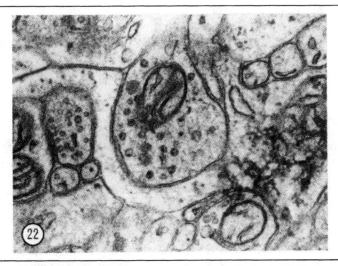

Figure 7-22. A monoaminergic nerve terminal containing small granular vesicles in the cerebellar molecular layer of a 1-year-old *pcd* mutant. Monoamine nerve terminals, which comprise those containing noradrenaline or serotonin, survive in the mutant long term, and their incidence per unit area increases due to their compression in the smaller, atrophic cortex. Potassium permanganate fixation. ×46,000 (L.C. Triarhou and B. Ghetti, unpublished micrograph).

1310 μm^2; at 3 months, 1 per 840 μm^2; at 6 months, 1 per 690 μm^2; at 9 months, 1 per 360 μm^2; and at 1 year of age, 1 per 290 μm^2 [21]. They are ensheathed by astroglial processes in most instances. They are also apposed to boutons that contain agranular vesicles and to stellate cells in the molecular layer. Clear synaptic specializations in the form of thickening of the synaptic membranes are not observed in either control or mutant mice.

It has been hypothesized that the survival of MA axons following the loss of their target cells can be attributed to the lack of an intimate adhesion to their target elements, to a possible functional interaction with the glia, or to the integrity of the extracerebellar terminal fields of the MA axon collaterals [21].

5. CHANGES IN POSTSYNAPTIC SYSTEMS

5.1. Deep cerebellar nuclei

The degeneration of Purkinje cells in *pcd* mutant mice results in a loss of presynaptic afferents to the deep cerebellar nuclei. In the mutants, the total volume of the deep cerebellar nuclei is reduced by 22% between 23 and 300 days of age. Neuronal populations in the deep cerebellar nuclei of control mice were 10,167 ± 949 at 23 days and 10,429 ± 728 at 300 days of age. The corresponding values in mutants were 9436 ± 1366 and 7424 ± 1324. There was a significant difference of 29% between 300-day-old mutants and age-matched controls ($p < 0.01$), and a 21% cell loss in 300-day-old mutants with respect to 23-day-old mutants ($p < 0.05$) [9].

Furthermore, neuronal diameters in control mice were 16.4 ± 0.72 μm (mean ±SD) at postnatal day 23 and 15.6 ± 0.64 μm at day 300. The corresponding values in *pcd* mutant mice were 15.7 ± 0.58 μm and 13.5 ± 0.24 μm. Diameters in 300-day-old mutants were significantly smaller than those in both age-matched controls and 23-day-old mutants ($p < 0.001$). These findings support the idea that the stability of deep cerebellar nuclei neurons in the mature cerebellum depends in part on the synaptic input from Purkinje cells.

6. NEUROCHEMICAL CHANGES

The degeneration of Purkinje cells induces several sets of neurochemical changes, namely, (1) loss of molecules contained in Purkinje cells, (2) modifications in glial markers, and (3) changes involving molecules contained in cells that degenerate or atrophy consequent to Purkinje cell loss.

6.1. GABA

All types of cerebellar cortical neurons use gamma-aminobutyrate (GABA) as a neurotransmitter, except granule cells, which use glutamate.

The concentration of GABA (nmol/mg wet weight) is normal in the cerebellar cortex and deep nuclei of *pcd* mutants before the onset of Purkinje cell degeneration. At 35 days of age, GABA concentration falls to 50% of normal values in the deep nuclei of *pcd*. The concentration in the cortical layers appears unchanged between 28 and 60 days of age [22].

Furthermore, the content of GABA (nmol/mg protein) was measured in the cerebellar vermis and hemispheres of normal and *pcd* mutant mice at 6, 9 and 12 months of age [23]. Relative to normal values, the GABA content in the vermis was 39% lower in *pcd* mutants at 6 months of age but did not differ from control values at 12 months. However, relative to data for normal mice, the GABA content in the hemispheres was consistently lower (20%) for all age groups. These results should be interpreted by keeping in mind that the value of mg protein decreases in *pcd* with the progression of cerebellar atrophy.

6.2. GABA and benzodiazepine receptors

Modifications in GABA receptors in *pcd* mutants have been monitored by examining muscimol and flunitrazepam binding. The GABA agonist muscimol binds to high-affinity GABA binding sites, located on the beta subunit of the $GABA_A$ receptor complex. Central benzodiazepine binding sites are located on the alpha subunit of the $GABA_A$ receptor-chloride channel macromolecular complex.

The distribution of cerebellar [^{3}H]muscimol binding sites has been studied in normal C57BL/6J mice and in pcd mutant mice by autoradiography [24]. In normal mouse cerebellum (80 days old), the highest concentration of [^{3}H]muscimol binding sites was observed in the granule cell layer. A much lower grain density was present over the Purkinje cell and molecular layers, and negligible numbers of binding sites were seen over the deep cerebellar nuclei and white matter. In 2-month-old pcd mutants, a 29% decrease in grain density over the granule cell layer was observed, while labeling was still present in the molecular layer. Neurochemical studies revealed a 50% decrease in benzodiazepine (BZ) receptors in 45-day-old pcd mutants after the degeneration of Purkinje cells. At 300 days there is an 80% decrease in BZ receptors, which occurs concomitantly with granule cell losses.

To determine the histological localization of these receptor changes, an in vitro autoradiographic technique was used to explore [H^3]flunitrazepam binding [25,26]. The highest concentration of [^{3}H]flunitrazepam binding sites was observed over the molecular layer. Intermediate grain density was present over the Purkinje cell layer and intermediate to high density was present over the deep cerebellar nuclei. Labeling over the granule cell layer was low. However, the difference in grain counts between control and mutant mice was not statistically significant in any of the layers.

When the appropriate mathematical correction factor was introduced for layer atrophy [27], there was a 60% decrease in grain counts in 45-day-old mutants in the molecular layer and a 84% decrease in 300-day-old mutants compared to controls. The initial decrease in total BZ receptors in 45-day-old mutant animals is associated with the selective loss of Purkinje cells. The amount of receptor binding that persists in 300-day-old mutants in the molecular layer would appear to reflect binding in the remaining parallel fibers of surviving granule cells. Because there is a significant decrease in granule cell layer width, especially in 300-day-old mutants, it has been proposed that other cerebellar elements present in this layer may contain BZ receptors. A substantial BZ binding in the deep cerebellar nuclei, both of 45-and 300-day-old animals, whether mutant or control, can be detected, and a substantial increase in grain density was observed over the deep cerebellar nuclei of pcd mutants when compared to littermate controls. These changes in grain density may be a manifestation of denervation supersensitivity subsequent to the loss of innervation by Purkinje cell axon terminals or to a reduction in the volume of the deep nuclei.

6.3. Amino acids

The content of glutamate (nmol/mg protein) was examined in the cerebellar vermis and hemispheres of normal and pcd mutant mice at 6, 9, and 12 months of age [23]. Relative to normal values, the glutamate content was approximately 50% lower in the vermis for the three age groups. In the hemispheres, glutamate content was also lower than control values and showed a progressive loss from 30% to 47% with age. The level of aspartate (the putative transmitter of olivocerebellar climbing fibers) was approximately 60% lower in the cerebellar vermis and 45–55% lower in the hemispheres of mutants with respect to controls for all age groups. Likewise, there was less alanine in the hemispheres (36–46%) and vermis (24%) in mutants relative to normal values at 6, 9, and 12 months of age. However, levels of glycine were 43–64% higher in the vermis and 77–100% greater in the hemispheres of mutants than in control mice. The highest values of glycine were observed at 9 and 12 months.

6.4. Noradrenaline

At 25–280 days of age, no significant changes in NA content (pmol/cerebellum) were detected during or after Purkinje cell degeneration [28]. However, since degeneration led to a reduction in cerebellar weight, NA concentration was increased in *pcd* mutants. These neurochemical data, in agreement with the fluorescence histochemical results, indicate that, in spite of the loss of a major postsynaptic target, the cerebellar NA input remains stable.

NA metabolism was also studied in brain regions of *pcd* mutants [29]. The purpose of those studies was to determine if cerebellar NA turnover is altered after Purkinje cell loss. The concentration of NA and of its major metabolite in mouse brain, 3-methoxy-4-hydroxy-phenylglycol (MHPG), were measured by liquid chromatography with electrochemical detection. The concentration of MHPG and the ratio of the concentrations of MHPG over NA (MHPG/NA) were taken as indices of NA turnover. Although the cerebellar NA content in *pcd* mice did not differ from controls, MHPG and the MHPG/NA ratio were slightly decreased in 3-month-old, and more decreased in 6- and 9-month-old mutants compared to controls. The MHPG content or the MHPG/NE ratio was not decreased in younger mutants, who were 22 or 45 days old. The accumulation of L-DOPA (3,4-dihydroxy-phenylalanine) after the administration of NSD 1015 to inhibit aromatic L-amino acid decarboxylase was measured as an index of NA synthesis. The accumulation of L-DOPA in *pcd* cerebellum was decreased to 69%, 53%, and 37% of control values at 3, 6, and 12 months, respectively; it did not change in the brain stem, while it was slightly decreased in the hypothalamus. MHPG content and concentration and the MHPG/NA ratio did not change in the brain stem. In the hypothalamus, the NA concentration and content were slightly increased in *pcd* mice at all ages, and the MHPG/NA ratio was decreased in 6- and 9-month-old-mice. In summary, although NA axons in the cerebellum are maintained in *pcd* mice after Purkinje cell degeneration, NA turnover is decreased, suggesting that the synthesis and release of neurotransmitter does not continue at normal rates at stages concomitant with the absence of target cells.

6.5. Serotonin

Both Purkinje and granule cells represent target cells for cerebellar 5HT axons originating in the raphé nuclei and other brain areas. Furthermore, both of these cell types degenerate in the *pcd* mutant: Purkinje cells as an effect of the mutation and granule cells probably trans-synaptically. Therefore, the content and turnover of 5HT were determined in the cerebellum of *pcd* mutant mice at 3–15 months of age. The serotonin content did not decrease in *pcd* mouse cerebellum but tended to increase slightly after 7 months [30]. The ratio of 5-hydroxyindoleacetic acid (5-HIAA) to 5HT was significantly decreased in cerebellum at 7–15 months, but not at 3 or 6 months. The decrease in this ratio is indicative of a decreased 5HT turnover. Similar changes were not seen in the brain stem or hypothalamus in mice up to 14 months of age, but slight decreases were observed at 15 months. Another index of turnover, the accumulation of 5-HIAA after the administration of probenecid to block its efflux from brain, was decreased by 46% in 7-month-old *pcd* mice in the cerebellum, but not in the brain stem or hypothalamus. The decrease in 5HT turnover in the cerebellum of *pcd* mutant mice occurs subsequent to and perhaps due to the loss of the target Purkinje and granule cells.

6.6. Second messengers

Experiments carried out on tissue slices of *pcd* cerebellum showed that NA causes an accumulation of cyclic AMP in the cerebellum of *pcd* that is far greater than that in normal mice [11,31]. The greatest elevation in cyclic AMP (300%) occurred between 30 and 128 days of age, which is the time period when a severe astrocytic response is found throughout the cerebellar cortex. NA elicited a smaller increase in cyclic AMP from 155-day-old mice than at earlier ages, and the response continued to decrease with age; at 270 days, an equal accumulation, and at 365 days, a lower accumulation, of cyclic AMP was detected in *pcd* cerebella. During this time, the Purkinje cell debris had been removed, the granule cell layer was depleted of granule cells, and the molecular layer was deprived of a large number of parallel fibers. However, a large number of astrocytic processes was still seen in the neuropil, even after phagocytosis of neuronal debris was completed.

Biochemical experiments in vitro established that the exaggerated accumulation of cyclic AMP in the presence of NA was not due to lower catabolism of cyclic AMP, a synergistic interaction with adenosine, or lower protein in the *pcd* cerebellum. The significance of the exaggerated accumulation of cyclic AMP in cerebella undergoing degeneration is unknown. The correlates of heightened NA-stimulated accumulation of cyclic AMP with neuronal loss and the glial cell reaction might indicate that cyclic nucleotides might play a role in controlling glial cell functions, such as proliferation, migration, and phagocytosis.

6.7. Glycolipids

The distribution of cerebellar gangliosides has been studied in *pcd* mutant mice at postnatal days 25, 30, 50, and 150. Purkinje cell loss is associated with significant reductions in cerebellar weight and ganglioside concentration. The neuronal loss is also correlated with reductions of gangliosides GT1a/LD1 and GT1b, and with elevations of GD3. It appears that GTa/LD1 and GT1b are concentrated in Purkinje cells and that GD3 is enriched in reactive glial cells. A slight, but significant, reduction in GD1a concentration occurs only in older *pcd* mice, consistent with the idea that GD1a is enriched in mature granule cells [32].

A significant reduction in the content of two members of the sulfoglucuronyl-neolacto-series of glycolipids (SGGLs), 3-sulfoglucuronyl-lacto-N-neotetraosylceramide (SGGL-1), and 3-sulfoglucuronyl lacto-N-norhexasoylceramide (SGGL-2), in the cerebellum of the *pcd* has been detected. The expression of SGGLs was studied during the development of the *pcd* mutant cerebellum, and it was shown that the rate of decline in the level of SGGLs practically coincided with the loss of Purkinje cell perikarya, indicating that SGGLs are primarily localized in Purkinje cells and that initially, at least, there is no genetic defect in the biosynthesis of SGGLs in this mutant [33]. The precursor of SGGLs, lacto-N-neotetraosylceramide (paragloboside) and lacto-N-norhexaosylceramide, as well as other glycolipids derived from these precursors, such as X-determinant fucoglycolipids and disialosyl-lacto-N-neotetraosylceramide, are also present in normal cerebellum. Levels of paragloboside and its other derivatives, similar to SGGLs, are also significantly reduced in mutants with Purkinje cell abnormalities, including *pcd*, indicating that the entire paragloboside family of glycolipids is primarily associated with Purkinje cells in the cerebellum. Although levels of monoclonal antibody HNK-1-reactive glycolipids are reduced in the *pcd* mutant, HNK-1 reactive glycoproteins are not affected.

7. CONCLUDING REMARKS

As several neuronal populations degenerate in the CNS of *pcd* mutants, it is not always easy to distinguish between genetically determined and trans-synaptic degeneration. From a quantitative viewpoint, virtually all of the Purkinje cells, which are thought to be targets of the genetic mutation, degenerate over a short period of time. In contrast, only one fifth of the neurons in the deep cerebellar nuclei, one half of the neurons in the inferior olivary complex, and about nine tenths of granule cells are lost over much more protracted time periods. The severity and pattern of the degeneration, along with a comparative analysis of data from other cerebellar mutants, support the notion that those latter changes are most likely to be secondary. In all, it appears that secondary losses in *pcd* mutants proceed at a slower rate than losses that are under primary genetic control.

The reasons why secondary neuronal loss is severest in the granule cell layer and in the inferior olivary complex and why parallel and climbing fibers are more vulnerable to Purkinje cell loss than MA fibers are not fully understood.

The persistence of MA axons in *pcd* mice markedly contrasts with the loss of other types of axons presynaptic to Purkinje cells; this could be due to the different anatomical and functional properties of those systems, as opposed to the other types of neurons that are presynaptic to Purkinje cells. It is known that axon terminals and nerve cells that lose their postsynaptic targets undergo retrograde trans-synaptic degeneration in several anatomical systems. Thus, extensive loss of parallel fibers and granule cells is observed in *pcd* mutants, along with a marked fall in glutamate, their putative neurotransmitter. Similarly, the loss of inferior olivary neurons and their climbing fibers accounts, in all likelihood, for the drop in aspartate, the putative neurotransmitter of olivocerebellar climbing fibers.

Several possibilities can be considered in explaining the mechanisms of trans-synaptic cell losses in the *pcd* mutant cerebellum. As already stated, the different structural properties of the synaptic relationships of parallel or climbing-fiber endings with Purkinje cells, as opposed to those between MA axon terminals and Purkinje cells, may play a substantial role. As an alternative supposition, one may consider whether the excitatory amino-acid neurotransmitters of granule cells and inferior olivary neurons could become cytotoxic when they cannot be utilized properly by postsynaptic Purkinje cells. A final consideration could be a loss of trophic factors associated with Purkinje cell degeneration.

In all, the *pcd* mutant mouse provides a useful model for understanding the pathogenetic mechanisms underlying cerebello-olivary degeneration. There are extensive similarities between the degenerative events caused by that mutation and one familial form of human cerebellocortical degeneration, originally described by Holmes [34]. That entity is histopathologically characterized by a marked atrophy of the cerebellum, an almost complete disappearance of Purkinje cells, and an atrophy of the inferior olivary complex in association with the loss of olivocerebellar fibers. The olivary degeneration has been considered to be secondary to the atrophy of the cerebellar cortex [34,35]. Furthermore, neuronal depletion of the cerebellar dentate nucleus in humans associated with Purkinje cell loss is seen in cerebello-olivary atrophy of Holmes [36].

REFERENCES

1. Peduzzi J.D., Crossland W.J. (1983). Anterograde transneuronal degeneration in the ectomamillary nucleus and ventral lateral geniculate nucleus of the chick. J. Comp. Neurol. 213:287–300.
2. Larramendi L.M.H. (1969). Analysis of synaptogenesis in the cerebellum of the mouse. In: R. Llinás (ed): Neurobiology of Cerebellar Evolution and Development. Chicago: Institute for Biomedical Research AMA/ERF, pp. 803–843.

3. Mullen R.J., Eicher E.M., Sidman R.L. (1976). Purkinje cell degeneration, a new neurological mutation in the mouse. Proc. Natl. Acad. Sci. USA 73:208–212.
4. Landis S.C., Mullen R.J. (1978). The development and degeneration of Purkinje cells in *pcd* mutant mice. J. Comp. Neurol. 177:125–144.
5. Green M.C. (1981). Genetic Strains and Variants of the Laboratory Mouse. Stuttgart: Gustav-Fischer-Verlag.
6. Wassef M., Sotelo C., Cholley B., Brehier A., Thomasset M. (1987). Cerebellar mutations affecting the postnatal survival of Purkinje cells in the mouse disclose a longitudinal pattern of differentially sensitive cells. Dev. Biol. 124:379–389.
7. Chang A.C., Triarhou L.C., Alyea C.J., Low W.C., Ghetti B. (1989). Developmental expression of polypeptide PEP-19 in cerebellar cell suspensions transplanted into the cerebellum of *pcd* mutant mice. Exp. Brain Res. 76:639–645.
8. Wassef M., Simons J., Tappaz M.L., Sotelo C. (1986). Non-Purkinje cell GABAergic innervation of the deep cerebellar nuclei: A quantitative immunocytochemical study in C57BL and in Purkinje cell degeneration mutant mice. Brain Res. 399:125–135.
9. Triarhou L.C., Norton J., Ghetti B. (1987). Anterograde transsynaptic degeneration in the deep cerebellar nuclei of Purkinje cell degeneration (*pcd*) mutant mice. Exp. Brain Res. 66:577–588.
10. Sotelo C., Alvarado-Mallart R.M. (1988). Integration of grafted Purkinje cells into the host cerebellar circuitry in Purkinje cell degeneration mutant mouse. Prog. Brain Res. 78:141–154.
11. Ghetti B., Truex L., Sawyer B., Strada S., Schmidt M.J. (1981). Exaggerated cyclic AMP accumulation glial cell reaction in the cerebellum during Purkinje cell degeneration in *pcd* mutant mice. J. Neurosci. Res. 6:789–801.
12. Schwartz J.P., Ghetti B., Truex L., Schmidt M.J. (1982). Increase of a nerve growth factor-like protein in the cerebellum of *pcd* mutant mice. J. Neurosci. Res. 8:205–211.
13. Ghetti B., Alyea C.J., Muller J. (1978). Studies on the Purkinje cell degeneration (*pcd*) mutant: Primary pathology and transneuronal changes. J. Neuropathol. Exp. Neurol. 37:617.
14. Triarhou L.C., Norton J., Alyea C., Ghetti B. (1985). A quantitative study of the granule cells in the Purkinje cell degeneration (*pcd*) mutant. Ann. Neurol. 18:146.
15. Sotelo C., Changeux J.-P. (1974). Transsynaptic degeneration "en cascade" in the cerebellar cortex of staggerer mutant mice. Brain Res. 67:519–526.
16. Sotelo C., Triller A. (1979) Fate of presynaptic afferents to Purkinje cells in the adult nervous mutant mouse: A model to study presynaptic stabilization. Brain Res. 175:11–36.
17. Caddy K.W.T., Biscoe T.J. (1979). Structural and quantitative studies in the normal C3H and Lurcher mutant mouse. Phil. Trans. R. Soc. Lond. (Biol) 287:167–201.
18. Ghetti B., Norton J., Triarhou L.C. (1987). Nerve cell atrophy and loss in the inferior olivary complex of "Purkinje cell degeneration" mutant mice. J. Comp. Neurol. 260:409–422.
19. Triarhou L.C., Ghetti B. (1988). Stabilisation of neurone number in the inferior olivary complex of aged *pcd* mutant mice. Eur. J. Neurosci. (Suppl. 1):25.
20. Felten D.L., Felten S.Y., Perry K.W., Fuller R.W., Nurnberger J.I. Sr., Ghetti B. (1986). Noradrenergic innervation of the cerebellar cortex in normal and in Purkinje cell degeneration mutant mice: Evidence for long-term survival following loss of the two major cerebellar cortical neuronal populations. Neuroscience 18:783–793.
21. Triarhou L.C., Ghetti B. (1986). Monoaminergic nerve terminals in the cerebellar cortex of Purkinje cell degeneration mutant mice: Fine structural integrity and modification of cellular environs following loss of Purkinje and granule cells. Neuroscience 18:795–807.
22. Roffler-Tarlov S., Beart P.M., O'Gorman S., Sidman R.L. (1979). Neurochemical and morphological consequences of axon terminal degeneration in cerebellar deep nuclei of mice with inherited Purkinje cell degeneration. Brain Res. 168:75–95.
23. McBride W.J., Ghetti B. (1988). Changes in the contents of glutamate and GABA in the cerebellar vermis and hemispheres of the Purkinje cell degeneration (*pcd*) mutant. Neurochem. Res. 13:121–125.
24. Rotter A., Gorenstein C., Frostholm A. (1988). The localization of $GABA_A$ receptors in mice with mutations affecting the structure and connectivity of the cerebellum. Brain Res. 439:236–248.

25. Rotter A., Frostholm A. (1986). Cerebellar benzodiazepine receptor distribution: An autoradiographic study of the normal C57BL/6J and Purkinje cell degeneration mutant mouse. Neurosci. Lett. 71:66–71.
26. Rotter A., Frostholm A. (1988). Cerebellar benzodiazepine receptors: Cellular localization and consequences of neurological mutations in mice. Brain Res. 444:133–146.
27. Vaccarino F.M., Ghetti B., Nurnberger J.I. Sr. (1985). Residual benzodiazepine (BZ) binding in the cortex of *pcd* mutant cerebella and qualitative BZ binding in the deep cerebellar nuclei of control and mutant mice: An autoradiographic study. Brain Res. 343:70–78.
28. Ghetti B., Fuller R.W., Sawyer B.D., Hemrick-Luecke S.K., Schmidt M.J. (1981) Purkinje cell loss and the noradrenergic system in the cerebellum of *pcd* mutant mice. Brain Res. Bull. 7:711–714.
29. Ghetti B., Perry K.W., Fuller R.W. (1987). Norepinephrine metabolism in the cerebellum of the Purkinje cell degeneration (*pcd*) mutant mouse. Neurochem. Int. 10:39–47.
30. Ghetti B., Perry K.W., Fuller R.W. (1988). Serotonin concentration and turnover in cerebellum and other brain regions of *pcd* mutant mice. Brain Res. 458:367–371.
31. Schmidt M.J., Ghetti B. (1980). Exaggerated norepinephrine-stimulated accumulation of cyclic AMP *in vitro* in cerebellar slices from *pcd* mutant mice following Purkinje cell loss. J. Neur. Transmiss. 48:49–56.
32. Seyfried T.N., Yu R.K. (1990). Cerebellar gangioside abnormalities in *pcd* mutant mice. J. Neurosci. Res. 26:105–111.
33. Chou D.K.H., Flores S., Jungalwala F.B. (1990). Loss of sulfoglucuronyl and other neolactoglycolipids in Purkinje cell abnormality murine mutants. J. Neurochem. 54:1589–1597.
34. Holmes G. (1907). A form of familial degeneration of the cerebellum. Brain 30:467–489.
35. Holmes G., Stewart T.G. (1908). On the connection of the inferior olives with the cerebellum in man. Brain 31:125–137.
36. Schoene W.C. (1985). Degenerative diseases of the central nervous system. In: Davis R.L., Robertson D.M. (eds). Textbook of Neuropathology. Baltimore: Williams and Wilkins, pp. 788–823.

II. CLINICAL NEUROSCIENCES OF THE CEREBELLUM

8. CLASSIFICATION AND EPIDEMIOLOGY OF CEREBELLAR DEGENERATIONS

ANDREAS PLAITAKIS

1. DEFINITIONS

The cerebellar degenerations encompass a large number of heterogenous neurological disorders that are characterized pathologically by degeneration and atrophy of the cerebellum and its connections and clinically by disturbances of motor coordination or *ataxia*. Because involvement of the spinal cord may also occur in these cerebellar disorders, the term *spinocerebellar degeneration* has often been used interchangeable with the above broader term *cerebellar degeneration*. In other types, cerebellar atrophy can be found in combination with atrophic lesions of the basal ganglia, brain stem, spinal cord, and/or cerebral cortex. Such disorders may be best categorized under *multiple system atrophy*, although some authors will use this term only when they refer to disorders with coexisting progressive autonomic failure.

Another unsettled issue concerns the nosologic relationship between the above-defined cerebellar degenerations and a wide variety of metabolic encephalopathies in which ataxia is a clinical feature. In contrast to the classic cerebellar degenerations, which are characterized by localized CNS atrophic changes, most of these encephalopathies show widespread pathologic changes of the nervous system. However, as described in Chapter 13, a notable exception is the ataxic form of hexosaminidase deficiency, which can be associated with cerebellar atrophic lesions, as shown by neuroimaging studies, which are indistinguishable from those seen in patients with classic cerebellar degenerations.

Because the majority of the cerebellar degenerative disorders are genetically transmitted, they are often called *hereditary ataxias*. Many of these ataxias are known as *eponymic syndromes*, usually named after the authors who first described them. Some of these, such as the Friedreich's ataxia, proved to be homogeneous by clinical, genetic, and pathologic criteria, and their eponymic definition has endured the test of time. Others, however, such as Marie's ataxia, turned out to be quite heterogeneous. Hence, their eponymic definition proved not to be useful in the study of these disorders and has largely been abandoned by most authors.

2. CLASSIFICATION

The nosology of cerebellar degenerative disorders has been studied extensively over the past century, during which time many classification schemes have been proposed. As with the other human neurodegenerations of unknown etiology, the classification of cerebellar ataxias has been based on clinical, pathologic, and genetic criteria. However, because these features can vary considerably, even within members of the same family, these studies are subject to substantial limitations, and as such, they should be considered as provisional until the gene-dependent fundamental biochemical defects of these disorders are determined [1]. Given the rapid progress made recently in the molecular genetics of inherited human disorders, this goal is expected to be achieved in the not too distant future.

Meanwhile, a working classification is needed both for the clinicians who care for ataxia patients as well as for the clinical and the basic investigators who study these disorders. The clinician, by narrowing the diagnosis to a particular subtype of cerebellar degeneration, will often be able to predict its evolution and to provide genetic counseling to the patient and his or her family. Such information could help the patient in the planning of his or her family, and generally in making long-term life plans. With respect to the study of the inherited ataxias, there is currently a major effort toward characterizing these disorders at the molecular level using the so-called "reverse-genetics" approach. The initial step in these investigations is to establish a linkage between the presence of the disease and of DNA markers of known chromosomal localization. Therefore, for such an approach to be successful a precise definition of the clinical disorder is a necessary prerequisite.

We have previously proposed a classification scheme [2] that was based on conventional clinicopathological concepts [1–7] and on our own experience with these disorders. In this classification, we have, in addition, used data from neuroimaging investigations [7], which permit the study of brain morphological changes in living patients. This has allowed clinicopathological correlations during the various phases of disease evolution, rather

Table 8-1. Classification of cerebellar degenerations

A. Childhood or juvenile onset
   I. Cerebellar disorders with areflexia or hyporeflexia
      a. Friedreich's ataxia
      b. Ataxia-telengiectasia
      c. Ataxia-oculomotor apraxia
      d. Xeroderma pigmentosum
      d. Dissynergia cerebellaris myclonica
      e. Degeneresce systematisee optico-cochleo-dentelee
      f. Refsum's disease
      g. Abetalipoproteinemia
      h. Vitamin E deficiency states
   II. Cerebellar disorders with normal or increased reflexes
      a. Olivo-ponto-cerebellar atrophy
         1. Autosomal dominant
         2. Autosomal recessive
         3. X-linked
      b. Marinesco-sjogren Syndrome
      c. Hexosaminidase deficiency
      d. Adrenoleukodystrophy
B. Adult onset
   I. Cerebello-olivary atrophy or cortical cerebellar degeneration
      a. Autosomal dominant
      b. Autosomal recessive (Holmes)
      c. Sporadic
   II. Olivopontocerebellar atrophy
      a. Autosomal dominant
         1. Linked to human chromosome 6p
             Menzel's type
         2. Not-linked to human chromosome 6p
             with slowed saccades (Wadia's type)
             with retinal degeneration and other features
             Spinopontine variety
             Variants with deafness or other features
      b. Autosomal recessive
      c. X-linked
      d. Sporadic
         1. With autonomic failure
         2. Without autonomic failure

than at the end stages seen at autopsy. Here a revised version of this classification is presented (Table 8-1). It has been expanded to include rare and obscure entities, which, though first described decades ago, still attract interest, as evidenced by recent publications related to these subjects. The disorders shown in Table 8-1 primarily represent the classic, idiopathic cerebellar degenerations, which, as already mentioned, are characterized by localized CNS atrophic changes and for which no specific therapy is currently available. On the other hand, when ataxia occurs as a clinical feature of disorders with known etiology, it is often amenable to specific treatments. These treatable ataxic encephalopathies are shown in Table 8-2.

Because afflictions that start in childhood can differ substantially from those that begin in later life, the various cerebellar degenerations were

Table 8–2. Treatable ataxic syndromes*

Clinical syndrome	Laboratory tests	Treatment
Ataxic encephalopathies caused by specific metabolic disorders		
Amino acid metabolism		
Hartnup disease	Neutral monoamino carboxylic amino acids and indole products in the urine	Nicotinamide, protein supplement
Maple syrup urine disease	Branched-chain amino acids and branched-chain keto acids in blood and urine	Dietary, thiamine, peritoneal dialysis
Urea cycle enzyme deficiencies	Blood ammonia, urea cycle substrates in blood and urine, organic acids in urine, enzyme assays	Dietary, sodium benzoate, sodium phenylacetate
Lipid metabolism		
Abetalipoproteinemia and hypobetalipoproteinemia	Cholesterol, triglycerides, lipoproteins, vitamin E in plasma, acanthocytosis	Dietary, vitamins A, E, K
Refsum's disease	Phytanic acid in plasma	Dietary, plasmapheresis
Metal metabolism		
Wilson's disease	Serum ceruloplasmin; copper in serum, urine, and liver; incorporation of radioactive copper into ceruloplasmin	Penicillamine, triethylene-tetramide hydrochloride
Ataxic syndrome associated with dysproteinemia	Plasma protein electrophoresis, immunoelectrophoresis	Plasmapheresis, chemotherapy
Ataxic syndromes caused by vitamin deficiencies		
Alpha-tocopherol	Vitamin E in plasma	Vitamin E
Biotin-responsive encephalopathy	Biotin in plasma, organic acids in urine, enzyme assays	Biotin
Thiamine-deficiency encephalopathy	Thiamine in blood, TDP effect on red cell transketolase	Thiamine
Cerebellar dysfunction in endocrinopathies		
Myxedema-ataxia	Thyroid function studies	Thyroid replacement
Cerebellar dysfunction in posterior fossa lesions or cerebellar anomalies		
Arnold-Chiari malformation, neoplasms, abscesses, or vascular malformations	Angiograms, CT scan, MRI, skull and spine x-rays	Operation, radiotherapy
Drug or toxin-induced cerebellar ataxia		
Anticonvulsants (diphenylhydantoin, phenobarbital, carbamazepine)	Serum anticonvulsant levels	Adjustment of anticonvulsant dosage
Anticholinergic agents	Careful history of drug intake	Discontinuation of offending agent
Metals		
Lead	Lead in blood and urine	Chelating agents, hemodialysis
Thallium	Thallium in urine	
Lithium	Lithium in blood	
Familial paroxysmal ataxia	None	Acetazolamide
Gamma-glutamylcysteine synthase deficiency	Amino acids in urine, enzyme assays in liver tissue	Dietary

Modified from [5].

divided into two major groups: one with onset in childhood and another with onset in adulthood. Moreover, for reasons explained below, the childhood-onset ataxias were further subdivided into those characterized by areflexia and those with preserved or enhanced deep tendon reflexes.

3. CHILDHOOD-ONSET ATAXIA WITH AREFLEXIA OR HYPOREFLEXIA

Friedreich's ataxia is the most homogeneous and one of the rather commonly encountered spinocerebellar degenerations of childhood that is recessively inherited. The disorder usually starts at around puberty and shows a rather characteristic constellation of clinical symptoms and signs, which permit the diagnosis to be made by clinical criteria alone [6]. In the overwhelming majority of the cases, the presenting symptom is progressive ataxia, with gait being predominantly affected, as well as dysarthria, areflexia, proprioceptive sensory loss, corticospinal deficits (muscle weakness, extensor plantar responses), skeletal deformities (kyphoscoliosis, pes cavus). The pathological changes of FA are largely limited to the spinal cord [8] and, for this reason, the disease is considered to be the prototype of **spinal ataxia**. Neuroimaging studies have, accordingly, shown atrophy of the spinal cord with minimal cerebellar changes (see Chapter 13).

Other forms of childhood-onset ataxia with hyporeflexia or areflexia are relatively rare and include a variety of systemic disorders associated with specific metabolic, immunologic, and/or DNA repair defects, such as ataxia-telengiectasia, xeroderma pigmentosum, abetalipoproteinemia, Refsum's disease, and vitamin E deficiency states. Also, certain obscure idiopathic degenerations, such as dyssynergia cerebellaris myoclonica, dégénérescence systématisée optico-cochléo-dentelée, and ataxia-oculomotor apraxia, a disorder clinically mimicking ataxia telengiectasia, can often be associated with depressed or absent deep-tendon reflexes. In spite of their areflexia or hyporeflexia, these disorders can be distinguished from Friedreich's disease by their specific clinical and laboratory features, which are described below.

Ataxia-telengiectasia, first described by Louis Bar [9], is usually manifested in early childhood by progressive ataxia of gait, titubation, choreoathetosis, dysarthria, decreased facial expression, and a variety of eye-movement abnormalities, termed *oculomotor apraxia* [10,11]. In advanced stages, muscle weakness, areflexia, and proprioceptive sensory loss may occur. The characteristic feature of the disease is the progressive development of conjuctival, oral, and cutaneous telengiectasias, with the latter found primarily in the face, ear lobes, knees, and elbows [10,11]. The disease is recessively inherited and is characterized by a variable degree of immunodeficiency and an inability to repair radiation-induced DNA damage (see Chapter 19). Also, increased levels of a-fetoprotein and carcinoembryonic antigen have been detected in such patients [11]. Recently, Gatti et al. [12] performed a genetic linkage analysis of 31 families affected by ataxia-telengiectasia and succeeded in mapping the disease's gene locus to human chromosome 11 (region q22–23).

Aicardi et al. [13] recently described 14 patients affected by a slowly progressive syndrome characterized by ataxia, choreoathetosis, and oculomotor apraxia, similar to that of ataxia telengiectasia, but these patients did not show any of the systemic abnormalities of the disorder. Hence, this disorder seems to be distirct from ataxia-telengiectasia.

Xeroderma pigmentosum is another rare disorder, some forms of which can be associated with defective DNA repair mechanisms and neurologic involvement [14,15]. The latter includes progressive cerebellar ataxia, motor and sensory neuropathy, dysarthria, choreoathetosis, pyramidal weakness, proprioceptive sensory loss, and areflexia. Pathological studies have revealed atrophic changes in the cerebellum, brain stem, susbtantia nigra, and spinal cord (posterior and lateral columns). The characteristic feature of the disease is a marked sensitivity to ultraviolet radiation, leading to the development of skin lesions, such as sunlight burns, freckles, telengiectasias, keratoses, and a variety of skin and other malignancies [14].

Abetalipoproteinemia is a recessively inherited disorder that is characterized by early-onset steatorrhea, retinitis pigmentosa, progressive cerebellar ataxia, areflexia, and proprioceptrive sensory loss [16,17]. Some patients may also show skeletal deformities similar to those seen in FA. Acanthocytosis is present in the majority of these patients. The concentration of cholesterol in the plasma is markedly reduced due to a virtual absence of apoprotein B, a component of low-density lipoproteins (LDL), very low-density lipoproteins, and chylomicrons. Because of this defect, the absorption and transport of fat-soluble vitamins (A, D, E, and K) are markedly compromised, and these changes appear to be implicated in the pathogenesis of the neurologic complications of the illness. Indeed, the administration of large doses of vitamin E, if given in early stages, can prevent or ameliorate the neurologic manifestations of the disease [18]. Morphological abnormalties of hepatic peroxisomes have been recently described in a 17-year-old boy with spinocerebellar degeneration associated with abetalipoproteinemia and vitamin E deficiency [19] and, as such, the pathogenesis of this disorder may be linked with that of the peroxisomal afflictions (see below).

Other disorders causing steatorrhea, such as chronic liver disease, small-bowel resection, blind loop, intestinal lymphagiectasis, or transport defects that are specific for vitamin E [20], can also lead to a deficiency of this vitamin, with resultant progressive neurologic dysfunction [2]. Clinical features of vitamin E deficiency include ataxia, retinitis pigmentosa, dysarthria, oculomotor disturbances, areflexia, proprioceptive sensory loss, and postive Babinski signs. The diagnosis can be established by documenting deficient levels of vitamin E in the serum or plasma of these patients.

Refsum's disease is a recessively inherited metabolic disorder that usually starts before age 20 [21], although late adult onset can also occur. There is a characteristic constellation of clinical symptoms and signs, which include retinitis pigmentosa, motor and sensory neuropathy with diminished or absent deep-tendon reflexes, progressive cerebellar ataxia, nerve deafness, skin icthyosis, and skeletal abnormalities. Elevation of the spinal fluid protein without pleocytosis is a consistent finding. The diagnosis is confirmed by demonstrating a marked increase in the plasma levels of phytanic acid [25].

Dyssynergia cerebellaris myoclonica was first described by Rumsay Hunt [22], who considered it to represent a primary atrophy of the dentate nucleus. Clinical manifestations include ataxia, hyporeflexia or areflexia, proprioceptive sensory loss, and myoclonic epilepsy. The genetic transmission is autosomal recessive. The presence of myoclonus is the main feature that distinguishes this from other forms of ataxia, although it may develop several years after the onset of cerebellar dysfunction [22]. Some authors have suggested [23] that the term *myoclonic encephalopathy* be applied to this disorder, as well as to the progressive myoclonic epilepsy of Unverricht-Lundborg [24], since both are characterized by myoclonus, ataxia, and system degeneration.

The combination, however, of cerebellar ataxia and myoclonus is by no means specific for these syndromes, since it can occur in a variety of metabolic encephalopathies [25–27] and in certain forms of idiopathic cerebellar degeneration, such as **olivopontocerebellar atrophy** (OPCA) (see below) and a rare disorder known by its descriptive name, **dégénérescence systématisée optico-cochléo-dentelée** [28,29]. The latter disorder is characterized pathological by atrophy of the optic and auditory pathways, cerebellum (dentate nucleus), medial lemniscus, basal ganglia, thalamus, oculomotor nuclei, and spinal cord. Clinically, there is blindness, deafness, psychomotor retardation, tremor, cerebellar dysfunction, corticospinal deficits, and myoclonus starting in infancy and early childhood. Finally, patients with Jakob-Kreutzfeld disease can, sometimes, present with progressive ataxia and myoclonic jerks, thus masquerading as cerebellar degeneration (Table 8-3). The often coexisting progressive dementia should alert the clinician to the possibility of this disorder, although ataxia associated with dementia can also occur in some forms of OPCA and in the Gerstmann-Sträussler syndrome [30].

4. CHILDHOOD-ONSET ATAXIA WITH ACTIVE OR INCREASED REFLEXES

Ataxic disorders with normal or enhanced reflexes that start in childhood are clearly heterogenous, and most of them remain poorly understood. The majority of these patients suffer from idiopathic forms of cerebellar degeneration, while in others ataxia may be a feature of a metabolic encephalopathy, such as hexosaminidase deficiency or adrenoleukodystrophy.

Regarding the idiopathic cases, it has been proposed [6] that early-onset ataxia with retained tendon reflexes represents an entity among those of non-Friedreich's childhood disorders, but this has not been defined pathologically.

On the other hand, some patients with childhood-onset cerebellar degeneration and active deep-tendon reflexes have been found pathologically to have olivopontocerebellar atrophy (OPCA) [31–33]. Although this is traditionally regarded as an adult-onset disease, there is increasing awareness that OPCA can also occur in infancy or childhood. The genetic transmission may be autosomal dominant [31], recessive [32], or X-linked [33]. Clinical features include cerebellar ataxia, corticospinal deficits, oculomotor disturbances, bulbar dysfunction (dysarthria and dysphagia), extrapyramidal manifestations, and peripheral neuropathy occurring in various combinations. Also, variants showing retinal degeneration, optic atrophy, mental retardation, peripheral neuropathy, deafness, and myoclonus can occur in childhood [31]. Brain CT scans and MRI may be of essential help in establishing the diagnosis in living patients by revealing atrophy of the cerebellum and brain stem (for details see Chapter 13).

Marinesco-Sjogren syndrome is another rare form of childhood-onset cerebellar degeneration with preserved reflexes. This recessively inherited neurological disorder is characterized by early-onset progressive cerebellar ataxia, gaze-evoked nystagmus, dysarthria, mental retardation, myopathy, and cataracts, with the latter often found in infancy or early childhood [1,34]. Skeletal anomalies, such as microcephaly, brachydactyly, and kypkoscoliosis, are common [1]. Deep-tendon reflexes are often increased or are normal and the Babinski sign is positive.

Of the metabolic disorders, **hexosaminidase deficiency** has been recognized during the past 15 years as a cause of progressive ataxia with onset in childhood, adolescence, or even adulthood [35–37]. In addition to cerebellar dysfunction, most patients usually show normal or increased reflexes, anterior horn cell signs (muscle weakness, amyotrophy, and fasciculations), oculomotor disturbances, and mental changes (progressive dementia or recurrent psychosis). Neuroimaging studies may show a characteristic pattern of cerebellar atrophy (see Chapter 13).

Adrenoleukodystrophy is best known as an X-linked disorder that affects boys in early childhood, causing a progressive dementing encephalopathy, with the white matter being predominantly involved. In addition to the classic form of the disease, a few cases have been reported in the literature that showed a predominant ataxia, thus masquerading as spinocerebellar degeneration [38,39]. Moreover, female heterozygotes can become symptomatic, usually in their adult life, and may show progressive spinal cord dysfunction and ataxia [46]. MRI of the brain may reveal white matter changes, including the degeneration of certain neuronal pathways [41]. The disease has been shown to result from peroxisomal dysfunction and is associated with increased plasma and tissue levels of the very long-chain fatty acids, which can be used as a specific diagnostic test for these disorders [40].

5. ADULT-ONSET ATAXIA

The majority of patients with progressive ataxic syndromes that start in adult life will have pathologically either cortical cerebellar degeneration or olivopontocerebellar atrophy. Although these may be distinguished by certain clinical criteria, this is often difficult. The advent, however, of the CT scan and MRI, which allow the study of the living pathology of the CNS structures, have permitted the use morphological criteria in the differential diagnosis of these disorders, as described in detail in Chapter 13.

5.1. Cerebello-olivary atrophy or cortical cerebellar degeneration

This disorder is characterized pathologically by atrophy of the cerebellar cortex and inferior olives, and clinically by rather pure cerebellar ataxia [42,43]. Neurologic manifestations are those of a predominant middle-line cerebellar dysfunction and include a slowly progressive gait ataxia, dysarthria, and oculomotor disturbances, such as gaze-evoked nystagmus, broken-up pursuit, ocular dysmetria, and inability to suppress the vestibular-ocular reflex. Appendicular ataxia is less prominent than gait ataxia, with the legs being more affected than the arms. Although Holmes [42] originally described a recessive form of cerebello-olivary atrophy associated with hypogonadism, subsequent reports have indicated that the disorder, as defined above, is often transmitted in an autosomal dominant fashion [43]. Sporadic cases may also occur, and these are known as late or acquired cortical cerebellar atrophy [44,45]. Although most such cases are idiopathic and are probably genetic in origin, others may represent neurologic complications of systemic diseases, such as neoplasms (paraneoplastic cerebellar degeneration), immunological disorders, dyshormonal states, vitamin deficiencies, or chronic alcoholism [45].

5.2. Olivo-ponto-cerebellar atrophies

These are heterogenous disorders that share common pathologic features, such as atrophy of the cerebellum, pons, middle cerebellar peduncles, and inferior olives. These are often associated with atrophic lesions affecting the spinal cord, basal ganglia, and cerebral cortex. Clinically, they show evidence for a multisystemic involvement, which can distinguish them from the above-described cerebello-olivary atrophy.

5.2.1. Previous attempts to classify the OPCAs

Greenfield [3] was the first to separate the OPCAs into sporadic (Dejerine-Thomas) and hereditary (Menzel) types. Later on, Konigsmark and Weiner [46] divided the familial OPCAs into five main types (I–V). Type I included 10 pedigrees with autosomal dominant OPCA showing clinical and pathologic features that were similar to those originally described by Menzel [47]. Type II included two families with autosomal recessive transmission and variable age of disease onset. Type III included five families with autosomal dominant OPCA, characterized by progressive ataxia, visual loss, and sometimes with ophthalmoplegia and upper-motor neuron deficits. Type IV included the family described by Schut and Haymaker [48,49], believed to be distinct from other dominant OPCA kindreds, due to its marked clinical and pathologic variation. Type V included two dominant kindreds in which afflicted members developed cerebellar ataxia, dysarthria, rigidity, ophthalmoplegia, and dementia.

There are a number of problems related to Konigsmark and Weiner's classification. The authors relied on clinical and pathologic descriptions reported in the literature, which are known to vary considerably from author to author. This led to such errors as classifying the same family under two different categories. Thus, the disease described by Gray and Oliver [50] was classified as OPCA I (Menzel's type), while that reported by Schut-Haymaker was placed under OPCA IV, when, in fact, both papers reported the same family [51].

Many of the studies used by Konigsmark and Weiner were decades old, and as such, they probably lacked the sophistication of contemporary neurology and neurophysiology. This is particularly true for a number of modern neuroophthalmologic concepts related to the different types of eye movements and their control by cerebellar and brain-stem structures, which were not known at the time most of the above reports were written. Also, a number of electrophysiologic methods have been increasingly used in recent years in the study of cerebellar degenerations, and these have furthered our understanding of these disorders. Such electrophysiologic studies have documented the presence of motor or sensory neuropathy or both in the majority of dominant OPCA, including Schut-Haymaker disease [52–55]. Despite this neuropathy, however, the deep-tendon reflexes are retained or become abnormally hyperactive due to coexisting pyramidal tract degeneration. In some instances, however, peripheral neuropathy may become particularly prominent, leading to hyporeflexia or areflexia. This seemed to have been the case for a patient from the Schut-Haymaker kindred, which was described by Schut [48] as a FA phenotype occurring in a family afflicted by dominant OPCA. The undue emphasis placed on such a rare instance led to the conclusion that a marked clinical variability is characteristic for this kindred [46]. By more recent accounts [54,55], however, the Schut-Haymaker pedigree does not show a greater degree of clinical variation than other dominant OPCAs, and as such, its separate categorization is no longer justified.

5.2.1. Present concepts of OPCA nosology
Since a hereditary disorder is defined by its primary genetic abnormality [4], it seems prudent to classify all cases, which can be traced to a single OPCA family, under that particular kindred. This will assure, with a reasonable

certainty, that all such cases are related to the same genetic mutation. It is further reasonable to consider that families showing similar features are probably afflicted by the same disorder. Such an organization could be helpful in the study of the diseases, because it may permit the investigation of closely related, if not identical, entities. Unfortunately, many of the clinical and pathologic manifestations found in OPCA patients are nonspecific deficits of cerebellar and other system dysfunctions, and can occur in various combinations, even within families, thus greatly complicating efforts to classify these disorders on the basis of signs complex. On the other hand, certain clinical manifestations, such as slowed saccadic eye movements and/or blindness, can occur with consistency within families, and as such they can be used in the classification of these disorders, as described below.

With these considerations in mind, it is proposed that dominant OPCA be divided into the following five main categories. The first is to include the pedigrees that show the general OPCA features of progressive cerebellar ataxia, dysarthria, nystagmus, corticospinal deficits, peripheral neuropathy, and/or choreiform movements but that lack the specific clinical deficits found in the variants described below. As such, this category encompasses dominant pedigrees previously classified under **Menzel's** OPCA [56–58] and under type I OPCA of the Konigsmark-Weiner classification, as well as the Schut-Haymaker kindred, for the reasons stated above. Moreover, recent studies have suggested that these diseases may be genetically heterogenous, with some, but not others, being linked to the HLA locus on the short arm of the sixth chromosome (6p region) (see Chapter 17). Hence, these diseases can be further subdivided into those mapping to human chromosome 6p and those not mapping to this chromosomal region. Also, to this author's knowledge, the gene loci for all the categories of dominant OPCA described below have not been linked to the sixth chromosome.

The second category is to include the variant of dominant OPCA that is associated with **slowed saccadic eye movements**. Wadia and Swami [59] provided the first clear description of slowed saccadic eye movements in patients with a dominant OPCA associated with peripheral neuropathy. Although this is indeed a characteristic disorder of ocular motility, as discussed below, some authors have not been willing to accept its specific occurrence in a form of OPCA by pointing out that "ophthalmoplegia" can occur in diverse forms of cerebellar and basal ganglia degenerative disorders [6,60]. Wadia and Swami [59] were, however, careful to separate slowed saccades from other disturbances of ocular motility that can be seen in patients with cerebellar disorders, such as oculomotor nerve paralysis due to brain-stem or peripheral nerve lesions, pseudo-ophthalmoplegia due to conjugate gaze palsy, and progressive external ophthalmoplegia, resulting from ocular myopathy.

We have also studied [61, and upublished data] three dominant OPCA kindreds afflicted by this disorder. Clinical features include progressive ataxia, dysarthria, dysphagia, weakness of eye closure, peripheral neuropathy, corticospinal deficits, and lower motor neuron signs. Visual acuity is preserved, although in some patients disc pallor was detected on fundoscopy. In early disease stages, the volitional saccades are selectively affected (slowed) and these changes can be quantitated by electro-oculography [5]. Spontaneous and gaze-evoked nystagmus, which is commonly seen in other types of cerebellar degeneration, are characteristically absent. Moreover, the normal oculovestibular and opto-kinetic nystagmus cannot be elicited in such patients, but instead the eye movements generated are pendular, lacking the quick phase of this nystagmus. On the other hand, the smooth-pursuit eye movements remain characteristically intact and at no time are broken up by catch-up saccades, as often seen in patients with other types of cerebellar disorders. These abnormalities were clearly distinct from those occurring in progressive supranuclear palsy (Steele-Richardson-Ozewski syndrome), in which the vertical gaze is characteristically affected. Moreover, in this disorder the large-amplitude horizontal saccades are slowed, whereas the small-range saccades are relatively preserved. This, however, is not the case for the OPCA patients, in which all amplitude saccades are uniformly affected.

Neuroretinal degeneration in association with slowed saccades can also occur as a consistent finding in affected members of other dominant OPCA pedigrees [62–65], thus justifying the placement of these pedigrees under a third category of dominant OPCA. The disorder is characterized by progressive cerebellar ataxia, bulbar dysfunction, peripheral neuropathy, lower motor neuron signs, blindness, and slowed eye movements. In advanced disease stages, the eyes may show little motility, thus explaining the term *ophalmoplegia* that has been used by several authors to refer to the oculomotor abnormalities of these patients. In addition, some patients may show myoclonic movements and dementia or psychosis. The majority of the reported cases have occurred in black families [62–66] in which there was a considerable variation in the mode of presentation and the age of disease onset, with frequent childhood occurrence [31]. Neuroimaging studies in a few patients have revealed CNS changes that were similar to those seen in patients with OPCA associated with slowed saccades (see Chapter 13).

It should, however, be noted that cerebellar degeneration associated with blindness can occur in a variety of metabolic and idiopathic disorders [6–66], most of which appear to be inherited in an autosomal recessive manner. As such, this should be considered as being disease specific, only within the spectrum of dominant OPCA.

The fourth category is the disorder described by Boller and Seggara [67], which has been termed **spino-pontine atrophy**. These authors reported a large dominant kindred affected by a progressive neurologic disorder, characterized clinically by ataxia of gait and limb, oculomotor nerve palsies causing diplopia, nystagmus, bulbar dysfunction, corticospinal deficits, and proprioceptive sensory loss, and pathologically by a predominant atrophy of the pons and spinal cord, with minor involvement of the inferior olives and cerebellum. Additional families showing similar clinical pathologic changes, but with a pleomorphic expression, have been reported [68–70]. Linkage analysis in one these families [70] produced strongly negative results with the HLA locus on the sixth chromosome, thus suggesting that this disorder is genetically distinct from the HLA-linked forms of dominant OPCA.

We have previously described [7] morphologic changes of the brain observed in CT of one patient from the kindred described by Boller and Seggara [67]. They revealed characteristic alterations in the ventral pons, probably resulting from degeneration of the descending pyramidal tracts, as well as atrophy of the cerebellar hemispheres and vermis. The corresponding neurologic features were cerebellar ataxia, bulbar dysfunction, oculomotor nerve palsy, and a marked spasticity. Since these morphological and clinical features are, in many respects, distinct from those occurring in other forms of OPCA, a separate categorization of this disorder seems to be justified. However, it is as yet unclear as to whether spinopontine atrophy, diagnosed on the basis of its neuropathologic characteristics, represents a homogeneous entity, since clinical features, such as the deep-tendon reflexes, have been variable, ranging from marked spasticity and hyperreflexia [67] to depressed or absent reflexes [70].

The fifth category is OPCA with **deafness** and other features [6]. Although deafness can be found in a variety of recessively inherited cerebellar disorders [2,6,32–34], it may also occur as a consistent clinical feature in kindreds affected by dominant cerebellar degeneration [71,72]. Additional clinical features described are myoclonus and/or sensory neuropathy. In a large pedigree described by Treft et al. [73], deafness occurred in early life (first and second decades), along with blindness, in 23 affected family members. In later life, some of these patients evidenced cerebellar dysfunction, ptosis, ophthalmoplegia, and/or mild myopathy. As such, this variant of dominant cerebellar degeneration associated with sensorineural deafness seems to be heterogenous.

5.3. Sporadic olivopontocerebellar atrophy

In contrast to the above-described dominant OPCA, which usually starts during the second or fourth decade of life, sporadic OPCA cases often occur in late adult life and can be associated with signs of Parkinsonism. We have separated the late-onset OPCA into two groups on the basis of the presence or absence of progressive autonomic failure, including orthostatic hypotension. As described in the Chapter 13, these groups were found to differ not only in their clinical features, but also in their natural history, brain morphological changes, and perhaps in their etiology. Thus, although the late-onset OPCA associated with autonomic failure was always found to occur sporadically, the form of the disease that was not associated with autonomic failure was found in several cases to affect siblings and, as such, it may be genetically transmitted.

5.4. Recessive cerebellar degeneration

Some authors have considered adult-onset cerebellar degeneration to be extremely rare [6,74]. However, as discussed in Chapters 13 and 14, the frequency of recessive OPCA seems to have been markedly underestimated. Similarly to FA and other recessively inherited disorders, many cases of recessive OPCA may appear to occur sporadically. Their recognition may often depend on how thoroughly the family tree of each patient with seemingly sporadic ataxia has been investigated.

Because of our particular interest in the biochemical defects of these disorders, the families of our patients with ataxia have been investigated thoroughly. In 24 out of 182 patients with adult-onset cerebellar degeneration studied, the genetic data obtained were consistent with recessive inheritance (Table 8-3). Four of these patients evidenced rather pure cerebellar dysfunction and were classified under cortical cerebellar degeneration or cerebello-olivary atrophy [43], while the remaining 20 patients showed additional neurologic features, and as such, they were categorized under OPCA.

Corticospinal deficits were the most common associated neurologic features in these recessive cases. In some families downbeat nystagmus was a consistent neurological finding, while in others extrapyramidal deficits (Parkinsonism) occurred. There appeared to be an intrafamilial correlation of the age at onset and the associated clinical features, thus possibly suggesting that recessive cerebellar degeneration is genetically heterogenous. In about half of the families studied, only males were affected, thus raising the possibility of X-linked recessive inheritance (see Chapter 13). The study of brain morphological changes by MRI or CT revealed a particular type of severe cerebellar atrophy, which appeared to include the subcortical white matter (see Chapter 13). This was associated with brain stem changes that were much less pronounced than those seen in patients with dominant OPCA. As such, these morphologic features suggest that recessive cerebellar degeneration occupies its own place between the classic olivopontocerebellar atrophy and cortical cerebellar degeneration (cerebello-olivary atrophy).

6. DIAGNOSIS

When the clinician is confronted with a case of ataxia, the diagnosis should be established only after a careful consideration of the constellation of clinical features, the presence of specific clinical deficits, natural history of the disease, family history, neurologic findings of other affected relatives, and if available, the pathologic diagnosis of deceased family members. Neuroimaging studies, such as computed tomography (CT) and magnetic resonance imaging (MRI), may reveal atrophy of the cerebellum, brain stem, spinal cord, and/or cerebral cortex, and as such, they have become essential confirmatory tests for most forms of cerebellar degeneration. Not uncommonly, the presence of cerebellar atrophy on CT or MRI may first draw attention to the possibility of a degenerative cerebellar disease.

Neuroimaging studies seem to be particularly valuable in the differential diagnosis of childhood spinocerebellar degenerations. Thus, the presence of significant cerebellar and brain-stem atrophy virtually excludes the possibility of Friedreich's ataxia, which, as already described, is primarily associated with atrophy of the spinal cord. Also, in the evaluation of adult-onset ataxias, MRI and CT may be of value in separating the cortical cerebellar degenerations from the olivopontocerebellar atrophies.

Electrophysiologic studies (see Chapter 10) may also be of help in the study of these disorders by revealing aspects of function and dysfunction of the peripheral nervous system and certain central pathways. Finally, routine and specialized hematologic and biochemical investigations are of value, primarily for excluding other diagnoses or revealing treatable forms of ataxia. A search for the specific etiologic factors, shown in Table 8-2, should be undertaken in every case of cerebellar ataxia that cannot be categorized with a reasonable certainty as belonging to one of the idopathic cerebellar degenerations.

7. EPIDEMIOLOGY

Schoenberg [75] has pointed out that the epidemiologic study of inherited ataxia poses major problems and challenges, which are primarily due to the rarity by which they occur and the difficulties surrounding their diagnosis and classification. A few such studies have been published in the past [75–81], but these suffer from a lack of clear diagnostic criteria [75] and the use of diagnostic categories, such as Marie's ataxia, which are now considered obsolete.

Despite these limitations, it has been generally accepted that the overall prevalence of hereditary ataxia is less than 6 cases/100,000 [75]. Skre's studies [80,81] in Western Norway have further indicated a prevalence rate of about 3 cases/100,000 for autosomal dominant ataxia and 1 case/100,000 for Friedreich's ataxia. The latter figure agrees with the prevalence of this disorder in Iceland, as reported by Gudmundson [77]. However, in isolated, inbred populations higher prevalence rates have been reported [82–86]. In this regard, recent studies have shown the existence of forms of OPCA in Cuba [85] and in the Iakut population of Eastern Siberia, with a prevalence rate of 13–184 cases/100,000 [86]. Hence, some forms of cerebellar degeneration show such an appreciable prevalence rate in the population, while others are extremely rare and may occur in single families. Hence, the likelihood of encountering these afflictions in neurologic practice or even in referral centers is quite low.

For over a decade, we have been conducting biochemical investigations in our laboratory on patients with cerebellar degenerative disorders. During

Table 8-3. Cerebellar disorders diagnosed in 218 personally evaluated patients with ataxia

A. Childhood onset (n = 36)
   I. Areflexic/hyporeflexic syndromes (n = 23)
      1. Friedreich's ataxia — 14
      2. Other idiopathic disorders — 9
   II. Normoreflexic/hypereflexic syndromes (n = 13)
      1. Idiopathic unclassified — 6
      2. Olivopontocerebellar atrophy — 4
      3. Juvenile hexosaminidase deficiency — 2
      4. Juvenile Nieman-Pick disease — 1

B. Adult onset (n = 182)
   I. Pure cerebellar syndrome (cortical cerebellar degeneration or cerebello-olivary atrophy) (n = 16)
      1. Dominant — 6
      2. Recessive — 4
      3. Sporadic — 6
   II. Cerebellar dysfunction associated with additional neurologic features (olivopontocerebellar degeneration and other multisystem atrophies) (n = 140)
      1. Dominant (n = 29)
         a. Menzel's type — 14
         b. With slowed saccades — 11
         c. With retinal degeneration — 2
         d. Other — 2
      2. Recessive (n = 20)
         a. With corticospinal deficits — 13
         b. With Parkinsonism and other features — 5
         c. With peripheral neuropathy — 2
      3. Sporadic without autonomic failure (n = 61)
         a. With corticospinal deficits — 26
         b. With extrapyramidal and other features — 17
         c. With areflexia, amyotrophy, and/or sensory neuropathy — 13
         c. With dementia and other features — 5
      4. Sporadic with progressive autonomic failure (n = 30)
         a. With Parkinsonism and other features — 10
         b. Without Parkinsonism — 20
   III. Cerebellar ataxia associated with systemic disorders (n = 17)
      1. Malignancy — 5
      2. Chronic alcoholism — 3
      3. Vitamin B12 deficiency — 3
      4. Hypothyroidism (myxedema ataxia) — 1
      5. Monoclonal gammopathy — 1
      6. Mitochondrial myopathy — 1
      7. Paget's disease — 1
      8. Scleroderma — 1
      9. Ulcerative colitis — 1
   IV. Ataxia associated with various conditions (n = 9)
      1. Communicating hydrocephalus — 3
      2. Cerebellar form of spongiform encephalopathy — 2
      3. Brain-stem arteriovenous anomaly — 1
      4. Chronic dilantin therapy — 1
      5. Postanoxic cerebellar-myoclonic syndrome — 1
      6. Postinfectious cerebellar syndrome — 1

this time, we had the opportunity to evaluate a rather large number of ataxia patients referred to us from around the country and abroad. Over 200 patients with various types of cerebellar degenerations have been personally examined. In the majority of these patients, detailed clinical, genetic, electrophysiological, biochemical, and neuroimaging evaluations were performed. Table 8-3 shows the various diagnoses established in these individuals according to the classification strategy described above. Although this does not constitute a population-based sample of ataxia patients, it allows some estimation of the frequency by which the various types of ataxia can be encountered in a referral center.

ACKNOWLEDGMENT

This work was supported by NIH grant NS-16871 and RR-71, Division of Research Resources, General Clinical Research Center Branch.

REFERENCES

1. Refsum S., Skre H. (1978). Neurological approaches to the inherited ataxias. Adv. Neurol. 21:1–13.
2. Plaitakis A. (1978). Cerebellar degenerations. Curr. Neurol. 7:159–192.
3. Greenfield J.G. (1954). The Spino-Cerebellar Degenerations. Oxford: Blackwell.
4. Currier R.D. (1984). A classification for ataxia. In: R.C. Duvoisin, A. Plaitakis (eds): Advances in Neurology, Vol. 41. The Olivopontocerebellar Atrophies. New York: Raven Press, pp. 1–12.
5. Plaitakis A., Gudesblatt M. (1984). The hereditary ataxias. Curr. Neurol. 5:471–509.
6. Harding A.E. (1984). The hereditary ataxias and related disorders. In: Clinical Neurology and Neurosurgery Monographs. London: Churchill Livingstone, pp. 1–226.
7. Huang Y.P., Plaitakis A. (1984). Morphological changes of olivopontocerebellar atrophy in computed tomography and comments on its pathogenesis. Adv. Neurol. 41:39–85.
8. Oppenheimer D.R. (1984). Diseases of the basal ganglia, cerebellum and motor neurons. In: J. Hume Adams, J.A.N. Corsellis, L.W. Duchen (eds): Greenfield's Neuropathology. New York: John Wiley & Sons, pp. 699–747. 1959: 62:364–369.
9. Louis-Bar. (1941). Sur un syndrome progressif comprenant des telangiectasies capillaires, cutannes et conjoctivales symetriques, a disposition naevoide et des troubles cerebelleux. Confinia Neurologica (Basel) 4:32–42.
10. Smith J.L., Cogan D.G. (1959). Ataxia-telengiectasia. Arch. Ophthalmol.
11. Boder E. (1985). Ataxia-telengiectasia: An overview. In: R.A. Gatti, M. Swift (eds): Ataxia-Telengiectasia: Genetics, Neuropathologuy, and Immunology of a Degenerative Disease of Childhood. New York: Alan R. Liss 1985:1–63.
12. Gatti R.A., Berkel I., Boder E., et al. (1988). Localization of an ataxia-telengiectasia gene to chromosome 11q22–23. Nature 336:577–580.
13. Aicardi J., Barbosa C., Andermann E. (1988). Ataxia-ocular motor apraxia: A syndrome mimicking ataxia-telengiectasia. Ann. Neurol. 24:497–502.
14. Robbins J.H., Kraemer K.H., Lutzner M.A. et al. (1974). Xeroderma pigmentosum. An inherited disease with sun sensitivity, multiple cutaneous neoplasms and abnormal DNA repair. Ann. Intern. Med. 80:221–248.
15. Tomimoto H., Hirose S., Akiguchi I. (1989). Xeroderma pigmentosum presenting clinical features of spinocerebellar degeneration. Rinsho Shinkeigaku 29:172–176.
16. Kornzweig A.L., Bassen F.A. (1957). Retinitis pigmentosa, acanthocytosis and herodegenerative neuromuscular disease. Arch. Ophthalmol. 58:183–187.
17. Salt H.B., Wolff O.H., Fosbrooke A.S., Cameron A.H., Hubble D.V. (1960). On having no beta-lipoprotein: A syndrome comprising abetalipoproteinemia, acanthocytosis and steatorrhoea. Lancet II:325–329.

18. Muller D.P.R., Lloyd J.K., Wolf O.H. (1983). Vitamin E and neurological function. Lancet I:209–214.
19. Collins J.C., Scheinberg I.H., Giblin D.R., Sternlied I. (1989). Hepatic peroxisomal abnormalities in abetalipoproteinemia. Gastrenterology 97:766–770.
20. Harding A.E., Mathews S., Jones S., et al. (1985). N. Engl. J. Med. 313:32–35.
21. Steinberg, D. (1989). Refsum disease. In: Scriver C.R., Beaudet A.L., Sly W.S., Valle D. (eds): The Metabolic Basis of Inherited Disease. New York: McGraw-Hill, II:1533–1550.
22. Hunt J.R. (1921). Dyssynergia cerebellaris myoclonica-primary atrophy of the dentate system. Brain 44:490–538.
23. Lance J.W. (1986). Action myoclonus, Rumsay-Hunt syndrome, and other myoclonic syndromes. Adv. Neurol. 43:33–55.
24. Koskiniemi M., Donner M., Majuri H., et al. (1974). Progressive myoclonus epilepsy. A clinical and histopathological study. Acta Neurol. Scandin. 50:307–332.
25. Spiro A., Moore C.L., Prineas J.W., et al. (1970). A cytochrome related inherited disorder of the nervous system and muscle. Arch. Neurol. 23:103–112.
26. Thomas P.K., Adams J.D., Swallow D., Stewart G. (1979). Sialidosis type 1; cherry red spot-myoclonus syndrome with sialidase deficiency and altered electrophoretic mobilities of some enzymes known to be glycoproteins. J. Neurol. Neurosurg. Psychiatry 42:873–880.
27. Janeway R., Ravens J.R., Pearce L.A., et al. (1967). Progressive myoclonus epilepsy with Lafora inclusion bodies. I. Clinical, genetic, histopathologic and biochemical aspects. Arch. Neurol. 16:565–582.
28. Nyssen R., Van Bogaert I. (1934). La degenerescence systematisee optico-cochleo-dentelee. Rev. Neurol. 2:321–345.
29. Ferrer I., Campistol J., Tobena L., et al. (1987). Dégénérescence systématisee optico-cochleodentelée. J. Neurol. 234:416–420.
30. de Courten-Mayers G., Mandybur T.I. (1987). Atypical Gerstmann-Straussler syndrome or familial spinocerebellar ataxia and Alzheimer's disease. Neurology 37:269–275.
31. Colan R.V., Snead O.C., Ceballos R. (1981). Olivopontocerebellar atrophy in children: A report of seven cases in two families. Ann. Neurol. 10:355–363.
32. Harding B.N., Dunger D.B., Grant D.B., Erdohazi M. (1988). Familial olivopontocerebellar atrophy with neonatal onset: A recessively inherited syndrome with systemic and biochemical metabolic abnormalities. J. Neurol. Neurosurg. Psychiatry 51:385–390.
33. Lutz R., Bodensteiner J., Gray C. (1898). X-linked olivopontocerebellar atrophy. Clin. Genet. 35:417–422.
34. Superneau D.W., Wertelecki W., Zellweger H., Bastian F. (1987). Myopathy in Marinesco-Sjogren syndrome. Eur. Neurol. 26:8–16.
35. Rapin I., Suzuki K., Valsamis M.P. (1976). Adult (chronic) GM2-gangliosidosis-atypical spinocerebellar degeneration in a Jewish sibship. Arch. Neurol. 33:120–130.
36. Johnson W.G., Choutorian A., Miranda A. (1977). A new juvenile hexosaminidase deficiency disease presenting as cerebellar ataxia: Clinical and biochemical studies. Neurology 27:1012–1018.
37. Willner J.P., Grabowski G.A., Gordon R.E., et al. (1981). Chronic GM2 gangliosidosis masquarading as atypical Friedreich ataxia: Clinical, morphologic and biochemical studies of nine cases. Neurology 787–798.
38. Marsden C.D., Obeso J.A., Lang A.E. (1982). Adrenoleukodystrophy presenting as spinocerebellar degeneration. Neurology 32:1031–1032.
39. Nakazato T., Takeshi S., Nakamura T. (1989). Adrenoleukodystrophy presenting as spinocerebellar degeneration. Eur. Neurol. 29:229–234.
40. Moser W.H., Moser A.B. (1989). Adrenoleukodystrophy (X-linked). In: Scriver C.R., Beaudet A.L., Sly W.S., Valle D. (eds): The Metabolic Basis of Inherited Disease. New York: McGraw Hill, pp. 1511–1532.
41. Kumar A.J., Rosenbaum A.E., Naidu Saeta (1987). Adrenoleukodystrophy: Correlating MR imaging with CT. Radiology 165:497–504.
42. Holmes G. (1907). A form of familial degeneration of the cerebellum. Brain 30:466–489.
43. Eadie M.J. (1975). Cerebelo-olivary atrophy (Holmes type). In: P.J. Vinken, G.W. Bruyn (eds): Handbook of Clinical Neurology. Amsterdam: North Holland Publishing, 21:403–414.

44. Marie P., Foix C., Alajouanine T. (1922). De l'atrophie cerebelleuse tardive a predominance corticale. Rev. Neurol. 2:849–885.
45. Mancall E.L. (1975). Late (acquired) cortical cerebellar atrophy. In: P.J. Vinken, G.W. Bruyn (eds). Handbook of Clinical Neurology. Amsterdam, North Holland, 21:477–508.
46. Konigsmark B.W., Weiner L.P. (1970). The olivopontocerebellar atrophies: A review. Medicine 49:227–241.
47. Menzel P. (1891). Beitrag zur Kenntniss der hereditaren ataxie und kleinhirnatrophie. Arcv. Psychiatrie Nervenkr. 22:160–190.
48. Schut J.W. (1950). Hereditary ataxia: Clinical study through six generations. Arch. Neurol. Psychiatry 63:535–568.
49. Schut J.W., Haymaker W. (1951). Hereditary ataxia: A pathologic study of five cases of common ancestry. J. Neuropathol. Clin. Neurol. 1:183–213.
50. Gray R.C., Oliver C.P. (1941). Marie's hereditary cerebellar ataxia (olivopontocerebellar atrophy). Minnesota Med. 24:327–335.
51. Kark R.A.P., Rosenberg R.N., Schut L.J. (1978). The inherited ataxias. Biochemical viral and pathological studies. Adv. Neurol. 21:107–112.
52. Bennett R.H., Ludvigson P., DeLeon G., et al. (1984). Large-fiber sensory neuropathy in autosomal dominant spinocerebellar degeneration. Arch. Neurol. 41:175–178.
53. Carenini L., Finocchiaro G., DiDonato S., et al. (1984). Electromyography and nerve conduction study in autosomal dominant olivopontocerebellar atrophy. J. Neurol. 231:34–37.
54. Landis D.M., Rosenberg R.N., Landis S.C., et al. (1974). Olivopontocerebellar degeneration. Clinical and ultrastructural abnormalities. Arch. Neurol. 31:295–307.
55. Haines J.L., Schut L.J., Weitkamp L.R., et al. (1984). Spinocerebellar ataxia in a large kindred: Age at onset, reproduction, and genetic linkage studies. Neurology 34:1542–1548.
56. Currier R.D., Glover G., Jackson T.F., et al. (1984). Spinocerebellar ataxia: Study of a large kindred: I. General information and genetics. Neurology 34:1542–1548.
57. Jackson J.F., Currier R.D., Terasaki P.I., et al. (1977). Spinocerebellar ataxia and HLA linkage. N. Engl. J. Med. 296:1138–1141.
58. Nino H.E., Noreen H.T., Dufey D.P., et al. (1980). A family with hereditary ataxia: HLA typing. Neurology 30:12–20.
59. Wadia N.H., Swami R.K. (1971). A new form of heredofamilial spinocerebellar degeneration with slow eye movements (nine families). Brain 94:359–374.
60. Lepore F.E. (1984). Disorders of ocular motility in the olivopontocerebellar atrophies. Adv. Neurol. 41:97–103.
61. Plaitakis A., Huang Y.P., Rudolph S. (1983). Clinical, electrophysiological and CT findings in dominant olivopontocerebellar atrophy with slow saccades. Neurology 33(Suppl. 2):218.
62. Jampel R.S., Okazaki H., Bernstein H. (1961). Ophthalmoplegia and retinal degeneration associated with spinocerebellar ataxia. Arch. Ophthalmol. 66:247–259.
63. Carpenter S., Schumacher G.A. (1966). Familial infantile cerebellar atrophy associated with retinal degeneration. Arch. Neurol. 14:82–94.
64. Weiner L.P., Konigsmark B.W., Stoll J. Jr., Magladery J.W. (1967). Hereditary olivopontocerebellar atrophy with retinal degeneration. Arch. Neurol. 16:364–376.
65. Traboulsi E.I., Maumenee I.H., Green R. (1988). Olivopontocerebellar atrophy with retinal degeneration. Arch. Ophthalmol. 106:801–806.
66. Wisniewski K.E., Madrid R.E., Dambska M., et al. (1987). Spinocerebellar degeneration with polyneuropathy associated with ceroid lipofuscinosis in one family. J. Child. Neurol. 2:33–41.
67. Boller F., Segarra J.M. (1969). Spino-pontine degeneration. Eur. Neurol. 2:356–373.
68. Taniguchi R., Konigsmark B.W. (1971). Dominant spino-pontine atrophy. Report of a family through three generations. Brain 94:349–358.
69. Pogacar S., Amber M., Conlin W.J., et al. (1978). Dominant spinopontine atrophy: Report of two additional members of family W. Arch. Neurol. 35:156–162.
70. Bale A.E., Bale S.J., Schlesinger S.L., McFarland H.F. (1987). Linkage analysis in spinopontine atrophy: Correlations of HLA linkage with phenotypic findings in hereditary ataxia. Am. J. Med. Genetics 27:595–602.
71. May D.L., White H.H. (1990). Familial myoclonus, cerebellar ataxia, and deafness. Specific genetically-determined disease. Arch. Neurol. 19:331–338.

72. Baraitser M., Goody W., Halliday A.M., et al. (1984). Autosomal dominant late onset cerebellar ataxia with myoclonus, peripheral neuropathy and sensoneural deafness: Aclinicopathological report. J. Neurol. Neurosurg. Psychiatry 47:21–25.
73. Treft R.L., Sanborn G.E., Carey J., et al. (1984). Dominant optic atrophy, deafness, ptosis, ophthalmoplegia, dystaxia, and myopathy. Ophthalmology 91:908–915.
74. Koeppen A.H., Barron K.D. (1984). The neuropathology of olivopontocerebellar atrophy. Adv. Neurol. 41:13–38.
75. Schoenberg B.S. (1978). Epidemiology of the inherited ataxias. Adv. Neurol. 21:15–32.
76. Brewis M., Poskanzer D.C., Rolland C., Miller H. (1966). Neurological disease in an English city. Acta Neurol. Scand. 42(Suppl 24):9–89.
77. Gudmundsson K.R. (1969). Prevalence and occurrence of some rare neurological diseases in Iceland. Acta Neurol. Scand. 45:114–118.
78. Kurland L.T. (1958). Descriptive epidemiology of selected neurologic and myopathic disorders with particular reference to a survey in Rochester, Minnesota. J. Chronic Dis. 378–418.
79. Refsum S., Skre H. (1978). Nosology, genetics and epidemiology of hereditary ataxias. Adv. Neurol. 19:497–508.
80. Skre H. (1974). Spinocerebellar ataxia in Western Norway. Clin. Genet. 6:265–288.
81. Skre H. (1975). Friedreich's ataxia in Western Norway. Clin. Genet. 287–298.
82. Barbeau A., Roy M., Sadibelouiz M., Wilensky M.A. (1984). Recessive ataxia in Acadians and "Cajuns." Can. J. Neurol. Sci. 11:526–544.
83. Keats B.J.B., Ward L.J., Shaw J., et al. (1989). "Acadian" and "classical" forms of Friedreich's ataxia are most probably caused by mutations at the same locus. Am. J. Med. Genet. 33:266–268.
84. Dean G., Chamberlain S., Middleton L. (1988). Friedreich's ataxia in Kathikas-Arodhes, Cyprus. Lancet I:587.
85. Orosco B., Estrada R., Perry T.L., et al. (1989). Dominantly inherited olivopontocerebellar atrophy from eastern Cuba. Clinical, neuropathological and biochemical findings. J. Neurol. Sci. 93:37–50.
86. Goldfarb L.G., Chumakov M.P., Petrov P.A., et al. (1989). Olivopontocerebellar atrophy in a large kinship in Eastern Siberia, Neurology 39:1527–1530.

9. THE CEREBELLAR CORTEX AND THE DENTATE NUCLEUS IN HEREDITARY ATAXIA

ARNULF H. KOEPPEN, AND DAVID I. TUROK

1. INTRODUCTION

The hereditary ataxias constitute a heterogeneous group of neurological disorders. Clinical classification has been difficult because of substantial interfamily variation of symptoms and signs. Holmes [1] expressed the hope that morphological observations would ultimately provide some order. However, as clinicoanatomical reports grew in number, the pathological data failed to offer the expected clarification, and numerous eponymic designations appeared. Genetic heterogeneity may now be added as the chromosomal location of the defective gene(s) is becoming known (chromosome no. 9 for Friedreich's disease [2]; chromosome no. 6 for some forms of olivopontocerebellar atrophy [OPCA] [3]). The International Classification of Diseases (1980) [4] lists *spinocerebellar disease* and recognizes hereditary and sporadic forms. It retains some eponymic designations but excludes OPCA, though the familial form of this condition was described in 1891 [5]. A simple approach to a practical classification of hereditary ataxia is the recognition of the age at onset and the mode of transmission (whether autosomal dominant or recessive, or X-linked recessive). The prototype of the autosomal recessive ataxias is Friedreich's disease. Olivopontocerebellar atrophy [5] and familial cortical cerebellar degeneration [6] are typically autosomal dominant. Friedreich's ataxia begins in youth [7], whereas many types of

A. Plaitakis (ed.), CEREBELLAR DEGENERATIONS: CLINICAL NEUROBIOLOGY. Copyright © 1992.
Kluwer Academic Publishers, Boston. All rights reserved.

dominant ataxia are adult-onset conditions. In the forms classified as *spastic ataxia* by Bell and Carmichael [8], autosomal dominant transmission and juvenile onset characterize the clinical syndrome. Friedreich [9–13] gave extensive clinical and neuropathological descriptions of his cases, and neuropathological findings have been part of innumerable subsequent case reports. Despite the known heterogeneity of the morphological alterations in hereditary ataxia, trends can, nevertheless, be recognized: The autosomal dominant forms, perhaps with the exception of "spastic ataxia," tend to have severe lesions of the cerebellar cortex, while the dentate nucleus is commonly unaffected. In contrast, the cerebellar cortex is variably involved in Friedreich's ataxia or is entirely normal (reviews in [14–16]). The dentate nucleus may be quite seriously affected. Ataxia-telangiectasia adds to the heterogeneity of the hereditary ataxias. It is a recessive disorder but is characterized by prominent loss of Purkinje cells. Machado-Joseph disease is an autosomal dominant form of ataxia in which the dentate nucleus is more severely affected than the cerebellar cortex.

This report is not based on large numbers of cases. Instead, an effort was made to obtain autopsy tissues as soon as possible and to achieve optimal fixation by perfusion of the brain with fixatives. Whenever possible, sections were prepared with the vibrating microtome (vibratome) to preserve antigenic properties for immunocytochemistry. The description is restricted to the cerebellar cortex and the dentate nucleus, though the brain stem and spinal cord were also examined. Staining procedures are listed as standard histological stains, Golgi impregnation, immunocytochemical methods, and lectin affinity cytochemistry.

Previous neuropathological reports often emphasized Purkinje cell depletion. Dendritic alterations were not examined until recently [17]. Changes occurring in stellate and basket cells, and in Golgi neurons, were difficult to determine due to the lack of a suitable stain. Descriptions of glial changes did not extend beyond "Bergmann gliosis," and reactions of microglia were totally disregarded. Current methods permit a much more detailed assessment of all of the described cell types.

2. THE CEREBELLAR CORTEX

2.1. Standard histological stains

The cell bodies of Purkinje cells and Golgi neurons are readily revealed by such stains as hematoxylin and eosin, and cresyl violet (Figures 9-1 and 9-2). The nuclei of granule cells are also clearly shown. In the molecular layer, staining is essentially restricted to nuclei, and the distinction of stellate and basket cells, microglial cells, and endothelial cells is difficult. The nuclei of Golgi epithelial cells are readily shown, but their processes toward the surface of the cerebellar cortex (the Bergmann glia) are not revealed. Hematoxylin and eosin, and cresyl violet do not permit adequate visualization of dendrites. Only occasionally are the primary and secondary shafts shown. Silver stains,

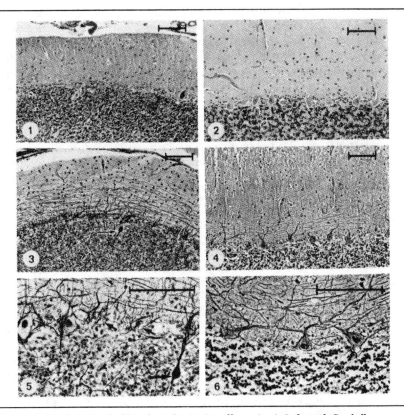

Figures 9-1 to 9-6. Standard histological stains (paraffin sections). Left panel: Cerebellar cortex in autosomal dominant cortical cerebellar atrophy. Right panel: Age-matched normal cerebellum. Stains: Figures 9-1 and 9-2, hematoxylin and eosin; Figures 9-3 to 9-6, Bodian silver technique. Both stains reveal Purkinje cell loss in cortical cerebellar atrophy. The Bodian stain shows empty baskets (short arrows in Figure 9-5) and torpedoes (long arrows, Figures 9-3 and 9-5). The relative retention of the parallel fibers in the molecular layer of the patient with hereditary ataxia is shown in Figures 9-3 and 9-5. Magnification markers: 100 µm.

such as the Bodian technique (Figures 9-3 to 9-6), reveal the parallel fibers of the molecular layer and baskets around the bodies of Purkinje cells. The soma of the Purkinje cell and some dendrites are also visualized.

Hereditary cortical cerebellar degeneration is characterized by loss of Purkinje cells (Figure 9-1). Hematoxylin and eosin and cresyl violet show vast stretches of cerebellar cortex that are devoid of Purkinje cells. The loss is not uniform, and the paleocerebellum is somewhat resistant [18]. Axonal expansions of Purkinje cells (torpedoes) are readily seen on hematoxylin and eosin stains and stand out on silver stains (Figures 9-3 and 9-5). In areas of Purkinje cell depletion, the nuclei of Golgi epithelial cells seem more abundant, but the term *Bergmann gliosis* should be used with caution. It cannot be easily decided whether Bergmann glia or Golgi epithelial cells actually proliferate in response to Purkinje cell loss. The Bodian stain is quite useful in that it shows the collapsed baskets, which seem to persist for extended periods after the disappearance of the Purkinje cells (Figure 9-5). This silver stain was recently examined by modern neurochemical methods and appears to be a true stain for neurofilament protein [19]. Axons constituting the empty baskets derive from parallel fibers. These fibers commonly persist in hereditary cerebellar cortical atrophy (Figure 9-5). It is of interest that the Bodian technique generally stains basket fibers more vigorously than the parallel fibers from which they arise. This observation can also be made by immunocytochemical methods with antisera to neurofilament proteins (see below). The Bodian method stains torpedoes very well, and often the continuity of these fusiform axonal swellings with a delicate axon can be visualized (Figure 9-5).

Cell and silver stains provide insufficient detail of Purkinje cell dendrites, Golgi neurons, and basket and stellate cells.

In the recessive forms of hereditary ataxia, such as Friedreich's ataxia, the cerebellar cortex is occasionally quite seriously affected [20], but none of the cases available in this study showed Purkinje cell depletion. Sections from one case revealed many bizarre dendrites (see Figure 9-38).

2.2. Golgi impregnation

In this study, the method of Braitenberg et al. [21] was used to achieve Golgi impregnation. Dimethylsulfoxide (1%) was added to the potassium dichromate and silver nitrate solutions to enhance penetration. Also, the silver nitrate solution was renewed every other day until 6 days had elapsed. Successful Golgi impregnation of normal human cerebellar cortex showed the true expanse of Purkinje cell dendrites, ranging from the primary shaft(s) to the delicate terminal, "spiny," branchlets (Figures 9-8 and 9-12). Occasionally, stellate, basket, and Golgi neurons were visualized at the same time. Golgi stains generally stained only a small percentage of nerve cells, and the method may have excluded severely diseased Purkinje cells. The impregnated cells may only represent the relatively intact neurons, and the dendritic lesion may not have been adequately revealed. Incomplete impregnation was checked by microscopy with Nomarski interference optics to prevent confusion between incomplete impregnation and true dendritic loss. Despite the described drawbacks, Golgi techniques have been successfully applied to the cerebellum of patients with hereditary ataxia [17,22,23]. The static images suggested the loss of spiny branchlets and the progressive fragmentation of larger dendrites (Figures 9-7, 9-9, and 9-11). Occasionally, a cluster of spiny branchlets arose from a primary or secondary dendrite, suggesting remodeling of the dendritic tree (Figure 9-11). The orderly division into branches of diminishing caliber, and ultimately into the terminal branches,

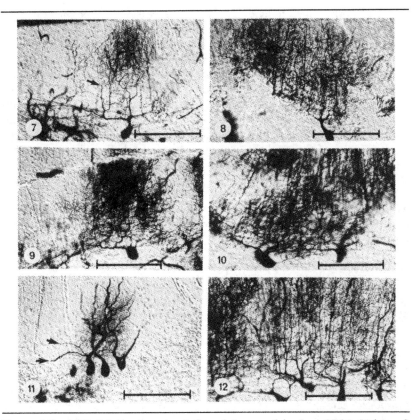

Figures 9-7 to 9-12. Golgi impregnation. Left panel: Autosomal dominant OPCA. Right panel: Normal cerebellum. The dendritic tree is much less elaborate in hereditary ataxia (Figure 9-7) than in the normal cerebellum (Figure 9-8), and some dendrites appear broken off (arrows in Figures 9-7 and 9-11). More than one primary dendrite may arise from an affected Purkinje cell body (Figure 9-9), but similar observations may be made in a normal aged person (Figure 9-10). Remodeling of Purkinje cell arborization may occur, and spiny branchlets may arise from proximal dendritic shafts (Figure 9-11). The overall height of the Purkinje cell expanse is reduced compared to a control specimen (Figure 9-12). Nomarksi interference optics; Magnification markers: 100 µm.

was replaced by an abrupt transition of thick shafts into delicate branches. The disturbed arborization also resulted in an overall reduction of Purkinje cell height, accounting for the thinning of the molecular layer. The narrow molecular layer can be appreciated on routine stains, but the study of dendritic loss required documentation by special techniques. Golgi stains also suggested a reduction in the size of Purkinje cell bodies and occasionally visualized their axonal expansions. Multiple primary dendrites may arise from some Purkinje cell bodies (Figures 9-9 and 9-11), but the cerebellar cortex of normal-aged individuals also may reveal two, three, or even four primary dendrites [24] (Figures 9-10 and 9-12).

A successful Golgi stain may be used to quantitate the dendritic tree of nerve cells [25,26], but the described techniques may be difficult to apply to "idiodendritic" neurons, such as Purkinje cells. The Golgi stain sometimes revealed the cell bodies of Golgi neurons and their long apical dendrites. Golgi impregnation proved somewhat unsatisfactory for the study of cortical cerebellar neurons other than Purkinje cells. In a total of eight cases of hereditary ataxia, the authors could not find a consistent alteration of the small neurons of the cerebellar cortex, whereas Purkinje cell changes were seen quite readily. For stellate, basket, and Golgi neurons, immunocytochemistry proved superior to the Golgi method (see below).

Golgi stains were prepared on blocks from the cerebellum of four patients with recessive ataxia. In two, no Purkinje cell staining occurred at all. In the other two specimens, Purkinje cell visualization was satisfactory and revealed no abnormalities.

2.3. Immunocytochemistry and lectin affinity cytochemistry

Table 9-1 lists the antigens that were examined on tissue sections from the cerebellum by the avidin-biotin peroxidase complex (ABC) method [27]. Microglia were visualized by affinity cytochemistry with the biotinylated lectin *Ricinus communis* agglutinin I (RCA-I) [28]. The lectin was visualized by ABC as for immunocytochemistry.

2.4. Glutamic acid decarboxylase (GAD) (E.C. 4.1.1.15)

The reaction product for GAD is a suitable marker for neurons that biosynthesize gamma-aminobutyric acid (GABA) and hence are thus called *GABA-ergic*. Only very prompt fixation of human postmortem cerebellum by perfusion permitted immunocytochemical visualization of GAD in Purkinje cells, stellate and basket neurons, parallel fibers, and Purkinje cell baskets [17]. The methods of purification and application of the antiserum used in this study have been published [29].

In normal adult human cerebellum, GAD reaction product showed the Purkinje cell body and GAD-reactive basket (Figures 9-14 and 9-16). Primary and secondary dendrites were also revealed. Dense reaction product decorated parallel fibers. For unknown reasons, Golgi cells (which are also GABAergic) failed to show reaction product. In cortical cerebellar atrophy, as illustrated in Figure 9-13, Purkinje cell perikarya were devoid of GAD reaction product, while dense staining characterized Purkinje cell baskets. Empty baskets were abundant and indicated the site of total collapse of Purkinje cell somata (Figures 9-13 and 9-15). Torpedoes were well visualized by anti-GAD (Figure 9-15). It was not surprising that granule cells did not react with anti-GAD. However, numerous GABAergic axons traverse the granular layer on their way from Purkinje cells to the dentate nucleus.

Table 9-1. Antisera and antibodies in the study of the cerebellum and the dentate nucleus in hereditary ataxia and in normal controls

Antigen(s) or epitopes (abbreviation)	Description of antisera or antibody	Cells or tissue constituent visualized
Glutamic acid decarboxylase (GAD)	Sheep polyclonal	Purkinje cells, stellate and basket cells; parallel fibers and their collaterals; dentate afferents from Purkinje cells
Microtubule-associated protein 1 (MAP1A)	Mouse monoclonal (IgG)	Most neurons of the cerebellar cortex and dentate nucleus with variable intensity; Golgi neurons not well visualized
Microtubule-associated protein 2 (MAP2)	Rabbit polyclonal	Most neurons of the cerebellar cortex and dentate nucleus; Purkinje cells not well visualized
Neuron-specific enolase (NSE)	Rabbit polyclonal	Most neurons of the cerebellar cortex and the dentate nucleus; Purkinje cells commonly not well visualized; axons
Neurofilament proteins (nonphosphorylated)	Mouse monoclonal (IgG) (SMI-33)	Purkinje cell somata and dendrites, Lugaro cells; neurons of the dentate nucleus
Neurofilament proteins (phosphorylated)	Mouse monoclonal (IgG) (SMI-31)	Axons; prominent staining of Purkinje cell baskets; dentate afferents
Cyclic GMP-dependent protein kinase (cGK)	Rabbit polyclonal	Purkinje cell bodies and larger dendrites
Synaptic protein(s) (P38; synaptophysin)	Mouse monoclonal (IgG)	Synaptic terminals
Glial fibrillary acidic protein (GFAP)	Rabbit polyclonal	Bergmann glia and other astrocytes; astrocytic processes stain more readily than cell bodies
Non-neuronal enolase (NNE)	Rabbit polyclonal	Actrocytic cytoplasm

These axons were not visualized with anti-GAD, presumably because of a relatively low concentration of the enzyme in the axon when compared to axon terminals.

2.5. Microtubule-associated proteins 1 and 2 (MAP 1 and MAP 2)

Antisera to MAPs gave interesting differential staining of the neurons in the cerebellar cortex, provided fixation was not unduly delayed. The results of immunocytochemical reactions with anti-MAPs on normal human cerebellum were quite similar to the observations in optimally fixed animal brain [30–33].

Anti-MAP2 readily visualized granule, stellate, basket, and Golgi cells (Figures 9-17 to 9-22). Purkinje cell dendrites were also shown, but Purkinje cell perikarya were too pale. In contrast, anti-MAP1 (in this study a mouse

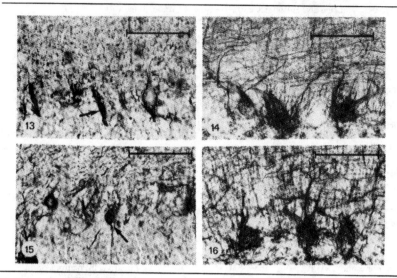

Figures 9-13 to 9-16. Immunocytochemical visualization of glutamic acid decarboxylase (Vibratome sections). Left panel: Autosomal dominant OPCA. Right panel: Normal cerebellum. In OPCA, the perikarya of Purkinje cells either fail to stain with the antiserum or have completely disappeared (Figrue 9-13). The loss of Purkinje cells causes the well-known appearance of empty baskets (arrow in Figure 9-13). Torpedoes are readily visualized with the antiserum (arrow in Figure 9-15). While Purkinje cell baskets are prominently revealed, parallel fibers in OPCA seem less abundant (Figures 9-13 and 9-15) than in normal cerebellum (Figures 9-14 and 9-16). The perikarya of stellate and basket neurons are rarely visualized with anti-GAD. Magnification marker: 100 µm.

monoclonal anti-MAP1A) gave dense staining of Purkinje cell somata and larger dendrites (Figures 9-23 to 9-28). Stellate and basket cells were also clearly revealed by anti-MAPs (Figures 9-21 and 9-22, 9-27 and 9-28). The application of anti-MAP2 and anti-MAP1A in adjacent sections proved quite informative in cortical cerebellar atrophy (Figures 9-17 to 9-22 and 9-23 to 9-28). Neither stain revealed terminal dendrites or spiny branchlets. The long apical dendrites of Golgi neurons could sometimes be traced through the molecular layer. In hereditary ataxia with predominantly cortical cerebellar atrophy, there was a general reduction in the size of the Purkinje cell perikaryon (Figure 9-23), and the dendritic profiles were less abundant (Figure 9-17, MAP2; Figure 9-23, MAP1A). MAP2 immunocytochemistry suggested that Golgi cells were more abundant (Figure 9-19), but a more likely explanation is crowding due to the loss of granule cells. This loss may be retrograde due to Purkinje cell depletion, but the density of granule cells did not permit an assessment of their dendritic or perikaryonal changes. Retrograde atrophy may also be responsible for the loss of reaction product

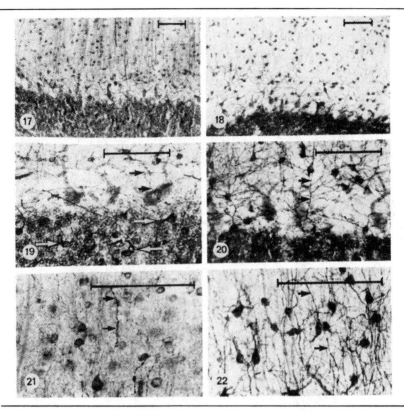

Figures 9-17 to 9-22. Immunocytochemical visualization of microtubule-associated protein 2 (MAP2) (Vibratome sections). Left panel: Autosomal dominant OPCA. Right panel: Normal cerebellum. Anti-MAP2 readily visualizes stellate, basket, and Golgi neurons, whereas Purkinje cell bodies reveal only a somewhat hazy reaction product (Figures 9-19 and 9-20). Circular reaction product characterizes granule cells (Figure 9-19), but the density of these neurons in the granular layer commonly obscures individual cells. Dense reaction product also does not permit full resolution of Golgi neurons in the normal cerebellum (Figure 9-20). In the case of OPCA, Golgi neurons are visible in the granular layer (Figures 9-17 and 9-19) and appear numerically increased (white arrows in Figure 9-19). This crowding of Golgi cells is thought to occur due to the loss of granule cells. The black arrows (Figures 9-19 to 9-22) point to the long apical dendrites of Golgi neurons, which pass to the surface of the molecular layer. Golgi neurons appear unaffected by the disease process. Higher magnification of the molecular layer reveals that stellate cells in OPCA are rounded off and devoid of longer dendrites (Figure 9-22). In the normal molecular layer, MAP2-reactive dendrites of stellate cells may reach a length of nearly 50 μm. Magnification marker: 100 μm.

in the dendrites of stellate and basket cells, and for their abnormal rounded appearance (Figures 9-21 and 9-27).

No adequately preserved tissue for MAP immunocytochemistry was available from patients with recessive ataxia.

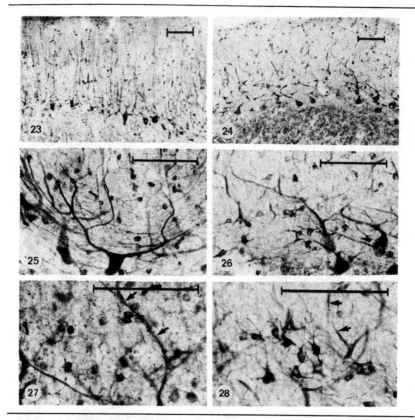

Figures 9-23 to 9-28. Immunocytochemical visualization of microtubule-associated protein 1 (MAP1A) (Vibratome sections). Left panel: Autosomal dominant OPCA. Right panel: Normal cerebellum. Purkinje cell bodies, major dendrites, stellate, and basket neurons are well shown by anti-MAP1A. Golgi neurons are not stained. In OPCA, Purkinje cell perikarya are generally smaller and often give rise to two or more primary dendrites (Figure 9-25). Stellate and basket cells in OPCA (Figures 9-25 and 9-27) are frequently rounded off and lack the stellate appearance of these neurons in the normal molecular layer (Figures 9-26 and 9-28). The appearance is similar to the result after the application of anti-MAP2 (Figures 9-19 and 9-21). The arrows in Figures 9-27 and 9-28 indicate Purkinje cell dendrites. Magnification markers: 100 μm.

2.6. Neuron-specific enolase (NSE) (E.C. 4.2.1.11, gamma)

In the adult mammalian nervous system, NSE is a good immunocytochemical marker for nerve cells [34]. In animal cerebellum, Purkinje cells are readily visualized by anti-NSE. Interestingly, stellate and basket cells are reactive for both NSE and NNE [35]. Granule cells are shown by a characteristic ringlike reaction product. Axons are also stained. In human autopsy tissue, Purkinje cell immunocytochemistry with anti-NSE is fre-

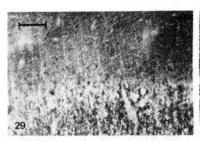 

Figures 9-29 to 9-30. Immunocytochemical visualization of neuron-specific enolase (Vibratome sections). Left: Autosomal dominant OPCA. Right: Normal cerebellum. Reaction product is finely granular in the molecular layer and coarse in the granular layer. Purkinje cells, Golgi neurons, and stellate and basket neurons are not visualized. There are no major differences in the staining intensity of the cerebellar cortex of OPCA and control cerebellum. Magnification marker: 100 µm

quently unsatisfactory due to autolytic changes and/or inherently poor staining of these cells in human cerebellum [36].

The present study confirmed the relative lack of Purkinje cell staining in normal perfusion-fixed cerebellum. The molecular layer showed a finely granular or diffuse reaction product, and the stellate and basket cells were often indistinct. In some specimens, Golgi and Lugaro cells were well revealed by anti-NSE. Reaction product in the human granular layer was quite similar to the appearance in animals and strongly resembled the staining pattern with anti-MAP2. Anti-NSE was very useful for the visualization of nerve cells in the dentate nucleus (see below). Comparing hereditary cortical cerebellar atrophy and control cases (Figures 9-29 and 9-30), only thinning of the molecular layer and perhaps some attenuation of the granular layer were shown.

2.7. Neurofilament proteins

Monoclonal antisera raised against neurofilament proteins [37–39] reveal an immunocytochemical heterogeneity of nerve cells [39], which is based, in large measure, on protein phosphorylation [37]. Neurofilament protein in axons reacts best with antibodies directed against phosphorylated determinants. Axonal injury leads to the accumulation of phosphorylated neurofilament protein in somata and dendrites of nerve cells, and Purkinje cells are no exception [40].

In this study, two commercially available antibodies against neurofilament proteins (SMI-31 and SMI-33; Sternberger-Meyer, Jarrettsville, MD) were used on vibratome sections of cerebellar cortex. SMI-33 showed normal human Purkinje cells with great cytological detail, which approached the resolution of a successful Golgi preparation (Figures 9-31 to 9-34).

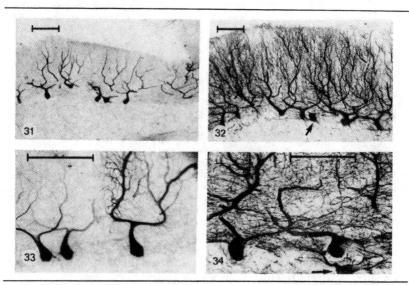

Figures 9-31 to 9-34. Immunocytochemical visualization of nonphosphorylated neurofilament protein with monoclonal antibody SMI-33 (Vibratome sections). Left panel: Autosomal dominant cortical cerebellar atrophy. Right panel: Normal cerebellum. In hereditary ataxia (Figures 9-31 and 9-33), the overall height of Purkinje cells is reduced, and the dendritic tree is nearly devoid of small dendrites and spiny branchlets. The elaborate dendritic expansions shown in normal Purkinje cells (Figures 9-32 and 9-34) are similar to a successful Golgi impregnation, and Lugaro cells are shown (arrows in Figures 9-32 and 9-34). Lugaro cells are sparse or absent in this form of cortical cerebellar atrophy (Figures 9-31 and 9-33). Magnification markers: 100 μm.

Interestingly, Lugaro cells were also shown (Figures 9-32 and 9-34), whereas Golgi, stellate, and basket neurons, and granule cells were not stained. SMI-33 did not stain axons in the molecular and granular layers. The supplier reports that staining is not abolished by dephosphorylation so that reaction product in human material is also considered indicative of nonphosphorylated neurofilament proteins.

Immunocytochemistry with SMI-31 readily revealed parallel fibers, their collaterals, and many other axons (Figures 9-35 to 9-38). The density of axons in the internal third of the molecular layer was especially well shown. Reaction product in Vibratome sections was very abundant, and the high concentration of antigen in baskets and pinceaux was illustrated (Figures 9-36 and 9-38). The application of SMI-31 and SMI-33 to human cerebellum in hereditary ataxia confirmed many observations previously available only after the much more laborious Golgi method. In hereditary cortical cerebellar atrophy, SMI-33 revealed a considerable reduction in the overall height of the remaining Purkinje cells and depletion of the terminal branch-

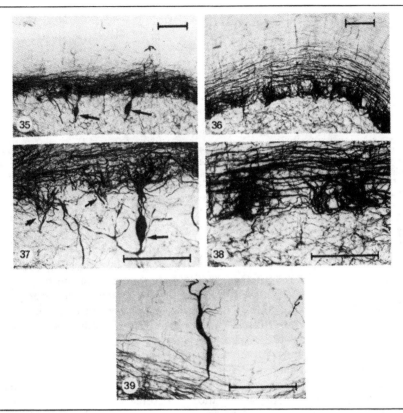

Figures 9-35 to 9-39. Immunocytochemical visualization of phosphorylated neurofilament protein with monoclonal antibody SMI-31 (Vibratome sections). Left panel: Autosomal dominant cortical cerebellar atrophy. Right panel: Normal cerebellum. The sections in Figures 9-35 to 9-39 were obtained from the same block as those illustrated in Figures 9-31 to 9-34. Figures 9-35, 9-37, and 9-39 can be directly compared with Figures 9-31 and 9-33. Figures 9-32, 9-34, 9-36, and 9-38 also came from the same block of normal cerebellar cortex. In contrast to the dramatic loss of Purkinje cells and dendritic arborization in the affected cerebellar cortex (Figures 9-31 and 9-33), parallel fibers and baskets are well preserved. Empty baskets are abundant (short arrows in Figure 9-37), and torpedoes contain a dense reaction product (long arrows in Figures 9-35 and 9-37). SMI-31 does not stain normal Purkinje cell bodies or dendrites. However, occasional bizarre dendritic structures are present in the affected molecular layer (Figure 9-39), implying the accumulation of phosphorylated neurofilament protein in Purkinje cell dendrites. Magnification markers: 100 μm.

lets, which sometimes reached an extreme degree (Figures 9-31 and 9-33). The Purkinje cell somata were reduced in size. Though quantitative counts of Lugaro cells have not been performed, these cells were rarely seen in affected cerebellar cortex.

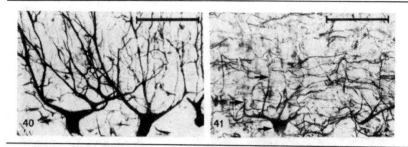

Figures 9-40 to 9-41. Immunocytochemical visualization of nonphosphorylated and phosphorylated neurofilament protein in the cerebellar cortex of a patient with Friedreich's ataxia (Vibratome sections). In this case, alterations of Purkinje cells were severe (SMI-33, Figure 9-40). SMI-31 (Figure 9-41) revealed a serious loss of parallel fibers and abnormal reaction product in cell bodies and dendrites of Purkinje cells (arrows). Magnification markers: 100 μm.

SMI-31 showed a surprising degree of retention of parallel fibers (Figures 9-35 and 9-37), though the internal third of the molecular layer was reduced in thickness, giving the impression of at least some attenuation of parallel fibers. Many collapsed baskets remained where Purkinje cells were lost (Figures 9-35 and 9-37). SMI-31 showed torpedoes very clearly (Figures 9-35 and 9-37). Occasionally, bizarre expansions of Purkinje cell dendrites were revealed by this antibody to phosphorylated neurofilament protein (Figure 9-39), but no widespread accumulation of reaction product occurred in the remaining cell bodies and dendrites.

SMI-31 and SMI-33 were useful in the immunocytochemistry of human cerebellum, even if autopsy and fixation were somewhat delayed. Prolonged fixation, rather than autopsy delay, proved detrimental, and more so for the nonphosphorylated epitopes than the phosphorylated determinants. Only immersion-fixed cerebellum was available from cases of Friedreich's disease, and SMI-33 revealed no reaction product on most specimens. However, sections from one case showed a surprising degree of Purkinje cell loss and dendritic modification (Figure 9-40). In this case, SMI-31 stained some cell bodies (Figure 9-41), suggesting the accumulation of phosphorylated neurofilament protein. Parallel fibers were reduced in number. This patient had a characteristic family history and typical clinical findings during life, which included diabetes mellitus and cardiomyopathy.

2.8. Cyclic guanosine monophosphate-dependent protein kinase (cGK) (E.C. 2.7.1.37 [cyclic GMP])

Though cGK was isolated in good purity from bovine lung [41], anti-cGK proved useful in the immunocytochemical staining of Purkinje cells. Stellate, basket, and granule cells, and Golgi neurons revealed no immunoreactivity [42,43]. Vascular smooth muscles in the brain are quite rich in cGK immu-

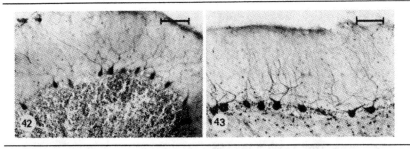

Figures 9-42 to 9-43. Immunocytochemical visualization of cyclic GMP-dependent protein kinase (Vibratome sections). Left: Autosomal dominant cortical cerebellar atrophy. Right: Normal cerebellum. The antiserum reveals Purkinje cell bodies and major dendrites. The affected cerebellar cortex (Figure 9-42) shows focal loss of Purkinje cells and smaller than normal Purkinje cell somata. Magnification markers: 100 μm.

noreactivity, and some astrocytic staining may also occur. Cyclic GMP-dependent protein kinase in cerebellum appears to have a specific substrate, G substrate, which has been well characterized and is also Purkinje cell specific [44–46]. cGK immunoreactivity increases with the development of the cerebellar cortex and is specifically linked to the appearance of maturing Purkinje cells [47]. It is not surprising that cGK protein is extremely low in mouse mutants with Purkinje cell loss, i.e., the homozygous *nervous mouse* (nr/nr) and the homozygote with *Purkinje cell degeneration* (Pcd/Pcd) [48]. The restriction of cGK to Purkinje cells made anti-cGK attractive in the study of cerebellar cortex of patients with hereditary ataxia. Normal Purkinje cells stained well, though spiny branchlets were not visualized (Figure 9-43). In hereditary cortical cerebellar atrophy, Purkinje cell alterations, such as impoverished dendritic trees (Figure. 9-42), were shown, though not with the clarity of antineurofilament proteins (SMI-33). No adequately fixed tissue was available from patients with recessive ataxia.

2.9. Synapse-specific protein (P38; synaptophysin)

Antisera raised against two synapse-specific proteins (Ia and Ib) have been used with some success in the visualization of synaptic terminals of the cerebellar cortex and other tissues of the central and peripheral nervous systems [49]. A monoclonal antibody raised against another synapse-specific protein (designated P38 [50] or synaptophysin [51]) was tried on Vibratome sections of normal human cerebellum and the cerebellum of patients with hereditary ataxia. In the relatively thick sections (40 μm), reaction product gave finely granular staining of the molecular layer (Figures 9-44 to 9-47), but the thicker Purkinje cell dendrites appeared studded with terminals (Figure 9-47). Reaction product in the granular layer was coarse and likely represented mossy fiber terminals (Figures 9-46 and 9-47). In hereditary

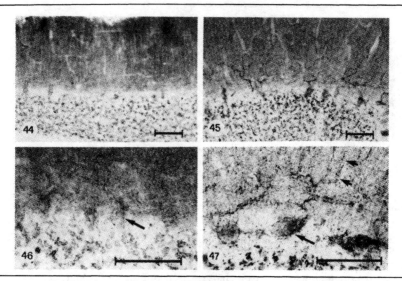

Figures 9-44 to 9-47. Immunocytochemical visualization of a synapse-specific protein (P38) (Vibratome sections). Left panel: Autosomal dominant cortical cerebellar atrophy. Right panel: Normal cerebellum. In affected and normal cerebellar cortex, reaction product in the molecular layer is diffuse or finely granular. It is relatively coarse in the granular layer. Purkinje cell bodies are recognizable due to studding with synaptic terminals (long arrows in Figures 9-46 and 9-47). However, only the normal cerebellar cortex shows the outline of primary, secondary, and tertiary dendrites (short arrows in Figure 9-47). Figure 9-46 suggests a numerical reduction of mossy fiber terminals. Magnification markers: 100 μm.

cortical cerebellar atrophy, there was serious loss of synaptic terminals on cell body and proximal dendrites of Purkinje cells (Figure 9-46). Reaction product in the molecular layer was not substantially reduced. However, mossy fiber terminals were reduced in number in the illustrated case (Figure 9-46). No suitable material was available from patients with recessive ataxia.

2.10. Glial fibrillary acidic protein (GFAP)

A comprehensive account of the history of cerebellar glial cells can be found in Palay and Chan-Palay [52]. For reasons of historical accuracy, the glial cells located near the somata of Purkinje cells should be called *Golgi epithelial cells* and their slender studded processes *Bergmann glia*. The cell type attributed to Fañanás [53] should be viewed as a Golgi epithelial cell. Bergmann glia have often been considered as smooth apical processes of Golgi epithelial cells, a concept that should be revised [52]. In addition to Bergmann glia, the cerebellar cortex contains astrocytes with a more common appearance. They are frequent only in the granular layer and have been divided further into smooth and velate types, addressing the microscopic anatomy of their processes [52]. Immunocytochemistry for GFAP readily reveals Bergmann glia, and velate and smooth astrocytes, establishing their common astrocytic nature. However, Bergmann glia in adult cerebellum normally show vimentin immunoreactivity [54,55] and can also be induced to biosynthesize ferritin [56]. The linear shafts of the Bergmann glia may be viewed as scaffolding for the molecular layer. There is compelling morphological evidence that these shafts guide the external granule cells in their migration to their final destination in the internal granular layer [57,58]. Though much more abundant than Purkinje cells, Bergmann glia are not necessarily distributed at random [59,60], but many mimic a second espalier at right angles to that of Purkinje cell dendrites.

It is quite likely that alterations of the cerebellar cortex, and especially loss and dendritic alteration of Purkinje cells in the molecular layer, affect Bergmann glia. Secondary changes in the granular layer can be expected to affect smooth and velate astrocytes.

Since immunocytochemistry for GFAP works well on formalin-fixed paraffin-embedded tissues (Figures 9-48 to 9-56), many cases of hereditary ataxia could be examined for glial changes. An effort was made to achieve tangential sections of the folial crest to obtain Bergmann glia in cross section. This task proved more difficult than in the rat and rabbit, because the main shafts of Bergmann glia in the human cerebellum are more delicate than in smaller animals. Also, side branches were found to be quite abundant and obscured true cross sections. Nevertheless, rows of these glial shafts were found in hereditary ataxia and normal controls (Figures 9-52 and 9-53). The spacing between these rows of glial processes was greatly reduced in some cases of cortical cerebellar atrophy (Figure 9-52). In human paraffin-embedded tissues, simultaneous staining of Purkinje cell dendrites and Bergmann glia was not possible, and no information was obtained on the physical relationship of dendrites and glial processes, as reported in animals [59]. Though difficult to quantify, GFAP reaction product in Bergmann glia appeared increased in many cases. It was not apparent whether additional Golgi epithelial cells, as a sign of true glial hyperplasia, occurred as well. The disappearance of Purkinje cells likely increased the apparent number of nuclei of Golgi epithelial cells per unit length without true proliferation.

Astrocytes in the granular layer were not abundant, even though the immediately subjacent folial white matter was truly gliotic, Bergmann "gliosis" also characterized those cases of recessive ataxia where Purkinje cell depletion was more serious (Figure 9-54).

2.11. Non-neuronal enolase (E.C. 4.2.1.11, alpha; NNE)

Only sections of cerebellar cortex from dominant ataxia and normal controls were available for staining with anti-NNE (Figures 9-57 to 9-60). In contrast to anti-GFAP, anti-NNE gave more reaction product in the cell bodies of astrocytes. This observation is not surprising, because NNE is a soluble

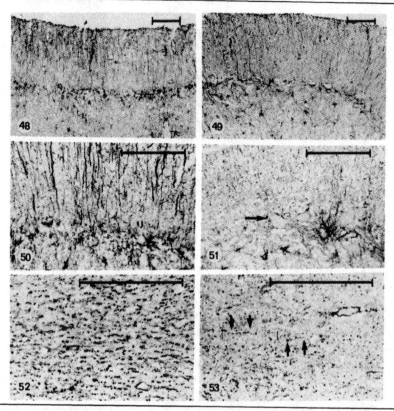

Figures 9-48 to 9-53. Immunocytochemical visualization of glial fibrillary acidic protein (GFAP) (paraffin sections). Left panel: Autosomal dominant cortical cerebellar atrophy. Right panel: Normal cerebellum. Reaction product is more prominent in Bergmann glia of the affected molecular layer (Figures 9-48 and 9-50). The more typical astrocytes of the granular layer do not appear numerically increased and do not contain denser than normal reaction product. In the normal cerebellar cortex, glial reaction product surrounds Purkinje cell bodies and thus gives a faint outline of the perikaryon (arrow in Figure 9-51). In the affected cerebellum (Figure 9-50), the loss of Purkinje cells leads to relatively dense spacing of Golgi epithelial cells in the Purkinje cell layer. A tangential section through the crest of a normal cerebellar folium shows a punctiform reaction product in transversely sectioned Bergmann glia (Figure 9-53). The dots are commonly arranged in rows (arrows in Figure 9-53). In cortical cerebellar atrophy (Figure 9-52), the shafts of Bergmann glia appear much thicker, the linear arrangement is much more prominent, and spacing between the rows is tighter. Magnification marker: 100 µm.

protein and is not associated with glial fibrils. In sections from normal cerebellum, Purkinje cell bodies and some larger dendrites were shown as negative images (Figure 9-60). These voids were absent in the sections from cases with Purkinje cell depletion (Figure 9-59).

9. The cerebellar cortex and the dentate nucleus   223

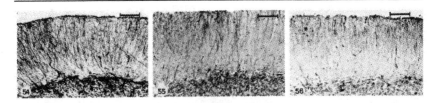

Figures 9-54 to 9-56. Immunocytochemical visualization of glial fibrillary acidic protein (GFAP) in Friedreich's ataxia (paraffin sections). The section illustrated in Figure 9-54 is derived from a patient with Friedreich's disease, characterized by severe Purkinje cell alterations. This case is identical to the one illustrated in Figures 9-40 and 9-41. Figure 9-55 illustrates the cerebellar cortex from a patient with Friedreich's disease in whom no Purkinje cell alteration had occurred. Figure 9-56 illustrates the normal cerebellar cortex. Only the patient with Purkinje cell lesions (Figure 54) showed increased GFAP reaction product in Bergmann glia and in the astrocytes of the Purkinje cell layer. Magnification markers: 100 µm.

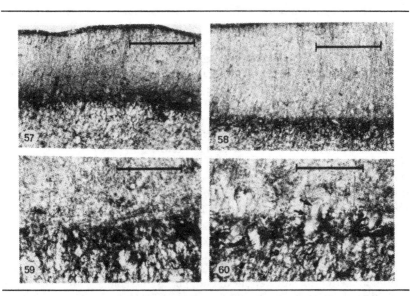

Figures 9-57 to 9-60. Immunocytochemical visualization of non-neuronal enolase (NNE) (Vibratome sections). Left panel: Autosomal dominant OPCA. Right panel: Normal cerebellum. Anti-NNE is a suitable antiserum for the visualization of Golgi epithelial cells and other astrocytes. The shafts of Bergmann glia are not well revealed, but their expansions near the surface of the molecular layer are shown. In hereditary ataxia, the thinning of the molecular layer is apparent (Figure 9-57). Condensation of the reaction product in the Purkinje cell layer is caused, in part, by the loss of Purkinje cells (Figure 9-59). In the normal cerebellar cortex, reaction product provides a negative image of Purkinje cell bodies (arrows in Figure 9-60). Magnification markers: 100 µm.

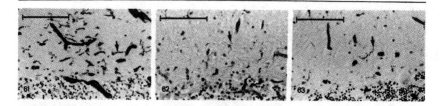

Figures 9-61 and 9-63. Visualization of microglia with biotinylated *Ricinus communis* agglutinin-1 (paraffin sections; hematoxylin counterstain). Figure 9-61. Autosomal dominant cortical cerebellar atrophy. Figure 9-62. Friedreich's ataxia without pathological changes in the molecular layer. Figure 9-63. Normal cerebellum. Typical microglial cells are indicated by arrows. In the case of dominant ataxia (Figure 9-61), near total Purkinje cell loss was associated with a numerical increase of microglial cells and hypertrophy of microglial cytoplasm. Friedreich's ataxia (Figure 9-62) did not differ from the normal state (Figure 6-63). Magnification markers: 100 µm.

2.12. Affinity cytochemistry with *Ricinus communis* agglutinin 1 (RCA-1)

Biotinylated RCA-1 is a useful cytochemical marker for microglia in sections of human brain [28]. The lectin also attaches readily to vessel walls, but the distinct shape of microglial cells makes RCA-1 suitable for their recognition and for the examination of pathological reactions (Figures 9-61 to 9-63). The staining procedure is effective on paraffin sections, and 15 cases of dominant and recessive ataxia were available. With the described staining procedure, the normal adult human cerebellar cortex revealed scattered microglial cells with delicate processes. They were readily found in the molecular layer due to the relative lack of other nuclei (Figures 9-61 to 9-63). They were less abundant in the granular layer, but the white matter showed many microglial cells between fibers. In hereditary ataxia with cortical cerebellar atrophy, the microglial response was quite variable. In two very aggressive cases of OPCA (mother and daughter), no differences from the normal state were found. In other cases of dominant ataxia and Purkinje cell loss, microglial cells were more frequent and displayed hypertrophy of their cytoplasmic processes (Figure 9-61). Occasionally, only focal microglial proliferation occurred. In some cases of Friedreich's disease, microglial hypertrophy was present in the molecular layer but did not necessarily coincide with regions of Purkinje cell depletion. In the illustrated case of Friedreich's disease, RCA-1 gave a normal appearance of microglia in the molecular layer (Figure 9-62).

3. THE DENTATE NUCLEUS

The dentate nucleus is commonly well preserved in OPCA and hereditary cortical cerebellar atrophy, but this exemption is not invariable. Preservation may apply only to neurons, while glial nuclei may be more abundant than normal. In the familial form of dyssynergia, cerebellaris myoclonica (the "Ramsay Hunt syndrome") [61], neuronal loss of the dentate nucleus is

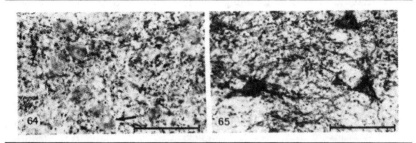

Figures 9-64 and 9-65. Immunocytochemical visualization of glutamic acid decarboxylase (GAD) in the dentate nucleus (Vibratome sections). Left: Autosomal dominant OPCA. Right: Normal dentate nucleus. In the normal dentate nucleus (Figure 9-65), neuronal cell body and main dendrites are visualized by dense surface reaction product. In the section from the case of OPCA (Figure 9-64), the stellate outline of dentate neurons is not apparent, since the reaction product is sparse. Some cell bodies are devoid of GAD-reactive terminals (arrow in Figure 9-64). Magnification markers: 100 µm.

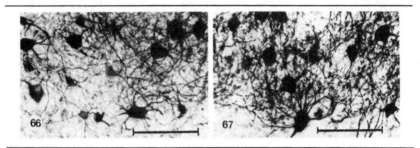

Figures 9-66 and 9-67. Immunocytochemical visualization of microtubule-associated protein 2 in the dentate nucleus (vibratome sections). Left: Autosomal dominant OPCA. Right: Normal dentate nucleus. There are no differences in cell bodies and dendrites, nor in the density of reaction product between normal dentate nucleus and the dentate nucleus of the patient with OPCA. Magnification markers: 100 µm.

severe and is associated with degeneration of the spinocerebellar tracts and dorsal columns. Neuronal loss in the dentate nucleus is a relatively frequent *brain* lesion of Friedreich's disease [reviews in refs. 14–16], and degeneration of the superior cerebellar peduncle is an expected change.

At first glance, the normal dentate nucleus appears less complex than the cerebellar cortex, as it lacks the variety of neurons that make up the molecular, Purkinje cell, and granular layers. However, measurements of dentate neurons in humans have revealed at least three groups of nerve cells, including two large types and one small type [62]. Also, the dendrites of dentate neurons are quite elaborate, as shown in Figures 9-66 and 9-67. They do not arise from the entire circumference of the soma but are heavily concentrated on one side streaming toward the center of the ribbonlike gray matter. In this manner, synaptic circuitry is most efficient, since dentate afferents massively converge on a relatively small way station. With advancing age, the dentate nucleus loses some of its volume, and neuronal cell density increases. Absolute nerve cell counts remain nearly constant [62]. Due to these observations, caution is advised in the interpretation of neuronal loss, but depletion of myelinated fibers in the hilum strongly supports the diagnosis of neuronal atrophy of the dentate nucleus in Friedreich's ataxia.

Golgi impregnation of the dentate nucleus in animals (rat and monkey) shows neurons of considerable complexity in size, shape, and dendritic expansion [63]. It is quite likely that the human dentate nucleus has a similar assortment of nerve cells. Furthermore, the distribution of various neuronal types is not diffuse or random, possibly raising difficulties in the interpretation of pathological changes when single sections are examined.

It would be of obvious interest to examine the fastigial, globose, and emboliform nuclei in hereditary ataxia. They are rarely seen on sections, despite careful dissection of the cerebellum. Sections of the dentate nucleus occasionally reveal some neurons of the emboliform nucleus.

In OPCA and familiar cortical cerebellar atrophy, the dentate nucleus is subject to deafferentation by the loss of Purkinje cell axons. In the recessive forms of hereditary ataxia, such as Friedreich's disease, relatively few afferents from the cerebellar cortex are lost, while there is a reduction of the numerous collateral fibers of extracerebellar origin. Friedreich's disease is less common now than the dominant forms of hereditary ataxia, and no perfusion-fixed cerebellum was available. However, some immersion-fixed material was provided by other physicians, and several antigens remained detectable.

Golgi stains confirmed the presence of large and small neurons but did not reveal pathological changes. There was no hint that dentate neurons in dominant or recessive ataxia responded by an impoverished dendritic tree, as is so commonly seen with Purkinje cell atrophy.

Sufficient case material was available from dominant ataxia for the immunocytochemical visualization of GAD, MAP2, MAP1A, NSE, phosphorylated, and nonphosphorylated neurofilament proteins, and GFAP. Microglial cells were studied by lectin affinity cytochemistry. Routine cell stains confirmed the integrity of the dentate nucleus in the autosomal dominant forms of hereditary ataxia and the common loss of these nerve cells in Friedreich's disease.

3.1. Glutamic acid decarboxylase (GAD)

In the dentate nucleus, reaction product for GAD was restricted to the axon terminals that outlined the soma and primary dendrites (Figure 9-65). In a case of OPCA, the number of GAD-reactive terminals was greatly reduced (Figure 9-64), confirming a loss of afferent fibers from Purkinje

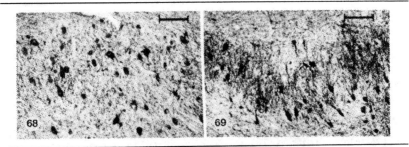

Figures 9-68 and 9-69. Immunocytochemical visualization of neuron-specific enolase in the dentate nucleus (Vibratome sections). Left: Autosomal dominant OPCA. Right: Normal dentate nucleus. Neuronal cell bodies are well visualized. A somewhat more elaborate dendritic tree is suggested by the reaction product on the section from the normal cerebellum (Figure 9-69), but neuronal sizes are identical for OPCA and normal dentate nucleus. Magnification markers: 100 μm.

cells. Adequately prepared material from a patient with recessive ataxia was not available.

3.2. Microtubule–associated proteins (MAP1A and MAP2)
Cell bodies and dendrites were readily visualized by antibodies to MAP, provided the tissue was obtained promptly. The reaction product with polyclonal anti-MAP2 is illustrated in Figures 9-66 and 9-67. The interdigitating dendrites of the neurons are shown. No differences between normal and dominant ataxia are apparent. Due to the great sensitivity of microtubule-associated proteins to postmortem changes, no adequately fixed material was available from a recessive case.

3.3. Neuron-specific enolase (NSE)
Reaction product normally occurred in the cell bodies and larger dendrites of dentate neurons, and staining intensity varied somewhat. The reaction product was slightly granular and thus differed from that of MAP2 and MAP1A. There were no apparent differences between the dentate nucleus of patients with OPCA or cortical cerebellar atrophy and the normal dentate nucleus (Figures 9-68 and 9-69). However, in three specimens from patients with Friedreich's ataxia, neuronal loss was quite severe (Figure 9-70). The remaining nerve cells were smaller than normal (Figure 9-71). Some were revealed only by a counterstain because they had lost their NSE immunoreactivity.

3.4. Neurofilament proteins (antibodies SMI-31 and SMI-33)
Only the results of staining with SMI-31 are illustrated. SMI-33 provided neuronal visualization that strongly resembled NSE (Figures 9-68 and 9-69).

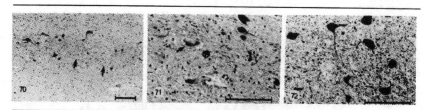

Figures 9-70 and 9-72. Immunocytochemical visualization of neuron-specific enolase in the dentate nucleus of Friedreich's ataxia (paraffin sections; hematoxylin counterstain). Figures 9-70 and 9-71 are from the same case. Figures 9-72 is from a normal dentate nucleus. Focal neuronal loss is present in the dentate nucleus (between the arrows in Figure 9-70). Many remaining nerve cells in Figure 9-71 show only sparse reaction product, and one neuron is devoid of immunoreactivity (arrow). Magnification markers: 100 μm.

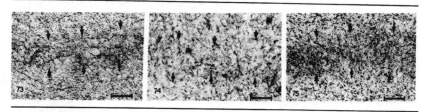

Figures 9-73 and 9-75. Immunocytochemical visualization of phosphorylated neurofilament protein in the dentate nucleus with monoclonal antibody SMI-31 (Vibratome sections). Figure 9-73. Autosomal dominant cortical cerebellar atrophy. Figure 9-74. Friedreich's disease. Figure 9-75. Normal dentate nucleus. The gray matter of the dentate nucleus is outlined by arrows (Figures 9-73 to 9-75). In the case of dominant ataxia, the density of SMI-31 reaction product on both sides of the gray matter ribbon appears reduced (Figure 9-73). In Friedreich's disease, the loss of reaction product in axons and axon terminals is more serious (Figure 9-74). Magnification markers: 100 μm.

The dendrites were not shown in the same detail as by MAP2 (Figures 9-66 and 9-67). There were no differences between the normal dentate nucleus (Figure 9-75) and the dentate nucleus from cerebellar atrophy or OPCA. Due to prolonged fixation in formalin, the available tissue from patients with Friedreich's ataxia did not give a satisfactory reaction product with SMI-33. However, the better resistance of phosphorylated epitopes to fixation and/or autolysis permitted the study of the dentate nucleus with SMI-31. In all sections, the neurons were surrounded by a network of SMI-31-reactive fibers, which were more delicate than the basket and parallel fibers. The number of these fibers about dentate neurons was reduced in dominant *and* recessive ataxia (Figures 9-73 and 9-74). In the cases of dominant ataxia with loss of Purkinje cell input to the dentate nucleus, the lack of reaction product about the cell bodies of dentate neurons is not surprising. The degree of axonal loss illustrated for the recessive ataxias, such as Friedreich's disease (normal Purkinje cells), implies considerable loss of terminal axons from

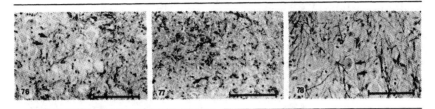

Figures 9-76 to 9-78. Immunocytochemical visualization of glial fibrillary acidic protein (GFAP) in the dentate nucleus (paraffin sections; hematoxylin counterstain). Figure 9-76. Autosomal dominant cortical cerebellar atrophy. Figure 9-77. Friedreich's disease. Figure 9-78. Normal dentate nucleus. Anti-GFAP reveals delicate astrocytic processes that traverse the gray matter of the dentate nucleus (Figure 9-78). Only minor augmentation of these processes appears in the dentate from the case with dominant ataxia (Figure 9-76). In contrast, the dentate nucleus of Friedreich's disease (which reveals neuronal loss), shows small-cell gliosis and increased cytoplasmic GFAP reaction product. The arrows indicate glial processes (Figures 9-76 and 9-78) and small astrocytes (Figure 9-77), respectively. Magnification markers: 100 µm.

other, presumably extracerebellar, sources. Collaterals from climbing and mossy fibers are obvious candidates.

3.5. Synapse-specific protein (P38; synaptophysin)

Anti-P38 was successfully applied to sections from several normal cerebella and in one case of dominant ataxia with severe Purkinje cell loss but good preservation of the dentate nucleus. Also, a single suitably fixed specimen from a patient with recessive ataxia was available. The findings in the normal dentate nucleus were similar to those obtained with anti-GAD (Figure 9-65): Neuronal cell bodies and proximal dendrites were outlined by immunoreactive terminals. In the section from the patient with cortical cerebellar atrophy, synaptic density in the dentate nucleus was *not* visibly reduced. In the case of recessive ataxia, the gray matter of the dentate nucleus was narrowed due to neuronal loss, and P38-reaction product did not reveal the normal negative image of nerve cell bodies. Nerve terminals appeared crowded, undoubtedly the result of lost dentate nerve cells and their neuropil.

3.6. Glial fibrillary acidic protein (GFAP)

Deafferentation of dentate neurons in cortical cerebellar atrophy and the presumed primary loss of dentate neurons in recessive ataxia are both likely to generate an astrocytic response (Figures 9-76 and 9-77). In the normal dentate nucleus, anti-GFAP revealed little reaction product in the astrocytes of the gray matter but numerous delicate processes in the neuropil (Figure 9-78). In deafferentation due to Purkinje cell loss, such as in OPCA or cortical cerebellar atrophy, only a modest increase of these processes occurred (Figure 9-76). Fibrous gliosis was rather prominent in the white matter of the hilus and the amiculum. In Friedreich's ataxia with neuronal loss in the dentate, GFAP was readily shown in the cytoplasm of gray matter astrocytes (Figure 9-77), though the astrocytes did not reach the large size of gemistocytes. In the recessive cases, the dentate hilus was gliotic.

3.7. Lectin affinity cytochemistry with RCA-1
The human dentate nucleus contains abundant microglial cells, which are readily revealed by affinity cytochemistry with RCA-1. In contrast to the cerebellar cortex and the white matter of the folia, they have more cytoplasm and coarser projections, giving the appearance of some hypertrophy. The dentate nucleus was available for study in cases of dominant and recessive ataxia, but none of the sections revealed a microglial response that exceeded the abundance and appearance of these cells under normal conditions.

4. ELECTRON MICROSCOPY OF THE CEREBELLAR CORTEX IN HEREDITARY ATAXIA
Electron microscopic studies of the cerebellum in hereditary ataxia remain extremely limited. The reasons are obvious. Biopsies often are not performed because there is little hope that tissue diagnosis will lead to effective therapy. Promptly fixed autopsy tissue is suitable for ultrastructural examination, but all the requirements for prompt removal, fixation by intracisternal injection or perfusion are generally not met. Nevertheless, two electron microscopic studies of familial OPCA have been reported (three biopsies, three autopsies) [64,65]. Many complex ultrastructural abnormalities were observed that gave little insight into the cause and pathogenesis of OPCA.

Petito et al. [64] reported a reduction in the number of spiny branchlets, but the remaining terminal dendrites appeared to have a normal complement of synapses. In contrast, synapses on smooth Purkinje cell dendrites were few, and hypertrophic glial processes surrounded these dendrites. The Purkinje cell dendrites themselves contained an excess of membrane-bound vacuoles, some unusual inclusions, and abnormal mitochondria. Basket cells had fewer synapses than normal. In the granular layer, a considerable reduction of mossy fiber terminals was observed. This layer also contained torpedos with neurofilamentous hypertrophy and an admixture of mitochondria and other electron-dense particles.

Landis et al. [65] described numerous degenerative changes in Purkinje cell dendrites, but also bizarre inclusions. Glial hypertrophy was present in regions of Purkinje cell loss. Stellate and basket cells were relatively normal, and hypertrophic basket fibers could be found as long as Purkinje cell debris was present. Synaptic connections to stellate and basket cells were deemed normal. Macrophages were reported in all layers of the cerebellar cortex. Both reports confirmed the preservation of Golgi cells.

A comparison of light and electron microscopic observations shows several similarities. They include loss of Purkinje cells, torpedos, hypertrophic glial processes, preservation of Golgi cells, and limited alterations of stellate and

Table 9-2. Pathogenesis of cerebellar lesions in hereditary ataxia

Postulate	Supportive evidence
Dendritic atrophy precedes death of Purkinje cells	Golgi impregnation and immunocytochemistry for nonphosphorylated neurofilament protein (SMI-33) shows intact Purkinje cell bodies but reduced dendritic arborization
Purkinje cells are the primary target of the disease process	Torpedoes are readily demonstrated by immunocytochemistry for GAD and phosphorylated neurofilament protein. Retrograde changes affect stellate and basket cells (MAP1A and MAP2). "Empty baskets" are shown by silver staining of axons and immunocytochemistry for GAD and phosphorylated neurofilament protein (SMI-31). Dentate neurons show loss of GAD-reactive terminals but normal cell bodies and dendrites (NSE, MAP1A, MAP2). Parallel fibers persist after Purkinje cells have disappeared, implying the survival of granule cells, stellate cells, and basket neurons.
Hereditary cortical cerebellar atrophy is unrelated to GABA biosynthesis	While Purkinje cells disappear, other GABAergic neurons persist (stellate, basket, and Golgi neurons) (MAP1A, MAP2, GAD)
Dentate neurons undergo primary atrophy in recessive ataxia	Afferent fibers from the cerebellar cortex are largely intact due to preservation of Purkinje cells in most cases. Efferent fibers of the dentate nucleus are depleted (standard myelin and axon stains; SMI-31).

basket cells. Purkinje cell dendrites showed many ultrastructural changes, and the profusion of membranous and tubular inclusions stands out in both reports. None of the described changes are necessarily specific for OPCA. Degenerating mitochondria are common to many other disorders.

5. PATHOGENESIS

The static images presented in this work permit only a limited interpretation of the pathogenesis of the cerebellar cortical and dentate lesions in hereditary ataxia. Pathogenesis is a dynamic process that is only inadequately explored by the gross and microscopic study of tissues. Patients who die from hereditary ataxia generally reveal the morphological end stage of their illness, though early death by intercurrent illness or suicide has come to attention from time to time [66]. The reported case [66] of an 18-year-old man who died from a self-inflicted gunshot wound to the chest may be cited. While not ataxic prior to death, he was thought to be at a very high risk for dominant ataxia due to his HLA genotype. The cerebellar cortex revealed considerable Purkinje cell loss, suggesting that neuronal atrophy antedates the onset of ataxia and that a certain degree of neuronal depletion must be reached for the first symptoms to become manifest.

Certain reasonable assumptions may be made on the basis of Golgi impregnation and immunocytochemical observations (Table 9-2). In the adult central nervous system, retrograde and anterograde transneuronal degenerations cannot be ignored. The death of Purkinje cells can be expected to affect granule, stellate, and basket neurons because these cells send synaptic input to Purkinje cell dendrites and cell bodies. It is well known that Purkinje cell atrophy of variable hereditary and nonhereditary causes leads to retrograde atrophy of the inferior olivary nuclei. The interpretation is the loss of climbing fiber connectivity, though Koeppen et al. [17] also suggested an independent alteration of inferior olivary neurons in OPCA. The loss of Purkinje cells clearly deprives the neurons of the dentate nucleus of major synaptic input and should lead to transneuronal degeneration. However, the dentate nucleus proved remarkably resistant to change, despite serious Purkinje cell depletion. The loss of dentate neurons in recessive ataxia remains quite enigmatic, because afferents from sources other than Purkinje cells are incompletely understood. Immunocytochemical observations with an antibody to phosphorylated neurofilament protein (SMI-31) actually suggest more deafferentation of the dentate nucleus in recessive ataxia than offered by other neuroanatomical techniques. Since neuronal degeneration may affect neurons across several synapses, distant morphological alterations may mimic primary involvement. Nevertheless, available evidence strongly favors primary Purkinje cell atrophy in OPCA and other forms of dominant cerebellar cortical atrophy. In the recessive forms, a primary lesion of the dentate nucleus may only be suggested, and stronger neuropathological support is needed.

Vinters and associates [67] suggested a novel pathogenesis for the severe Purkinje cell loss in ataxia-telangiectasia. They proposed that Purkinje cells in ataxia-telangiectasia are subject to improper interaction between their dendrites and parallel fibers. In support of this postulate, they illustrated abnormal dendrites of Purkinje cells, and displacement of their cell bodies into the molecular layer. This attractive hypothesis also implies that Purkinje cells are fatally flawed during gestation. In the dominant ataxias that have been studied in this investigation, misplaced Purkinje cells were not more frequent than in patients of comparable age who had no ataxia during life. Anomalously located Purkinje cells without argyrophilic baskets are found with some frequency in routine autopsy material. The ectopic location of Purkinje cells and their connections with parallel fibers have been studied in great detail [68,69]. Ectopia may occur in association with Purkinje cell loss, and the patient may be ataxic during life [69]. Cretinism is also known to lead to ectopic location and loss of Purkinje cells [70], and there is experimental support for the pathogenesis in hypothyroid experimental animals. Vinters et al. [67] suggested that major dendritic abnormalities, as observed in telangiectasia, could not have occurred if Purkinje cells had acquired a proper temporal and spatial relationship to parallel fibers. The immunocytochemical observations reported here appear to contradict this postulate. Parallel fibers remain largely intact, while Purkinje cells undergo extensive modification. Golgi impregnation of Purkinje cells in aged individuals also support the concept that these neurons undergo major remodeling during adult life [24]. For the stated reasons, it is assumed here that Purkinje cells in hereditary cortical cerebellar atrophy are initially quite normal but then undergo dendritic loss and ultimate cell death.

Certain monoclonal antibodies allow selective staining of Purkinje cells, even though the nature of the protein is not always fully known [72-74]. It can be postulated that a mutation produces the absence of a minor Purkinje cell-specific protein or an amino acid substitution in a major protein. Obviously, the defect cannot be of such magnitude that Purkinje cells are grossly altered in early life, since some patients with hereditary ataxia survive for decades. The alteration or absence of a specific protein conceivably initiates the relentless morphological process. While this causative mechanism is attractive and has proved rewarding in other heritable diseases, there is no current evidence that points toward the absence of a Purkinje cell-specific protein or a protein that is unique for the neurons of the dentate nucleus. Biochemical findings must also be reconciled with the many lesions that occur outside the cerebellum in OPCA and other forms of hereditary ataxia. It is also unlikely that immunocytochemistry will detect minor modifications of a protein. The absence of a major structural protein, for example, is not compatible with the late onset and prolonged survival that characterize some dominant ataxias. Nevertheless, proteins that are restricted or heavily concentrated in Purkinje cells or neurons of the dentate nucleus may be sequenced and offer the opportunity to develop oligonucleotide probes. Probes may, in turn, be used to examine complementary deoxyribonuleic acid libraries of human cerebellum, enhancing the likelihood of finding an important mutation.

ACKNOWLEDGMENTS

This work was supported by the Veterans Administration, the National Ataxia Foundation, the Alexander-von-Humboldt Foundation, and the Human Genetics Institute of Humburg University (Germany). Neurochemical Research, Inc. provided funds for the purchase of supplies. Several antisera (antibodies) were generously supplied by other scientists: Dr. Richard B. Vallee, Shrewsbury, MA, USA (MAP 1 and MAP 2); Dr. Donald E. Schmechel, Durham, NC, USA (GAD and NNE); Drs. Ulrich Walter and Suzanne Lohmann, Würzburg, Germany (cGK); and Dr. Charles Ouimet, New York, NY, USA (P38). Specimens were made available by Prof. Hans-Joachim Colmant, Hamburg, Germany, Dr. Philip V. Best, Aberdeen, Scotland; Drs. Robert D. Currier and Jose Bebin, Jackson, MS, USA; Dr. Horst P. Schmitt, Heidelberg, Germany; and Prof. Reinhard L. Friede, Göttingen, Germany.

REFERENCES

1. Holmes G. (1907). An attempt to classify cerebellar disease, with a note on Marie's hereditary cerebellar ataxia. Brain 30:545-567.
2. Chamberlain S., Shaw J., Rowland A., Wallis J., South S., Nakamura Y., von Gabain A.,

Farrall M., Williamson R. (1988). Mapping of mutation causing Friedreich's ataxia to human chromosome 9. Nature 334:248–250.
3. Jackson J.F., Currier R.D., Terasaki P.I., Morton N.E. (1977). Spinocerebellar ataxia and HLA-linkage. N. Engl. J. Med. 296:1138–1141.
4. The International Classification of Diseases. (1988). ICD 9 CM, 2nd ed. USDHHS. Vol. 1.
5. Menzel, P. (1891). Beitrag zur Kenntnis der Hereditären Ataxie und Kleinhirnatrophie. Arch. Psychiat. Nervenkrankh. 222:160–190.
6. Marie P., Foix C., Alajouanine T. (1922). De l'atrophie cérérebelleuse tardive à prédominance corticale. Rev. Neurol. 38:849–885; 38:1082–1111.
7. Harding A.E. (1981). Friedreich's ataxia: A clinical and genetic study of 90 families with an analysis of early diagnostic criteria and intrafamilial clustering of clinical features. Brain 104:589–620.
8. Bell J., Carmichael E.A. (1939). On hereditary ataxia and spastic paraplegia. In: R.A. Fisher (ed): The Treasury of Human Inheritance. Vol. 4. Cambridge: University Press, pp. 140–281.
9. Friedreich N. (1863). Ueber degenerative Atrophie der spinalen Hinterstränge. Virchows Arch. Pathol. Anat. Physiol. Klin. Med. 26:391–419.
10. Friedreich N. (1863). Ueber degenerative Atrophie der spinalen Hinterstränge. Virchows Arch. Pathol. Anat. Physiol. Klin. Med. 26:433–459.
11. Friedreich N. (1863). Ueber degenerative Atrophie der spinalen Hinterstränge. Virchows Arch. Pathol. Anat. Physiol. Klin. Med. 27:1–26.
12. Friedreich N. (1876). Ueber Ataxie mit besonderer Berücksichtigung der hereditären Formen. Virchows Arch. Pathol. Anat. Physiol. Klin. Med. 68:145–245.
13. Friedreich N. (1877). Ueber Ataxie mit besonderer Berücksichtigung der hereditären Formen. Virchows Arch. Pathol. Anat. Physiol. Klin. Med. 70:140–152.
14. Greenfield J.G. (1954). The Spino-Cerebellar Degenerations. Oxford: Blackwell.
15. Tyrer J.H. (1975). Friedreich's Ataxia. In: P.J. Vinken, G.W. Bruyn (eds): Handbook of Clinical Neurology, Vol. 21. Amsterdam: North Holland, pp. 319–364.
16. Oppenheimer D.R. (1979). Brain lesions in Friedreich's ataxia. Can. J. Neurol. Sci. 6:173–176.
17. Koeppen A.H., Mitzen E.J., Hans M.B., Barron K.D. (1986). Olivopontocerebellar atrophy: Immunocytochemical and Golgi observations. Neurology 36:1478–1488.
18. Berciano J. (1982). Olivopontocerebellar atrophy. J. Neurol. Sci. 53:253–272.
19. Autilio-Gambetti L., Crane R., Gambetti P. (1986). Binding of Bodian's silver and monoclonal antibodies to defined regions of human neurofilament subunits: Bodian's silver reacts with a highly charged unique domain of neurofilaments. J. Neurochem. 46:366–370.
20. Guillain G., Bertrand I., Mollaret P. (1932). Les lésions susmedullaires dans la maladie de Friedreich. C. R. Soc. Biol. 111:965–967.
21. Braitenberg J., Guglielmotti V., Sada E. (1967). Correlation of crystal growth with the staining of axons by the Golgi procedure. Stain Technol 42:277–283.
22. Fujisawa K., Nakamura A. (1982). The human Purkinje cell: A Golgi study in pathology. Acta Neuropathol. (Berlin) 56:255–264.
23. Ferrer I., Kulisevski J., Vasquez J., Gonzalez G., Pineda M., (1988). Purkinje cells in degenerative diseases of the cerebellum and its connections: A Golgi study. Clin. Neuropath. 7:22–28.
24. Kato T., Hirano A., Llena J.F. (1985). Golgi study of the human Purkinje cell soma and dendrites. Acta Neuropathol. (Berlin) 68:145–148.
25. Sholl D.A. (1953). Dendritic organization in the neurons of the visual and motor cortices of the cat. J. Anat. 87:387–406.
26. Huttenlocher P.R. (1975). Synaptic and dendritic development and mental defect. In: N.A. Buchwald (ed): Mechanisms in Mental Retardation. New York: Academic Press. pp. 123–140.
27. Hsu S.M., Raine L., Fanger H. (1981). Use of avidin-biotin-peroxidase complex (ABC) in immunoperoxidase techniques: A comparison between ABC and unlabeled antibody (PAP) procedures. J. Histochem. Cytochem. 29:577–580.
28. Mannoji H., Yeger H., Becker L.E. (1986). A specific histochemical marker (lectin *Ricinus communis* agglutinin 1) for normal human microglia, and application to routine histopathology. Acta Neuropathol. (Berlin) 71:341–343.

29. Oertel W.H., Schmechel D.E., Mugnaini E., Tappaz, M.L., Kopin I.J. (1981). Immunocytochemical localization of glutamate decarboxylase in rat cerebellum with a new antiserum. Neuroscience 6:2715-2735.
30. DeCamilli P., Miller P.E., Navone F., Theurkauf W.E., Vallee R.B. (1984). Distribution of microtubule-associated protein 2 in the nervous system of the rat studied by immunofluorescence. Neuroscience 11:819-846.
31. Bloom F.S., Schoenfeld T.A., Vallee R.B. (1984). Widespread distribution of the major polypetide component of MAP 1 (microtubule-associated protein 1) in the nervous system. J. Cell Biol. 98:320-330.
32. Huber G., Matus A. (1984). Immunocytochemical localization of microtubule-associated protein 1 in rat cerebellum using monoclonal antibodies. J. Cell Biol. 98:777-781.
33. Bernhardt R., Matus A. (1984). Light and electron microscopic studies of the distribution of microtubule-associated protein 2 in rat brain: A difference between dendritic and axonal cytoskeleton. J. Comp. Neurol. 226:203-221.
34. Marangos P.J., Schmechel D. (1980). The neurobiology of the brain enolases. In: M.B.H. Youdim, W. Lovenberg, D.F. Sharman, R. Lagnado, (eds): Essays in Neurochemistry and Neuropharmacology, Vol. 4. New York: John Wiley, pp. 211-247.
35. Schmechel D.E., Brightman M.W., Marangos P.J. (1980). Neurons switch from non-neuronal enolase to neuron-specific enolase during differentiation. Brain Res. 190:195-214.
36. Royds J.A., Parsons M.A., Taylor C.B., Timperley W.R. (1982). Enolase isoenzyme distribution in the human brain and its tumors. J. Pathol. 137:37-49.
37. Sternberger L.A., Sternberger N.H. (1983). Monoclonal antibodies distinguish phosphorylated and nonphosphorylated forms of neurofilaments in situ. Proc. Natl. Acad. Sci. USA 80:6126-6130.
38. Lee V., Wu H.L., Schlaepfer W.W. (1982). Monoclonal antibodies recognize individual neurofilament triplet protein. Proc. Natl. Acad. Sci. USA 79:6089-6092.
39. Goldstein M.E., Sternberger L.A., Sternberger N.H. (1983). Microheterogeneity ("neurotypy") of neurofilament proteins. Proc. Natl. Acad. Sci. USA 80:3101-3105.
40. Shiurba R.A., Eng L.F., Sternberger N.H., Sternberger L.A., Urich H. (1987). The cytoskeleton of the human cereberllar cortex: An immunohistochemical study of normal and pathological material. Brain Res. 407:205-211.
41. Walter U., Miller P., Wilson F., Menkes D., Greengard P. (1980). Immunological distinction between guanosine 3':5'-monophosphate-dependent and adenosine 3':5'-monophosphate-dependent protein kinases. J. Biol. Chem. 255:3757-3762.
42. Lohmann S., Walter U., Miller P.E., Greengard P., DeCamilli P. (1981). Immunohistochemical localization of cyclic GMP-dependent protein kinase in mammalian brain. Proc. Natl. Acad. Sci. USA 78:653-657.
43. DeCamilli P., Miller P.E., Levitt P., Walter U., Greengard P. (1984). Anatomy of cerebellar Purkinje cells in the rat determined by a specific immunohistochemical marker. Neuroscience 11:761-817.
44. Aswad G., Greengard P. (1981). A specific substrate from rabbit cerebellum for guanosine-3':5' monophosphate-dependent protein kinase. J. Biol. Chem. 256:3487-3493.
45. Aswad D., Greengard P. (1981). A specific substrate from rabbit cerebellum for guanosine-3':5' monophosphate-dependent protein kinase. J. Biol. Chem. 256:3494-3500.
46. Aitken A., Bilham T., Cohen P., Aswad D., Greengard P. (1981). A specific substrate from rabbit cerebellum for guanosine-3':5'-monophosphate-dependent protein kinase. J. Biol. Chem. 256:3501-3506.
47. Levitt P., Rakič P., DeCamilli P., Greengard P. (1984). Emergence of cyclic guanosine 3':5'-monophosphate-dependent protein kinase immunoreactivity in developing rhesus monkey cerebellum: Correlative immunocytochemical and electron microscopic analysis. J. Neurosci. 4:2553-2564.
48. Schlichter D.J., Detre J.A., Aswad D.W., Chehrazi B., Greengard P. (1980). Localization of cyclic GMP-dependent protein kinase and substrate in mammalian cerebellum. Proc. Natl. Acad. Sci. USA 77:5537-5541.
49. DeCamilli P., Cameron R., Greengard P. (1983). Synapsin I (protein I), a nerve terminal-specific phosphoprotein. I. Its general distribution in synapses of the central and peripheral nervous system demonstrated by immunofluorescence in frozen and plastic sections. J. Cell Biol. 96:1337-1354.

50. Jahn R., Schiebler W., Ouimet C., Greengard P. (1985). A 38,000-dalton membrane protein (P38) present in synaptic vesicles. Proc. Natl. Acad. Sci. USA 82:4137–4141.
51. Wiedenmann B., Franke W.W. (1988). Identification and localization of synaptophysin, and integral membrane protein of M_r 38,000 characteristic of presynaptic vesicles. Cell 41:1017–1028.
52. Palay S.L., Chan-Palay V. (1974). Cerebellar Cortex. New York: Springer.
53. Fañanás J.R. (1916). Contribucion al estudio de la neuroglio del cerebelo. Trab. Lab. Invest. Biol. Madrid 14:163–179.
54. Shaw G., Osborn M., Weber K. (1981). An immunofluorescence microscopical study of the neurofilament triplet proteins, vimentin, and glial fibrillary acidic protien within the adult rat brain. Eur. J. Cell Biol. 26:68–82.
55. Bovolenta P., Liem R.K.H., Mason C.A. (1984). Development of cerebellar astroglia: transitions in form and cytoskeletal content. Develop. Biol. 102:248–259.
56. Koeppen A.H., Dentinger M.P. (1988). Brain hemosiderin and superficial siderosis of the central nervous system. J. Neuropath. Exper. Neurol. 47:249–270.
57. Rakic P. (1971). Neuron-glia relationship during granule cell migration in developing cerebellar cortex. A Golgi and electron microscopic study in *Macacus rhesus*. J. Comp. Neurol. 141:283–312.
58. Sidman R.L., Rakic P. (1973). Neuronal migration with special reference to developing human brain: A review. Brain Res. 62:1–35.
59. De Blas A. (1984). Monoclonal antibodies to specific astroglial and neuronal antigens reveal the cytoarchitecture of the Bergmann glia fibers in the cerebellum. J. Neurosci. 4:265–273.
60. De Blas A., Cherwinski H.M. (1985). The development of the Bergmann fiber palisades in the cerebellum of the normal rat and in the weaver mouse. Brain Res. 342:234–241.
61. Hunt J.R. (1921). Dyssynergia cerebellaris myoclonica-primary atrophy of the dentate system. Brain 44:490–538.
62. Höpker W. (1951). Das Altern des Nucleus dentatus. Zeitschr. Altersforsch. 5:256–277.
63. Chan-Palay V. (1977). Cerebellar Dentate Nucleus. New York: Springer Verlag.
64. Petito C.K., Hart M.N., Porro R.S., Earle K.M. (1973). Ultrastructural studies of olivopontocerebellar atrophy. J. Neuropath. Exper. Neurol. 32:503–522.
65. Landis D.M.D., Rosenberg R.N., Landis S.C., Schut L., Nyhan W.L. (1974). Olivopontocerebellar degeneration: Clinical and ultrastructural abnormalities. Arch. Neurol. 31:295–307.
66. Subramony S.H., Smith E.E., Currier R.D., Jackson J.F., Bebin J., McCormick G.M., Fain D.B. (1986). Purkinje cell loss before clinical onset in dominant cerebellar ataxia. Neurology 36 (Suppl.):302.
67. Vinters H.V., Gatti R.A., Rakic P. (1985). Sequence of cellular events in cerebellar ontogeny relevant to expression of neuronal abnormalities in ataxia-telangiectasia. In: R. Gatti, M. Swift (eds): Ataxia-Telangiectasia: Genetics, Neuropathology, and Immunology of a Degenerative Disease in Childhood. New York, Alan R. Liss, pp. 233–255.
68. Von Santha K. (1930). Über die Entwicklungsstörungen der Purkinjeneurone. Arch. Psychiat. Nervenheilk 91:373–410.
69. Kirschbaum W., Eichholz A. (1932). Uber primäre Kleinhirnrindenatrophie. Deutsch Zeitschr. Nervenheilk 125:21–44.
70. Lotmar F. (1931). Entwicklungsstörungen in der Kleinhirnrinde beim endemischen Kretinismus. Zeitschr. ges Neurol. Psychiat. 130:412–435.
71. Mikoshiba K., Huchet M., Changeux J.-P. (1979). Biochemical and immunological studies on the P400 protein, a protein characteristic of the Purkinje cell from mouse and rat cerebellum. Dev. Neurosci. 2:254–275.
72. Weber A., Schachner M. (1982). Development and expression of cytoplasmic antigens in Purkinje cells recognized by monoclonal antibodies. Cell Tiss. Res. 227:659–676.
73. Yamakuni R., Usni H., Iwanaga T., Kondo H., Odani S., Takahashi Y. (1984). Isolation and immunohistochemical localization of a cerebellar protein. Neurosci. Lett. 45:235–240.
74. Hawkes R., Leclerc N. (1987). Antigenic map of the rat cerebellar cortex: the distribution of parasagittal bands as revealed by monoclonal anti-Purkinje cell antibody mabQll3. J. Comp. Neurol. 256:29–41.

10. CLINICAL NEUROPHYSIOLOGY IN OLIVOPONTOCEREBELLAR ATROPHY

S. CHOKROVERTY

1. INTRODUCTION

Olivopontocerebellar atrophy (OPCA) defines a chronic progressive degenerative neurological disorder in which morphological alterations are found, in the pontine, arcuate, lateral reticular, and olivary nuclei, and in the cerebellar cortex [1]. The term *OPCA* was introduced by Dejerine and Andre-Thomas in 1900 [2], who described under this heading two patients, a 53-year-old woman and a 44-year-old man, presenting with progressive gait difficulty due to incoordination and dysarthria resulting from a progressive cerebellar syndrome without any history of similar affection in the family members. Dejerine-Thomas performed a postmortem examination in one of these patients and found the morphological lesions as described above. Since their description of OPCA, numerous cases have appeared in the literature over the years. Eadie [1] reviewed 35 pathologically confirmed cases of OPCA, and later on Berciano [3] included 117 pathologically verified cases. Both the clinical and pathological findings have gradually evolved over the years since the original description and included several other systems [1]. Particularly noteworthy is the involvement of the basal ganglia, as well as structural alterations in the corticospinal tract and the anterior horn cells, and sometimes the posterior column of the spinal cord [1]. The term *OPCA* continues to generate controversy since its original description. Before Dejerine-

A. Plaitakis (ed.), *CEREBELLAR DEGENERATIONS: CLINICAL NEUROBIOLOGY*. Copyright © 1992.
Kluwer Academic Publishers, Boston. All rights reserved.

Thomas [2], similar clinical-pathological findings have been described in a neurodegenerative disease with autosomal dominant inheritance by Menzel and is now included under the heading of dominant OPCA (Menzel type) [1]. A review of the literature clearly shows the heterogeneity of the clinical and pathological, and even biochemical, findings in OPCA.

Classification of OPCA has been difficult because of lack of clear-cut diagnostic criteria. Greenfield [4], in his monograph on spinocerebellar degeneration, classified OPCA as a mixed form of spinocerebellar degeneration and mentioned that it was not possible to classify OPCA based on the clinical phenomenology. Koningsmark and Weiner [5] sought to classify OPCA into five distinct types and also included two other types. However, their classification is controversial because of inclusion of the same Schut family members into two different categories. Also, their type 2 (Fickler-Winkler type) may not fulfill all the criteria of OPCA. A classification based on the clinical pattern recognition, pathological findings, and genealogical data has been attempted by various authors, but no uniform agreement has been reached. An attempt has been made to identify a specific biochemical defect in OPCA, but even in that endeavor it was noted that OPCA is a heterogeneous disorder [6,7]. Recently, Plaitakis et al. [8] noted partial deficiency of the enzyme glutamate dehydrogenase (GDH) in leukocytes and skin fibroblasts in three patients with OPCA. Their reports were followed by other reports [7,9–16] showing a similar deficiency in OPCA patients. We [10,11] were unable to identify a distinctive clinical-physiological pattern in these cases. Moreover, a serious question has recently been raised by us [17] and others [18] regarding the specificity of partial GDH deficiency in OPCA. We [17] and others [18,19] have found such a partial GDH deficiency in other neurodegenerative diseases and in many normal individuals. Therefore, we [17] revised our original opinion and agreed with Aubbey et al. [18] that low GDH deficiency does not correlate with a specific clinical type of OPCA. Our suggestion has further been strengthened by the biochemical and postmortem findings in our second OPCA patient [20].

2. ROLE OF CLINICAL NEUROPHYSIOLOGY IN OPCA

Thus, a brief survey of the literature suggests that the diagnosis of OPCA remains problematical from the clinical pattern recognition only [21]. Furthermore, this disorder closely resembles several other neurodegenerative disease, such as Parkinson's disease, striatonigral degeneration, progressive supranuclear palsy, corticobasal degeneration, the Shy-Drager syndrome, other types of multiple system atrophy, sometimes even amyotrophic lateral sclerosis, as well as the common demyelinating disease (multiple sclerosis). An attempt has, therefore, been made to find laboratory tests that might be specific for the diagnosis of OPCA. Neuroimaging studies may help in many cases by showing the pontine and cerebellar atrophy, but these findings may not be specific for OPCA. A variety of neurophysiologic tests have been developed and applied with a view to finding some diagnostic abnormalities in OPCA. Again, it was found that there was no specific test available, but a combination of neurophysiologic tests may help in differentiating OPCA from some of the diseases mentioned above. Herein lies the importance of clinical neurophysiology in OPCA. In this chapter I will review the role of neurophysiologic tests in the diagnosis and differential diagnosis of OPCA. This review is based on a survey of the literature, as well as our experience with 37 cases of OPCA [11,22] diagnosed on the basis of clinical pattern recognition and neuroimaging findings, and postmortem observations in three patients. For the purpose of this review I will define OPCA as a chronic slowly progressive neurodegenerative disorder, with or without other affected family members, starting with cerebellar or extrapyramidal symptoms (atypical Parkinsonism) or a mixed cerebellar-Parkinsonian syndrome, but always with some cerebellar dysfunction on clinical examination or on neuroimaging study (CT or MRI) showing cerebellar atrophy. Subacute to chronic cerebellar degeneration or dysfunction resulting from a variety of known causes must be excluded by an appropriate history, physical findings, and laboratory investigations before making a diagnosis of OPCA.

3. PHYSIOLOGICAL TESTS IN OPCA

A variety of physiological tests may be performed to document dynamic functional deficits in many patients with OPCA. Neurophysiologic tests to evaluate various systems in OPCA will be discussed under the following headings.

3.1. Clinical neurophysiology of the peripheral somatic nervous system

Tests that may document dysfunction of the afferent (sensory) and efferent (motor) somatic fibers of the peripheral nervous system may include electromyography (EMG) of the proximal and distal muscles; a sensory-motor nerve conduction study, including an F response study to document proximal motor conduction; and the H reflex to test the afferent, efferent, and spinal segmental reflex arc [11].

Peripheral neuropathy has not been recorded in the apparently sporadic forms of OPCA [1]. Impaired nerve conduction has been reported in patients with dominantly inherited OPCA [23,24] and in dominantly inherited spinocerebellar degeneration, who also had peroneal atrophy and pas cavus, in addition to pyramidal tract signs [25]. In cases of different categories of spinocerebellar degeneration, including OPCA, Dyck and Ohta [26] found impairment of sensory nerve conduction. EMG findings in 1 of 3 GDH-deficient OPCA by Plaitakis et al. [8] also suggested a neuropathic process. McLeod and Evans [27] found abnormalities of sensory nerve conduction in 9 of 19 patients with hereditary OPCA or cerebello-olivary degeneration. They also observed a significant impairment of sensory conduction and mild slowing of motor conduction in the lateral popliteal nerve in the whole group as compared with controls.

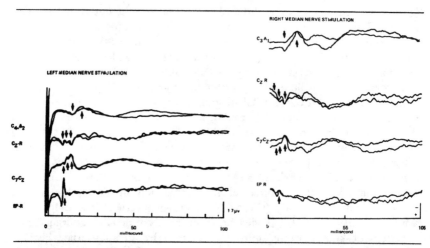

Figure 10-1. Median nerve short latency somatosensory evoked response. Left side: Normal control. Right side: A patient with OPCA; note prolonged N13/14–N20 and EP-N20. C4-A2/C3-A1: Arrows point to P13–14 and N20. Cz-R: Arrows point to P9, P11, and P13–14. C7-Cz: Arrows point to N9, N11, and N13–14. EP-R: Arrow points to Erb's point potential (peak). (Reprinted with permission from Chokroverty et al. [11].)

We obtained EMG and nerve-conduction study in 30 patients with OPCA [11,22]. Family history suggested a dominant mode of inheritance in two patients, and in the remaining 28 the disease was either sporadic or recessive. EMG showed evidence of diffuse and widespread anterior horn cell dysfunction in two patients, resembling the findings noted in patients with amyotrophic lateral sclerosis. Nerve conduction study, including the F response and H reflex, showed evidence of a sensory-motor, predominantly sensory, axonal type of neuropathy distally in the legs in 16 patients. Eight of the 16 patients had purely sensory, four had mixed sensory-motor neuropathy with equal affection, and four had mixed sensory-motor but predominantly sensory neuropathy. Of the eight patients with pure sensory neuropathy, four also had absent H reflex and, therefore, in the remaining four patients, the only abnormality was reduced to absent sural nerve action potentials. These four patients were all older individuals (more than 60 years of age), and these findings may possibly be age related in these patients, rather than indicating a sensory neuropathy related to OPCA.

3.2. Clinical neurophysiology of the central somatic nervous system

These tests may document dysfunction of the afferent and the efferent fibers in the central nervous system. The *afferent conduction* may be studied by obtaining somatosensory evoked responses (SER) after stimulation of the median nerve (MN) [11] at the wrist and the peroneal (PN) or the tibial nerve in the lower limbs. Specific afferent conduction may be observed in the visual pathways by obtaining pattern-reversal visual evoked response (PVER) and in auditory pathways by obtaining brain-stem auditory evoked responses (BAER) [11]. For details regarding the technique of obtaining the evoked potentials, the readers may consult the monograph by Chiappa [28]. Recently, central motor conduction has also been studied by applying high-voltage percutaneous electrical stimulation [29] or magnetic stimulation [30] over the surface of the head and obtaining the compound muscle action potentials in the abductor digiti minimi or the abductor pollicis brevis muscles in the hands or the tibialis anterior and gastrocnemius-soleus muscles in the legs.

The anatomical structures conducting MN-SER, PN-SER, and BAER are in close contiguity to be structures thought to be degenerated in OPCA. Therefore, in theory, the MN-SER, PN-SER, and BAER findings may be abnormal in these patients, but the reported observations have been contradictory. The available data are consistent with the generally accepted contention that OPCA is a heterogeneous group of disorders. Another reason for the conflicting observations may have been the lack of a clear definition of OPCA and the consequent inclusion of patients with a variety of hereditary and sporadic cerebellar and spinocerebellar ataxias under the category of OPCA. Furthermore, the findings may also indicate that there may be selective degeneration of one system of fibers, with sparing of contiguous fibers belonging to another system.

Nuwer et al. [31] described normal N20, and Erb's point to N20 interpeak latencies in six patients with OPCA. On the other hand, abnormal MN-SER findings (prolonged central conduction from the spinal cord to the cerebral cortex) were described by Pedersen and Trojaborg [32] in 6 of 11 patients with hereditary ataxia. Anziska and Cracco [33] observed abnormal MN-SER (absent N20 in all, 1 with absent P13–14, and another with prolonged interpeak latencies) in four patients with dominant cerebellar ataxia. The clinical features of these patients were not provided. Hammond and Wilder [34] also found abnormal MN-SER in one and impaired later components in two OPCA patients. In our own study [11,22] of MN-SER in 18 patients with OPCA, we found a prolonged central conduction time (N13–N20) in six patients (Figure 10-1). We [22] obtained somatosensory evoked response studies in 11 patients after bilateral stimulation of the peroneal nerves at the knees and found normal absolute and interpeak latencies of the responses in all patients.

Nuwer et al. [31] found abnormal BAER in all five OPCA patients they tested. Their findings agreed with those of Gilroy and Lynn [35], who reported abnormal BAER in three patients with OPCA. Hammond and Wilder [34] also had abnormal BAER in 1 of 2 OPCA patients. Pedersen and Trojaborg [32] found abnormal BAER results in only 3 of 11 patients with hereditary cerebellar ataxia. But Fujita et al. [36] reported normal BAER

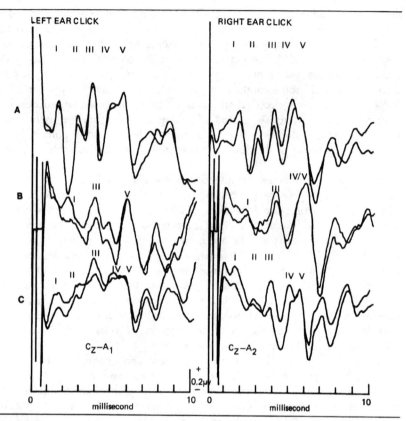

Figure 10-2. Brain-stem auditory evoked response. A. Normal control. B. A patient with prolonged waves I, III, and V but normal interpeak latencies. C. A patient with OPCA showing a normal response. (Reprinted with permission from Chokroverty et al. [11].)

results in 20 patients with spinocerebellar degeneration, consisting of 11 parenchymatous cerebellar and 9 with multiple system atrophy not sufficiently defined to be called OPCA. Satya-Murti et al. [37] also reported normal BAER findings in two patients with OPCA. In our own study [11,22] in 28 patients, we found normal BAER findings in 20 and abnormal findings in eight patients. In 2 of these 8 patients, prolongation of wave I but normal interpeak latencies of wave I–III and wave III–V suggested peripheral auditory nerve dysfunction (Figure 10-2). Six of these eight patients showed evidence of pontomedullary auditory pathway dysfunction (prolonged interpeak latency of waves I–III in four and prolonged interpeak latency of waves III–V in two).

Nuwer et al. [31] stated that most patients with inherited ataxias other than

Friedreich's ataxia had prolonged P100 latencies, but no details about OPCA patients were given. One of their six OPCA patients had normal PVER. Bird and Crill [38] reported normal P100 latencies in three OPCA patients. Pedersen and Trojaborg [32] found abnormal PVER in 5 of 11 patients with hereditary cerebellar ataxia. Asselman et al. [39] noted abnormal PVER in 3 of 5 patients with spinocerebellar degeneration, but no clinical information was given. Moller et al. [40] found delayed visual evoked responses in members of a family with dominant spinocerebellar ataxia, but these authors also provided no clinical details. Hammond and Wilder [34] reported abnormal PVER in two OPCA patients. Visual evoked potentials, studied in 2 of 3 GDH-deficient patients by Plaitakis et al. [8], were normal in one and borderline abnormal in the other. We [11,22] studied PVER in 26 patients with OPCA. We found mildly prolonged latency of the major positive wave (P100) in five, implying mild optic nerve dysfunction (Figure 10-3).

Conduction in central motor pathways
Conduction in the central motor pathways by transcranial electrical or magnetic stimulation of the motor cortex from the surface [29,30] may evaluate the function of the corticospinal tract, which may be degenerated in many patients with OPCA. Claus et al. [41] studied central motor conduction to small hand muscles using magnetic stimulation of the motor cortex and electrical stimulation of the proximal cervical roots over the cervical spinal cord in 13 patients with late-onset cerebellar degeneration, including patients with OPCA. They observed prolonged central motor conduction in 38% of patients with late-onset cerebellar degeneration. They concluded that central motor conduction abnormalities were not specific to any individual disorder, but in patients with various degenerative ataxic disorders, they found the highest abnormality in patients with Friedreich's ataxia and the lowest in patients with late-onset cerebellar degeneration, including OPCA.

What is the cause of central and peripheral axonal degeneration in OPCA? Greenfield [4] in 1954 introduced the term *dying back* to describe the degeneration in spinocerebellar degeneration. The general concept of the dying back neuropathy was later elaborated by Cavanagh [42]. Later the term *central-peripheral distal axonopathy* was introduced by Spencer and Schaumburg [43] to illustrate the earliest changes in peripheral and central distal axons. These authors [43] and Dyck and Ohta [26] stated that central-peripheral distal axonopathy is an important manifestation in spinocerebellar degeneration, including OPCA. Thomas [44] also speculated that a similar pattern of degeneration may possibly occur in OPCA but has not been adequately characterized, either electrophysiologically or histopathologically.

3.3. Autonomic neurophysiology in OPCA

Studies based on stimulation and ablation experiments have confirmed that the cerebellum plays an important role in the control of the autonomic

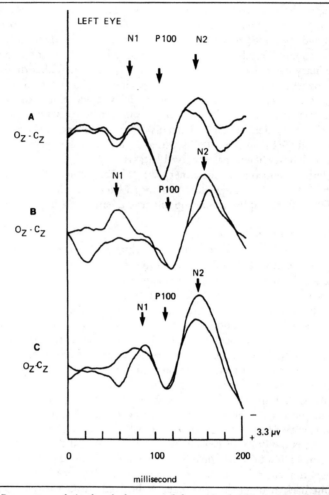

Figure 10-3. Pattern-reversal visual evoked response (left eye stimulation). A. A patient with OPCA showing normal P100 latency. B. A patient with OPCA showing mild prolongation of P100 latency. C. Normal control. (Reprinted with permission from Chokroverty et al. [11].)

nervous system (ANS) functions [45]. Mitchell [46] and Ingram [47] summarized their past observation on the role of the cerebellum in ANS function.

Miura and Reis [48] and Achari and Downman [49] described the pressor responses by electrical stimulation of the rostral fastigial nucleus, called the *fastigial pressor responses*, which resulted from widespread sympathetic activation. Doba and Reis [50] found that stimulation of the rostral fastigial nucleus in cats caused an increase in blood pressure, tachycardia, and an increase in peripheral vascular resistance. The responses noted by Doba and Reis [50] closely approximated the cardiovascular reflex responses evoked by the assumption of the upright posture. Doba and Reis [50] later produced bilateral rostral fastigial electrolytic lesions in cats and noted that there was an impairment of the compensatory reflex response on tilting. The lesions in the nucleus interpositus did not impair these responses. These findings suggested that cerebellum actively participated in the regulation of orthostatic cardiovascular responses. Experiments in cats by Lutherer et al. [51] proved that the cell bodies within the rostral fastigial nucleus, and not the fibers of passage, were implicated in the neural control of the cardiovascular system.

Cerebellar modulation of gastrointestinal and urinary bladder function has been documented, but the results have been inconsistent. Martner [52] concluded that a restricted fastigial area in the cerebellum linked the parasympathetic and sympathetic systems and affected virtually all types of autonomic functions, including reflexes concerned with bulbospinal homeostatic mechanisms. All these would result in cardiovascular, gastrointestinal, or urinary bladder adjustments. Thus, there is substantial evidence pointing to cerebellar participation in the ANS. Hence, it may be possible to document autonomic dysfunction in patients with cerebellar degeneration showing lesions of the fastigial nucleus or the fastigiobulbar projections.

A number of tests are available for evaluation of the function of the ANS [53–55]. It is possible to localize the lesions to the sympathetic or parasympathetic division and to the preganglionic or postganglionic sites to assess the integrity of the entire ANS. Many of these tests can be performed in the clinical neurophysiology laboratory.

It is possible to test the entire baroreceptor reflex arc function by performing tilt-table study and the Valsalva maneuver. The Valsalva maneuver is an invasive test, but the Valsalva ratio (longest R-R/shortest R-R interval in the electrocardiograph) is a simple noninvasive test of cardiac autonomic function [56].

Sympathetic efferent pathways can be tested by performing the following tests: cold pressure and mental arithmetic tests; pharmacological tests, such as norepinephrine infusion, plasma norepinephrine levels in the supine and erect positions, plasma renin activity in the supine and erect positions, and intraocular epinephrine (1:1000), 1% hydroxyamphetamine, and 4% cocaine to test pupillary sympathetic function; sympathetic skin response [57]; and sudomotor function.

Tests to determine the integrity of the parasympathetic efferent pathways consist of the following: atropine and neostigmin tests; beat-to-beat variation of the heart rate; intraocular pilocarpine (0.062%); insulin-induced gastric acidity; radiological study for gastrointestinal motility; and cystometrogram.

The involvement of the ANS is rarely mentioned in the literature on spinocerebellar degeneration, including OPCA [58–60]. In those types with involvement of the peripheral nerves, ANS dysfunction has been mentioned sporadically [59,60]. Vasomotor disturbances resulting from preferential involvement of the postganglionic sympathetic nerve fibers have been described in some patients with Charcot-Marie-Tooth disease [59]. Miyazaki [61] evaluated 45 patients with cerebellar degeneration and found orthostatic hypotension in 20 patients, four of whom had the Shy-Drager syndrome. Two of the four patients with an abnormal Valsalva response had the Shy-Drager syndrome. The author did not separate the Shy-Drager syndrome from OPCA and used the term *multiple system atrophy* for these two conditions, in agreement with some other authors. One should, however, note that in his Table 10-2 Miyazaki listed autonomic symptoms preceding cerebellar and other neurological features in the Shy-Drager syndrome by 1–5 years, while this sequence was reversed in other patients with cerebellar degeneration.

Structural lesions in the locus ceruleus (catecholaminergic neurons), nucleus ambiguus, and dorsal motor nucleus of the vagus (parasympathetic neurons) have been described in OPCA [1,3,4,62,63]. Staal et al. [64] examined neuropathologically three siblings in one generation of five patients with an unusual form of dominant OPCA and described a unique observation of abundant intrafascicular calcification in the sympathetic nerve fibers and their ganglia, in addition to abnormalities of the peripheral nerves, atrophy of the spinal cord, and the typical findings of OPCA. Clinically their patients did not have any symptoms attributable to autonomic nerve dysfunction. Whether there are lesions in the intermediolateral neurons of the spinal cord and the sympathetic ganglia or other autonomic structures in classical OPCA cannot be stated with certainty.

The central autonomic and somatic neurons, which are known to degenerate in OPCA, are in close proximity, and there is morphological evidence of degeneration of some central autonomic neurons in OPCA. It is, therefore, logical to expect degeneration of the autonomic and somatic structures in parallel fashion in OPCA. It is, however, not known how commonly or severely the ANS is involved in OPCA, becuase a systematic study of ANS has not been undertaken in a large number of OPCA patients. Patients with OPCA do not usually complain of symptoms related to orthostatic hypotension, such as dizziness or faint feeling in the erect posture. This may be due to the fact that the autonomic deficits are mild. It should be noted that urinary sphincter dysfunction, which may be a manifestation of autonomic neuropathy, has been described by Dejerine and Andre-Thomas, and by Menzel in some cases of OPCA [1,2].

Clinical expression of dysautonomia may be mild and evident only in formal tests of autonomic function. In the author's personal experience, autonomic function tests demonstrate deficits in only some OPCA patients. Recently the author [53,65], along with coinvestigators, studied autonomic function in 10 OPCA patients; five patients demonstrated a fall of 20–50 mmHg systolic blood pressure in the erect position, accompanied by

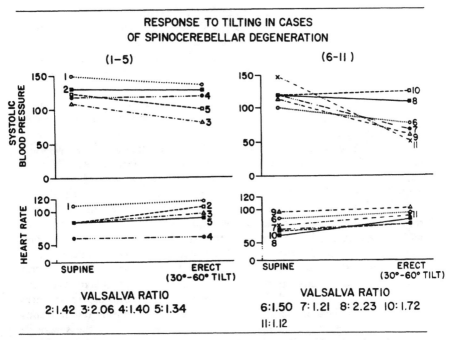

Figure 10-4. Effect of head-up tilt on systolic blood pressure (BP) and heart rate in patients with spinocerebellar degeneration (OPCA).

postural symptoms of dizziness, faintness, or sweating in five patients and a fixed heart rate in two (Figure 10-4). These findings indirectly suggested a critical reduction of cerebral blood flow and impaired circulatory homeostasis in the erect posture. One patient had a low Valsalva ratio and absence of post-Valsalva reflex bradycardia, indicating interruption of the total baroreceptor reflex arc. We suggested that postural hypotension without impairment of the Valsalva response in four patients may be consistent with a central defect above the level of the medulla. Impaired cold pressor and mental arithmetic tests and fixed heart rate in the erect position in two patients implied efferent sympathetic dysfunction. On the other hand, normal cold pressor and mental arithmetic tests and tachycardia on tilting in the remaining patients indicated intact efferent cardiac sympathetic pathways.

In five patients we observed low supine plasma renin or its failure to rise adequately in the upright position in the presence of normal urinary sodium, suggesting a reduction of impulse traffic from supraspinal centers to the sympathetic nerve terminals. A normal supine level of plasma norepinephrine and its failure to rise adequately in the erect position in one patient without postural hypotension may also indicate reduced impulse traffic at the sympathetic nerve endings due to reduced descending impulses from the supraspinal centers, which were not directly concerned with mediation of circulatory reflexes. A similar suggestion was made by Aminoff and Wilcox [66] to explain norepinephrine supersensitivity without postural hypotension in cases of paralysis agitans. Thus, autonomic dysfunction in some patients with OPCA, as manifested by orthostatic hypotension, impaired valsalva response, cold pressor and mental arithmetic tests, and a subnormal rise of plasma renin and norepinephrine in the erect posture suggested a sympathetic defect.

The exact site of the responsible central lesion causing autonomic dysfunction in OPCA remains speculative. Lesions in the nucleus tractus solitarius, where the baroreceptor neurons are presumed to be located, the supramedullary controlling pathways to the nucleus tractus solitarius, the fastigial nucleus of the cerebellum, fastigiobulbar fibers, and the intermediolateral columns of the spinal cord may all be implicated [53]. In one of the patients with autonomic dysfunction, neuropathological examination [65] at necropsy confirmed the clinical diagnosis of OPCA and showed marked neuronal loss and gliosis in the substantia nigra, pons, cerebellum, locus ceruleus, inferior olives, and lateral reticular nucleus of the medulla [65]. There was widespread loss of Purkinje cells in the cerebellum, along with severe degeneration of the fastigial nucleus. The nucleus tractus solitarius did not show significant morphologic changes. Thus, in this patient [65] the severe degeneration of the fastigial nucleus appears to be the most likely site of lesion responsible for the autonomic deficit. Adequate data regarding the extent and frequency of involvement of the parasympathetic nervous system in OPCA are not available. Thus, patients with OPCA (familial or sporadic) may manifest mild to moderate dysautonomia. It is often subclinical but may be elicited by special tests of autonomic function. A discussion of the autonomic dysfunction in OPCA cannot be complete without a brief review of the Shy-Drager variant of primary orthostatic hypotension syndrome [53,54,67], a multisystem neurodegenerative disease with prominent dysautonomia. There are striking similarities in the clinical and morphological findings between the Shy-Drager syndrome and the classical OPCA. We believe that the severity and initial presentation of dysautonomic manifestations, the shorter duration of the illness, and the findings of severe loss of neurons of the intermediolateral cell column of the thoracic spinal cord separate the Shy-Drager syndrome from the usual case of OPCA [53]. Whether there are distinct biochemical or neurochemical differences between these conditions remains to be determined. There are also certain differences between those patients with Parkinson's disease, OPCA, and striatonigral degeneration who may later develop evidence of dysautonomia and those presenting initially with dysautonomia who later show somatic neurological manifestations (cerebellar or Parkinsonian or both) [53]. The course of the illness, the biochemical findings in general, and the response to various therapeutic modalities appear to be different. Generally the course is relentlessly progressive and is shorter in multiple system atrophy with progressive autonomic failure (Shy-Drager syndrome) than in other degenerative diseases [53]. The data presently available appear to distinguish the Shy-Drager syndrome from classical Parkinson's disease, striatonigral degeneration, and OPCA, all of which share some common clinical and morphological findings [53]. The patients with the Shy-Drager syndrome and striatonigral degeneration and OPCA do not respond as well to levodopa treatment as do patients with Parkinson's disease. The Shy-Drager syndrome appears to be a distinct entity rather than simply representing one end of the spectrum of neurodegenerative disease.

3.4. Clinical neurophysiology of abnormal movements

Some patients with OPCA may have abnormal involuntary movements, e.g., tremor, myoclonus, and dystonia.

3.4.1. Tremor

The physiological characteristics of cerebellar and other types of tremors may be understood by accelerometric study with spectral analysis of average epochs to quantitate tremors [68–71]. Some of the physiological characteristics of the various types of tremors may be summarized as follows: (1) Rest tremor: The Parkinsonian rest tremor has a frequency of 3–5 Hz and the EMG shows characteristic alternating agonist and antagonist bursts. (2) Physiological tremor, including exaggerated physiological tremor, has a frequency in the range of 8–12 Hz. The EMG bursts are usually synchronous in agonist-antagonist muscles. (3) Essential tremor: The frequency is faster than the rest tremor but is slower than the physiological tremor and is in the range of 5–8 Hz. EMG bursts are mostly synchronous but can be alternating. (4) Intention tremor: This is associated with cerebellar disease and shows rhythmic alternating EMG bursts as the target is approached. (5) Neuropathic tremor: Both synchronous and alternating activities are seen in the agonist-antagonist EMG at a frequency of 6–8 Hz. This may be seen in patients with peripheral neuropathy. (6) Hysterical tremor: Mostly this tremor resembles action tremors with alternating agonist-antagonist EMG bursts and varying tremor frequency.

3.4.2. Myoclonus [68,72–78]

Myoclonus may be defined as sudden shocklike involuntary muscle jerks occurring synchronously or asynchronously, symmetrically or asymmetrically, diffusely or focally, rhythmically or arhythmically, and originating generally in the central nervous system. Myoclonus may be seen in some patients with OPCA. Physiologically the myoclonic EMG bursts are brief, 10–100 msec in duration, seen synchronously in the agonist and antagonist muscles in the reflex type of myoclonus. The true myoclonus generally does not last more than 250–300 msec and sometimes up to 500 msec. Prolonged bursts up to 500–1000 msec or longer may be seen in patients with myoclonic dystonia or dystonic myoclonus. According to the site of origin of the discharge, myoclonus can be physiologically classified into following three types: cortical, brain stem or subcortical, and spinal myoclonus. The myoclonus in cerebellar disease, such as what may be noted in OPCA, is generally subcortical in type, and its characteristics may be summarized as follows: (1) The jerks are generalized and more proximal than distal. (2) The muscles are activated up the brain stem and down the spinal cord. (3) The burst duration is long, usually 100–150 msec and sometimes longer. (4) EEG spikes may be associated with muscle jerks, but these are not time locked to the EMG event. (5) The cortical evoked potential amplitude is normal. (6) The transcortical reflex is not hyperactive, except in reticular reflex myoclonus, where this reflex is hyperactive.

The four most important tests to study myoclonus consist of polymography, jerk-locked backaveraging (EEG-EMG correlate), transcortical reflex, and somatosensory evoked potential amplitude. In stimulus-sensitive myoclonus, the study of the startle reflex may be helpful in some cases.

3.4.3. Dystonia [68,79–81]

Dystonia in the form of torsion dystonia or torticollis may be noted in some patients with OPCA. In fact, Menzel's original case developed spasmodic torticollis with facial contortions [1]. Physiologically there are certain characteristics noted that may be helpful in the diagnosis of dystonia and are summarized below [79–81]: (1) EMG of the involuntary movements in dystonia may show one or all of the following three patterns: (a) continuous EMG activities lasting from 2 to 30 sec followed by a brief silence, (b) rhythmic repetitive cocontractions lasting from 1–2 sec, followed by a relative EMG silence of 1–2 sec, (c) brief irregular muscle jerks, similar to myoclonic bursts of 50–100 msec in duration. (2) EMG characteristics of the voluntary movements in dystonia [79–81] show preservation of normal reciprocal activation of agonist-antagonist muscles during rapid flexion-extension movements. However, during voluntary movements requiring precision and delicate maneuvers, these bursts are prolonged and show excessive cocontractions in the agonist and antagonist muscles, with overflow to other muscles. In some patients after ballistic movements, the first agonist burst is prolonged. (4) The transcortical reflex may be hyperactive in some patients [68], but this may be due to the fact that the patients could not relax the muscles. (5) Median somatosensory evoked potentials are normal. (6) The backaveraging technique shows no time-locked EEG activity preceding the dystonic muscle spasms. (7) Reciprocal inhibition: In dystonia there is impairment of reciprocal inhibition between agonist and antagonist muscles [79–81]. This may be found both in focal and generalized dystonia. This impairment may cause excessive cocontraction between agonist and antagonist muscles. Reciprocal inhibition may be studied by applying radial

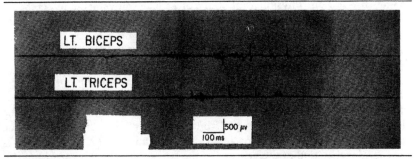

Figure 10-5. Characteristic triphasic EMG bursts in the agonist (left biceps) and antagonist (left triceps) muscles in a normal subject after ballistic elbow flexion.

nerve "conditioning" stimuli in the spiral groove before, during, and after the delivery of median "test" stimuli at the elbow and recording the H reflex amplitudes in the flexor carpi radialis muscles [79–83]. The H-reflex amplitude alterations are noted and are expressed as a percentage of the control amplitude. Day et al. [82] found three periods of inhibition (0, 10, and 75 msec intervals between the "conditioning" and the "test" stimuli) while studying reciprocal inhibition between agonist and antagonist muscles of the forearm in normal human subjects. Dystonic patients show impairment of all three periods of inhibition.

3.5. Dysfunction of the voluntary movements in OPCA

The cerebellum plays an important part in the control of voluntary movements by interacting with the cortical and subcortical circuits [68,84]. The cerebellum and basal ganglia have two-way connections with various regions of the cerebral cortex. The cerebellum also has direct connections with the brain stem and indirectly connects with the spinal cord, and these two subcortical inputs greatly influence the cerebral cortex in the control of voluntary movement. The cerebellum participates in the initiation and coordination of the movements; the basal ganglia helps in influencing the direction, force, and amplitude of the movements; and the cerebral cortex, via the supplementary motor area, plans and programs the movements. The corticospinal system then commands and executes the movements, and the spinal segmental motor apparatus finalizes these movements. A breakdown in any of these circuits may cause impairment of the voluntary movements.

One type of voluntary movement that may show impairment in cerebellar disease, such as OPCA, is the ballistic movement [85]. In normal individuals ballistic movements (e.g., rapid elbow flexion or extension) show the characteristic triphasic EMG bursts (Figure 10-5): an initial agonist burst followed by the antagonist burst, and later the final agonist burst. The duration of each burst is usually from 50 to 150 msec, and the antagonist burst is a little bit longer in duration. Physiological study of ballistic movements in cerebellar ataxia may show derangement of the normal triphasic EMG pattern: prolongation of the initial agonist and subsequent antagonist bursts.

3.6. Clinical neurophysiology of sleep and breathing in OPCA

The nature of alterations of sleep and breathing in neurological disorders generally depends upon the anatomical location of the lesions [86]. Experimental studies in animals have shown distinct cerebellar influence on the sleep-wakefulness mechanisms [87], but the role of the cerebellum on the respiratory control mechanisms in sleep is not known. In OPCA there is primary degeneration of the neurons of the lateral reticular nuclear group [1], which has projections to the medial reticular formation. Pathological alterations in OPCA are also found in the locus ceruleus and other neighboring structures in the pons and medulla [1,3]. The respiratory and hypnogenic neurons are located in these areas [86]. Therefore, there may be concomitant degeneration of these neurons in OPCA.

There are only isolated reports of alterations of sleep and breathing in patients with OPCA [88–95]. Osorio and Daroff [88] reported absent REM sleep in two patients with OPCA with slow eye movements and speculated that degeneration of the structures in the region of the paramedian pontine reticular formation was responsible for the slow eye movements and absent REM. On the other hand, Masdeu et al. [89] observed the presence of REM sleep in a patient with OPCA and slow saccadic eye movements, but they suggested that the presence and absence of REM sleep may depend upon the course and severity of the illness. Neil et al. [90] described several abnormalities of sleep in two cases of familial OPCA after performing an all-night polysomnographic study. These authors found reduced REM, decreased slow-wave sleep, and increased stage I non-REM sleep. These authors further stated that the phasic components of REM sleep were more severely affected than the tonic components. The authors suggested that these findings may have resulted from lesions of the locus ceruleus and pontine tagmentum that have been observed in OPCA. Salva and Guilleminault [91] reported an abnormal control of muscle tone after polygraphic study during sleep in two patients with OPCA. They noted bursts of EMG activity throughout both non-REM and REM sleep, and a progressive disappearance of muscle atonia (that is, abnormally high muscle tone was present during a large segment of REM sleep) during REM. This type of REM sleep without muscle atonia has been described following lesions in the pontine tagmentum. Shimizu et al. [92] and Schenck et al. [93] also described cases of OPCA showing the absence of REM sleep muscle atonia and accompanied by dream-enacting behavior. Thus OPCA patients may show a REM sleep behavior disorder.

In a daytime polygraphic study, Chokroverty et al. [65] documented repeated episodes of central, upper airway obstructive and mixed apneas

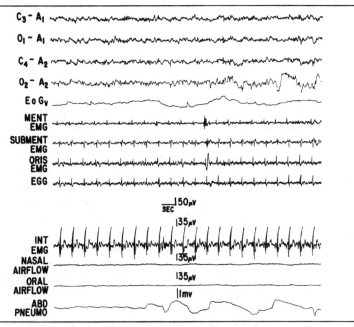

Figure 10-6. Polygraphic recording in a patient with OPCA showing four channels of EEG (C3-A1; O1-A1; C4-A2; O2-A2); vertical electrooculogram (EOGv); electromyograms (EMGs) of mental (MENT), submental (SUBMENT), orbicularis oris (ORIS), and intercostal (INT) muscles; electroglossogram (EGG); nasal and oral airflow and abdominal pneumogram (ABD PNEUMO). Note the portion of an episode of mixed apnea (initial central followed by upper airway obstructive) during stage 2 non-rapid eye-movement sleep. (Reprinted with permission from Chokroverty et al. [65].

(Figure 10-6) during sleep in 5 of 10 patients with OPCA. The apneas ranged from 10 to 62 sec in duration, with an apnea index of 30–55. Three patients had pure central apnea and two had all three types of apneas. Most of the apneas occurred during non-REM sleep stage II. In one of these patients at necropsy, the nucleus tractus solitarius was found to be normal, but the lateral reticular nucleus showed neuronal loss and gliosis. Later we [22] described two more patients with OPCA showing sleep apnea, and thus in a total of 12 patients, seven patients documented sleep apnea. Adelman et al. [94] also found mixed central and obstructive sleep apnea in a case of OPCA. Salazar-Grueso et al. [95] documented abnormal sleep architecture and central sleep apnea in a patient with autosomal-dominant OPCA who improved on long-term trazodone treatment. It should be noted that sleep apnea is well documented in patients with multiple system atrophy with progressive autonomic failure (the Shy-Drager syndrome), which clinically and pathologically resembles OPCA. The two most important tests to document various sleep abnormalities, daytime alertness and sleep-related respiratory dysrhythmias, consist of all-night and daytime polysomnography [96] and the multiple sleep latency test [97]. For sleep staging and scoring in adults, the guidelines and criteria developed by Rechtschaffen and Kales [98] are currently being followed.

4. CONCLUSIONS

OPCA thus shows heterogeneity not only in the clinical, pathological, and biochemical findings, but also in the clinical neurophysiologic deficits. This heterogeneity is in part a reflection of difficulty in defining the entity succinctly, but also in part is due to varying involvement of the different systems of the body. Some authors [99–102], therefore, suggested that we should discard the term OPCA in favor of *multiple system atrophy* (MSA). I do not subscribe to this view. We are using a nonspecific term, *MSA*, to "lump" many neurodegenerative disorders of unknown etiology, and in my opinion this does not advance our understanding of the disease. On the other hand, excessive "splitting" will also not help understand the pathogenesis of the disorder. I would, therefore, suggest that we retain the term OPCA, which has withstood the test of time since its introduction by Dejerine-Thomas [2] at the beginning of this century.

As discussed previously, despite its heterogeneity, it is important to perform a combination of neurophysiologic tests, particularly in cases of OPCA without the "classical" presentation. Many neurodegenerative diseases and multiple sclerosis may resemble OPCA, at least in the early stage. From the standpoint of clinical neurophysiologic tests, generally normal peripheral neurophysiologic functions; normal central conduction, except sometimes impairment of the visual evoked potentials; mild ANS dysfunction; and sleep abnormalities will favor a diagnosis of Parkinson's disease. Severe ANS dysfunction preceding somatic neurological disorder, normal peripheral and central somatosensory and motor conduction, and the presence of sleep-related breathing abnormalities will suggest a diagnosis of the Shy-Drager syndrome. In corticobasal degeneration and striatonigral degeneration, sufficient studies have not been performed, but no particular neurophysiologic abnormalities have been noted, except for nonspecific mild diffuse EEG background slowing. In progressive supranuclear palsy, sometimes EEG [103] abnormalities (nonspecific slowing and synchronous rhythmic delta or focal epileptiform abnormalities in the temporal region) may be noted, in addition to sleep abnormalities. However, the ANS function and the other peripheral and central neurophysiologic tests are normal. In amyotrophic lateral sclerosis, EMG findings suggest diffuse anterior horn cell dysfunction in the presence of normal or mild slowing of motor conduction; generally the somatosensory evoked potential is normal but the central motor conduction may be impaired. The sleep and autonomic neurophysiologic tests in amyotrophic lateral sclerosis usually do not show any particular abnormality, although in some patients there may be evidence of mild ANS defects on laboratory testing. In multiple sclerosis, the peripheral neurophysiologic tests should be completely normal. There may be abnormalities in the afferent and efferent central conduction, suggesting the involvement of multiple systems. Some patients may have sleep and sleep-related breathing abnormalities, and sometimes ANS deficits have been noted on laboratory tests.

In conclusion, an understanding of the clinical neurophysiology may help us in clearly defining the entity of OPCA, and neurophysiologic tests may be important in monitoring the progression of the disease and in assessing the prognosis and efficacy of a possible drug treatment in the future.

REFERENCES

1. Eadie M.J. (1975). Olivopontocerebellar atrophy. In: P.J. Vinken, G.W. Bruyn (eds): Handbook of Clinical Neurology, Vol. 21. Amsterdam: North-Holland Publishing, pp. 415–457.
2. Dejerine J., Thomas A. (1900). L'atrophie olivo-ponto-cerebelleuse. Nouv Lcon de la Salpet 13:330–370.
3. Berciano J. (1982). Olivopontocerebellar atrophy: A review of 117 cases. J. Neurol. Sci. 53:253–272.
4. Greenfield J.G. (1954). The Spinocerebellar Degeneration. Springfield, IL: Charles C. Thomas.
5. Konigsmark B.W., Weiner L.P. (1970). The olivo-ponto-cerebellar atrophies: A review. Medicine (Baltimore) 49:227–241.
6. Perry T.L. (1984). Four biochemically different types of dominantly inherited olivopontocerebvellar atrophy. In: R.C. Duvoisin, A. Plaitakis (eds): The Olivopontocerebellar Atrophies, Advances in Neurology, Vol. 41. New York: Raven Press, pp. 205–216.
7. Maruyama S., Yamaguchi T. (1984). Glutamate and pyruvate dehydrogenase deficiency in spinocerebellar degeneration. In: R.C. Duvoisin, A. Plaitakis (eds): The Olivopontocerebellar Atrophies, Advances in Neurology, Vol. 41. New York: Raven Press, pp. 255–265.
8. Plaitakis A., Nicklas W.J., Desnick R.J. (1980). Glutamate dehydrogenase deficiency in 3 patients with spinocerebellar syndrome. Ann. Neurol. 7:297–303.
9. Duvoisin R.C., Chokroverty S., Lepore F., Nicklas W. (1983). Glutamate dehydrogenase deficiency in patients with olivopontocerebellar atrophy. Neurology 33:1322–1326.
10. Duvoisin R.C., Chokroverty S. (1984). Clinical expression of gluitamate dehydrogenase deficiency. In: R.C. Duvoisin, A. Plaitakis (eds): The Olivoponocerebellar Atrophies, Advances in Neurology, Vol. 40. New York: Raven Press 40:267–279.
11. Chokroverty S., Duvoisin R.C., Sachdeo R., Sage J., Lepore F., Nicklas W. (1985). Neurophysiologic study of olivopontocerebellar atrophy with or without glutamate dehydrogense deficiency. Neurology 35:652–659.
12. Yamaguchi I., Hayashi K., Murakami H., Ota K., Maruyama S. (1982). Glutamate dehydrogenase deficiency in spinocerebellar degeneration. Neurochem Res. 7:627–636.
13. Konagaya Y., Konayaga M., Mano Y., Takayanagi T. (1986). Glutamate dehydrogenase and its isoenzyme activity in olivopontocerebellar atrophy. J. Neurol. Sci. 74:231–236.
14. Sorbi S., Tonini S., Giannini E., Piacenti S., Marinin P., Amaducci L. (1986). Abnormal platelet glutamate dehydrogenase activity and activation in dominant and nondominant olivopontocerebellar atrophy. Ann. Neurol. 19:239–245.
15. Finocchiaro G., Taroni F., Di Donato S. (1986). Glutamate dehydrogenase in olivopontocerebellar atrophies: Leucocytes, fibroblasts, and muscle mitochondria. Neurology 36:550–553.
16. Kajiyama K., Veno S., Tatsumi T., Yorifuji S., et al. (1988): Decreased glutamate dehydrogenase protein in spinocerebellar degeneration. J. Neurol. Neurosurg. Psychiatry 51:1078–1080.

17. Duvoisin R.C., Nicklas W.J., Ritchie V., Sage J., Chokroverty S. (1988). Low leucocyte glutamate dehydrogenase activity does not correlate with a particular type of multiple system atrophy. J. Neurol. Neurosurg. Psychiatry 51:1508–1511.
18. Aubby D., Saggu H.K., Jemmer P., Quinn N.P., Harding A.H., Marsden C.D. (1988). Leucocyte glutamate dehydrogenase activity in patients with degenerative neurological disorders. J. Neurol. Neurosurg. Psychiatry 51:893–902.
19. Hugon J., et al. (1989). Glutamate dehydrogenase (GDH) and aspartate aminotransferase (AAT) in leucocytes of patients with motoneuron disease. Neurology 39:956–958.
20. Chokroverty S., et al. (1990). Multiple system degeneration with glutamate dehydrogenase deficiency: Pathology and biochemistry. J. Neurol. Neurosurg. Psychiatry 53:1099–1101.
21. Duvoisin R.C. (1987). The olivopontocerebellar atrophies. In: C.D. Marsden, S. Fahn (eds): Movement Disorders 2. London: Butterworths, pp. 249–269.
22. Chokroverty S. (1990). Neurophysiological study in olivopontocerebellar atrophy. Electroencephalogr. Clin. Neurophysiol. 76:91 page abstract.
23. Currier R.D., Glover G., Jackson J.F., et al. (1972). Spinocerebellar ataxia: Study of large kindred. Part I. General information and genetics. Neurology 22:1040–1043.
24. Subramony S.H., Currier R.D. (1983). Peripheral nerve involvement in late-onset familial ataxias. Muscle and Nerve 6:537 abstract.
25. Frey H.J., Frey M.L., Riekkinen P.J., et al. (1973). A family with autosomal dominant spinocerebellar ataxia, with electrophysiological finding. Ann. Clin. Res. 5:163–167.
26. Dyck P.J., Ohta M. (1984). Neuronal atrophy and degeneration predominantly affecting peripheral sensory neurons. In: P.J. Dyck, P.K. Thomas, E.H. Lambert (eds): Peripheral Neuropathy, Vol. II. Philadelphia: W.B. Saunders, pp. 791–824.
27. McLeod J.G., Evans W.A. (1981). Peripheral neuropathy in spinocerebellar degenerations. Muscle and Nerve 4:51–61.
28. Chiappa K. (1990). Evoked Potentials in Clinical Medicine. New York: Raven Press.
29. Merton P.A., Morton H.B. (1980). Stimulation of the cerebral cortex in the intact human subject. Nature 285:227.
30. Barker A.T., Jalinous R., Freeston I.L. (1985). Non-invasive magnetic stimulation of the human motor cortex. Lancet 2:1106–1107.
31. Nuwer M.R., Perlman S.L., Packwood J.W., Kark A.P. (1983). Evoked potential abnormalities in the various inherited ataxias. Ann. Neurol. 13:20–27.
32. Pederson L., Trojaborg W. (1981). Visual, auditory and somatosensory pathway involvement in hereditary cerebellar ataxia, Friedreich's ataxia and familial spastic paraplegia. Electroencephalogr. Clin. Neurophysiol. 52:283–297.
33. Anziska B.J., Cracco R.Q. (1983). Short-latency somatosensory evoked potentials to median nerve stimulation in patients with diffuse neurologic disease. Neurology 33:989–993.
34. Hammond E.J., Wilder B.J. (1983). Evoked potentials in olivopontocerebellar atrophy. Arch. Neurol. 40:366–369.
35. Gilroy J., Lynn G.E. (1978). Computerized tomography and auditory evoked potentials: Use in the diagnosis of olivopontocerebellar degeneration. Arch. Neurol. 35:143–147.
36. Fujita M., Hosoki M., Miyazaki M. (1981). Brainstem auditory evoked responses in spinocerebellar degeneration and Wilson disease. Ann. Neurol. 9:43–47.
37. Satya-Murti S., Cacace T., Hanson P.A. (1979). Abnormal auditory evoked potentials in hereditary motor-sensory neuropathy. Ann. Neurol. 5:445–448.
38. Bird T.D., Crill W.E. (1981). Pattern-reversal visual evoked potentials in the hereditary ataxias and spinal degenerations. Ann. Neurol. 9:243–250.
39. Asselman P., Chadwick D.W., Marsden C.D. (1975). Visual evoked responses in the diagnosis and management of patients suspected of multiple sclerosis. Brain 98:261–282.
40. Moller E., Hindfelt B., Olsson J. (1978). HLA-determination in families with hereditary ataxia. Tissue Antigens 12:357–366.
41. Claus D., Harding A.E., Hess C.W., Mills K.R., Murray N.M.F., Thomas P.K. (1988). Central motor conduction in degenerative ataxic disorders: A magnetic stimulation study. J. Neurol. Neurosurg. Psychiatry 51:790–795.
42. Cavanagh J.B. (1964). The significance of the "dying-back" process in experimental and human neurological diseases. Int. Rev. Exp. Pathol. 3:219.
43. Spencer P.S., Schaumburg H. (1976). Central-peripheral distal axonopathy—the pathology

of dying-back polyneuropathies. In: H.M. Zimmerman (ed): Progress in Neuropathology, Vol. III, New York: Grune and Stratton, pp. 253–295.
44. Thomas P.K. (1982). Selective vulnerability of the centrifugal and centripetal axons of primary sensory neurons. Muscle and Nerve 5:S117–121.
45. Dow R.S., Moruzzi G. (1958). The Physiology and Pathology of the Cerebellum. Minneapolis, MN: University of Minnesota Press.
46. Mitchell G.A.G. (1953). Anatomy of the Autonomic Nervous System. Edinburgh: Livingstone.
47. Ingram W.R. (1960). Central autonomic mechanisms. In: Handbook of Physiology, Section I, Neurophysiology, Vol. 2, p. 951. Washington DC: American Physiology Society.
48. Miura M., Reis D.J. (1970). A pressor response from fastigial nucleus and its related pathway in brainstem. Am. J. Physiol. 219:1330–1336.
49. Achari N.K., Downman C.D.D. (1970). Autonomic effector responses to stimulation of nucleus fastigius. J. Physiol. 210:637–659.
50. Doba N., Reis D.J. (1972). Cerebellum: Role in reflex cardiovascular adjustment to posture. Brain Res. 13:495–500.
51. Lutherer L.O., Williams J.L., Everse S.J. (1989). Neurons of the rostral fastigial nucleus are responsive to cardiovascular and respiratory challenges. J. Autonom. Nerv. Syst. 27:101–112.
52. Martner J. (1975). Cerebellar influence on autonomic mechanisms. Acta Physiol. Scand. 425(Suppl.):8–35.
53. Chokroverty S. (1984). Autonomic dysfunction in olivopontocerebellar atrophy. In: R.C. Duvoisin, A. Plaitakis (eds): The Olivopontcerebellar Atrophies, Advances in Neurology, Vol. 41. New York: Raven Press, pp. 105–141.
54. Chokroverty S. (1986). The Shy-Drager Syndrome. Neurology and Neurosurgery Update Series 7(2):1–8.
55. Bannister R. (1988). Autonomic Failure. Oxford: Oxford University Press, 2nd ed.
56. Levin A.B. (1966). A simple test of cardiac function based upon the heart rate changes induced by the valsalva maneuver. Am. J. Cardiol. 18:90–99.
57. Shahani B.T., Halperin J.J., Boulu P., Cohen J. (1984). Sympathetic skin response—a method of assessing unmyelinated axon dysfunction in peripheral neuropathies. J. Neurol. Neurosurg. Psychiatry 47:536–542.
58. Bank W.A., Morrow G. (1972). A familial spinal cord disorder with hyperglycemia. Arch. Neurol. 27:136–144.
59. Jammes J.L. (1972). The autonomic nervous system in peroneal muscular atrophy. Arch. Neurol. 27:218–220.
60. Cartlidge N.E.F., Bone G. (1973). Sphincter involvement in hereditary spastic paraplegia. Neurology 23:1160–1163.
61. Miyazaki M. (1980). Shy-Drager syndrome: A nosological entity? The problem of orthostatic hypotension. In: I. Sobue (ed): Spinocerebellar Degeneration, Baltimore: University Park Press, pp. 35–43.
62. Oppenheimer D.R. (1976). Diseases of the basal ganglia, cerebellum and motor neurons. In: W. Blackwood, J.A.N. Corsellis (eds): Greenfield's Neuropathology, Chicago: Year Book Medical, pp. 608–651.
63. Lapresle J., Annabi A. (1979). Olivopontocerebellar atrophy with velopharyngolaryngeal paralysis—A contribution to the somatotopy of the nucleus ambiguus. J. Neuropathol. Exp. Neurol. 38:401–406.
64. Staal A., Stefanko S.Z., Busch H.F.M., Jennekens F.G.I., DeBruijn W.C. (1981). Autonomic nerve calcification and peripheral neuropathy in olivopontocerebellar atrophy. J. Neurol. Sci. 51:383–394.
65. Chokroverty S., Sachdeo R., Masdeo J. (1984). Autonomic dysfunction and sleep apnea in olivopontocerebellar degeneration. Arch. Neurol. 41:926–931.
66. Aminoff M.J., Wilcox C.S. (1971). Assessment of autonomic function in patients with a Parkinsonian syndrome. Br. Med. J. 4:80–84.
67. Shy G.M., Drager G.A. (1960). A neurological syndrome associated with orthostatic hypotension. Arch. Neurol. 2:511–527.
68. Chokroverty S. (1990). An approach to a patient with disorders of voluntary movements.

In: S. Chokroverty (ed): Movement Disorders. Costa Mesa, CA: PMA Publishing, pp. 1–43.
69. Lee R.G. (1987). The pathophysiology of essential tremor. In: C.D. Marsden, S. Fahn (eds): Movement Disorders 2. Boston: Butterworths, pp. 423–437.
70. Gresty N.A., Findley L.J. (1984). Definition, analysis and genesis of tremor. In: L.J. Findley, R. Capildeo (eds): Movement Disorders: Tremor. London: MacMillan, pp. 15–26.
71. Hallett M. (1986). Electrophysiological evaluation of tremor and central disorders of movement. In: M.J. Aminoff (ed): Electrodiagnosis in Clinical Neurology, 2nd ed. New York: Churchill Livingstone, pp. 385–401.
72. Chokroverty S., Manocha M., Duvoisin R.C. (1987). A physiologic and pharmacologic study in anticholinergic-responsive essential myoclonus. Neurology 37:608–615.
73. Hallett M., Chadwick D., Marsden C.D. (1979). Cortical reflex myoclonus. Neurology 29:1107–1125.
74. Marsden C.D., Hallett M., Fahn S. (1982). The nosology and pathophysiology of myoclonus. In: C.D. Marsden, S. Fahn (eds): Movement Disorders, Butterworths International Medical Reviews. London: Butterworths, pp. 196–248.
75. Obeso J.A., Rothwell J.C., Marsden C.D. (1985). The spectrum of cortical myoclonus. Brain 108:193–224.
76. Marsden C.D., Obeso J.A., Rothwell J.C. (1983). Clinical neurophysiology of muscle jerks: Myoclonus, chorea and tics. In: J.E. Desmedt (ed): Motor Control Mechanism in Health and Disease. New York: Raven Press, pp. 865–881.
77. Hallett M., Marsden C.D., Fahn S. (1986). Myoclonus. In: P.J. Vinken, G.W. Bruyn, H.L. Klawans (eds): Handbook of Clinical Neurology, Vol. 49. Amsterdam: Elsevier, pp. 609–625.
78. Hallett M., Chadwick D., Adam J., Marsden C.D. (1987). Reticular reflex myoclonus: A physiological type of human posthypoxic myoclonus. J. Neurol. Neurosurg. Psychiatry 40:253–264.
79. Rothwell J.C., Obeso J.A., Day B.C., Marsden C.D. (1983). Pathophysiology of dystonia. In: J. Desmedt (ed): Advances in Neurology, Vol. 39: Motor Control Mechanisms in Health and Disease. New York: Raven Press, pp. 851–864.
80. Rothwell J.C., Obeso J.A. (1987). The anatomical and physiological basis of torsion dystonia. In: C.D. Marsden, S. Fahn (eds): Movement Disorders 2. Stoneham, MA: Butterworths, pp. 313–331.
81. Marsden C.D. (1984). The pathophysiology of movement disorders. In: J. Jankovic (ed): Neurologic Clinics, Vol. 2, No. 3. Philadelphia: W.B. Saunders, pp. 435–459.
82. Day B.L., Marsden C.D., Obeso J.A., Rothwell J.C. (1984). Reciprocol inhibition between the muscles of the human forearm. J. Physiol. 349:519–534.
83. Baldisseri F., Campodelli P., Cavalleri P. (1983). Inhibition from radial group Ia afferents of H-reflex in wrist flexors. Electromyogr. Clin. Neurophysiol. 23:187–193.
84. Alexander G.E., DeLong M.R. (1986). Organization of supraspinal motor systems. In: A.K. Asbury, G.M. McKhann, W.I. McDonald (eds): Disease of the Nervous System, Vol. I. Philadelphia: W.B. Saunders Company, pp. 352–369.
85. Hallett M. (1983). Analysis of abnormal voluntary and involuntary movements with surface electromyography. In: J.E. Desmedt (ed): Motor Control Mechanisms in Health and Disease, Advances in Neurology, Vol. 39. New York: Raven Press, pp. 907–914.
86. Chokroverty S. (1986). Sleep and breathing in neurological disorders. In: N.H. Edelman, T.V. Santiago (eds): Breathing Disorders of Sleep. New York: Churchill Livingston, pp. 225–264.
87. Cunchillos J.D., DeAndre's I. (1982). Participation of the cerebellum in the regulation of the sleep-wakefulness cycle. Results in cerebellectomized cats. Electroencephalogr. Clin. Neurophysiol. 53:549–558.
88. Osorio I., Daroff R.D. (1980). Absence of REM and altered NREM sleep in patients with spinocerebellar degeneration and slow saccades. Ann. Neurol. 7:277–280.
89. Masdeu J., Chokroverty S., Gorelick P. (1981). Absence of REM sleep in patients with spinocerebellar degeneration and slow saccades. Ann. Neurol. 9:95–96.
90. Neil J.F., Holzer B.C., Spiker D.G., et al. (1980). EEG sleep alterations in olivopontocerebellar degeneration. Neurology 30:660–662.

91. Salva M.A.Q., Guilleminanlt C. (1986). Olivopontocerebellar degeneration, abnormal sleep and REM sleep withouth atonia. Neurology 36:576–577.
92. Shimizu, Sugita Y., Teshima Y., Hishikaws Y. (1981). Sleep study in patients with spinocerebellar degeneration and related diseases. In: W.P. Koella (ed): Sleep 1980. Basel: S. Karger, pp. 435–437.
93. Schenck Ch., Bundlie S.R., Ettinger M.G., Mahowald M.W. (1986). Chronic behavioral disorders of human REM sleep: A new category of parasomnia. Sleep 9:293–308.
94. Adelman S., Dinner D.S., Goren H., Little J., Nickerson P. (1984). Obstructive sleep apnea in association with posterior fossa neurologic disease. Arch. Neurol. 41:509–510.
95. Salazar-Grueso E.F., Rosenberg R.S., Roos R.P. (1988). Sleep apnea in olivopontocerebellar degeneration: Treatment with trazodone. Ann. Neurol. 23:399–401.
96. Chokroverty S., Sharp J.T. (1981). Primary sleep apnea syndrome. J. Neurol. Neurosurg. Psychiatry 44:970–982.
97. Carskadon M.A., et al. (1986): Guidelines for multiple sleep latency test (MSLT): A standard manner of sleepiness. Sleep 9(4):519–524.
98. Rechtschaffen A., Kales A. (1968). A Manual of Standardized Terminology, Techniques and Scoring Systems for Sleep Stages of Human Subjects. Los Angeles: Brain Information Service/Brain Research Institute.
99. Harding A.E. (1987). Commentary: Olivopontocerebellar atrophy is not a useful concept. In: C.D. Marsden, S. Fahn (eds): Movement Disorders 2. London: Butterworths, pp. 269–271.
100. Graham J.G., Oppenheimer D.R. (1969). Orthostatic hypotension and nicotine sensitivity in a case of multiple system atrophy. J. Neurol. Neurosurg. Psychiatry 32:28–34.
101. Oppenheimer D.R. (1980). Multiple system atrophy and the Shy-Drager syndrome. In: Spinocerebellar Degeneration, I. Sobue (ed): Baltimore: University Park Press, pp. 165–170.
102. Takei Y., Mirra S.S. (1973). Striatonigral degeneration: A form of multiple system atrophy with clinical Parkinsonism. In: H.M. Zimmerman (ed): Progress in Neurolopathology. New York: Grune & Stratton, pp. 217–251.
103. Nygaard T.G., Duvoisin R.C., Manocha M., Chokroverty S. (1989). Seizures in progressive supranuclear palsy. Neurology 39:138–140.

11. PATHOPHYSIOLOGY OF ATAXIA IN HUMANS

H.-C. DIENER AND J. DICHGANS

1. INTRODUCTION

Disorders of the cerebellum result in clinical signs and symptoms that were comprehensively described and summarized by Holmes in 1917 [1], 1922 [2–5], and 1939 [6], and later by Dow and Moruzzi [7], as well as Gilman et al. [8]. Holmes described that lesions of the lateral parts of the posterior cerebellar hemisphere cause dys-synergia, dysmetria, dysdiadochokinesia, and dysarthria. Damage to the cerebellar vermis and the anterior lobe result in ataxia of stance and gait. Lesions of the vestibulocerebellum cause deficits in retinal image stabilization and damage of the dorsal vermis causes deficient eye saccades (see Chapter by Fetter and Dichgans). Consequently, despite its homogeneous intrinsic anatomical structure, the cerebellum may be subdivided into functional compartments, and specific motor functions can be ascribed to these subunits [8–11]. The specification obviously is due to the specific afferent and efferent connections of each of the functional subdivisions. Despite considerable knowledge from animal and human physiology and pathophysiology [9,11], we still lack a widely accepted hypothesis of the principle way, in which the uniform anatomical structure of the intrinsic cerebellar network contributes to the control of movement.

This chapter (1) will describe the features of pathological vs. normal limb movements and body posture in cerebellar disorders and (2) will attempt to

A. Plaitakis (ed.), CEREBELLAR DEGENERATIONS: CLINICAL NEUROBIOLOGY. Copyright © 1992.
Kluwer Academic Publishers, Boston. All rights reserved.

262 II. Clinical neurosciences of the cerebellum

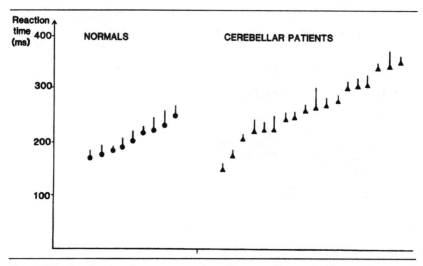

Figure 11-1. Reaction times in an arm extension task measured in terms of the onset latency of EMG activity in the anterior deltoid muscle in nine normal subjects (median and standard deviations) and 18 patients with diffuse cerebellar atrophy. Note the overlap in reaction times in normal subjects and cerebellar patients.

describe the physiology and pathophysiology of motor disturbances in as far as they are understood up to now.

The discussion of the role of the cerebellum in both short- and long-term motor learning, including adaptation of basic reflexes, such as the vestibulo-ocular reflex and conditioned reflexes, as well as disturbances of eye movements, will not be addressed in this chapter. The reader is referred to the chapters by Ito and Fetter and Dichgans in this volume.

2. PATHOPHYSIOLOGY AND CLINICAL SYMPTOMS OF CEREBELLAR DISORDERS OF LIMB MOVEMENTS

2.1. Delay of movement initiation

Delayed initiation of movement is best tested clinically by having the patient touch the index finger of the examiner with his index finger (outstretched arm). The patient is instructed to follow with his finger the abruptly initiated horizontal arm movements of the examiner as quickly as possible. This procedure also visualizes almost all of the other parametric abnormalities involved in ataxia of the arm (discussed below), as well as difficulties of postural stabilization of the trunk during arm movement.

The delay in the initiation of a movement has been observed both in humans with cerebellar disorders and in animals with cerebellar lesions [1,12–18]. The delay in movement onset can be seen in proximal (Figure 11-1) and distal joints and in fast elbow, wrist, and finger movements [26], as well as in slow visuo-motor tracking tasks [19]. Lesions of the dentate nucleus (the output relay of the cerebellar hemispheres) in monkeys resulted in increased reaction times (RT) when triggered by visual and auditory stimuli [20]. Increased RT were seen after cooling of the dentate nucleus, irrespective of whether simple or complex movements were performed [17]. This result suggests that the cortico-ponto-neocerebellar loop, with its efferents through the dentate and thalamus back to the motor cortex, takes part in the initiation of fast movements. In accordance with this hypothesis, it has been observed that neurons in the neocerebellar cortex and dentate nucleus change their discharge frequency prior to activation of cortical motoneurons [23,25] and that movement-related responses of cortical neurons are delayed when there is dentate dysfunction [13,20–22]. Cerebellar lesions do not cause a global depression of motor-cortex neurons, but lead to a decrease in phasic motor-cortex neural discharge in some neurons [23–25]. Changes in the tonic background activity of spinal interneurons due to cerebellar lesions that could in theory result in hypotonia are unlikely to explain the increased reaction time. We have observed delayed reaction times in patients with cerebellar atrophy and normal muscle tone [26].

One should also keep in mind that delayed reaction times are not specific for cerebellar damage and have been observed in patients with Parkinson's disease with some [27] but not other experimental conditions [28] and in primates with pyramidal tract lesions [29].

2.2. Disorders of movement termination (dysmetria) and of movement velocity and acceleration (dyskinesia)

According to our clinical observations, hypermetria can most often be observed in fast movements, hypometria occurs with slow movements of small amplitude. In a recent series of experiments we studied fast, goal-directed movements at the elbow, wrist, and finger in a reaction-time paradigm [18]. The investigation of cerebellar patients almost always showed an excessive extent of movement (hypermetria). This was seen in all three joints investigated. Hypermetria was more marked for aimed movements of small (5°) amplitudes. A characteristic of cerebellar disordered movements, which was present at all amplitudes (5°,30°,60°), at all joints, and in all six patients was an asymmetry with decreased peak accelerations and increased peak decelerations compared to normal movements [18]. Elbow flexions in monkeys performed during reversible cooling of the cerebellar nuclei, when compared with control movements of the same peak velocity, showed smaller magnitudes of acceleration and larger magnitudes of deceleration [30].

A similar asymmetry of movement parameters in cerebellar patients has also recently been reported by others [31]. However, in this case movements were characterized by short accelerations and long decelerations. Results of the two studies cannot easily be compared because the patients from the second study [31] had mild cerebellar impairments, duration rather than magnitude was studied, and the asymmetry was seen for slow but not fast movements. Hallett et al. described a prolonged acceleration time in simple rapid elbow flexions in patients with cerebellar deficits [32].

Fast movements of the arm are performed with a triphasic EMG pattern (Figure 11-2A) [33–34]. Under normal conditions, the first change observed is an inhibition of tonic antagonistic activity. The movement starts with a large initial burst of activity in the agonist. This burst is followed by a silent period and then a second burst of agonist activity. In the antagonist, there is a single burst that occurs at about the time of the agonist silent period and acts to brake the movement. The general opinion is that this pattern is centrally preprogrammed, released by the motor cortex, but that it can be influenced by peripheral feedback.

The disorder of acceleration in hypermetric movements was associated with agonist EMG activity that was less abrupt in onset, smaller in magnitude, and more prolonged in duration (Figure 11-2E) [18]. The disorder of deceleration was associated with delayed onset of phasic antagonistic EMG activity (Figure 11-2D) [18]. This was also true for monkeys with cooling of the dentate nucleus [30]. A delay in antagonistic activity within the three-burst pattern of human elbow and thumb movements was also observed in humans by Marsden et al. [12] and by Hallett and coworkers [32]. The delay of antagonist activity is not restricted to fast monoarticular limb movements, but can also be observed during multijoint movements such as throwing [35].

The cerebellum possibly specifies the cortical movement command and sends it back to the motor cortex [36]. This would be consistent with the delayed initiation, the decrease in phasic components of cortical motoneuron discharge, and the decreased accelerations found during cerebellar cooling in monkeys [25].

Conceptually hypometria can result from two different mechanisms: (1) Prolongation of the braking antagonist activity (Figure 11-2B) or (2) delayed persistence of antagonistic activity beyond movement initiation. Hallett et al. [32] observed a delay in onset of the initial antagonistic inhibition when patients with cerebellar lesions were asked to perform ballistic elbow flexions against tonic triceps activity. The consequence was a cocontraction of biceps and triceps at movement onset that resulted in hypometria (Figure 11-2C).

Lack of *rebound* conceptually is a form of reflexive hypermetria. This function is assessed by resisting the forceful intent of the patient to flex his elbow. In normal subjects, a sudden release of the opposing force results in a short movement of the forearm, which comes immediately to a stop and returns to its initial position (rebound). In patients with cerebellar lesions, the movement continues and the unchecked movement of the forearm can even strike against the patient's chest or nose. EMG recordings from elbow muscles in normal subjects showed a silent period in biceps and activity in triceps after a latency of about 50 msec following the release of the

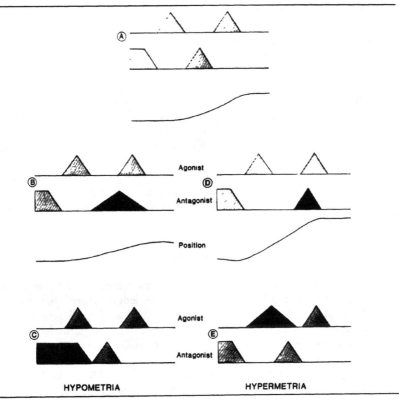

Figure 11-2. **A**: EMG pattern in agonist and antagonist in a fast goal-directed limb movement depicted schematically according to Hallett et al. [34]. The initial inhibition of tonic antagonistic activity is followed by a triphasic pattern of EMG activity in agonist and antagonist. **B,C**: Hypometria may result from either prolonged braking activity in the antagonist or delayed inhibition of the antagonist at movement onset. **D,E**: Hypermetria can result from delayed braking activity in the antagonist or prolonged activity in the agonist accelerating the limb.

isometrically contracted biceps muscle. This reciprocally organized pattern could not be observed in three patients with cerebellar lesions. EMG activity in the biceps continued after the release, and triceps activity was substantially delayed [37].

Studies of the sequential activation of agonists and antagonists involved in a multijoint movement [35] and of its coordination with the postural stabilization point to the only partially understood complexity of motricity to which the cerebellum contributes in terms of the temporal coordination of selected muscular activation and inhibition as well as the determination of force. *Decomposition of movement* (see below) is the result of cerebellar dysfunction in this respect.

2.3. Abnormalities of force

The few experiments performed so far indicate that the cerebellum is involved in the control of maintaining constant force, but not in the production of maximal force. Maintaining a low isometric force between the thumb and index finger, e.g., when holding a pen while writing, requires a cocontraction of nearly all arm and hand muscles [38]. The task of prehension is accompanied by decreased firing of Purkinje cells in the intermediate zone of the cerebellar cortex during the static phase of maintaining force [39]. Experiments performed by Mai et al. indicate that patients with cerebellar lesions have difficulties maintaining constant force (Figure 11-3) and that this deficiency is not correlated with the severity of either dysdiadochokinesia or tremor [40]. The maximal force of grip is comparable to normals. This observation, made in patients with a chronic deficiency, is in contrast to Holmes's [1] observation of asthenia in acute cerebellar lesions (with hypotonia).

2.4. Slowing in performance of rapid alternating movements (dysdiadochokinesia)

Diadochokinesia is tested by patting with one hand, rapidly alternating between palm up and palm down [41,42]. Dysdiadochokinesia theoretically could be explained by the mechanisms described so far. Slowness at the turning points could be due to delays in movement initiation and/or dysmetria at the end of the movement. Abnormalities of movement velocity and acceleration could also contribute.

In a recent experiment we investigated together with Hore and Wild rapid alternating movements at the elbow, wrist, and finger in cerebellar patients. On the affected side, the movements were slower and were irregular in movement amplitude and frequency. EMG activity on the affected side was prolonged, again irregular, and exceeded the start of movement at the turning points (unpublished observation).

2.5. Decomposition of movements (dys-synergia)

Dys-synergia describes the inability to perform movements in three-dimensional space involving multiple sets of agonistic and antagonistic muscles acting at different joints. This deficit can be observed if the patient is asked to move his arm from an outstretched horizontal position to touch the tip of his nose (finger-nose-test).

With the new techniques for recording two- and three-dimensional movements, Becker et al. [35] studied throwing movements in normals and cerebellar patients. Elbow-wrist coordination and the coordination of hand opening with activation of more proximal arm muscles were preserved in patients. Patients, however, were unable to coordinate the muscles so as to produce the same hand direction from trial to trial.

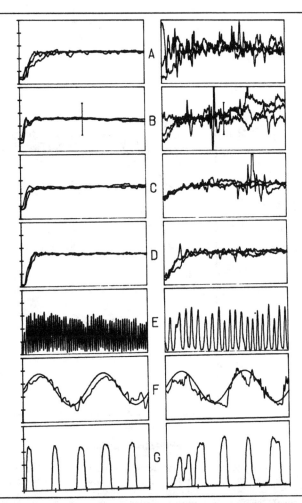

Figure 11-3. Isometric force control of a patient with diffuse cerebellar atrophy (on the right) and a normal subject (on the left). **A–D**: Maintenance of constant force; three superimposed trials are shown for each condition. Target force was 2.5 N for A–C and 12.5 N for trial D. Target force was indicated by a horizontal line on a monitor; the actual force exerted was fed back by the length of a vertical bar (continuous feedback) in A, B, and D. The visual feedback was withdrawn in condition B for the last 10 sec. In condition C, a discrete feedback was used, which signaled only one of three states (correct force, too much, too less force). **E**: Fast repetitive force changes; the changes were required between 6.25 N and 18.73 N. **F**: Force tracking of a sinusoidal target. **G**: Maximum grip force measured only between the thumb and index finger (From Mai et al. [41], with permission.)

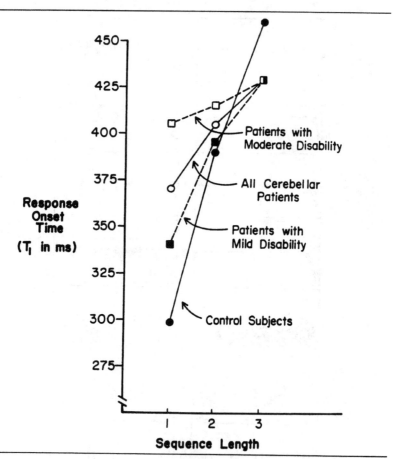

Figure 11-4. Reaction time to initiate the first response (T1) in the sequence as a function of sequence length: L = 1 for index finger responses, L = 2 for index and ring finger response sequences, and L = 3 for index, ring, and middle finger response sequences. All cerebellar patients (solid line, open circles) are compared with normal control subjects (solid line, closed circles). In addition, a subgroup of patients with moderate disability (dashed line, open squares) is compared with a subgroup of patients with mild disability (dashed line, closed squares). (From Inhoff et al. [44] with permission).

2.6. Disturbances of motor programming

Fast complex movements are supposed to be preprogrammed prior to execution. The concept of a motor program [43] originally derived from the observation that well-trained fast actions are executed too fast to be dominated by sensory feedback. This assumption may only partially be true, since most movements take longer than 100 msec and allow spinal and transcortical feedback to contribute. It is, however, still supposed that a neural representation of an entire complex action is prepared before response onset. The action is then executed as a single unit when triggered. One of our basic hypotheses is that the cerebellum helps to coordinate the timing between the single components of a movement, scales the size of muscular action, and coordinates the sequence of agonists and antagonists. The basic structure of a motor program (e.g., muscle groups involved, sequence of muscles activated) is obviously not generated within the cerebellum. Cerebellar patients, as well as monkeys with cerebellar dentate cooling, are still able to perform complex motor tasks [12,19,23,30,32,35,40], although the execution is slow and ataxic.

Motor programming was investigated in an experiment by Inhoff et al. [44]. Normal subjects and 22 patients with various cerebellar disorders were instructed to execute sequences of finger movements in a simple reaction-time paradigm. Evidence for anticipatory motor programming in normal subjects was reflected in the pattern of onset times and inter-key-press times. Normal subjects and patients with mild cerebellar dysfunction showed increases in response onset time as the sequence length increased from a single to two or three consecutive finger movements. Patients with severe cerebellar symptoms showed abnormally long reaction times and no further increase in their initial reaction time with the number of finger movements to be performed. Furthermore, cerebellar dysfunction was associated with slower inter-key-press reaction times (Figure 11-4). This result indicates that anticipatory motor programming critically depends on cerebellar integrity.

Motor preparation and associated postural adjustments can also be considered as complex movements. They prevent significant shifts of the center of gravity when the arms or the body's axis move. We recently investigated motor preparation in human subjects who were asked to rise on their toes [45]. This task requires an active shift of the body forward through tibialis anterior activity (preparation) prior to the execution of the task by triceps surae activation. The basic pattern of preparatory and executional motor activity was preserved in cerebellar patients. Cocontraction of antagonistic muscles and changes in the relative timing of preparation and execution could be seen (Figures 11-5A and 11-5B).

2.7. Disturbances of feedback

In contrast to rapid preprogrammed movements, precise and slow limb movements require continuous guidance by proprioceptive and/or visual feedback. The cerebellum receives abundant sensory information and could well provide the correcting signals for deviations from the intended movement path. Holmes [6] stated that slow movements in cerebellar patients are jerky, intermittent, or clonic. Studies by Beppu et al. described disturbances in slow limb movements [19,46], and studies by Mai et al. described disturbances in slow force changes [40]. Access to visual feedback does not improve disturbed limb movements in the clinical examination [10] or in the

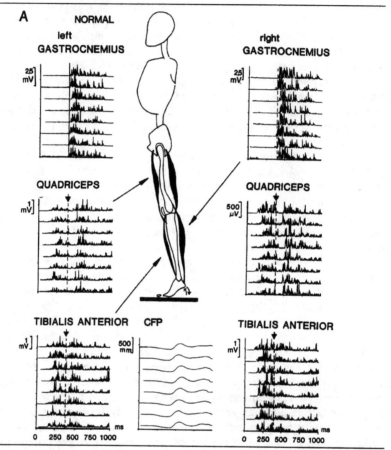

Figure 11-5. Original recordings of rectified EMGs on a single trial basis prior to and during the task of rising on tip toes. The single traces in the prime mover, the gastrocnemius muscle, are aligned in time, irrespective of their initial latency. All the other muscles are accordingly shifted in time. **A**: In a normal subject, the complex movement starts with premotor activity in the tibialis anterior, and shortly later in the quadriceps. The focal motor activity in the gastrocnemius muscle follows. The recording of the center of foot pressure (CFP) shows the two phases of shift of the body axis forward (downward deflection), followed by raising on the tip toes (upward deflection). **B**: EMG recordings in a patient with cerebellar atrophy. Note the increased duration and the increased scatter of premotor activity in the tibialis anterior on the right more than on the left and the increased delay between preparatory activity in this muscle and the start of the movement through gastrocnemius activity. Preparatory activity in the quadriceps is missing.

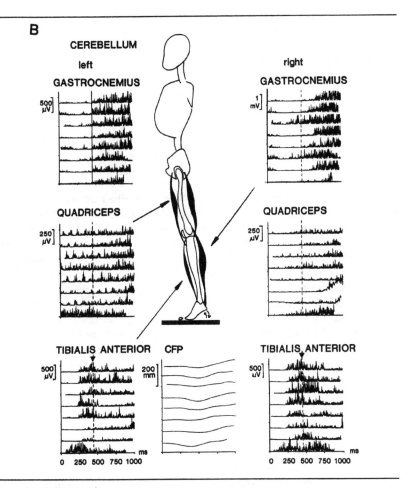

Figure 11-5. Continued

control of force [40,47]. In elbow and wrist tracking movements, patients perform even better without visual feedback [46,48].

2.8. Disturbances of time perception

Slow movements not only require feedback but also feedforward control. Recent experiments by Ivry et al. [49,50] indicate that cerebellar patients have disturbances in the acoustic and visual perception of time intervals. Disturbed perception of time intervals makes feedforward control extremely difficult. This finding agrees with the observations by Miall et al. [51] that monkeys in visuo-motor tracking tasks use perceived target velocity to correct their movements, a task that requires feedforward computation. With cooling of the interpositus nucleus of the cerebellum, the amplitudes of correction movements were correlated with the error between the target and the joystick, a fact suggesting a feedback mode of action with the reversible lesion. Timing is an essential feature of repetitive movements, such as finger tapping. Studies in patients with circumscribed lesions indicate that the cerebellar hemispheres are critically involved in the operation of the timing process, whereas the medial regions of the cerebellum are associated with the execution of these responses [50].

2.9. Hypotonia and hyporeflexia

Hypotonia, hyporeflexia, and asthenia were described as typical cerebellar symptoms of an acute lesion by Holmes [1]. Hypotonia is not exclusively associated with lesions of the cerebellum. It may also occur with lesions of the thalamic relay of cerebello-cortical afferents, with lesions of the striatum (like in Huntington's chorea), and in recent lesions of the pyramidal pathways. In our experience, hypotonia can only be observed in acute cerebellar damage (hemorrhage, surgery) and usually disappears within a few days or weeks. Patients with cerebellar atrophies usually present with normal muscle tone and normal tendon reflexes. Gilman [8,52] ascribed the depressed excitability of stretch reflexes in the affected limbs and the hypotonia in cats with acute cerebellar lesions to a reduction of proprioceptive inflow due to a decrease in the resting discharge of both static and dynamic fusimotor fibers. The latter mechanism, i.e., depressed fusimotor activity, has also been considered responsible for the delayed onset and slowed acceleration of movement [53]. However, this explanation is unlikely to explain comparable symptoms, i.e., the delayed onset and slowed acceleration seen in humans.

2.10. Tremor

Cerebellar tremor occurs as an oscillatory movement that becomes more prominent as the moving limb approaches a target. It is commonly described as intention tremor. However, the terms *kinetic, goal-directed tremor*, or *terminal tremor* better describe the clinical appearence. Powerful but brief involuntary movements at the beginning of a movement are due to intention myoclonus, not tremor, and occur in diseases involving the dentate nucleus or the superior cerebellar peduncle. Kinetic cerebellar tremor in arm movements can be observed in the finger-nose test. Leg tremor is seen during the heel-shin test. Postural body tremor is tested during the Romberg test with closed eyes.

Originally some components of cerebellar tremor were attributed to conscious efforts to correct dysmetric movements [7,9,54,55]. During clinical examination, however, one can clearly separate the rhythmical tremor from irregular correcting movements.

Currently, there exist two theories to explain cerebellar tremor. The first theory postulates a central oscillator disinhibited by cerebellar dysfunction; the second postulates a disorder of long loop reflexes. A strong argument for the existence of a central oscillator is the preservation of cerebellar tremor, despite deafferentation in monkeys with dorsal root section [56,57]. The fact that patients with Friedreich's ataxia partly deafferented by severe peripheral and spinal afferent neuropathy may also exhibit "cerebellar" tremor cannot, however, be used as evidence for the existence of a central oscillator. If residual afferent conduction is present, cerebellar tremor can also result from abnormal conduction in the supposed transcortical feedback loop.

The second theory ascribes cerebellar tremor to a sequence of enhanced and/or delayed long-latency transcortical reflexes [24,25,58,59,62]. Arguments for the significance of peripheral against central factors are the influence of proprioceptive in contrast to the visual feedback on tremor and the dependency of tremor amplitude and frequency on the mechanical state of the limb [58–61].

Long loop reflexes in leg muscles of standing subjects have been shown to play an important role in postural control. Long loop reflexes in leg muscles of standing humans evoked through sudden tilts of a supporting platform prevent the body from falling and keep it upright [63–66]. Increased long-latency reflexes result in an overcompensation of this postural task and lead to a regular anterior-posterior body sway with a frequency of 3 Hz, which is best seen in patients with anterior lobe atrophy due to chronic alcoholism [67–70].

3. CONTROL OF POSTURE AND POSTURAL MOVEMENT STRATEGIES IN CEREBELLAR PATIENTS

With respect to body posture, two regions of the cerebellum are of particular interest with regard to their types of afferents and reciprocal target regions. These are the vestibulocerebellum and the spinocerebellum (mostly the anterior lobe). The analysis of human stance on a force-measuring platform quantifies the clinical Romberg test (Figure 11-6A) and has allowed lesions at different locations to be distinguished according to their kind of postural instability [71–73]. Lesions of the spinocerebellar part of the anterior lobe are mainly observed in chronic alcoholics and lead to anterior-posterior body sway, with a frequency of about 3 Hz (Figure 11-6D). The tremor is provoked by eye closure. Patients rarely fall. This is due to the fact that the body tremor is opposite in phase in the head, trunk, and legs, thus resulting in a minimal shift of the center of gravity. Visual stabilization of posture is preserved.

Lesions of the lower vermis, mainly due to tumors or hemorrhage, cause postural ataxia of the head and trunk while sitting, standing, and walking. Postural sway is omnidirectional and contains frequency components below 1 Hz. Visual stabilization, as evaluated by comparing sway with eyes closed

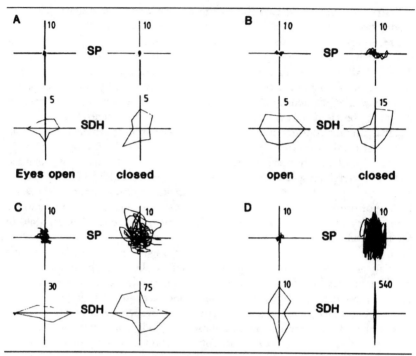

Figure 11-6. Recording of sway path (SP) in anterior-posterior and lateral direction and the calculated sway direction histogram (SDH). **A**: Normal subject. **B**: Increased omnidirectional sway in a patient with hemorrhage of the vestibulo-cerebellar vermis. **C**: Predominantly lateral sway in a patient with Friedreich's ataxia. **D**: Predominantly anterior-posterior sway in a chronic alcoholic with atrophy of the anterior lobe. Note the different scalings (A vs. B–D) of the axes of the sway direction histogram, summing the instantaneous sway directions within each of eight directional bins.

and sway with eyes open, is less than in the other groups of cerebellar patients (Figure 11-6B).

Recent experiments in patients with cerebellar disease by Bronstein et al. [74] investigated the control of balance with eyes open, eyes closed, and in response to visual stimuli generated by lateral displacements of a moveable room. Cerebellar lesions spared the visuopostural loop and also spared the ability to shift from visual to proprioceptive control of postural sway.

Long-latency responses can easily be recorded with surface electrodes and offer the opportunity to quantify cerebellar dysfunction in limb muscles. Early case reports of patients with cerebellar lesions by Marsden et al. described a loss or diminution of the initial component of the long-latency stretch reflex in the long flexor of the thumb, with a relative preservation or

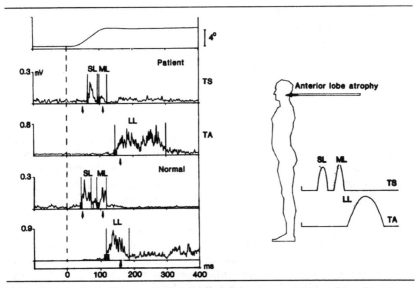

Figure 11-7. Postural reflexes evoked through a platform tilt toe-up in a standing subject; rectified and averaged EMG responses from the triceps surae (TS) and anterior tibial muscles (TA). Short- (SL) and medium- (ML) latency responses in TS and stabilizing long-latency response (LL) in TA. The patient with anterior lobe atrophy exhibited a prolonged activity in the anterior tibial muscle. The right side of the figure shows mean latencies, durations, and integrals of SL, ML, and LL in a population of normals (dotted lines) and in a population of 42 patients with anterior lobe atrophy due to chronic alcoholism. Note the increased duration and integral of LL in TA.

delay of the later component in four patients with a unilateral cerebellar lesion [75].

Later studies were unable to replicate these results. We studied, together with Friedemann and Noth, electromyographic responses to the stretch of hand muscles (first dorsal interosseus) and leg muscles (triceps surae, tibialis anterior) in 41 patients with cerebellar disorders [76–78]. Latencies of all early and late EMG responses were normal. The cerebellum is, therefore, unlikely to be the primary generator of these late EMG responses. However, duration and intensity of long-latency responses were increased in both upper and lower extremities (Figure 11-7). The intensity and duration of short- and medium-latency responses were normal in the cerebellar patients of all three studies mentioned above. These results indicate that the cerebellum has a modulatory influence on the size and duration, and therefore the force of long-latency reflexes, though not on their latency.

Variations in the amplitude and/or velocity of platform displacement and measurements of the size of the related EMG responses from leg and trunk muscles have shown abnormal gain control in cerebellar patients [79]. Reflex responses had normal latencies, and postural synergies were basically normal. EMG response vs. stimulus amplitude gains were increased and EMG response vs. stimulus velocity gains were reduced [80]. It may be hypothesized that the abnormal control of amplitude and the duration of long-loop reflexes is at least one of the constituents of postural tremor (and limb tremor?) in cerebellar patients [68].

4. COMMENT

The present understanding of the physiological mechanisms of ataxia and other cerebellar symptoms in humans is still very preliminary. It must largely resort to the results of animal experiments. In view of the differences between animal species, one should only apply those obtained from monkeys for the possible interpretation of human pathophysiology. The paucity of experimental studies in humans is partly due to the rarity of well-delineated and purely cerebellar lesions in patients who are able and willing to serve as subjects. Cerebellar atrophies mostly result in diffuse loss of Purkinje cells and deterioration of afferent and efferent cerebellar pathways. Many patients initially present with, or later on develop, multisystem atrophies, such as OPCA. Vascular lesions almost routinely also involve the brain stem. Tumors may compress the neighboring structures.

Most textbooks still rely on Gordon Holmes's descriptions, which imply that most cerebellar symptoms occur together. However, quantitative studies on populations of patients reveal a wide variety of performances and clearly prove that the occurrence and severity of the disturbed ability to control movement and force are variable from one patient to the other and are usually not correlated [40]. Lack of correlation may partly be ascribed to the topographical segregation of cerebellar function [10].

ACKNOWLEDGMENTS

The authors thank J. Hore and D. Burke for helpful comments. Many experiments reported in this chapter were supported by the Deutsche Forschungsgemeinschaft, SFB 307/A3.

REFERENCES

1. Holmes G. (1917). The symptoms of acute cerebellar injuries due to gunshot injuries. Brain 40:461–535.
2. Holmes G. (1922). Clinical symptoms of cerebellar disease and their interpretation. The Croonian lectures I. Lancet 1:1177–1182.
3. Holmes G. (1922). Clinical symptoms of cerebellar disease and their interpretation. The Croonian lectures II. Lancet 1:1231–1237.
4. Holmes G. (1922). Clinical symptoms of cerebellar disease and their interpretation. The Croonian lectures III. Lancet 2:59–65.
5. Holmes G. (1922). Clinical symptoms of cerebellar disease and their interpretation. The Croonian lectures IV. Lancet 2:111–115.
6. Holmes G. (1939). The cerebellum of man. (The Hughlings Jackson memorial lecture). Brain 62:1–30.

7. Dow R.S., Moruzzi G. (1958). The Physiology and Pathology of the Cerebellum. Minneapolis: University of Minnesota Press.
8. Gilman S., Bloedel J., Lechtenberg R. (1981). Disorders of the Cerebellum. Philadelphia: Davis.
9. Brooks V.B., Thach W.T. (1981). Cerebellar control of posture and movement. In: J.M. Brookhart, V.B. Mountcastle (eds): Handbook of Physiology, Section 1, Volume 2, Part 2. Bethesda: American Physiological Society, pp. 877–946.
10. Dichgans J., Diener H.C. (1984). Clinical evidence for functional compartmentalization of the cerebellum. In: J.R. Bloedel, J. Dichgans, W. Precht (eds): Cerebellar Functions. Berlin: Springer, pp. 126–147.
11. Ito M. (1984). The Cerebellum and Neural Control. New York: Raven Press.
12. Marsden C.D., Merton P.A., Morton H.B., Hallett M., Adam J., Rushton D.N. (1977). Disorders of movement in cerebellar disease in man. In: F. Rose (ed): The Physiological Aspect of Clinical Neurology. Oxford: Blackwell, pp. 179–199.
13. Meyer-Lohmann J., Hore J., Brooks V.B. (1977). Cerebellar participation in generation of prompt arm movements. J. Neurophysiol. 40:1038–1050.
14. Lamarre Y., Bioulac B., Jacks B. (1978). Activity of precentral neurons in conscious monkeys: Effects of deafferentation and cerebellar ablation. J. Physiol. (Paris) 74:253–264.
15. Lamarre Y., Jacks B. (1978). Involvement of the cerebellum in the initiation of fast ballistic movement in the monkey. Electroenceph. Clin. Neurophysiol. 34(Suppl.):441–447.
16. Miller A.D., Brooks V.B. (1982). Parallel pathways for movement initiation in monkeys. Exp. Brain Res. 45:328–332.
17. Beaubaton D., Trouche E., Legallet E. (1984). Neocerebellum and motor programming: Evidence from reaction-time studies in monkeys with dentate nucleus lesions. In: S. Kornblum, J. Requin (eds): Preparatory—States & Processes. London: Lawrence Erlbaum Assoc. Publ., pp. 303–320.
18. Hore J., Wild B., Diener H.C. (1991). Cerebellar dysmetria at the elbow, wrist and fingers. J. Neurophysiol. 65:563–571.
19. Beppu H., Suda M., Tanaka R. (1984). Analysis of cerebellar motor disorders by visually guided elbow tracking movement. Brain 107:787–809.
20. Spidalieri G., Busby L., Lamarre Y. (1983). Fast ballistic arm movements triggered by visual, auditory and somesthetic stimuli in the monkey. II. Effects of unilateral dentate lesion on discharge of precentral cortical neurons and reaction time. J. Neurophysiol. 50:1359–1379.
21. Thach W.T. (1975). Timing of activity in cerebellar dentate nucleus and cerebral motor cortex during prompt volitional movement. Brain Res. 88:233–241.
22. Thach W.T. (1978). Correlation of neural discharge with pattern and force of muscular activity, joint position, and direction of the intended movement in motor cortex and cerebellum. J. Neurophysiol. 41:654–676.
23. Meyer-Lohmann J., Conrad B., Matsunami K., Brooks V.B. (1975). Effects of dentate cooling on precentral unit activity following torque pulse injections into elbow movements. Brain Res. 94:237–251.
24. Vilis T., Hore J. (1980). Central neural mechanisms contributing to cerebellar tremor produced by limb perturbations. J. Neurophysiol. 43:279–291.
25. Hore J., Flament D. (1988). Changes in motor cortex neural discharge associated with the development of cerebellar limb ataxia. J. Neurophysiol. 60:1285–1302.
26. Wild B. (1990). Schnelle zielgerichtete Bewegungen der oberen Extremität bei Normalpersonen und cerebellären Patienten. M.D. Thesis, Tübingen.
27. Heilman K.M., Bowers D., Watson R.T., Greer M. (1976). Reaction times in Parkinson's disease. Arch. Neurol. (Chicago) 33:139–140.
28. Sheridan M.R., Flowers K.A., Hurrell J. (1987). Programming and execution of movement in Parkinson's disease. Brain 110:1247–1271.
29. Hepp-Reymond M.C., Wiesendanger M. (1972). Unilateral pyramidotomy in monkeys; effects on force and speed of a conditioned precision grip. Brain Res. 36:117–131.
30. Flament D., Hore J. (1986). Movement and electromyographic disorders associated with cerebellar dysmetria. J. Neurophysiol. 55:1221–1233.
31. Brown S.H., Hefter H., Mertens M., Freund H.-J. (1990). Disturbances in human arm movement trajectory due to mild cerebellar dysfunction. J. Neurol. Neurosurg. Psychiatry 53:306–313.

32. Hallett M., Berardelli A., Matheson J., Rothwell J., Marsden C.D. (1991). Physiological analysis of simple rapid movements in patients with cerebellar deficits. J. Neurol. Neurosurg. Psychiatry 53:124–133.
33. Wacholder K., Altenburger H. (1926). Beiträge zur Physiologie der willkürlichen Bewegung. 10. Einzelbewegungen. Pflügers Arch. Physiol. Mensch. Tiere 214:642–661.
34. Hallett M., Shahani B.T., Young R.R. (1975). EMG analysis of stereotyped voluntary movements in man. J. Neurol. Neurosurg. Psychiatry 38:1154–1162.
35. Becker W.J., Kunesch E., Freund H.-J. (1990). Coordination of a multi-joint movement in normal humans and in patients with cerebellar dysfunction. Can. J. Neurol. Sci. 17:264–274.
36. Eccles J. (1977). Cerebellar function in the control of movement. In: F.C. Rose (ed): Physiological Aspects of Clinical Neurology. Oxford: Blackwell, pp. 157–178.
37. Terzuolo C.A., Viviani P. (1973). Parameters of motion and EMG activities during some simple motor tasks in normal subjects and cerebellar patients. In: J.S. Cooper, M. Riklan, R.S. Snider (eds): The Cerebellum, Epilepsy and Behavior. New York: Plenum Press, pp. 173–215.
38. Smith A.M., Wetts T., Kalaska J.F. (1985). Activity of dentate and interpositus neurons during maintained isometric prehension. In: A.W. Goodwin, I. Darion-Smith (eds): Hand Function and the Neocortex. Berlin: pp. 248–258.
39. Smith A.M., Bourbonnais D. (1981). Neuronal activity in cerebellar cortex related to the control of prehensile force. J. Neurophysiol. 45:286–303.
40. Mai N., Bolsinger P., Avarello M., Diener H.C., Dichgans J. (1988). Control of isometric finger force in patients with cerebellar disease. Brain 111:973–998.
41. Babinski J. (1899). De l'asynergie cérébelleuse. Rev. Neurol. 7:806–816.
42. Babinski J. (1902). Sur le rôle du cervelet dans les actes volitionnels necessitants une succession rapide de mouvements (diadochocinésie). Rev. Neurol. 10:1013–1015.
43. Keele S.W. (1968). Movement control in skilled motor performance. Psycholog. Bull. 77:155–158.
44. Inhoff A.W., Diener H.C., Rafal R.D., Ivry R. (1989). The role of cerebellar structures in the execution of serial movements. Brain 112:565–581.
45. Diener H.C., Dichgans J., Guschlbauer B., Bacher M., Rapp H., Langenbach P. (1990). Associated postural adjustments with body movement in normal subjects and patients with Parkinsonism and cerebellar disease. Rev. Neurol. (Paris) 146:555–563.
46. Beppu H., Nagaoka M., Tanaka R. (1987). Analysis of cerebellar motor disorders by visually guided elbow tracking movement. 2. Contribution of the visual cues on slow ramp pursuit. Brain 110:1–18.
47. Mai N., Diener H.C., Dichgans J. (1989). On the role of feedback in maintaining constant grip force in patients with cerebellar disease. Neurosci. Lett. 99:340–344.
48. Morrice B.-L., Becker W.J., Hoffer J.A., Lee R.G. (1990). Manual tracking performance in patients with cerebellar incoordination: Effects of mechanical loading. Can. J. Neurol. 17:275–285.
49. Ivry R., Keele S.W. (1989). Timing functions of the cerebellum. J. Cogn. Neurosci. 1:136–152.
50. Ivry R.B., Keele W.S., Diener H.C. (1988). Differential contributions of the lateral and medial cerebellum to timing and to movement execution. Exp. Brain Res. 73:167–180.
51. Miall R.C., Weir D.J., Stein J.F. (1987). Visuo-motor tracking during reversible inactivation of the cerebellum. Exp. Brain Res. 65:455–464.
52. Gilman S. (1969). The mechanism of cerebellar hypotonia. An experimental study in the monkey. Brain 92:621–638.
53. MacKay W.A., Murphy J.T. (1979). Cerebellar modulation of reflex gain. In: G.A. Kerkut, J.W. Phyllis (eds): Progress in Neurobiology, Vol. 13. Oxford: Pergamon, pp. 361–417.
54. Goldberger M.E., Growdon J.H. (1973). Pattern of recovery following cerebellar deep nuclear lesions in monkeys. Exp. Neurol. 39:307–322.
55. Growdon J.H., Chambers H.W., Liu C.N. (1967). An experimental study of cerebellar dyskinesia in the rhesus monkey. Brain 90:603–632.
56. Gilman S., Carr D., Hollenberg J. (1976). Kinematic effects of deafferentation and cerebellar

ablation. Brain 99:311–330.
57. Liu C.N., Chambers W.W. (1971). A study of cerebellar dyskinesia in the bilaterally deafferented forelimbs of the monkey (Macaca mulatta and Macaca speciosa). Acta Neurobiol. Exp. 31:263–289.
58. Vilis T., Hore J. (1977). Effects of changes in mechanical state of limb on cerebellar intention tremor. J. Neurophysiol. 40:1214–1224.
59. Hore J., Flament D. (1986). Evidence that a disordered servolike mechanism contributes to tremor in movements during cerebellar dysfunction. J. Neurophysiol. 56:123–136.
60. Flament D., Vilis T., Hore J. (1984). Dependence of cerebellar tremor on proprioceptive but not visual feedback. Exp. Neurol. 84:314–325.
61. Sanes J.N., LeWitt P.A., Mauritz K.-H. (1988). Visual and mechanical control of postural and kinetic tremor in cerebellar system disorders. J. Neurol. Neurosurg. Psychiatry 51:934–943.
62. Flament D., Hore J. (1987). Comparison of cerebellar intention tremor under isotonic and isometric conditions. Brain Res. 439:179–186.
63. Diener H.C., Bootz F., Dichgans J., Bruzek W. (1983). Variability of postural "reflexes" in humans. Exp. Brain Res. 52:423–428.
64. Diener H.C., Dichgans J., Bootz F., Bacher M. (1984). Early stabilization of human posture after sudden disturbances: Influence of rate and amplitude of displacement. Exp. Brain Res. 56:126–134.
65. Nashner L.M. (1976). Adapting reflexes controlling the human posture. Exp. Brain Res. 26:59–72.
66. Nashner L.M. (1981). Analysis of stance posture in humans. In: A.L. Towe, E.S. Luschei (eds): Handbook of Behavioural Neurobiology, Vol. 5. New York: Plenum Press, pp. 527–561.
67. Dichgans J., Diener H.C. (1987). The use of short- and long-latency reflex testing in leg muscles of neurological patients. In: A. Struppler, A. Weindl (eds): Clinical Aspects of Sensory Motor Integration. Berlin: Springer, pp. 165–175.
68. Mauritz K.H., Schmitt C., Dichgans J. (1981). Delay and enhanced long-latency reflexes as the possible cause of postural tremor in late cerebellar atrophy. Brain 104:97–116.
69. Diener H.C., Dichgans J., Bacher M., Guschlbauer B. (1984). Characteristic alterations of long loop "reflexes" in patients with Friedreich's ataxia and late atrophy of the anterior cerebellar lobe. J. Neurol. Neurosurg. Psychiatry 47:679–685.
70. Dichgans J., Diener H.C. (1985). Postural ataxia in late atrophy of the cerebellar anterior lobe and its differential diagnosis. In: M. Igarashi, F.O. Black (eds): Vestibular and Visual Control on Posture and Locomotor Equilibrium. Basel: Karger, pp. 282–289.
71. Dichgans J., Mauritz K.H., Allum J.H.J., Brandt Th. (1976). Postural sway in normals and ataxic patients: Analysis of the stabilizing and destabilizing effects of vision. Aggressologie 17C:15–24.
72. Dichgans J., Diener H.C. (1986). Different forms of postural ataxia in patients with cerebellar diseases. In: W. Bles, Th. Brandt (eds): Disorders of Posture and Gait. Amsterdam: Elsevier, pp. 207–213.
73. Mauritz K.H., Dichgans J., Hufschmidt A. (1979). Quantitative analysis of stance in late cortical cerebellar atrophy of the anterior lobe and other forms of cerebellar ataxia. Brain 102:461–482.
74. Bronstein A.M., Hood J.D., Gresty M.A., Panagi C. (1990). Visual control of balance in cerebellar and Parkinsonian syndromes. Brain 113:767–779.
75. Marsden C.D., Merton P.A., Morton H.B., Adam J. (1978). The effect of lesions of the central nervous system on long-latency stretch reflexes in the human thumb. In: J.E. Desmedt (ed): Progress in Clinical Neurophysiology, Vol. 5. Cerebral Motor Control in Man: Long Loop Mechanisms. Basel: Karger, pp. 334–341.
76. Friedemann H.H., Noth J., Diener H.C., Bacher M. (1987). Long latency EMG responses in hand and leg muscles: Cerebellar disorders. J. Neurol. Neurosurg. Psychiatry 50:71–77.
77. Diener H.C., Dichgans J., Bacher M., Guschlbauer B. (1984). Characteristic alterations of long loop "reflexes" in patients with Friedreich's ataxia and late atrophy of the anterior cerebellar lobe. J. Neurol. Neurosurg. Psychiatry 47:679–685.

78. Diener H.C., Dichgans J. (1986). Long loop reflexes and posture. In: W. Bles, Th. Brandt (eds): Disorders of Posture and Gait. Amsterdam: Elsevier, pp. 41–51.
79. Horak F.B., Diener H.C., Nashner L.M. (1986). Abnormal scaling of postural responses in cerebellar patients. Soc. Neurosci. Abstr 12:1419.
80. Nashner L.M., Horak F.B., Diener H.C. (1987). Scaling postural response amplitudes: Normals and patients with cerebellar deficits. Neurology 37 (Suppl. 1): 281.

12. OCULOMOTOR ABNORMALITIES IN CEREBELLAR DEGENERATION

M. FETTER AND J. DICHGANS

Despite the classical treatises by Holmes [1] and later by Cogan [2] on "specific" cerebellar eye signs, most clinicians during the past decades have been very cautious in relating observed eye-movement disorders to cerebellar lesions. This caution acknowledged the fact that frequently in cerebellar disease there is a coexisting involvement of brain-stem structures, making the clear correlation of signs and symptoms to cerebellar lesions rather difficult. However, over the last 25 years a host of clinical and neurophysiological data (the latter mostly derived from single-unit recordings and ablation studies [3], as well as anatomical tracing techniques), have cast some light on this dilemma. Meanwhile, we can at least try to attribute specific oculomotor subfunctions and related eye-movement abnormalities to distinct parts of the cerebellum [for more recent reviews see 4-7].

1. EXPERIMENTAL EVIDENCE OF THE CEREBELLAR ROLE IN EYE MOVEMENTS

The function of the cerebellum in the control of eye movements so far can be attributed to three distinct portions of the cerebellum: the flocculus with the paraflocculus, the nodulus and uvula, and the dorsal vermis with the underlying fastigial nuclei.

Westheimer and Blair [8] were the first to make it undoubtedly clear that cerebellar lesion alone, without the involvement of the brain stem, lead to abnormal eye movements. They reported abnormal oculomotor responses after complete one-step cerebellectomy, including the roof nuclei, in rhesus monkeys. The decerebellate oculomotor syndrome they described consisted of gaze-holding failure, i.e., the inability to maintain eccentric gaze (gaze-evoked nystagmus); pursuit failure, i.e., the almost complete absence of smooth-pursuit eye movements (cockwheeled pursuit); and transient failure of convergence. From the fact that they still could elicit conjugate smooth-pursuit eye movements by faradic stimulation of the pontine reticular formation, they concluded that after cerebellar ablation the motor apparatus is still capable of executing smooth-pursuit movements and pursuit failure in cerebellectomized monkeys may be due to failure of access to, or modulation of, the brain-stem motor mechanisms.

In a second ablation series, Westheimer and Blair [9] performed complete unilateral cerebellar lobectomies, including the underlying cerebellar nuclei. They found a constant conjugate drift of the eyes away from the side of the lesion as soon as fixation was prevented; a gaze-holding failure in the ipsilateral field of fixation; a unilateral pursuit deficit, i.e., the animal was unable to track with smooth-pursuit movements a target moving away from the midline towards the side of the lesion; the slow phase of the optokinetic response was slower for movement towards the side of the lesion; and, finally, asymmetry of the vestibular responses that have a faster slow-phase component and a longer duration whenever slow phases were induced away from the side of the lesion (rotation towards the side of the lesion). While juvenile monkeys recovered almost completely within a week, adult monkeys showed enduring deficits in gaze holding and smooth tracking. Similar results were reported by Burde and coworkers [10].

In a more recent study, Eckmiller and Westheimer [11] showed that, on the one hand, monkeys with extensive neonatal cerebellar ablations, if the intracerebellar nuclei were kept intact, did not show any discernable oculomotor dysfunction if tested as adult animals. On the other hand, if ablation also included the nuclei on one side, compensation was never complete, even several years later. While vestibulo-ocular and saccadic responses seemed normal, there were deficits in pursuit and gaze-holding performance. Since the cerebellar nuclei and cerebellar cortex probably receive matching sets of information, only removal of the cerebellar main output, i.e., the nuclei, leads to permanent dysfunctions. Systematic studies have not been performed to determine whether or not an adult individual can compensate at all for a cerebellar lesion without the cerebellar nuclei. Our own clinical experience is that lesions involving the cerebellar nuclei cause much stronger deficits that last permanently.

1.1. Flocculus/paraflocculus

In an attempt to further delineate the oculomotor function of specific cerebellar areas, Takemori and Cohen [12], and later Zee and coworkers [13], performed unilateral and bilateral ablations of the flocculus and paraflocculus in rhesus monkeys. They found that smooth tracking of small targets moving in space was impaired either with the head still (smooth pursuit) or moving (cancellation of the VOR by fixating a target rotating with the head). Smooth-pursuit gain amounted to 65%, affecting both horizontal and vertical pursuit [13]. While horizontal pursuit signals from the flocculus may be mediated through the ipsilateral medial and superior vestibular and perihypoglossal nuclei, vertical pursuit signals are probably mediated by a floccular projection to the rostral dentate nucleus and group y of the vestibular complex [14]. Additionally, Zee and coworkers [13] found in flocculectomized monkeys an impaired optokinetic response, with a decreased initial slow-phase velocity, an almost doubled rise time to a steady state, and a decreased steady-state gain, which became more pronounced with a stimulus velocity exceeding $60°/sec$. They concluded that these abnormalities can largely be attributed to the coexisting pursuit deficit, since optokinetic afternystagmus was normal. Waespe and coworkers [15] also performed bilateral and unilateral floccular lesions in monkeys and found that even unilateral flocculectomy affected the rapid rise in horizontal eye velocity during optokinetic stimulation to both sides, being more pronounced toward the ipsilateral, with a gain reduction of 50–70%, as compared to contralateral with a gain reduction of 30–65%.

Zee and coworkers [13] found that the vestibuloocular reflex (which acts during rotation of the head, producing compensatory eye movements with the same velocity but in the opposite direction as the head movement to maintain the gaze stable in space) was only mildly affected by bilateral flocculectomy. Only small changes in VOR gain (eye velocity/head velocity) and phase between eye velocity and head velocity appeared after the operation.

In contrast, Flandrin and coworkers [16] found acutely in cats, after unilateral flocculus lesions, a strongly increased vestibulo-ocular reflex gain of a factor of 1.5 to velocity steps exciting the labyrinth ipsilateral to the lesion (slow phase toward the contralateral side) and a decrease of about 60% in the opposite direction. However, this effect was transient. Responses became symmetric in about 20 days and were about 75% of the corresponding preoperative values. The authors also showed that the observed VOR gain asymmetries can not be solely attributed to the spontaneous nystagmus found after the floccular lesion, with slow phases away from the lesioned side.

The flocculectomized monkeys [13] showed some further findings. They had gaze-evoked nystagmus with exponentially decaying centripetal drift, downbeat nystagmus, and postsaccadic drift. All animals showed rebound nystagmus. Zee and coworkers [13] explained the latter results by a control function of the flocculus, and possibly the paraflocculus, on the time constant and stability of the proposed brain-stem oculomotor integrator, which is assumed to create a gaze position signal by integration, in the mathematical sense, of the eye-velocity signal. The integration is necessary, since all conjugate eye movements are initiated as eye-velocity commands, regardless of whether the eye movement is of vestibular, pursuit, or saccadic origin [17]. A decreased integrator time constant, making this integrator more "leaky," would produce gaze-holding deficits with exponentially decreasing slow-phase velocities. An inappropriate match between the phasic (pulse) and (its integration) the tonic (step) innervational changes that create a saccade would lead to what is called a *pulse-step mismatch*, i.e., a mismatch between saccade and the final eye position [18]. Recent evidence suggests that the brain-stem component of the neural integrator for horizontal eye movements is located within the rostral portion of the nucleus prepositus hypoglossi and/or the adjacent rostral part of the medial vestibular nuclei [19,20]. Both structures are interconnected with each other and with the cerebellar flocculus.

Further evidence underlining the role of the flocculus in oculomotor control was provided by microstimulation experiments [21] and by single-unit recordings from Purkinje cells in the flocculus [22–36].

The pursuit deficit found in flocculectomized monkeys was clearly less than that reported by Westheimer and Blair [8] and by Optican and Robinson [37] after total cerebellectomy (the latter found a pursuit gain of between 0 and 30%), suggesting that other parts of the cerebellum also play a considerable role in generating smooth tracking.

Recent evidence suggests that the dorsolateral pontine nuclei (dlpn) may be the source of afferent cerebro-cerebellar information to the flocculus for the generation of pursuit eye movements [38,39]. However, the bulk of the output from the dlpn is known to be directed to vermal lobules VI and VII, and the adjacent paravermal regions of the cerebellum [38,39]. Thus, also lobules VI and VII of the posterior vermis of the cerebellum, traditionally considered to be involved mainly in saccadic eye movements, may be a putative site for smooth-pursuit signal processing. This notion is supported by more recent neurophysiological findings [42–45].

The cerebellum is also important in the long-term maintenance of ocular motor accuracy, i.e., the "plastic" adaptive functions that keep ocular motor responses appropriate for sensory (e.g., visual and vestibular) stimuli [46–49]. Adaptive control of the amplitude, and even the direction, of the slow phases of the vestibulo-ocular reflex (the alterations being induced by optical devices) is impaired by floccular lesions [50–52]. There is further evidence that the smooth-pursuit system is also under adaptive control [53]. Whether the flocculus also mediates this adaptation remains to be proven. In the saccadic system the situation is a little more complicated. While lesions of the vermis and paravermis (lobules IV–IX) abolished adaptive control of the pulse of innervation, adaptive changes in the step of innervation (e.g., elimination of postsaccadic drift) still occurred. The latter was only abolished after floccular lesions [37,54].

Current hypotheses about the plastic adaptive functions of the cerebellum include the ideas that the cerebellum is a possible site of motor learning [46,52,55–59] and that the cerebellum calculates the necessary sensory-motor coordinate transformations so that movements will be accurate no matter what their direction or initial or final position [57] (see Chapter 2).

The neural organization of the flocculo-vestibulo-ocular system suggests three possible roles of the flocculus in controlling the vestibulo-ocular reflex (VOR) [46]: (1) as a side path to the major VOR arc, the flocculus may contribute to the dynamic characteristics of the VOR; (2) through visual feedback, the flocculus may rapidly correct the performance of the VOR so as to maintain constancy of retinal images (rapid readjustment of the VOR by vision) [13,61,62]; (3) when a correction is repeated, there is a progressive change in the internal parameters of the flocculus so that performance of the VOR will be improved (long-term adaptive modification of the VOR by vision).

The notion that the cerebellum is the only site of "motor learning," however, has been challenged by several authors [63–69].

1.2. Nodulus/uvula

Lesions of the nodulus produce predominantly abnormalities related to the vestibular "velocity storage" mechanism, which probably has its anatomical substrate in the caudal pons and medulla near or including the perihypoglossal and vestibular nuclei, and which receives additional input from the optokinetic system via the nucleus of the optic tract (NOT) and the nucleus prepositus hypoglossi (NPH). During sustained, constant-velocity rotations, the activity in the eighth nerve rapidly decreases exponentially with a time constant (time until 63% of the initial activity is lost) of about 5–6 sec. However, the vestibular nystagmus is longer and has a time constant of about 15 sec, because the canal signals are preserved by a presumed brainstem velocity storage mechanism.

Lesions of the nodulus in monkeys lead to an increase in the time constant of the VOR and OKAN to the level first encountered in the naive animal, losing any previous adaptation, as well as the inability to lower the time constant of the VOR upon functional demands, which usually happens when repetitive rotations in darkness are performed (vestibular habituation) or under conditions of vestibular-visual or intravestibular (otolith-canal) conflicts [70,71]. Tilt suppression of nystagmus was completely lost after nodulectomy and partial uvulectomy. The ability of the visual system to decrease the time constant of decay of stored central activity was not entirely lost but was markedly attenuated. The initial fast rise in the slow-phase velocity at the onset of OKN was unaffected by the lesions, which is consistent with the hypothesis that the rapid rise in OKN slow-phase velocity is mediated by the flocculus. The gain of the VOR was unchanged after nodulectomy and partial uvulectomy [71]. The same monkeys often showed periodic alternating nystagmus (a form of horizontal nystagmus that changes direction every few minutes) when placed in complete darkness. This periodic alternating nystagmus can be triggered by vestibular and optokinetic stimuli, being a further indication that the vestibular and optokinetic system shares the same velocity storage mechanism and that the nodulus and uvula are not only important in habituating but are also important in modulating and stabilizing the vestibulo-ocular reflex through inhibitory projections via the fastigial nuclei to the brain-stem velocity storage mechanism. Loss of this inhibitory modulation may lead to instability in the central velocity storage mechanism and consequently, for example, to the periodic alternating nystagmus [72].

Precht and coworkers [73] studied Purkinje cell activity in the nodulus and uvula of rabbits. They found vestibular, visual, and proprioceptive inputs, and sometimes convergent inputs from all these sensory systems, to Purkinje cells in the nodulus/uvula. The nodulus and uvula also project to the vestibular nuclei, but apparently to different areas than the flocculus [74].

1.3. Dorsal vermis/fastigial nuclei

Traditionally the dorsal vermis and underlying fastigial nuclei have been considered to be involved mainly in saccadic eye movements [42,75–79]. Systematic mapping with microstimulation disclosed that the region in the cerebellar vermis that yielded saccades with weak stimulus currents was mostly confined to lobule VII, but included a part of folium VIc in about 50% of the monkeys tested [80,81]. The direction of the saccade was topographically organized. Upward saccades were evoked from the anterior part, downward saccades from the posterior part, and horizontal saccades from the lateral part of the oculomotor vermis. Lesions in this region (dorsal vermis/fastigial nuclei) lead to saccadic dysmetria (usually being hypermetric for centripetal saccades and hypometric for centrifugal saccades) [82–85] and also abolish adaptive control of the pulse of innervation [37]. The saccadic dysmetria may be different between the left and right eyes [86,87]. For an overview of saccadic pulse and step disorders, see Figure 12-1.

Lesions involving the deep medial nuclei result in more pronounced overshoot dysmetria than cortical lesions alone, occasionally leading to large to-and-fro saccades with short intersaccadic intervals at about the position of the target, called macrosaccadic oscillations [88]. However, Zee and Robinson [89] suggested that another form of saccadic oscillations—ocular flutter and probably also opsoclonus (no intersaccadic interval)—may be due to another mechanism, since not infrequently this specific disturbance is seen without other cerebellar-type eye-movement disorders. They claimed that these oscillations may be due to impaired pause cell activity, which usually inhibits burst cells. Pause cells, lying near the midline within the pons and the central gray of the mesencephalon dorsal to the oculomotor nuclear complex, only cease firing during a saccade. The authors hypothesized that the saccadic pulse generator is inherently unstable because of a small delay in the eye-

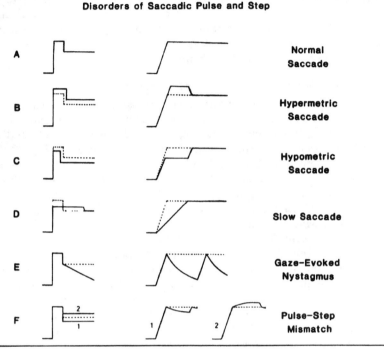

Figure 12-1. Disorders of the saccadic pulse and step. Innervation patterns for the initial saccade are shown on the left and theoretical eye movements (with eventual correction of the false end position) on the right. Dashed lines indicate the normal response. **A**: normal saccade; **B**: hypermetric saccade [the pulse amplitude (width × height) is too high but usually the pulse and step are matched appropriately]; **C**: hypometric saccade (the pulse amplitude is too small but the pulse and step are matched appropriately); **D**: slow saccade (decreased pulse hight but increased width and normal pulse-step match); **E**: gaze-evoked nystagmus [normal pulse, poorly sustained step (leaky integration) with exponential waveform of the drift back to primary position]; **F**: pulse-step mismatch, either with the step being relatively smaller than the pulse (backward postsaccadic drift, 1), or with the step being relativel higher than the pulse (onward postsaccadic drift or glissade, 2). (After Lee and Zee [6], with permission.)

position feedback loop. Consequently, the burst cells must be tonically inhibited to prevent saccadic oscillations during periods of fixation.

As mentioned in the chapter on flocculus, the dorsal vermis also contains neurons that discharge in relation to pursuit eye movements and to head movements [43–45,90]. This may mean that this area is also involved in the regulation of smooth-pursuit eye movements and vestibular information processing. Eventually this information may be used either to release and to determine, or just to perform, the fine tuning of catch-up saccades during combined eye-head tracking, or during smooth pursuit of the eye only when smooth pursuit lags target velocity. In this respect, Pierrot-Deseilligny and coworkers [91] recently demonstrated a patient with severe foveal pursuit and OKN deficits in all directions in whom magnetic resonance imaging showed an infarct involving the postero-inferior part of the vermis with preservation of the flocculus and brain stem.

2. CLINICAL SYMPTOMS CONFIRMING LESION EXPERIMENTS

Meanwhile, many of the experimental findings have also been shown in patients with structural lesions of the cerebellum. The most common findings were impaired smooth pursuit, impaired cancellation of VOR and caloric nystagmus, and reduced gain of optokinetic nystagmus, [4,5,92–105].

Von Reutern and Dichgans [106] found in patients with pontine angle tumors a bilateral diminution of horizontal optokinetic nystagmus that prevailed towards the side of the tumor and cockwheeled smooth pursuit toward the side of the lesion. They concluded that this syndrome is caused by pressure damage to the ipsilateral flocculus.

Less often were described increased VOR gain [4,98,100,104,105], gaze-evoked nystagmus [4,5,93,100,105,107], and rebound nystagmus [5,99,105].

Further clinical symptoms found with cerebellar dysfunction are downbeat nystagmus [93,104,105]; positional nystagmus, usually beating to the uppermost ear (ageotropic nystagmus) [108–111]; dysmetric saccades [5,92,105,112–115]; postsaccadic drift [116]; loss of suppression of VOR and OKAN by head tilt [117,118]; and instability of fixation [square wave jerks, saccadic intrusions, increased slow drift (spontaneous nystagmus), pendular oscillations, macrosaccadic oscillations, ocular flutter, and opsoclonus] [4,5,88,100,103,105,107,115,119–121].

One of the most impressive eye-movement abnormalities, opsoclonus or saccadomania, is preferably found in cerebellar encephalopathy in association with viral infections (Kinsbourne encephalitis) [122,123], neuroblastomas, or rarely, adenocarcinomas, but there are also idiopathic cases [124]. In some cases alterations of the dentate nucleus have been described [125–127]. A clear correlation of opsoclonus, however, to a localized structural lesion is not possible.

Some ocular motor signs observed in patients with cerebellar dysfunction do not as yet have an experimental correlate or a specific mechanistic explanation (e.g., divergent nystagmus [5,128], centripetal nystagmus [107], primary position upbeating nystagmus [129–131] and blink facilitation of saccades [132].

Eye-movement recordings of an 18-year-old patient who suffered from a pilocytic astrocytoma and was operated on several weeks prior to the recording, may exemplify some of the above-mentioned oculomotor signs of a cerebellar lesion (Figure 12-2). His left cerebellar hemisphere was subtotally removed.

Table 12-1 summarizes the current ideas about the topical localization of

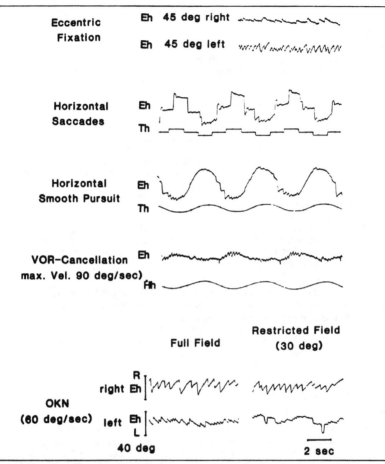

Figure 12-2. Original data of an 18-year-old patient with a left-sided pilocytic astrocytoma of the cerebellum. He had been operated on the tumor several weeks before, and a subtotal removal of the left cerebellar hemisphere had been performed. Eye movements were recorded with AC electrooculography. From top to bottom are shown eccentric fixation with massive gaze-evoked nystagmus predominantly to the left; horizontal saccades were hypermetric, especially for zentripetal saccades; smooth pursuit was cockwheeled only to the left (note the overlayed gaze-evoked nystagmus in the left hemifield), and VOR cancellation by fixating a small target while sinusoidal rotation with 0.2 Hz and 90°/sec maximum velocity was impaired predominantly to the left; OKN was better to the left than to the right, which becomes more pronounced with small-field stimulation. Eh = horizontal eye position; Th = horizontal target position; Ḣh = horizontal head velocity; R = right; L = left.

Table 12-1.

I. Flocculus (Paraflocculus)

Function: Retinal image stabilization

Deficits: A. Gaze-holding and fixation deficits
1. Gaze-evoked (occasionally centripetal) nystagmus, rebound nystagmus, downbeat nystagmus
2. Postsaccadic drift (glissades)
3. Square wave jerks (saccadic intrusions) (?)
4. Opsoclonus (?)

B. Pursuit-related deficits
(systems that stabilize targets moving in space)
1. Impaired smooth pursuit
2. Low gain of OKN at high stimulus velocities
3. Loss of fast initial rise at the beginning and initial drop at the end of optokinetic stimulation (direct pursuit component of OKN)
4. Impaired cancellation of the VOR and caloric nystagmus (direct pursuit component)

C. Plastic-adaptive functions
1. Defective long-term modulation of gain and direction of VOR resulting in inappropriate VOR (e.g., increased VOR gain)
2. Lack of adaptive adjustment of saccadic pulsestep ratio to suppress postsaccadic drift

II. Nodulus (Uvula)

Function: Control of the velocity storage mechanism (VOR and OKAN time constant, habituation, and intervestibular (otolith-canal) and visuo-vestibular interaction)

Deficits: A. Impaired attenuation of velocity storage
1. Loss of habituation of VOR and OKAN time constant
2. Loss of head-tilt and visually induced attenuation of the time constant of postrotatory and caloric nystagmus as well as OKAN
3. Partial reduction of fixation-suppression [fast initial (direct) component preserved]

B. Periodic alternating nystagmus

C. Positional nystagmus (?)

III. Dorsal vermis/fastigial nuclei

Function: Control of saccadic amplitude (and smooth-pursuit saccadic interactions?)

Deficits: A. Saccadic dysmetria, macrosaccadic oscillations; normal velocities and latencies

B. Lack of immediate and long-term adaptive adjustment of amplitude of saccadic pulse

C. Hypothetical: Low gain of smooth-pursuit and inadequate catch-up saccades during tracking

After Leigh and Zee [4], with permission.

---

specific cerebellar functions in oculomotor control and their pathology. The question mark indicates a questionable assignment.

3. DEGENERATIVE DISEASES OF THE CEREBELLUM AND SPINOCEREBELLAR ATAXIAS

In the remainder of this chapter we will review the literature on ocular motor findings caused by some specific degenerative disorders with prominent cerebellar eye-movement abnormalities. Even though numerous authors have tried to find specific patterns of oculomotor abnormalities confined to certain specific cerebellar diseases, diagnostic key patterns could not be defined. This is not surprising, since many degenerative disorders of the cerebellum involve extracerebellar structures as well, a fact that complicates conclusions as to the topological specificity of oculomotor symptoms.

Most publications dealing with oculomotor disturbances in cerebellar degenerations do not use a clear classification, and often several distinct degenerative syndromes are intermingled in such a way that a differential analysis is not possible. We are aware of the fact that the clinical classification of degenerative diseases frequently remains uncertain (see Chapter 8). We nevertheless felt obliged to invariably adopt the authors' classification, even if we were in doubt as to whether the assignments were justified. Recent clinical experience suggests a more symptom-oriented classification [133].

We feel that the knowledge of the kinds of oculomotor signs documented thus far as to a given syndrome may help the differential diagnosis.

3.1. Spinocerebellar ataxias

3.1.1. Friedreich ataxia

Friedreich ataxia is an autosomal recessive disease, with an onset usually before age 20. Degeneration involves predominantly the dorsal roots and spinal cord, and less extensively the cerebellum. Furman and colleagues [134] investigated 24 patients with well-documented Friedreich's ataxia. Using quantitative eye-movement recordings, they observed what they called a "characteristic" pattern of oculomotor abnormalities. The pattern included fixation instability [in all patients square-wave jerks, macro-square-wave jerks, or ocular flutter (occasionally more than one type in the same patient)], which seems to be the most specific sign; gaze-evoked nystagmus (in eight patients); rebound nystagmus (in three patients), inaccurate saccades (overshoots in 21 of 24 patients) with normal peak velocities; impaired smooth pursuit and optokinetic slow phases; decreased VOR gain (also described by Baloh and coworkers [88]), consistent with a study documenting atrophy of the vestibular nerves in two patients with Friedreich's ataxia [135]; and impaired visual-vestibular interaction (visual modification of the vestibulo-ocular reflex). The authors claimed that each of these findings can occur with other cerebellar atrophy syndromes, but the combination of findings appears to be highly specific for Friedreich's ataxia.

3.1.2. Abetalipoproteinemia (Bassen-Kornzweig)

Although the biochemical defects in lipid transport and metabolism have been well described, the cause of the neurologic defects, such as spinocerebellar degeneration, ophthalmoplegia, and peripheral neuropathy, remains uncertain (possibly vitamin E deficiency due to steatorrhoea). Yee and coworkers [136] described the oculomotor disturbances in three patients with abetalipoproteinemia: acquired exotropia, progressive paresis of the medial rectus muscles, and dissociated nystagmus on lateral gaze, with more nystagmus in the adducting eye (in contrast to dissociated nystagmus with internuclear ophthalmoplegia). One patient additionally showed abnormally slow voluntary saccades and slow or absent fast components of vestibular nystagmus and optokinetc nystagmus. All of these oculomotor symptoms most probably cannot be attributed to cerebellar damage, but rather to concomitant brain-stem lesions, such as the loss of saccadic "burst" neurons in the pontine reticular formation.

Slow saccades have also occasionally been found in patients with other unclassified spinocerebellar degenerations [137–140].

3.2. Pure cerebellar atrophies

Baloh and coworkers [94] investigated vestibulo-ocular function in four patients with pure cerebellar atrophy. They all had hyperactive caloric-induced and rotatory-induced nystagmus, abnormal smooth pursuit, abnormal optokinetic responses, and square wave jerks. Two patients had dysmetric saccades, and two patients had rebound nystagmus.

In a recent study of Yamamoto and coworkers [141], five patients had the "Holmes type of cerebellar atrophy." Only one patient had rebound nystagmus, no patient had fixation instabilities, they all had smooth pursuit disturbances, and all had dysmetric saccades.

Similar findings have been reported by Avanzini and coworkers [92]. In their group of patients with cerebellar pathology, there were two patients with "Holmes type cerebellar atrophy." Both had impaired smooth-pursuit eye movements, mainly normometric saccades for saccade amplitudes up to 15°, and partially hypometric, partially hypermetric saccades for saccade amplitudes between 15° and 30°. No fixation instabilities were observed in these patients.

Zee and coworkers [105] recorded and analyzed eye-movement abnormalities in patients from a kindred with Holmes-type late-onset, dominantly inherited, cerebellar ataxia. The patients showed almost the same deficits as monkeys with cerebellar lesions, including an inability to hold the eccentric gaze (gaze-paretic nystagmus), defective smooth pursuit, saccadic dysmetria, and defective visual suppression of vestibular nystagmus. Other findings included downbeating nystagmus, enhanced gain of the vestibulo-ocular reflex during rotation in darkness, rebound nystagmus, and post-saccadic drift (glissades).

Thurston and coworkers [142] reported a patient with a cerebellar degeneration (no further classification) who had a hyperactive vestibulo-ocular reflex. A twofold increase in VOR gain (peak eye velocity/peak head velocity) at high frequencies was associated with a VOR time constant of 6 sec. Visual cancellation of the VOR and smooth pursuit were also abnormal. They hypothesized that his high VOR gain was due to dysfunction of olivo-cerebellar projections. Physostigmine reduced the patients VOR gain, consistent with the hypothesis that the projections responsible for the effect are cholinergic.

The "Foix-Alajouanine type of cerebellar atrophy" mainly occurs in chronic alcoholics; it predominantly involves the more medial parts of the anterior lobe. Oculomotor symptoms are not obligatory. If they occur, they are largely of the type that one would ascribe to a floccular lesion [143]. Oculomotor findings of a 67-year-old male with advanced cerebellar cortical atrophy are depicted in Figure 12-3.

Avanzini and coworkers [92] studied three cases of "Foix-Alajouanine type cerebellar atrophy." Only one patient had square wave jerks; another patient had gaze-evoked nystagmus. All patients had impaired smooth pursuit and partially hypometric, partially hypermetric saccades for saccade amplitudes between 15° and 30°. Below 15° saccade amplitude they had predominantly normometric saccades.

Yamamoto and coworkers [141] found in seven patients with late cortical cerebellar atrophy in 100% smooth-pursuit disturbances, mixed hypermetric and hypometric saccades in three patients, hypometric saccades in three patients, and normal saccades in only one patient. There were no fixation disturbances, such as saccadic oscillations or square wave jerks.

3.3. Multiple systems atrophies

3.3.1. Machado-Joseph disease

This disease is classified as an autosomal dominant cerebellar atrophy (ADCA I) [152], but multiple systems are involved. The symptoms and signs are variable among the patients and are also variable in the same patient during the course of the disease [153,154]. The main neurological alterations are ataxia, akinesia, distal amyotrophy, progressive external ophthalmoplegia, facial and lingual fasciculations, and bulging eyes. Dawson and coworkers [156] investigated 26 patients with electrooculography. The patients had defects in caloric response (less than 6°/sec slow-phase velocity) (12 patients), sinusoidal tracking (12 patients), and optokinetic nystagmus (11 patients); dysmetric saccades (6 patients); and gaze paretic nystagmus (7 patients). None of the patients had spontaneous nystagmus.

3.3.2. Olivopontocerebellar atrophies

3.3.2.1. TYPE MENZEL I (CLASSIFICATION AFTER KONIGSMARK AND WEINER [168]). Yamamoto and coworkers [141] investigated 30 patients with Menzel-type olivopontocerebellar atrophy. Twenty percent had square wave jerks, 10% had saccadic oscillations, 30% had ocular flutter, and 36.7% had rebound nystagmus. All patients had smooth-pursuit disturbances, 16.7% had hypermetric saccades, 13.3% had mixed hypermetric and hypometric saccades, 53.3% had hypometric saccades, and four patients had normal saccades. In contrast, Avanzini and coworkers [92] found impaired smooth pursuit in only 1 out of 3 patients with Menzel-type OPCA. The same

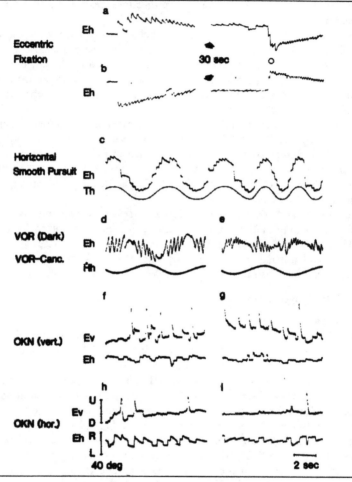

Figure 12-3. Gaze holding, smooth pursuit, and OKN deficits in advanced olivopontocerebellar atrophy. A 59-year-old female with severe ataxia of trunk and gait, more than in her arms and legs (AC recording). Gaze-evoked nystagmus during eccentric fixation (45°) to the right (a) and to the left (b), which slowly declines, after about 35-sec refixation of a target at the primary position (open circle). Note the immediate appearance of rebound nystagmus, which beats in the direction opposite to former gaze holding. Severe impairment of horizontal sinusoidal tracking (c). Vestibuloocular reflex (VOR) during sinusoidal chair rotation with a frequency of 0.2 Hz and a maximum velocity of 90°/sec. The gain (eye velocity/head velocity) is increased in both directions to about 1.2 (d). Impaired suppression of the VOR during attempted fixation of a head-stationary target attached to the rotating chair (e). Severe impairment of optokinetic nystagmus (OKN) in all directions (60°/sec velocity), slow phases upward (f), downward (g), to the left (h), and to the right (i). Eh = horizontal eye position; Ev = vertical eye position; Th = horizontal target position; Hh = horizontal head velocity; R = right; L = left; U = up, D = down.

patient had partially hypometric, partially hypermetric saccades for saccade amplitudes between 15° and 30°. The other two patients had normometric saccades. None of the patients had fixation abnormalities. Furthermore, Rondot and coworkers [155] found slow saccades in one patient with Menzel's ataxia.

3.3.2.2. SCHUT-HAYMAKER DISEASE (IV). Hutton and coworkers [144] described a group of the Schut-Haymaker kindred, an autosomal dominant form of olivopontocerebellar degeneration. The major features of this disease include marked incoordination, pyramidal tract involvement, and bulbar dysfunction with dysarthria and recurrent aspiration. Oculomotor performance was significantly impaired, with prolonged saccadic latencies, longer saccadic refixation times, reduced visual tracking performance, and saccadic hypermetria. These eye-movement abnormalities can precede the clinical diagnosis by at least 1–3 years and provide a useful tool for early diagnosis.

3.3.2.3. OLIVOPONTOCEREBELLAR ATROPHY WITH RIGOR AND DEMENTIA (V). Philcox and coworkers [145] described members of a South African kindred with autosomal dominant hereditary ataxia with rigor and dementia. In more severely affected individuals, voluntary shifts of gaze could only be made by turning the head. There was no diplopia or evident weakness of the ocular muscles. Blinking or jerking of the head to unfix the gaze or to initiate voluntary movement did not occur. Optokinetic responses were absent or reduced and smooth pursuit was impaired. In some patients, however vertical responses were lost while the horizontal responses were retained.

3.3.2.4. OLIVOPONTOCEREBELLAR ATROPHY WITH SLOW EYE MOVEMENTS AND PERIPHERAL NEUROPATHY. Singh and coworkers [146], Wadia [147,148], and Wadia and Swami [149] described a variety of olivopontocerebellar atrophy with slow eye movements. Characteristic eye findings in this syndrome were as follows: (1) saccadic and smooth-pursuit movements were extremely slow, particularly in the horizontal direction, with considerably less slowing in the vertical direction, particularly on downward gazing, and with no impairment of convergence. (2) There was a marked reduction in spontaneous scanning movements. (3) Head-jerking movements were used to help the relatively frozen eyes attempt to gaze in the lateral direction. Dysfunction of descending, efferent pathways in the brain stem was considered to be responsible for these oculomotor disturbances. In a recent review, Mizutani and coworkers [150] disclosed that all cases of olivopontocerebellar atrophy with slow saccades were hereditary.

3.3.2.5. NONHEREDITARY OLIVOPONTOCEREBELLAR ATROPHY (VI). In this group of 52 patients, Yamamoto and coworkers [141] found square wave jerks in 30.8%, saccadic oscillations in 11.5%, ocular flutter in 23.1%, and rebound nystagmus in 19.2% of the patients. They all had smooth-pursuit disturbances, 13.5% had hypermetric saccades, 21.2% had mixed hypermetric and hypometric saccades, 61.5% had hypometric saccades, and only two patients had normal saccades.

Recently, Baloh and coworkers [151] presented an interesting but rare finding in two (our of 150) patients with olivopontocerebellar atrophy (no further classification). Apart from gaze-evoked nystagmus, abnormal smooth pursuit and optokinetic nystagmus, and absent optokinetic-after-nystagmus, as well as hyperactive VOR gain, the patients' dominant VOR time constant amounted to 2 sec or less. This is even below the cupula time constant.

3.4. Ataxia-telangiectasia

Ataxia-telangiectasia is an autosomal recessive inherited, progressive disorder of unknown etiology, manifested by ataxia, telangiectasia, and sinopulmonary infection due to immune deficiency. A comprehensive treatise of the neurological abnormalities in ataxia-telangiectasia has been given by Sedgwick [157]; 84% of these patients had oculomotor abnormalities. They constitute one of the major hallmarks of the disease, particularly as it progresses. Major oculomotor findings are gaze-evoked nystagmus in all directions of gaze, difficulties in initiating and executing voluntary conjugate gaze, and overshooting of the head with preserved VOR, resembling congenital oculomotor apraxia. Baloh and coworkers [158] found abnormalities of voluntary saccades, with increased reaction times and marked hypometria of horizontal and vertical voluntary saccades. The saccade velocity remained normal. Vestibular and optokinetic fast components (involuntary saccades) had a normal amplitude and velocity, but the eyes deviated tonically in the direction of the slow phases. The authors felt that the difficulties in saccade initiation distinguish the eye movements from those with other familial cerebellar atrophy syndromes. Schmidt [159] found in two brothers with ataxia telangiectasia pathological smooth pursuit and hypometric saccades with preserved vestibulo-ocular reflex, voluntary saccade increased reaction times, gaze-holding failure, gaze nystagmus, absent optokinetic nystagmus, and convergence paresis. Rarely, an additional periodic alternating nystagmus was observed [160].

3.5. Congenital cerebellar hypoplasia

Cerebellar hypoplasia is a clinical syndrome with several causes, but it has many symptoms and signs in common. Hypoplasia of the human cerebellum occurs as a rare autosomal recessive disease [161] or can be associated with other progressive neurologic diseases, including Werdnig-Hoffmann disease [162], Tay-Sachs disease [163], and Menkes's disease [164]. It is occasionally seen with other developmental anomalies, such as the Arnold-Chiari type 2 malformation. Sarnat and Alcala [165] described seven cases. The most constant signs were generalized muscular hypotonia and developmental delay. Ataxia, dyssynergia, and intention tremor became evident during the first year. Jerky ocular movements were seen in the majority. Fixation- and gaze-evoked nystagmus were seen in most subjects; vertical or horizontal vestibular nystagmus could only be elicited in 1 of the 7 patients.

Furman and coworkers [130] described the oculographic features of a family whose members suffer from a dominantly inherited, early-onset, nonprogressive cerebellar syndrome, characterized by primary position spontaneous upbeating nystagmus and gait ataxia associated with cerebellar vermian atrophy. Additional findings were severely impaired horizontal and vertical smooth pursuit, impaired optokinetic nystagmus and visual-vestibular interaction [visual modification (augmentation and suppression) of the vestibulo-ocular reflex], symmetrical horizontal, but asymmetrical vertical, vestibulo-ocular reflex (higher velocity of the downward slow phases than the upward slow phases), and normal saccades. From the latter finding (normal saccades), the authors concluded that the deep cerebellar nuclei must have been spared in the disease.

3.6. Joubert's syndrome

This is an autosomal recessive familial syndrome with variable expressivity, presenting in early life with episodic hyperpnea (resembling Biot's respiration), ataxia, mental retardation, and abnormal eye movements associated with agenesis of the vermis (in one autopsy case, additionally, neuronal loss of the dentate nucleus was found) [166]. Eye-movement abnormalities consisted of irregular, usually horizontal but occasionally rotatory, spontaneous nystagmus (pendular, seesaw), incoordinate jerky eye movements and abnormalities of smooth pursuit, OKN, and saccades [167]. The syndrome is probably associated with malformation of the pontine and medullary structures, which leads to the respiratory abnormality.

REFERENCES

1. Holmes G. (1917). The symptoms of acute cerebellar injuries due to gunshot injuries. Brain 40:461–535.
2. Cogan D.G. (1956). Neurology of the Ocular Muscles, 2nd ed. Springfied, IL: Charles C. Thomas.
3. Dow R.S., Manni E. (1964). The relationship of the cerebellum to extraocular movements. In: M.B. Bender (ed): The Oculomotor System, New York: Harper and Row, pp. 280–292.
4. Dichgans J., Jung, R. (1975). Oculomotor abnormalities due to cerebellar lesions. In: G. Lennerstrand, P. Bach-y-Rita (eds): Basic Mechanisms of Ocular Motility and their Clinical Implications. Oxford: Pergamon Press, pp. 281–298.
5. Cogan D.G., Chu F.C., Reingold D.B. (1982). Ocular signs of cerebellar disease. Arch. Ophthalmol. 100:755–760.
6. Leigh R.J., Zee D.S. (1983). The Neurology of Eye Movements. Philadelphia: Davis Company.
7. Dichgans J. (1984). Clinical symptoms of cerebellar dysfunction and their topodiagnostical significance. Human Neurobiol. 2:269–279.
8. Westheimer G., Blair S.M. (1973). Oculomotor defects in cerebellectomized monkeys. Invest. Ophthalmol. 12:618–621.
9. Westheimer G., Blair S.M. (1974). Functional organization of primate oculomotor system revealed by cerebellectomy. Exp. Brain Res. 21:463–472.
10. Burde R.M., Stroud M.H., Roper-Hall G., Wirth F.P., O'Leary J.L. (1975). Ocular motor dysfunction in total and hemicerebellectomized monkeys. Br. J. Ophthalmol. 59:560–565.
11. Eckmiller R., Westheimer G. (1983). Compensation of oculomotor deficits in monkeys with neonatal cerebellar ablations. Exp. Brain Res. 49:315–326.

12. Takemori S., Cohen B. (1974). Loss of visual suppression of vestibular nystagmus after flocculus lesions. Brain Res. 72:213–224.
13. Zee D.S., Yamazaki A., Butler P.H., Guecer G. (1981). Effects of ablation of flocculus and paraflocculus on eye movements in primate. J. Neurophysiol. 46:878–899.
14. Chubb M.C., Fuchs A.F. (1982). Contribution of y group of vestibular nuclei and dentate nucleus of cerebellum to generation of vertical smooth eye movements. J. Neurophysiol. 48:75–99.
15. Waespe W., Cohen B., Raphan T. (1983). Role of the flocculus and paraflocculus in optokinetic nystagmus and visual-vestibular interactions: Effects of lesions. Exp. Brain Res. 50:9–33.
16. Flandrin J.M., Courjon J.H., Jeannerod M., Schmid R. (1983). Effects of unilateral flocculus lesions on vestibulo-ocular responses in the cat. Neuroscience 8:809–817.
17. Cannon S.C., Robinson D.A. (1986). The final common integrator is in the prepositus and vestibular nuclei. In: E.L. Keller, D.S. Zee (eds): Adaptive Processes in Visual and Oculomotor Systems, Oxford: Pergamon Press, pp. 307–311.
18. Optican L.M., Zee D.S., Miles F.A., Lisberger S.G. (1980) Oculomotor deficits in monkeys with floccular lesions. Soc. Neurosci. Abstr. 6:474.
19. Cannon S.C., Robinson D.A. (1985). Neural integrator failure from brain stem lesions in monkey. Invest. Ophthalmol. Vis. Sci. 26(Suppl.):47.
20. Cheron G., Godaux E., Laune J.M., Vanderkelen B. (1987). Disabling of the oculomotor neural integrator by kainic acid injections in the prepositus-vestibular complex of the cat. J. Physiol. 394:267–290.
21. Belknap D.B., Noda H. (1987). Eye movements evoked by microstimulation in the flocculus of the alert macaque. Exp. Brain Res. 67:352–362.
22. Lisberger S.G., Fuchs A.F. (1978). Role of primate flocculus during rapid behavioral modification of vestibuloocular reflex. I. Purkinje cell activity during visually guided horizontal smooth pursuit eye movements and passive head rotation. J. Neurophysiol. 41:733–763.
23. Lisberger S.G., Fuchs A.F. (1978). Role of primate flocculus during rapid behavioral modification of vestibuloocular reflex. II. Mossy fiber firing patterns during horizontal head rotation and eye movement. J. Neurophysiol. 41:764–777.
24. Kimura M., Maekawa K. (1981). Activity of flocculus Purkinje cells during passive eye movements. J. Neurophysiol. 46:1004–1017.
25. Miyashita Y. (1979). Interaction of visual and canal inputs on the oculomotor system via the cerebellar flocculus. Progr. Brain Res. 50:695–702.
26. Miyashita Y. (1984). Eye velocity responsiveness and its proprioceptive component in the floccular Purkinje cells of the alert pigmented rabbit. Exp. Brain Res. 55:81–90.
27. Miles F.A., Fuller J.H., Braitman D.J., Dow B.M. (1980). Long-term adaptive changes in primate vestibuloocular reflex. III. Electrophysiological observations in flocculus of normal monkeys. J. Neurophysiol. 43:1437–1476.
28. Miles F.A., Braitman D.J., Dow B.M. (1980). Long-term adaptive changes in primate vestibuloocular reflex. IV. Electrophysiological observations in flocculus of adapted monkeys. J. Neurophysiol. 43:1477–1493.
29. Noda H., Warabi T., Ohno M. (1987). Response properties and visual receptive fields of climbing and mossy fibers terminating in the flocculus of the monkey. Exp. Neurology 95:455–471.
30. Waespe W., Henn V. (1981). Visual-vestibular interaction in the flocculus of the alert monkey. II. Purkinje cell activity. Exp. Brain Res. 43:349–360.
31. Waespe W., Henn V. (1984). The primate flocculus in visual-vestibular interactions: Conceptual, neurophysiological, and antomical problems. In: Bloedel et al. (eds): Cerebellar Functions, Berlin: Springer-Verlag, pp. 109–125.
32. Waespe W., Henn V. (1985). Cooperative functions of vestibular nuclei neurons and floccular Purkinje cells in the control of nystagmus slow phase velocity: Single cell recordings and lesion studies in the monkey. In: A. Berthoz, J. Melvill Jones (eds): Adaptive Mechanisms in Gaze Control. Facts and Theories. Elsevier Science, pp. 233–250.
33. Waespe W., Rudinger D., Wolfensberger M. (1985). Purkinje cell activity in the flocculus of vestibular neurectomized and normal monkeys during optokinetic nystagmus (OKN) and smooth pursuit eye movements. Exp. Brain Res. 60:243–262.

34. Waespe W., Büttner U., Henn V. (1981). Visual-vestibular interaction in the flocculus of the alert monkey. I. Input activity. Exp. Brain Res. 43:337-348.
35. Markert G., Büttner U., Straube A., Boyle R. (1988). Neuronal activity in the flocculus of the alert monkey during sinusoidal optokinetic stimulation. Exp. Brain Res. 70:134-144.
36. Sato Y., Kawasaki T. (1990). Operational unit responsible for plane-specific control of eye movement by cerebellar flocculus in cat. J. Neurophys. 64:551-564.
37. Optican L.M., Robinson D.A. (1980). Cerebellar-dependent adaptive control of primate saccadic system. J. Neurophysiol. 44:1058-1076.
38. Langer T., Fuchs A.F., Scudder C.A., Chubb M.C. (1985). Afferents to the flocculus of the cerebellum in the rhesus macaque as revealed by retrograde transport of horseradish peroxidase. J. Comp. Neurol. 235:1-25.
39. Suzuki D.A., Keller E.L. (1984). Visual signals in the dorsolateral pontine nucleus of the alert monkey: Their relationship to smooth-pursuit eye movements. Exp. Brain Res. 53:473-478.
40. Brodal P. (1979). The pontocerebellar projection in the rhesus monkey: An experimental study with retrograde axonal transport of horseradish peroxidase. Neuroscience 4:193-208.
41. Brodal P. (1982). Further observations on the cerebellar projections from the pontine nuclei and the nucleus reticularis tegmenti pontis in the rhesus monkey. J. Comp. Neurol. 204:44-55.
42. Kase M., Miller D.C., Noda H. (1973). Discharges of Purkinje cells and mossy fibers in the cerebellar vermis of the monkey during saccadic eye movements and fixation. J. Physiol. 300:539-555.
43. Suzuki D.A., Noda H., Kase M. (1981). Visual and pursuit eye movement-related activity in posterior vermis of monkey cerebellum. J. Neurophysiol. 46:1120-1139.
44. Suzuki D.A., Keller E.L. (1988). The role of the posterior vermis of monkey cerebellum in smooth-pursuit eye movement control. I. Eye and head movement-related activity. J. Neurophysiol. 59:1-18.
45. Suzuki D.A., Keller E.L. (1988). The role of the posterior vermis of monkey cerebellum in smooth-pursuit eye movement control. II. Target velocity-related Purkinje cell activity. J. Neurophysiol. 59:19-40.
46. Ito M. (1982). Cerebellar control of the vestibulo-ocular reflex—around the flocculus hypothesis. Ann. Rev. Neurosci. 5:275-296.
47. Llinas R., Pellionisz A. (1986). Cerebellar function and the adaptive feature of the central nervous system. In: Adaptive Mechanisms in Gaze Control—Facts and Theories. A. Berthoz, J. Melvill Jones (eds): Elsevier Science, pp. 223-231.
48. Watanabe E. (1984). Neuronal events correlated with long-term adaptation of the horizontal vestibulo-ocular reflex in the primate flocculus. Brain Res. 297:169-174.
49. Nagao S. (1983). Effects of vestibulocerebellar lesions upon dynamic characteristics and adaptation of vestibulo-ocular and optokinetic responses in pigmented rabbit. Exp. Brain Res. 53:36-46.
50. Demer J.L., Echelman D.A., Robinson D.A. (1985). Effects of electrical stimulation and reversible lesions of the olivocerebellar pathway on Purkinje cell activity in the flocculus of the cat. Brain Res. 346:22-31.
51. Schultheis L.W., Robinson D.A. (1981). Directional plasticity of the vestibulo-ocular reflex in the cat. N.Y. Acad. Sci. 374:504-512.
52. Lisberger S.G., Miles F.A., Zee D.S. (1984). Signals used to compute errors in the monkey vestibulo-ocular reflex: Possible role of the flocculus. J. Neurophysiol. 52:1140-1153.
53. Optican L.M., Zee D.S., Chu F.C. (1985). Adaptive changes due to ocular muscle weakness in human pursuit and saccadic eye movements. J. Neurophysiol. 54:110-122.
54. Optican L.M., Zee D.S., Miles F.A. (1986). Floccular lesions abolish adaptive control of post-saccadic ocular drift in primates. Exp. Brain Res. 64:596-598.
55. Albus J.S. (1971). A theory of cerebellar function. Math. Biosci. 10:25-61.
56. Ito M. (1979). Adaptive modification of the vestibulo-ocular reflex in rabbits affected by visual inputs and its possible neuronal mechanisms. In: R. Granit, O. Pompeiano (eds): Progress in Brain Research, Vol. 50, Elsevier Science Publishers, pp. 757-761.
57. Ito M. (1980). Roles of the inferior olive in the cerebellar control of vestibular functions. In: Courville et al. (eds): The Inferior Olivary Nucleus: Anatomy and Physiology. New York: Raven Press, pp. 367-377.

58. Ito M. (1985). Synaptic plasticity in the cerebellar cortex that may underlie the vestibuloocular adaptation. In: Adaptive Mechanisms in Gaze Control—Facts and Theories. A. Berthoz, J. Melvill Jones (eds): Elsevier Science, pp. 213–221.
59. Marr D. (1969). A theory of cerebellar cortex. J. Physiol. 202:437–470.
60. Pellionisz A., Llinas R. (1982). Space-time representation in the brain. The cerebellum as a predictive space-time metric tensor. Neuroscience 7:2929–2970.
61. Robinson D.A. (1976). Adaptive gain control of vestibulo-ocular reflex by the cerebellum. J. Neurophysiol. 39:954–969.
62. Yagi T., Shimizu M., Sekine S., Kamio T., Suzuki J.I. (1981). A new neurological test for detecting cerebellar dysfunction. In: B. Cohen (ed): Vestibular and Oculomotor Physiology. Ann N.Y. Acad. Sci., 37:526–531.
63. Llinas R., Walton K., Hillman D.E. (1975). Inferior olive: Its role in motor learning. Science 190:1230–1231.
64. Llinas R., Walton K. (1977). Significance of the olivo-cerebellar system in compensation of ocular position following unilateral labyrinthectomy. In: Baker, and A. Berthoz (eds): Control of Gaze by Brain Stem neurons. Developments in Neuroscience, Volume 1. Elsevier pp. 399–408.
65. Courjon J.H., Flandrin J.M., Jeannerod M., Schmid R. (1982). The role of the flocculus in vestibular compensation after hemilabyrinthectomy. Brain Res. 239:251–257.
66. Haddad G.M., Friendlich A.R., Robinson D.A. (1977). Compensation of nystagmus after VIIIth nerve lesion in vestibulocerebellectomized cats. Brain Res. 135:192.
67. Judge S.J. (1987). Optically-induced changes in tonic vergence and AC/A ratio in normal monkeys and monkeys with lesions of the flocculus and ventral paraflocculus. Exp. Brain Res. 66:1–9.
68. Milder D.G., Reinecke R.D. (1983). Phoria adaptation to prisms—a cerebellar-dependent response. Arch. Neurol. 40:339–342.
69. Luebke A.E., Hain T.C. (1988). Phoria adaptation in patients with cerebellar dysfunction. Invest. Ophthalmol. Visual Sci. Abstr. 29:137.
70. Blair S.M., Gavin M. (1979). Modification of the macaque's vestibulo-ocular reflex after ablation of the cerebellar vermis. Acta Otolaryngol. 88:235–243.
71. Waespe W., Cohen B., Raphan T. (1985). Dynamic modification of the vestibulo-ocular reflex by the nodulus and uvula. Science 228:199–202.
72. Leigh R.J., Robinson D.A., Zee D.S. (1981). A hypothetical explanation of periodic alternating nystagmus: Instability in the optokinetic-vestibular system. Ann. N.Y. Acad. Sci. 374:619–635.
73. Precht W., Simpson J.I., Llinas R. (1976). Response of Purkinje cells in rabbit nodulus and uvula to natural vestibular and visual stimuli. Pflügers Arch. 367:1–6.
74. Haines D.E. (1977). Cerebellar corticonuclear and corticovestibular fibers of the flocculonodular lobe in a prosimian primate. J. Comp. Neurol. 174:607–630.
75. Aschoff J.C., Cohen B. (1971). Changes in saccadic eye movements produced by cerebellar cortical lesions. Exp. Neurol. 32:123–133.
76. Cohen B., Goto K., Shanzer S., Weiss A. (1965). Eye movements induced by electrical stimulation of the cerebellum in the alert cat. Exp. Neurol. 13:145–162.
77. McElligott J.G. (1979). Purkinje cell acitivity in the vermal cerebellum of the cat during trained saccadic eye movements and fixation. Soc. Neurosci. Abstr. 5:104.
78. Ron S., Robinson D.A. (1973). Eye movements evoked by cerebellar stimulation in the alert monkey. J. Neurophysiol. 36:1004–1022.
79. Keller E.L., Slakey D.P., Crandall W.F. (1983). Microstimulation of the primate cerebellar vermis during saccadic eye movements. Brain Res. 288:131–143.
80. Noda H., Fujikado T. (1987). Topography of the oculomotor area of the cerebellar vermis in macaques as determined by microstimulation. J. Neurophysiol. 58:359–378.
81. Noda H., Murakami S., Yamada J., Tamada J., Tamaki Y., Aso T. (1988). Saccadic eye movements evoked by microstimulation of the fastigial nucleus of macaque monkeys. J. Neurophysiol. 60:1036–1052.
82. Ritchie L. (1976). Effects of cerebellar lesions on saccadic eye movements. J. Neurophysiol. 39:1246–1256.
83. Vilis T., Hore J. (1981). Characteristics of saccadic dysmetria in monkeys during reversible lesions of medial cerebellar nuclei. J. Neurophysiol. 46:828–838.

84. Hepp K., Henn V., Jaeger J. (1982). Eye movement related neurons in the cerebellar nuclei of the alert monkey. Exp. Brain Res. 45:253–264.
85. Mackay W.A. (1988). Cerebellar nuclear activity in relation to simple movements. Exp. Brain Res. 71:47–58.
86. Vilis T., Snow R., Hore J. (1983). Cerebellar saccadic dysmetria is not equal in the two eyes. Exp. Brain Res. 51:343–350.
87. Snow R., Hore J., Vilis T. (1985). Adaptation of saccadic and vestibulo-ocular system after extraocular muscle tenectomy. Invest. Ophthalmol. Vis. Sci. 26:924–931.
88. Selhorst J.B., Stark L., Ochs A.L., Hoyt W.F. (1976). Disorders in cerebellar ocular motor control. II. Macrosaccadic oscillation—an oculographic, control system and clinicoanatomical analysis. Brain 99:509–522.
89. Zee D.S., Robinson D.A. (1979) A hypothetical explanation of saccadic oscillations. Ann. Neurol 5:405–414.
90. Suzuki D.A., Keller E.L. (1982). Vestibular signals in the posterior vermis of the alert monkey cerebellum. Exp. Brain Res. 47:145–147.
91. Pierrot-Deseilligny C., Amarenco P., Roullet E., Marteau R. (1990). Vermal infarct with pursuit eye movement disorders. J. Neurol. Neurosurg. Psychiatry 53:519–521.
92. Avanzini G., Girotti F., Crenna P., Negri S. (1979). Alterations of ocular motility in cerebellar pathology—an electro-oculographic study. Arch. Neurol. 36:274–280.
93. Baloh R.W., Spooner J.W. (1981). Downbeat nystagmus: A type of central vestibular nystagmus. Neurology 31:304–310.
94. Baloh R.W., Konrad H.R., Honrubia V. (1975). Vestibulo-ocular function in patients with cerebellar atrophy. Neurology 25:160–168.
95. Baloh R.W., Konrad H.R., Dirks D., Honrubia V. (1976). Cerebellar-pontine angle tumors—Results of quantitative vestibulo-ocular testing. Arch. Neurol. 33:507–512.
96. Baloh R.W., Jenkins H.A., Honrubia V., Yee R.D., Lau C.G.Y. (1979). Visual-vestibular interaction and cerebellar atrophy. Neurology 29:116–119.
97. Chambers B.R., Gresty M.A. (1983). The relationship between disordered pursuit and vestibulo-ocular reflex suppression. J. Neurol. Neurosurg. Psychiat. 46:61–66.
98. Dichgans J., Von Reutern G.M., Römmelt U. (1978). Impaired suppression of vestibular nystagmus by fixation in cerebellar and noncerebellar patients. Arch. Psychiat. Nervenkr. 226:183–199.
99. Hood J.D., Kayan A., Leech J. (1973). Rebound nystagmus. Brain 96:507–526.
100. Jung R., Kornhumber H.H. (1964). Results of electronystagmography in man: The value of optokinetic, vestibular, and spontaneous nystagmus for neurologic diagnosis and research. In: M. Bender (ed): The Oculomotor System. New York: Hoeber, pp. 428–488.
101. Larmande P., Delplace M.P., Autret A. (1980). Influence du cervelet sur la statique oculaire et les mouvements de poursuite visuelle. Rev. Neurol. 136:327–339.
102. Nemet P., Ron S. (1977). Cerebellar role in smooth pursuit movement. Doc. Ophthalmol. 43:101–107.
103. Potthoff P.C., Haustein M. (1970). Nystagmus und Elektro-nystagmogramm nach Kleinhirntumoroperationen. Neurochirurgia 13:174–188.
104. Zee D.S., Friendlich A.R., Robinson D.A. (1974). The mechanism of downbeat nystagmus. Arch. Neurol. 30:227–237.
105. Zee D.S., Yee R.D., Cogan D.G., Robinson D.A., Engel W.K. (1976). Ocular motor abnormalities in hereditary cerebellar ataxia. Brain 99:207–234.
106. von Reutern G.M., Dichgans J. (1977). Augenbewegungsstörungen als cerebelläre Symptome bei Kleinhirnbrückenwinkeltumoren. Arch. Psychiat. Nervenkr. 223:117–130.
107. Leech J., Gresty M., Hess K., Rudge P. (1977). Gaze failure, drifting eye movements, and centripetal nystagmus in cerebellar disease. Br. J. Ophthalmol. 61:774–781.
108. Allen G., Fernandez C. (1960). Experimental observation on postural nystagmus. I. Extensive lesions in posterior vermis of the cerebellum. Acta Otolaryngol. 51:2–12.
109. Fernandez C., Alzate L. (1960). Experimental observation on postural nystagmus. II. Lesions of the nodulus. Ann. Otol. Rhinol. Laryngol. 69:94–101.
110. Grand W. (1971). Positional nystagmus: An early sign in medulloblastoma. Neurology 21:1157–1159.
111. Sakata E., Ohtsu K., Shimura H., Sakai S. (1987). Positional nystagmus of benign

paroxysmal type (BPPN) due to cerebellar vermis lesions. Pseudo-BPPN. Auris, Nasus, Larynx 14:17–21.
112. Larmande P., Autret A. (1980). Influence du cervelet sur les mouvements oculaires volontaires. Rev. Neurol. 136:259–269.
113. Ron S., Nemet P. (1977). The cerebellar involvement in the generation of saccades. Doc. Ophthalmol. 43:109–114.
114. Selhorst J.B., Stark L., Ochs A.L., Hoyt W.F. (1976). Disorders in cerebellar ocular motor control. I. Saccadic overshoot dysmetria—an oculographic, control system and clinico-anatomical analysis. Brain 99:497–508.
115. Lesser R.L., Smith J.L., Levenson D.S., Susac J.O. (1973). Vertical ocular dysmetria. Am. J. Ophthalmol. 76:208–211.
116. Zee D.S. (1982). Ocular motor abnormalities related to lesions in the vestibulocerebellum in primate. In: G. Lennerstrand, D.S. Zee, E.L. Keller (eds): Functional Basis of Ocular Motility Disorders Oxford: Pergamon Press, pp. 423–430.
117. Hain T.C., Zee D.S., Maria B. (1988). Tilt suppression of the vestibulo-ocular reflex in patients with cerebellar lesions. Acta Otolaryngol. 105:15–20.
118. Heide W., Schrader V., Koenig E., Dichgans J. (1988). Impaired discharge of the eye velocity storage mechanism in patients with lesions of the vestibulo-cerebellum. Adv. Otorhinolaryngol. 41:44–48.
119. Dell'Osso L.F., Abel L.A., Daroff R.B. (1977). "Inverse latent" macro square-wave jerks and macro saccadic oscillations. Ann. Neurol. 2:57–60.
120. Hotson J.R. (1982). Cerebellar control of fixation eye movements. Neurology 32:31–36.
121. Leopold H.C. (1985). The syndrome of opsoclonus-myoclonus. Fortschr. Neurol. Psychiat. 53:42–54.
122. Ford F.R. (1966). Diseases of the Nervous System in Infancy, Childhood and Adolescence, 5th ed., Springfield, IL, Charles C. Thomas, p. 301.
123. Poullot B., Lauvin R., Baudet D., Chatel M. (1984). Encephalitis with opsoclonia and circumduction nystagmus. Rev. Otoneuroophthalmol. 56:335–346.
124. Hunter S., Kooistra C. (1986). Neuropathologic findings in idiopathic opsoclonus and myoclonus. Their similarity to those in paraneoplastic cerebellar cortical degeneration. J. Clin. Neuro-Ophthalmol. 6:236–241.
125. Ross A.T., Zemann W. (1967). Opsoclonus, occult carcinoma and chemical pathology in dentate nuclei. Arch. Neurol. 17:546–551.
126. Ellenberger C., Campa J.F., Netsky M.G. (1968). Opsoclonus and parenchymatous degeneration of the cerebellum. Neurology 18:1041–1046.
127. Giordana M.T., Soffietti R., Schiffer D. (1989). Paraneoplastic opsoclonus: A neuropathologic study of two cases. Clin. Neuropathol. 8:295–300.
128. Yee R.D., Baloh R.W., Honrubia V., Lau C.G.Y., Jenkins H.A. (1979). Slow build up of optokinetic nystagmus associated with downbeat nystagmus. Invest. Ophthalmol. 18:622.
129. Daroff R.B., Troost B.T. (1973). Upbeat nystagmus. JAMA 225:312.
130. Furman J.M.R., Baloh R.W., Yee R.D. (1986). Eye movement abnormalities in a family with cerebellar vermian atrophy. Acta Otolaryngol. 101:371–377.
131. Kase C.S., White J.L., Joselyn J.N. (1985). Cerebellar infarction in the superior cerebellar artery distribution. Neurology 35:705–711.
132. Zee D.S., Chu F.C., Leigh R.J., Savino P.J., Schatz N.J., Reingold D.B., Cogan D.G. (1983). Blink-saccade synkinesis. Neurology 33:1233–1236.
133. Klockgether T., Schroth G., Diener H.C., Dichgans J. (1990). Idiopathic cerebellar ataxia of late onset: Natural history and MRI morphology. J. Neurol. Neurosurg. Psych. 53:297–305.
134. Furman J.M.R., Perlman S. Baloh R.W. (1983). Eye movements in Friedreich's ataxia. Arch Neurol. 40:343–346.
135. Spoendlin H. (1974). Optic and cochleovestibular degenerations in hereditary ataxias: II. Temporal bone pathology in two cases of Friedreich's ataxia with vestibulo-cochlear disorders. Brain 97:41–48.
136. Yee R.D., Cogan D.G., Zee D.S. (1976). Ophthalmoplegia and dissociated nystagmus in abetalipoproteinemia. Arch. Ophthalmol. 94:571–575.
137. Zee D.S., Optican L.M., Cook J.D., Robinson D.A., Engel W.K. (1976). Slow saccades in spinocerebellar degeneration. Arch. Neurol. 33:243–251.

138. Osorio I. Daroff R.B. (1980). Absence of REM and altered NREM sleep in patients with spinocerebellar degeneration and slow saccades. Ann. Neurol. 7:277–280.
139. Usui S., Beppu H., Hirose K., Tanabe H., Tsukabi T. (1988). A family of spino-cerebellar degeneration with disturbances of ocular movement, choreoathetosis, amyotrophy and dementia. No. To. Shinkei. 40:953–961.
140. Al Din A.S., Al Kurdi A., Al Salem M.K., Al Nassar K.E., Al Zuhair A., Rudwan M.A., Ayish I., Barghouti J.A., Khaffaji S., Hamawi T. (1990). Autosomal recessive ataxia, slow eye movements, dementia and extrapyramidal disturbances. J. Neurol. Sci. 96:191–205.
141. Yamamoto H., Saito S., Sobue I. (1988). Bedside and electro-oculographic analysis of abnormal ocular movements in spinocerebellar degenerations: Effects of thyrotropin-releasing hormone. Neurology 38:110–114.
142. Thurston S.E., Leigh R.J., Abel L.A., Dell'Osso L.F. (1987). Hyperactive vestibulo-ocular reflex in cerebellar degeneration: Pathogenesis and treatment. Neurology 37:53–57.
143. Baloh R.W., Yee R.D., Honrubia V. (1986). Late cortical cerebellar atrophy. Clinical and oculographic features. Brain 109:159–180.
144. Hutton J.T., Albrecht J.W., Kuskowski M., Schut L.J. (1987). Abnormal ocular motor function predicts clinical diagnosis of familial ataxia. Neurology 37:698–701.
145. Philcox D.V., Sellars S.L., Pamplett R., Beighton P. (1975). Vestibular dysfunction in hereditary ataxia. Brain 98:309–316.
146. Singh BM., Ivamoto H., Strobos R.J. (1973). Slow eye movements in spinocerebellar degeneration. Am. J. Ophthalmol. 76:237–240.
147. Wadia N.H. (1973). An indigenous form of heredo-familial spinocerebellar degeneration with slow eye movements. Neurol. India Suppl. IV:561–580.
148. Wadia N.H. (1977). Heredo-familial spinocerebellar degeneration with slow eye movements—Another variety of olivopontocerebellar degeneration. Neurol. India 25:147–160.
149. Wadia N.H., Swami R.K. (1971). A new form of heredofamilial spinocerebellar degeneration with slow eye movements (nine families). Brain 94:359–374.
150. Mizutani T., Satoh J., Morimatsu Y. (1988). Neuropathological background of oculomotor disturbances in olivopontocerebellar atrophy with special reference to slow saccade. Clin. Neuropathol. 7:53–61.
151. Baloh R.W., Beykirch B.S., Tauchi B.S., Yee R.D., Honrubia V. (1988). Ultralow vestibular reflex time constant. Ann. Neurol. 23:32–37.
152. Harding A.E. (1984). The hereditary ataxias and related disorders. New York: Churchill Livingstone, pp. 153–158.
153. Pou-Serradell A., Russi A., Ferrer I., Galofre E., Escudero D. (1987). Maladie de Machado-Joseph dans une famille d'origine espagnole. Rev. Neurol. 143:520–525.
154. Hotson J.R., Langston E., Louis A.A., Rosenberg R.N. (1987). The search for a physiologic marker of Machado-Joseph disease. Neurology 37:112–116.
155. Rondot P., De Recondo J., Davous P., Vedrenne C. (1983). Menzel's hereditary ataxia with slow eye movements and myoclonus. J. Neurol. Sci. 61:65–80.
156. Dawson D.M., Feudo P., Zubick H.H., Rosenberg R., Fowler H. (1982). Electro-oculographic findings in Machado-Joseph disease. Neurology 32:1272–1276.
157. Sedgwick R.P. (1982). Neurological abnormalities in ataxia-telangiectasia. In: B.A. Bridges, D.G. Harnden (eds): Ataxia-Telangiectasia—A Cellular and Molecular Link Between Cancer, Neuropathology, and Immune Deficiency. John Wiley and Sons, pp. 23–35.
158. Baloh R.W., Yee R.D., Boder E. (1978). Eye movements in ataxia-telangiectasia. Neurology 28:1099–1104.
159. Schmidt D. (1973). Okulomotorische Symptome bei Kleinhirnerkrankung am Beispiel der Ataxia teleangiectatica (Louis-Bar). Klin. Mbl. Augenheilk. 173:329–333.
160. Stell R., Bronstein A.M., Plant G.T., Harding A.E. (1989). Ataxia telangiectasia: A reappraisal of the ocular motor features and the value in the diagnosis of atypical cases. Mov. Disord. 4:320–329.
161. Jervis G.A. (1950). Early familial cerebellar degeneration: Report of 3 cases in one family. J. Nerv. Ment. Dis. 111:398–407.
162. Weinberg A.G., Kirkpatrick J.B. (1975). Cerebellar hypoplasia and Werdnig-Hoffmann disease. Dev. Med. Child Neurol. 17:511–516.

163. Friede R.L. (1964). Arrested cerebellar development: A type of cerebellar degeneration in amaurotic idiocy. J. Neurol. Neurosurg. Psychiat. 27:41–45.
164. Williams R.S., Marshall P.C., Lott I.T. (1977). Dendritic abnormalities in "steely hair syndrome": A Golgi microscopic analysis, abstracted. Neurology 27:369.
165. Sarnat H.B., Alcala H. (1980). Human cerebellar hypoplasia. A syndrome of diverse causes. Arch. Neurol. 37:300–305.
166. Joubert M., Eisenring J.J., Robb J.P., Andermann F. (1969). Familial agensis of the cerebellar vermis. Neurology 19:813–825.
167. Lambert S.R., Kriss A., Gresty M., Benton S., Taylor D. (1989). Joubert syndrome. Arch. Ophthalmol. 107:709–713.
168. Konigsmark B.W., Weiner C.P. (1970). The olivopontocerebellar atrophies: A review. Medicine 49:277–241.

13. CLINICAL AND RADIOLOGIC FEATURES OF CEREBELLAR DEGENERATION

ANDREAS PLAITAKIS, SHOICHI KATOH, AND YUN PENG HUANG

The human cerebellar degenerations encompass a large number of heterogenous neurological disorders sharing common clinical and pathologic features. A detailed account of the nosology and epidemiology of these disorders is provided in Chapter 8. Some forms of cerebellar degeneration are obscure entities occurring with extreme rarity (even limited to single families), while others show an appreciable prevalence rate in the general population and, as such, they represent the primary hereditary ataxias.

This chapter focuses on these primary forms of cerebellar degeneration, i.e., those likely to be encountered in neurologic practice. Its main scope is to study the clinical and morphological aspects of these disorders and to emphasize recent progress that has advanced our understanding of these afflictions.

1. CLINICAL ASPECTS

Disordered motor coordination or ataxia is the cardinal neurologic manifestation of the majority of cerebellar degenerations. The severity of ataxia usually reflects the overall disease stage, because it results from degeneration and atrophy of the cerebellum and/or its connections with the spinal cord and brain stem. Depending on the particular form of cerebellar degeneration, other CNS structures may also undergo degeneration and atrophy, giving rise to corresponding neurologic deficits. Some cerebellar degenerations can

A. Plaitakis (ed.), CEREBELLAR DEGENERATIONS· CLINICAL NEUROBIOLOGY. Copyright © 1992.
Kluwer Academic Publishers, Boston. All rights reserved.

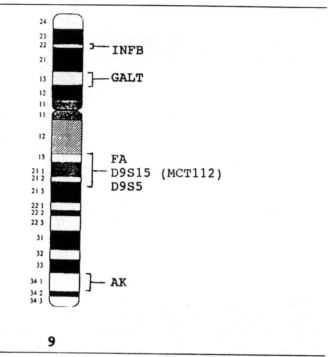

Figure 13-1. Schematic diagram of human chromosome 9 according to McKusisc [112]. The Friedreich's ataxia gene (FA) has been mapped to the 9 chromosome by the genetic linkage analysis of Chamberlain et al. [51]. According to Hanauer et al. [56], the FA gene and the markers D9S15 (MCT112) and D9S5 define a cluster of tightly linked loci that map to the proximal part of chromosome 9 (9q13-q21), as shown here. The physical location of interferon b (INFB), adenylate kinase (AK), and the galactose-1-phosphate uridylytransferase (GALT) are also shown.

be diagnosed on the basis of their symptom complex; others may require neuropathological data.

Of the childhood or juvenile-onset cerebellar degenerations, Friedreich's ataxia is probably the most commonly encountered and thoroughly studied hereditary ataxia. It can be diagnosed on the basis of its characteristic constellation of symptoms and signs, which include progressive cerebellar dysfunction, dysarthria, areflexia, muscle weakness, extensor plantar responses, proprioceptive sensory loss, cardiomyopathy, and skeletal abnormalities. The chromosomal locus of this disorder has been recently identified (Figure 13-1). Of the other forms of juvenile-onset ataxia, hexosaminidase deficiency is of particular interest, because it has been characterized at the molecular level and because the pattern of CNS atrophic changes resulting from this disorder is very similar to that seen in the classic cerebellar degenerations. Hence, the study of this enzymatic deficiency may provide clues to the understanding of the idiopathic ataxias and to factors determining the selective regional vulnerability that is characteristic of these disorders.

Of the adult-onset ataxias, the cortical cerebellar degenerations and the olivopontocerebellar atrophies are of particular importance because of the frequency with which they occur and because of recent progress made in understanding these disorders. The cortical cerebellar degenerations are often characterized pathologically by atrophy of the cerebellar cortex and inferior olives and, as such, are known as *cerebello-olivary atrophies*. In these disorders, cerebellar disturbances, primarily reflecting dysfuction of the middle cerebellar zone, are usually seen in the absence of any other associated clinical features (pure cerebellar syndrome). On the other hand, the OPCA are examples of disorders with *multiple system atrophy* in which cerebellar dysfunction is often associated with additional neurologic features reflecting the involvement of the brain stem, spinal cord, basal ganglia, and/or cerebral cortex. These disorders will be discussed in detail below.

2. NEUROIMAGING ASPECTS

Visualization of CNS structures by CT and MRI offers advantages over an examination of the brain at autopsy in that these techniques allow the study of the anatomical alterations in vivo or the living neuropathology of neurologic disorders at their different stages. In contrast, autopsy usually evaluates end-stage disease. We have previously studied the CT characteristics of five distinct types of OPCA [1] and have found clinicopathological correlations that proved to be of value in understanding these afflictions. Since then the advent of high-resolution MRI has permitted improved visualization of the CNS structures, i.e., the cerebellum, brain stem, and spinal cord, that degenerate in cerebellar ataxias. Bony artifacts, a major problem in the visualization of posterior fossa structures by CT, are neglible with the MRI technique. Hence, high-resolution MRI has become a major tool in the study of cerebellar degenerations. Accordingly, we have concentrated our efforts on studying the morphological changes of inherited ataxias using high-resolution MRIs. CT scans were used only if MRIs were not available or to compare the disease's progression. To make it easier for the reader to appreciate the anatomical alterations affecting the cerebellum in disease states, the normal anatomy of the cerebellum is presented diagramatically in Figure 13-2. In addition, the neuropathology of these disorders will be reviewed and compared to neuroimaging findings.

3. FRIEDREICH'S ATAXIA

3.1. Clinical findings

3.1.1. Neurologic features

Friedreich's ataxia was one of the first familial degenerative neurologic disorders to be identified as a distinct entity from other afflictions by Friedreich,

308 II. Clinical neurosciences of the cerebellum

A

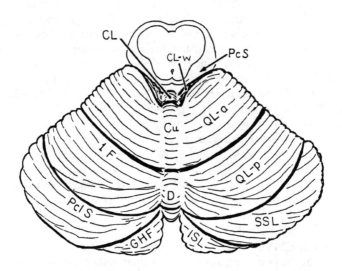

B

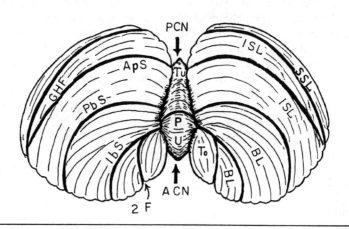

C

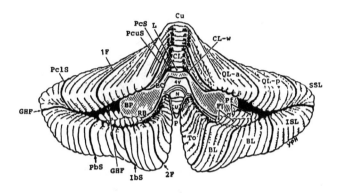

D

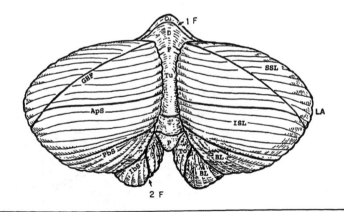

310 II. Clinical neurosciences of the cerebellum

E

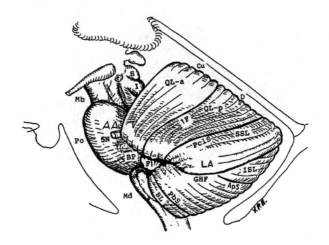

F

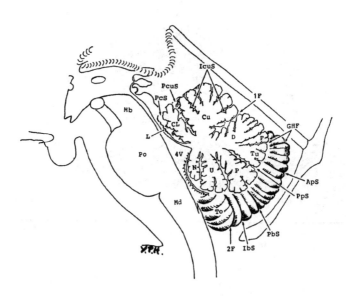

who, over a century ago [2], provided an accurate account of its clinical, pathologic, and genetic features. Even today, Friedreich's criteria used for the diagnosis of this disorder remain essentially unchanged [3].

The mean age at onset is around puberty (about 11 years), although in individual cases it can vary from 2 to 25 years [2–8]. Some authors have found evidence for a strong intrafamilial correlation of the age of onset of the disorder [3]. Hallmarks of the disease are gait and limb ataxia, dysarthria, areflexia, positive Babinski signs, proprioceptive sensory loss, oculomotor disturbances, kyphoscoliosis, and pes cavus deformities. In most cases the presenting neurologic manifestations are disturbances of equilibrium, dysarthria, and absent tendon reflexes, often with extensor plantar responses [5]. In rare cases, scoliosis or cardiomyopathy may be the presenting symptom [6].

The disease typically follows a chronic, progressive nonremtting course, with about 50% of the patients becoming wheelchair bound by age 25 [3]. As the disease progresses, ataxia of gait and limbs and dysarthria become more prominent. Also, skeletal deformities (kyphoscoliosis, pes cavus), sensory abnormalities (loss of joint position and vibration), and cardiomyopathy appear in the majority of the patients (55–80%) [5]. In about 10% of the cases, optic atrophy, oculomotor disturbances, and hearing loss are encountered [6]. Within a few years after the onset of symptoms, paraparesis gradually develops and is present in all patients with advanced disease. Also, wasting of the intrinsic muscles of the hands and feet, and of distal arms and legs, tends to develop in later stages of the illness [3,5]. Mentation is usually not affected in FA; this has been recently confirmed by psychometric studies

Figure 13-2. Diagram of the cerebellum showing its major lobules and sulci. (A) Superior surface, (B) inferior surface, (C) anterior surface, (D) posterior surface, (E) lateral surface, and (F) midsagittal section. Labeled structures are the anterior angle (AA), anterior cerebellar notch (ACN), ansoparamedian sulcus (ApS), brachium conjuctivum (BC), biventral lobule (BL), brachium pontis (BP), central lobule (CL), wing of the central lobule (CL-w), culmen (Cu), declive (D), flocculus (Fl), folium (F), great horizontal fissure (GHF), inferior colliculus (I), intrabiventral sulcus (IbS), intraculminate sulcus (IcuS), inferior semilunar lobule (ISL), lingula (L), alteral angle (LA), midbrain (Mb), medulla oblongata (Md), nodulus (N), pyramid (P), prebiventral sulcus (PbS), postclival sulcus (PclS), precentral cerebellar sulcus (PcS), preculminate sulcus (PcuS), posterior cerebellar notch (PCN), paraflocculus (Pf), pons (Po), prepyramidal sulcus (PpS), anterior part of the quadrangular lobule (QL-a), posterior part of the quadrangular lobule (QL-p), restiform body (RB), superior colliculus (S), superior semilunar lobule (SSL), cerebellar tonsil (To), tuber (Tu), and uvula (U). Also labeled are the primary fissure (1F), secondary fissure (2F), fourth ventricle (4V), and fifth nerve (5N). It should be noted that the term *fissure* is used to separate the anterior from the middle lobe (primary fissure) and the middle from the posterior lobe (secondary fissure). All other shallow grooves separating lobules are designated as *sulci*. The lobulus gracilis located below the ansoparamedian sulcus is included in the inferior semilunar lobule in this chapter. The following are synonyms that are frequently used in textbooks and other articles: primary fissure = anterior superior fissure; postclival sulcus = posterior superior sulcus; prepyramidal sulcus = suprapyramidal, prebiventral, or posterior inferior sulcus; secondary fissure = infrapyramidal, postpyramidal, or antherior inferior fissure.

[9] that revealed the absence of cognitive deficits, despite an impaired information processing speed [10].

3.1.2. Cardiac involvement

In as many as 40% of FA patients, signs of cardiomyopathy may appear at the time of neurologic presentation [7,11]. A higher percentage of patients will experience subjective and objective cardiac abnormalities during the course of the disease, and about 50% may die from cardiac disease [4]. Moreover, in rare cases, cardiac involvement may antedate the development of ataxia, as indicated by the report of Berg et al. [12], which described two siblings who presented with an acute cardiomyopathy only to develop neurologic signs of FA later. The cardiac abnormalities in FA do not seem to be transmitted independently of neurologic abnormalities; unaffected obligate heterozygotes and other unaffected first-degree relatives have normal electrocardiograms and echocardiograms.

There is no consensus among investigators concerning the specific nature of FA cardiomyopathy. Smith et al. [13] and Weiss et al. [14] have observed obstructive hypertrophic cardiomyopathy with asymmetric septal hypertrophy and systolic anterior motion of the mitral valve. However, others [7,16] have reported concentric left ventricular hypertrophy with an increase in the left atrial size and a decrease in the left ventricular size. More recently, Palagi et al. [17] described alterations in the timing of ventricular filling that were particularly evident with high heart rates and were associated with right ventricle hypokinesis.

EKG abnormalities are quite common but are nonspecific in FA patients and seem to be the best early indication of cardiomyopathy [7,15]. These include arrhythmias, changes in repolarization (negative T waves, ST depression in inferolateral leads), and the presence of Q waves, especially in leads I, II, and V 1–5. Cote et al. [15] have reported that while the ECG abnormalities tend to remain constant, the severity of the cardiomyopathy can progress independently of these changes and usually parallels that of the neurologic dysfunction.

3.1.3. Skeletal abnormalities

Musculoskeletal deformities, such as kyphoscoliosis and pes cavus, remain one of the major causes of disability in FA and can aggravate the progressive cardiomyopathy of the disease. Kyphoscoliosis occurs in about 80% of affected patients [6,18]. It tends to be most common in patients with disease onset prior to age 12 and tends to progress rapidly with growth spurts. Allard et al. [18,19] also noted that a profound increase in the Cobb angle measure of scoliosis occurred at about age 10. A relative torsion between the thoracic and lumbar segment was found to be most prominent, with non-ambulatory patients showing the most severe deformities. Changes in the lumbosacral angle have also been shown to coincide with abnormalities in posture and ambulation [19].

3.1.4. Autonomic dysfunction

Abnormalities of bladder function, such as urinary urgency and secondary incontinence, sometimes associated with nocturia, may occur in as many as 50% of patients with FA [20,21]. Cystometric alterations include prominent increases in bladder capacity and changes in inhibition and voluntary contraction. Moreover, external sphincter electromyograms have been shown to be abnormal in FA, demonstrating variations in the levels of electrical activity and in the coordination of vesicosphincter function [20,21].

3.1.5. Auditory system

Although from the clinical point of view the majority of FA patients do not have hearing loss, recent electrophysiologic studies have shown a profound subclinical dysfunction, involving both the peripheral and the central auditory pathways. Recordings of brain-stem auditory evoked responses (BAER) revealed the absence of completely normal wave forms. The sequential deterioration of evoked responses reflects both the disease duration and severity [22,25]. Changes in BAERs were shown to be gradual and did not precede the onset of the disease. As the disease progresses, waveforms tend to loose their morphology and may drop out entirely. However, the order in which the waves disappear remains a matter of controversy [8]. Satya-Murti [22] suggested that the auditory system dysfunction in FA was due to degeneration of the spiral ganglion, probably correlating with the pathologic disappearance of the inner spiral nerve, and that this could be similar to the degeneration of the large myelinated fibers in the dorsal root. However, Nuwer et al. [26] studied 20 patients with FA and reported normal BAER in all patients.

3.1.6. Visual system

Visual abnormalities appear to be common in FA. Livingstone et al. [27] and Carroll [28] have systematically evaluated FA patients and found clinical abnormalities in the majority of such patients, which included disc pallor, decreased visual acuity, reduced number of small vessels crossing the disc margin, or rarely, dyschromatopsia. The VEP showed reductions in amplitude and increases in temporal dispersion, with occasional interocular differences [27–30]. Pinto et al. [30] recently reported a moderate amplitude reduction in pattern electroretinogram that was thought to result from scattered fiber loss, as shown by red-free light retinography. The VEP abnormalities observed in FA are different from those seen in demyelinating disorders and are consistent with the findings of previous pathologic studies that suggested dropout of the large-diameter fast fibers in the pregeniculate visual system.

The oculomotor system is also commonly affected in FA. The main abnormalities are gaze-evoked nystagmus, impaired smooth pursuit, fixation instability, and inaccurate saccades [31]. Furman et al. [32] have suggested that these alterations, along with impaired visual-vestibular interaction and optokinetic slow phases, should be considered specific to FA; however, all these oculomotor disturbances reflect midline cerebellar dysfunction and can occur as nonspecific signs in lesions of the cerebellar vermis, flocculus, and brain stem [33].

Although nystagmus may be observed clinically in about 40% of FA patients, electronystamography has revealed the presence of nystagmus in as many as 70% of such patients [34,35]. Horizontal gaze-paretic nystagmus appears to be commonly encountered, while upbeat nystagmus is rarely seen [30,34,35]. Some authors [31] have noted rebound nystagmus after the eyes return to the midline from an eccentric position, but others [34] have brought attention to transient bursts of ocular myoclonus [33].

Abnormalities of smooth pursuit, consisting of jerky catch-up saccades at low eye velocity (as slow as 18°/sec) and fixation instability manifested clinically as square wave jerks, have been described in the majority of FA patients [31]. Also, failure of fixation suppression of the vestibulo-ocular reflex (decreased VOR gain) has been seen in more than three quarters of FA patients [31]. Optokinetic nystagmus is abnormal in 95% of the cases [35], with slow-phase velocity being particularly affected. Kirkham et al. [35] found that increasing the speed of the optokinetic stimulus failed to augment the slow-phase velocity and that there was a reduction in the maximum slow phase of caloric responses to 4–10°/sec.

3.1.7. Sensory and motor systems

Electrophysiologic abnormalities involving the neuromuscular system occur consistently in patients with FA; nondetectable sensory evoked responses are the hallmark of the disease [30,36]. These abnormalities, however, are not specific for FA, since they can occur in most neuronal degenerations affecting the peripheral sensory nerves, such as in hereditary sensory neuropathy. Somatosensory evoked potentials have shown delayed and dispersed cortical responses, suggesting abnormalities involving the central sensory pathways [38]. EMG usually reveals subtle signs of denervation, as evidenced by rare fibrillations at rest, increased numbers of polyphasic action potentials of long duration but low amplitude, and a decreased recruitment pattern with voluntary effort [36,37]. An inability to relax proximal muscles results in a high-frequency discharge pattern after needle insertion [36]. The distal motor latencies are normal.

3.1.8. Endocrine system

It has been known for several decades that diabetes mellitus occurs frequently in patients with FA. These findings have led to the suggestion that a defect in carbohydrate metabolism underlies the disease. In several studies the incidence of diabetes has varied from 8% to as high as 40% of FA patients [6,39–43]. Campanella et al. [41] reported overt clinical diabetes in 13% and abnormal glucose tolerance in 26% of 23 FA patients studied. Early- and late-onset diabetes mellitus has been described [6,39–43]. Early-onset diabetes can be (1) insulinopenic, characterized by frequent severe hyperglycemia and a tendency to develop ketoacidosis or (2) insulin resistant with modest hyperglycemia and elevated basal insulin levels. More recent studies [42,43] have indicated that diabetes mellitus in such patients is caused by the loss of islet cells, similar to type 1 diabetes, but without HLA association and without serologic evidence for autoimmune destruction of islet cells [43]. On the other hand, patients with late-onset diabetes mellitus tend to have a moderate degree of glucose intolerance that can be easily controlled with oral hypoglycemic agents [6]. In addition to diabetes mellitus, FA patients may have abnormalities of prolactin secretion and hypothyroidism [41].

3.2. Pathologic findings

FA is considered to be the prototype of the spinal form of ataxia. The most conspicuous CNS pathologic finding is spinal-cord atrophy [30,44,45]. There is degeneration of the posterior columns throughout the spinal cord; the gracile fasciculi are usually more severely affected than the cuneate fasciculi [45]. The pyramidal tracts show gradual loss as they descend down the spinal cord, but they tend to remain intact at the brain-stem level [45]. The motor neurons of the anterior horns may also show atrophic changes. The posterior spinocerebellar tracts are severely affected, while the anterior spinocerebellar tracts tend to be less markedly involved. Degeneration of Clarke's column is also consistently and severely involved [45].

Oppenheimer [45] noted that atrophy of the brain stem can occur in severe cases. The cranial sensory nuclei, such as the trigeminal, glossopharyngeal, and vagus nerves, are affected in most patients. However, the cranial motor nuclei were spared. In the motor cortex the giant pyramidal cells showed loss and/or shrinkage in all patients studied [45]. With respect to the cerebellar systems, the superior cerebellar peduncles and the accessory cuneate nuclei showed atrophy and degeneration, but the pontine nuclei and the middle cerebellar peduncles were not affected. Also, in the minority of the patients, cell loss was noted in the inferior olives. The cerebellar cortex is essentially normal, although some loss of Purkinje cells may occur [45]. Oppenheimer [45] described a severe cell loss in the dentate nuclei in 13 out of 15 FA patients studied. The same author noted that the white matter of the cerebellum was always gliotic, probably due to degeneration of afferent fibers. In about half the cases, neuronal loss was also found in the vestibular nuclei, cochlear nuclei, superior olives, optic tracts, geniculate bodies, and the basal nuclei (globus pallidus and subthalamic nucleus).

With respect to the peripheral nervous system, the most conspicuous and

316 II. Clinical neurosciences of the cerebellum

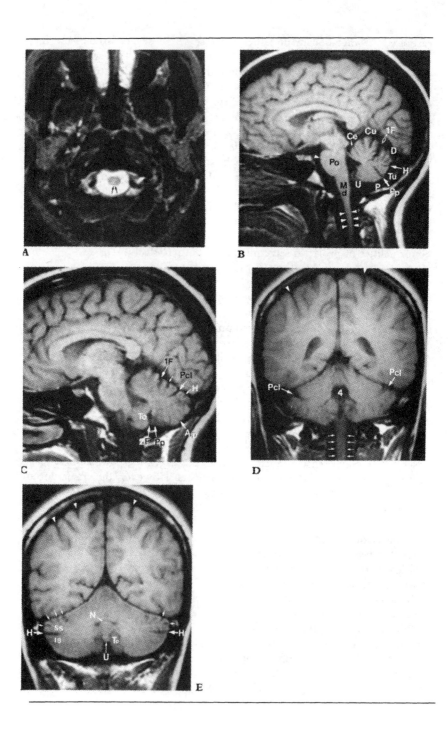

constant finding is degeneration of the large myelinated sensory fibers (loss of about 95% of the fibers), posterior roots, and sensory ganglia [30,44,45]. The small unmyelinated fibers of the peripheral nerves are preserved, whereas the interstitial fibrous tissue may increase [45]. The motor fibers are usually not involved.

3.3. Neuroimaging findings

Here we studied six patients with FA. Their age ranged from 18 to 50 years (mean age 29.2 years) and the disease's duration ranged from 2 to 32 years (mean duration 16.0 years). The age at onset of symptoms ranged from 7 to 18 years (mean 13.2 years). Brain CT scans were performed in all six patients and high-resolution were performed MRIs in two (Figures 13-3 to 13-5). In accord with the neuropathologic changes described above, the most conspicuous finding in CT and MRI was atrophy of the spinal cord (Figures 13-3 to 13-5). Sections through the cervical cord revealed flattening of the posterior aspect of the cord (Figures 13-3A and 13-4). In a patient with a disease duration of 32 years, the cervical cord appeared to be markedly shrunk in its anteroposterior direction (Figure 13-5). In addition, there was an alteration in the MRI signal, indicative of degenerative changes within the spinal cord (Figure 13-5).

The midbrain and pons were essentially normal, although the rostral part of basis pontis appeared to be slightly flattened in sagittal views (Figure 13-3B). The medulla oblongata was reduced in size (Figures 13-3B and 13-5). In the cerebellum, mild atrophic changes occurred in the superior vermian and paravermian regions. The primary fissure and, to a lesser extent, the

Figure 13-3. Friedreich's ataxia. Female, age 30, affected since age 11. Clinical features included progressive ataxia, dysarthria, areflexia, extensor plantar responses, kyphoscoliosis, pes cavus, and cardiomyopathy. She has been wheelchair bound since age 27. **A**: Transverse section at C1. The spinal cord is atrophic, especially posteriorly (arrows). **B**: Midsagittal section. The most conspicuous finding is atrophy of the spinal cord (row of unlabeled opposing arrowheads). The medulla oblongata (Md) is reduced in size. The anterosuperior aspect of the pons (Po) is minimally flattened (unlabeled single arrowhead). The visualized vermis shows atrophic changes, evidenced by the following: widening of the primary fissure (1F) and, to a lesser extent, of the sulci of the culmen (Cu) and declive (D), and the slightly prominent prepyramidal sulcus (Pp). The great horizontal fissure (H), central lobule (Ce), and the normal pyramid (P), tuber, (Tu), and uvula (U) are labeled. **C**: Parasagittal section showing slight atrophy of the paravermian area of the cerebellar hemisphere. There is widening of the primary fissure (1F), postclival sulcus (Pcl), and sulci within the posterior quadrangular lobule (unlabeled small arrow). The tonsil (To), secondary fissure (2F), prepyramidal sulcus (Pp), ansoparamedian sulcus (Ap), and great horizontal fissure (H) are labeled. **D**: Coronal section through the anterior portion of the fourth ventricle (4) showing atrophy of the cervical cord (row of opposing arrows), widening of the postclival sulci (Pcl), and slightly prominent parietal sulci (arrowheads). **E**: Coronal section through the roof of the posterior superior recesses of the fourth ventricle parallel to the brain stem. Note atrophy of the superior semilunar lobules (SS) (small arrows) and widening of the great horizontal fissures (H). The inferior semilunar lobule (IS), tonsil (Yo), nodulus (N), and uvula (U) are labeled. The intraparietal and adjacent sulci are widened (arrowheads).

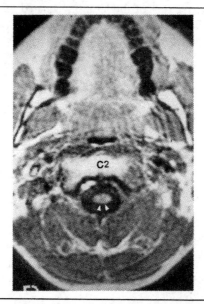

Figure 13-4. Friedreich's ataxia. Male, age 25, affected since age 16 by progressive ataxia, dysarthria, areflexia, corticospinal deficits, and proprioceptive sensory loss. There are also kyphoscoliosis and pes cavus deformities. A 20-year-old sister has been similarly affected since age 15. Transverse section of the spinal cord at the level of C2. The spinal cord is decreased in size anteroposteriorly. Its posterior aspect shows some flattening (arrowheads) due, presumably, to atrophy of the posterior and posterolateral portions of the spinal cord.

sulci of the culmen and declive, showed enlargement (Figure 13-3B). In the inferior vermis, the prepyramidal fissure was slightly widened (Figure 13-3B). The cerebellar hemispheres showed a slight degree of atrophy in the superior semilunar lobule (Figure 13-3E).

3.4. Genetic defect

Friedreich was first to note that the disease that bears his name occurs in families with siblings, but not parents or offsprings, of patients found to be similarly affected. Subsequent studies have confirmed these original observations and have established with a reasonable certainty that the disorder is inherited in an autosomal recessive manner. Analysis of pedigree data has revealed a calculated 0.25 segregation ratio [6], as expected in a recessively inherited disorder. In the United Kingdom, a heterozygote frequency of 1 in 110 individuals has been estimated [46]. Instances of vertical transmission have, however, been reported, but these are thought to represent examples of pseudodominant inheritance resulting from the mating of a homozygote with a heterozygote [46]. Although previous studies suggested a high con sanguinity rate among families afflicted by FA [47], more recent data revealed rates that were either slightly higher or the same as those found in the general population [6,48,49]. Harding [3] has suggested that these differences may reflect a reduction in consanguineous marriages or that more than one recessive gene may be involved.

3.4.1. Linkage of FA gene to human chromosome 9

In recent years DNA probes have been used successfully for establishing a linkage between restriction fragment length polymorphism (RFLP) and the chromosomal gene locus in several human genetic disorders [50]. (See Chapter 17 for the rationale and methodology used in genetic linkage studies.) These studies opened the way for the chromosomal mapping of the FA gene to human chromosome 9 by Chamberland et al. [51]. These investigators studied 22 FA families with three or more affected siblings. They assumed that the disease is single-gene disorder, despite the clinical indications for genetic heterogeneity suggested by Harding [3]. A total of 117 DNA markers and 36 polymorphic blood groups and protein markers were utilized. Analysis of these combined data excluded over 80% of the human genome, while suggesting three chromosomal regions as probable loci for the FA gene. These loci were intensively investigated and revealed a linkage between the FA genetic defect and two markers on chromosome 9. The first was the anonymous probe MCT112 (D9S15) and the second was the interferon-b gene (INFB). There was tighter linkage to the probe MCT112 than to the INFB probe, with the maximal lod score for the former being 6.41 and for the latter 2.98 [51]. Since the interferon-b gene had been mapped to the short arm of chromosome 9 (Figure 13-1) and the MCT112 (D9S15) probe to the centromeric region of this chromosome, Chamberlain et al. [51] tentatively assigned the FA gene to the 9 p22 centromere. These advances have made possible the prenatal [52] and presymptomatic diagnosis of Friedreich's ataxia on the basis of DNA analysis.

Following this original observation, Chamberlain et al. [53] studied 553 individuals from 80 families of distinct ethnic origin (English, German, French-Canadian, Acadian, and Spanish). All families were found to be linked to the D9S15 probe, with no recombination observed between the probe and the FA gene locus in any patient population studied. These results were thought to be consistent with a mutation at a single gene locus and strongly suggest that FA is genetically homogeneous [53].

Fujita et al. [54] recently confirmed the linkage of the FA gene to human chromosome 9. These authors found a second DNA marker (D9S5) to be also tightly linked to the FA gene without recombinants. They also reported a highly polymorphic microsatellite sequence at D9S15, which they used for further analysis [55]. The FA gene was found to show linkage disequilibrium (see Chapter 17 for an explanation of linkage disequilibrium) with D9S15, indicating the close proximity of the two loci. Based on these observations,

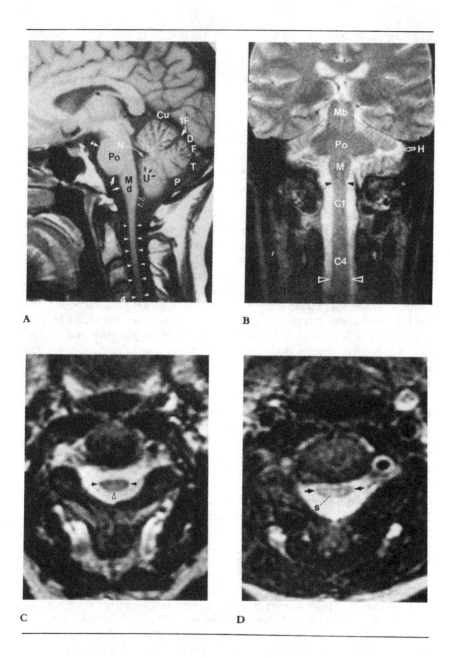

Figure 13-5. Another case of Friedreich's ataxia. Female, age 50, affected since age 18. The patient has been wheelchair bound since age 40. Clinical features include ataxia, dysarthria, quadriparesis with legs more involved than arms, areflexia, positive Babinski sign, proprioceptive sensory loss, and skeletal deformities. Two of the patient's nieces have been similarly affected. **A**: Midsagittal section. The most striking finding is tremendous atrophy of the spinal cord (row of opposing arrowheads), particularly its dorsal aspect. Note a semiarcuate depression of the posterior aspect of the upper cervical cord (double empty arrowheads) as one traces the posterior aspect of the pons (Po) and of the medulla oblongata (Md) downward to that of the spinal cord. The anteroposterior diameter of the cervical cord has shrunken down to 4 mm at the level of C4 (4), most likely due to atrophy of the posterior columns. The anterior aspect of the medulla oblongata (unlabeled white arrow) is retracted, presumably due to atrophy of the pyramids. Therefore, when the anterior aspect of the brain stem is traced inferiorly from the pons to the spinal cord, there is an abrupt depression at the pontomedullary junction, giving an impression of protrusion of the inferior part of the belly of the pons (unlabeled crossed arrow). In addition, the anterosuperior aspect of the pons is slightly retracted (upper crossed arrow). The belly of the pons is, therefore, more rounded than normal. The superior vermis is atrophic, with a prominent primary fissure (1F). The sulci within the culmen (Cu), declive (D), folium (F), and tuber (T) are also more clearly demonstrated. The concave anterior margin (small black arrows) of the uvula (U) is likely to be due to a slanted cut. The visualized mesencephalon and medial cerebral hemisphere are unremarkable. The normal pyramid (P) and nudulus (N) are labeled. **B**: Coronal section through the superior colliculi, aqueduct, pontine tegmentum, medulla oblongata, and spinal cord. The cervical cord is markedly atrophic. Its transverse diameter is smallest at its most superior part (8–9 mm) (black arrowheads), enlarges as it is traced inferiorly, and reaches 15 mm at the level of C5 (opposite empty arrowheads). In reality, the cord has collapsed in the anteroposterior direction, and therefore it has increased in transverse width (see Figure 13-5A). The medulla oblongata (M) is narrow transversely (in addition to anteroposteriorly), particularly in its inferior ends. The visualized anterior aspect of the cerebellum, the pons and its brachia pontis, and the midbrain (Mb) appear unremarkable. The spinal cord at C1 and C4, and the great horizontal fissure (H) are labeled. **C**: Transverse section of the spinal cord at the C2–C3 level. The spinal cord is atrophic. It measures 4.3 mm anteroposteriorly and 12 mm transversely (opposite arrowheads). The posterior part of the spinal cord shows minimally increased signal intensities, indicating degenerative changes. **D**: Transverse section of the spinal cord at the C4 level. The spinal cord is flat, with a markedly decreased anteroposterior and an increased transverse diameter (arrows), i.e., collapsed and flattened; anteroposteriorly the cord measures 3.7 mm. The central and posterior parts of the spinal cord show increased signal intensities (S), indicating degenerative changes, especially on the right (reader's left).

Fujita et al. [55] concluded that D9S5 and D9S15 are less than 1 centimorgan (cM) from the FA gene, and, as such, they may be used as starting points for isolating the disease gene. Hanauer et al. [56] recently mapped D9S5 by in situ hybridization to 9q13–q21 and assigned the FA gene, D9S5, and D9S15 cluster to the proximal area of the long arm of chromosome 9 (Figure 13-1).

4. HEXOSAMINIDASE DEFICIENCY (ATAXIC FORM)

4.1. Clinical findings

Hexosaminidase deficiency or GM2 gangliosidosis is a well-studied storage disorder that results from failure of this enzyme to cleave N-acetylgalactosamine from the GM2 ganglioside that accumulates in various body tissues, including the brain. Several forms of this disorder are known to result from distinct mutations involving the different isoforms of this enzyme. A predominant cerebellar syndrome has been described in several patients with juvenile-onset or childhood-onset disorders. Rapin et al. [57] first described an atypical spinocerebellar degeneration in two adult siblings with hexosaminidase A deficiency. Johnson et al. [58] reported a 4-year-old boy who since age $2\frac{1}{2}$ developed progressive intention tremor, gait and limb ataxia, slowing of eye movements, and decreased visual acuity, with pale discs and bilateral cherry-red spots. A deficiency of hexosaminidase A and B was documented in the patient's plasma. However, leukocytes and cultured skin fibroblasts from this patient showed small amounts of hexosaminidase S-like activity and greater amounts of hexosaminidase A-like activity. The parents had intermediate values.

Oonk et al. [59] described two adult sisters affected since age 20 by a progressive cerebellar ataxia, dysarthria, muscle weakness, and wasting, and corticospinal deficits. Biochemical studies showed a deficiency of hexosaminidase A and B in the patients' serum and leukocytes. Electron-microscopic examination of the skin showed dystrophic axons and small inclusions, reminiscent of membranous cytoplasmic bodies found in gangliosidoses. Willner et al. [60] reported a progressive spinocerebellar degenerative disorder in nine patients, aged 11–37 years, from four unrelated Ashkenazi Jewish families. All showed an early-onset cerebellar ataxia, with subsequent development of upper and lower motor deficits, marked dysarthria, and in some patients, dementia or recurrent psychotic episodes. Hexosaminidase A was markedly deficient in all sources analyzed. Rectal biopsy revealed membrane-bound lamellar cytoplasmic inclusions, consistent with lysosomal accumulations in rectal ganglia. Parents had enzyme activities consistent with heterozygosity, confirming the autosomal recessive transmission of the enzymatic deficiency.

4.2. Neuroimaging abnormalities
For neuropathologic alterations observed in various types of hexosaminidase deficiency, see Chapter 16. Here, we studied the CT scans of two adult patients (33 and 39 years of age) affected by progressive ataxia due to hexosaminidase A deficiency. The most striking CT abnormality was pancerebellar atrophy, associated with a marked dilatation of the fourth ventricle (Figure 13-6). Although the cerebellar atrophy was diffuse, the hemispheres were clearly more involved than the vermis. Interestingly, the brain stem showed minor changes (Figure 13-6C). The brachium conjuctivum was slightly atrophic, whereas the brachium pontis and restiform body were essentially normal (Figures 13-6C and 13-6D). There was also atrophy of the frontal and parietal lobes (figures not shown).

5. CEREBELLO-OLIVARY ATROPHY OR CORTICAL CEREBELLAR DEGENERATION

5.1. Clinical features

Holmes [61] is credited with providing the first clinicopathological description of this disorder. He reported a particular variety of this illness with an autosomal recessive transmission pattern and hypogonadism. Affected members of the family described by Holmes evidenced a rather pure cerebellar syndrome, starting in the third decade and evolving slowly over several decades. The presenting symptom was progressive gait ataxia, followed by incoordination of the upper extremities, dysarthria, nystagmus, and head tremor. Besides the cerebellar dysfunction, no additional neurological manifestations were observed. Pathologically, there was atrophy of the cerebellar cortex; the tonsils, uvula, and nodule were relatively spared. The inferior olives were markedly atrophic, and the restiform bodies were shrunken due to loss of the olivocerebellar fibers. Since the original description of Holmes, many cases have been described in the literature [62–64], showing similar pathologic features. However, the genetic transmission has often been dominant [62], rather than recessive, and sporadic cases have also been encountered.

Of 182 cases of adult-onset ataxia we studied (see Table 8-3 in Chapter 8), 16 patients showed progressive cerebellar dysfunction without associated neurologic features (pure cerebellar syndrome) and were, accordingly, classified as suffering from cortical cerebellar degeneration. In most of these patients, this clinical impression was substantiated by CT or MRI findings (see below). Six of these cases were dominant, four were recessive, and six were sporadic. In the dominant cases, the onset of symptoms ranged from 22 to 54 years of age (mean = 37.4 years) and the duration of the disease ranged from 5 to 35 years (mean = 17.8 years). In the recessive cases, the onset of disease varied widely from childhood to 72 years of age (mean = 48.0 years). The duration of the disease was also quite variable from 1 to 31 years (mean = 16 years). On the other hand, the majority of the sporadic cases tended to appear in late life, with the disease's onset ranging from 36 to 79 years (mean = 60.7 years) and the duration from 1 to 4 years (mean = 1.8 years).

As indicated above, these patients evidenced rather pure cerebellar dysfunction, with gait ataxia being the major manifestation of the disease. Limb ataxia was present mainly in the lower extremities, but it was generally less pronounced than gait ataxia. Dysarthric, dysrhythmic speech was observed in most patients. Dysarthria tended to appear at around the time at which incoordination of fine hand movements became noticeable. Oculomotor disturbances were observed in the majority of the cases and reflected primarily midline cerebellar dysfunctions. They consisted of gaze-evoked

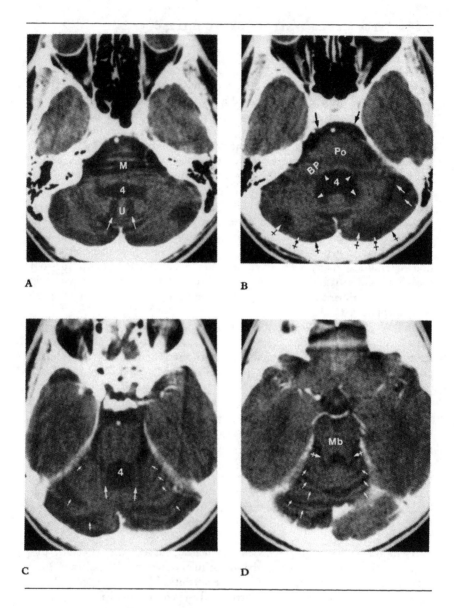

Figure 13-6. Hexosaminidase A deficiency (adult-onset ataxic form). Male, age 49, affected since age 31 by progressive gait and limb ataxia, dysarthria, paraparesis, muscle fasciculations, and corticospinal deficits. **A:** Transverse section through the lower part of the fourth ventricle. The paravermian subarachnoid spaces (arrows) and fourth ventricle (4) are widened. The medulla oblongata (M) and uvula (U) are labeled. **B:** Transverse section through the midportion of the fourth ventricle. The most striking finding is the markedly enlarged, ballooned (radiating arrowheads) fourth ventricle (4). The pons (Po) and brachia pontis (BP) show no definite atrophy (see Figure 13-6C). The great horizontal fissure (crossed arrows) and the postclival and intrasemilunar sulci (unlabeled white arrows) are widened. The pontine cisternal (unlabeled black arrows) are wider than usual for the patient's age. **C:** Transverse section through the midportion of the markedly enlarged fourth ventricle (4). Widened sulci are seen throughout the entire cerebellar hemispheres (including the anterior lobes). The vermis is comparatively spared. Atrophy is of a diffuse "fine comb" type and may be produced by uniform atrophy affecting the cerebellar folia. The brachia conjunctiva are thin bilaterally (double-crossed arrows), while the great horizontal fissures (crossed arrow) are widened. There is no pontine atrophy. **D:** Transverse section through the upper part of the cerebellum and the midbrain (Mb). The cerebellar sulci are diffusely and finely widened (unlabeled small arrows), while the folia appear to be uniformly atrophic ("fine comb" atrophy). The brachia conjunctiva (crossed arrows) are also slightly thin. The upper fourth ventricle is enlarged. The bases pedunculi and tegmentum of the mesencephalon are normal.

nystagmus, broken-up pursuit, inacurate saccades (ocular dysmetria), and inability to suppress the vestibulo-ocular reflex. There was no evidence of corticospinal deficits, peripheral neuropathy, or extrapyramidal features. A slight decrease in vibration in the toes or lower legs, observed in a few elderly patients, was attributed to age.

5.2. Neuropathologic findings

The pathologic features of cortical cerebellar degeneration or cerebello-olivary atrophy have been reviewed by Eadie [62] and Oppenheimer [45]. There is atrophy of the cerebellar cortex, with the white matter showing some degree of demyelination and gliosis. The upper cerebellum, and particularly the vermis, are affected. The dentate nucleus may show neuronal loss and gliosis. In the medulla, there is atrophy of the inferior olives and some wasting of the restiform body. Other brain-stem structures, including the middle cerebellar peduncles, are usually not involved. Some authors have observed atrophic changes in the frontal and temporal lobes.

5.3. Neuroimaging abnormalities

We have previously reported the CT findings of two patients with COA [1]. They revealed atrophic changes in the superior cerebellum (vermis and paravermian region) and less so in the lateral cerebellum (semilunar lobule). The brain stem was normal. Here, we studied the MRIs of four COA cases, including one case previously evaluated by CT [1]. In accord with the neuropathologic data described above and the CT findings reported pre-

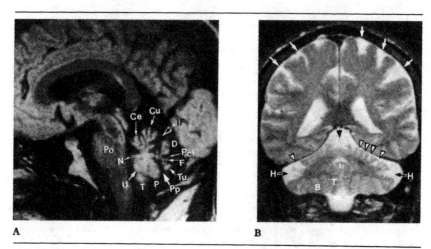

Figure 13-7. Dominant cortical cerebellar degeneration (cerebello-olivary atrophy). Male, age 57, affected since age 31 by slowly progressive cerebellar dysfunction not associated with other neurological deficits. Neurological features included gait ataxia (which is greater than limb ataxia), dysarthria, and oculomotor deficits that reflect midline cerebellar dysfunctions (broken-up pursuit occular dysmetria, inability to suppress the vestibulo-ocular reflex). There is autosomal dominant transmission with complete penetrance. **A:** Midsagittal section. The characteristic change is atrophy of the superior and middle part of the vermis. The declive (D), folium (F), tuber (Tu), culmen (Cu), and central lobule (Ce) are severely atrophic, while the pyramid (P), uvula (U), and nodulus (N) are much less affected. The postclival sulcus (Pcl), primary fissure (1F), and prepyramidal sulcus (Pp) are very wide. The pons (Po), medulla oblongata, and spinal cord are normal. **B:** Coronal section through the roof of the fourth ventricle. The great horizontal fissures (H) and, to a lesser extent, the sulci of the quadrangular lobules (empty arrowheads), are markedly widened. In other words, the lateral and posterior superior part of the cerebellar hemispheres are also involved. Superior vermian atrophy (black arrowhead) is obvious. The parietal sulci (white arrows) are also prominent. The fourth ventricle, adjacent inferior vermis (I), tonsils (T), and biventral lobules (B) are unremarkable.

viously [1], the major abnormality detected in MRI was atrophy of the superior cerebellum, with the vermis most severely involved (Figures 13-7 to 13-9). Of the vermian lobules, the declive, culmen, folium, and tuber were severely affected, while the uvula, nodule, and pyramid were relatively spared (Figure 13-7A). The cerebellar hemispheres were involved laterally, particularly in the region of the superior semilunar lobules (Figures 13-8B, 13-8C, 13-8D, 13-9B, 13-9D, 13-9F, and 13-9G). In the lower cerebellum, atrophic changes were noted in the inferior semilunar lobule and the lateral part of the biventral lobule, while the tonsil and the medial part of the biventral lobule were largely spared (Figures 13-8C, 13-8D, and 13-9D). The fourth ventricle was enlarged in three patients (Figures 13-8A and 13-9A), but it was normal in one (Figure 13-7). The brain stem showed no abnormalities, while the cerebral cortex showed atrophic changes in the parietal (Figures 13-7B, 13-8B, and 13-9F) and, to a lesser extent, in the frontal region.

6. THE OLIVOPONTOCEREBELLAR ATROPHIES

Menzel in 1891 [65] was the first to describe a hereditary form of cerebellar degenerative disorder that affected several members of a family in three generations. The patients presented with an adult-onset progressive cerebellar syndrome, characterized by gait and limb ataxia, dysarthria, head tremor, and choreiform movements of extremities. An autopsy of one patient revealed atrophy of the pons, cerebellum, inferior olives, spinal cord, and lower cranial nerve nuclei. Nine years later Dejerine Thomas [66] introduced the descriptive term *olivopontocerebellar atrophy* (OPCA) to refer to two sporadic cases of a late adult-onset cerebellar syndrome. In one case, autopsy revealed atrophy of the cerebellar cortex, inferior olives, the middle cerebellar peduncles, and the gray matter of the pons. Menzel's disease is considered to be the prototype of familial OPCA, while the disorder described by Dejerine and Thomas is the prototype of sporadic OPCA [44].

Since these original reports, numerous articles describing various types of OPCA, some of which have been limited to single families, have appeared in the literature. It is now clear that OPCA is not a single disease entity but is a group of genetically heterogenous neurologic disorders sharing common pathologic and clinical features. Their clinical manifestations include progressive cerebellar ataxia, cranial nerve dysfunction, oculomotor abnormalities, corticospinal deficits, extrapyramidal features, motor neuron signs, and/or peripheral neuropathy occurring in various combinations.

Pathologically, these disorders are characterized by degeneration and atrophy of the cerebellar cortex, inferior olives, olivocerebellar fibers, gray matter of the pons, pontocerebellar fibers, and middle cerebellar depuncles. As discussed below, these pathologic features can vary according to the subtype of OPCA and within members of the same family. They can extend to involve the spinal cord, substantia nigra, striatum, and areas of the cortex. Thus, the OPCAs are disorders with multiple system atrophy and, as such, they link the cerebellar degenerations with the extrapyramidal diseases and other diseases with system atrophy. Here, we studied four different types of dominantly inherited OPCA, as described below.

6.1. Menzel's type

6.1.1. Clinical features

There is considerable confusion surrounding the nosology of this disorder. While some authors regard all dominantly inherited OPCAs as Menzel's disease, others, such as Konigsmark and Weiner [67], have limited this definition to dominant pedigrees that show clinical and pathologic features similar to those originally described by Menzel [65]. As discussed in Chapter 8, there are problems with the Konigsmark-Weiner classification. This is

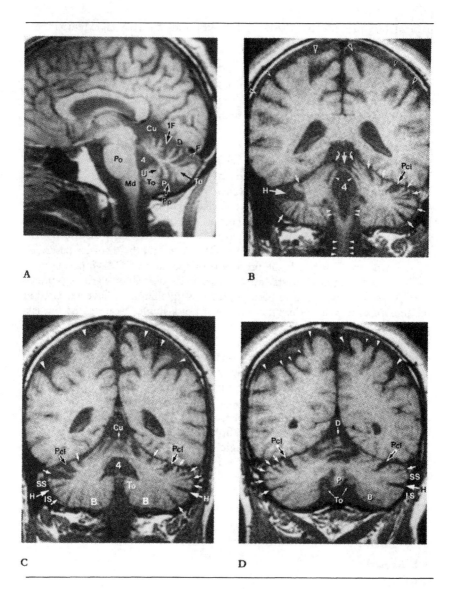

Figure 13-8. Another case (sporadic case) of cortical cerebellar degeneration (cerebello-olivary atrophy). Male, age 51, affected for 4 years by a progressive cerebellar disorder that is not associated with other neurologic deficits. Ataxia of gait, less so of limbs, and oculomotor deficits reflect midline cerebellar dysfunctions (broken-up pursuit, ocular dysmetria, and inability to suppress the vestibulo-ocular reflex). The family history was negaive. **A**: Sagittal section of the brain stem and cerebellar vermis. Inferiorly, the plane of section has deviated slightly, passing through a cerebellar tonsil (To). As a result, the uvula (U) and pyramid (P) are short. As in the previous case (Figure 13-8), there is atrophy of the culmen (Cu), declive (D), and folium (F). Evaluation of the inferior vermis is difficult due to a slightly oblique cut. The visualized pons (Po) and medulla oblongata (Md) appear unremarkable. The widened fourth ventricle (4), primary fissure (1F), the prepyramidal sulcus (Pp), and the tuber (Tu) are labeled. The spinal cord atrophy is difficult to assess. **B**: Coronal section through the fourth ventricle and upper cervical cord. There is a mild transverse atrophy of the spinal cord (opposing triple arrowheads), as well as shrinkage of the slightly arched superior and inferior cerebellar peduncles (opposite single and double arrowheads). The fourth ventricle (4) is enlarged. The culmen (unlabeled large arrow) and superior paravermian sulci (unlabeled curved arrows) are widened. The parietal sulci are slightly prominent (empty arrowheads). Aside from vermian atrophy, the sulci (unlabeled small arrows) of the semilunar and, to a lesser extent, biventral lobules and of the anterior parts of the quadrangular lobules are widened. The great horizontal fissures (H) and postclival sulci (Pcl) are very prominent, indicating lateral hemispheric atrophy. **C**: Coronal section through the fourth ventricle (4) just behind the brain stem. The sulci of the lateral part of the cerebellum (unlabeled arrows), particularly those located in the region of the superior (SS) and inferior semilunar lobules (IS), are widened. This is evidenced by the dilated postclival sulci (Pcl) and great horizontal fissures (H), and some sulci of the quadrangular lobules. The culmen (Cu) of the vermis shows marked atrophy, while the biventral lobules (B) and tonsils (To) are largely unaffected. The parietal sulci (arrowheads) are prominent. **D**: Coronal section of the cerebellum through the pyramid and declive. The superior vermian atrophy (the declival region, D) is obvious. The pyramid (P) is smaller than usual. There is considerable widening of the postclival sulci (Pcl) and great horizontal fissures (H), with the atrophic semilunar lobules (IS, SS, unlabeled arrows), while the biventral lobules (B) are clearly spared. The posterior cerebellar notch is slightly wide. The partially sectioned tonsils (To) are also labeled. The parietal sulci (arrowheads) are prominent.

particularly true for the large dominant kindred described by Schut and Haymaker [68], considered by Konigsmark and Weiner to represent a separate entity, although its clinical and pathologic features are similar to pedigrees that have been placed under Menzel's-type OPCA [67–79]. Recent studies have established a linkage between the gene locus of this disorder and the HLA complex on the sixth chromosome. As such, these data further suggest that the Schut-Haymaker family shares the same genetic abnormality as other dominant OPCA kindreds of Menzel's type (see Chapter 17).

We propose that all dominantly inherited disorders that show progressive cerebellar dysfunction (ataxia, dysarthria, nystagmus, and dysdiadochokinesis) in association with corticospinal deficits and/or peripheral neuropathy and that clearly lack the specific clinical features found in the OPCA variants described below (slowed saccades, oculomotor and other cranial nerve palsies, retinal degeneration, and deafness) be classified under Menzel's disease. Most likely this is a heterogenous subgroup of neurologic disorders,

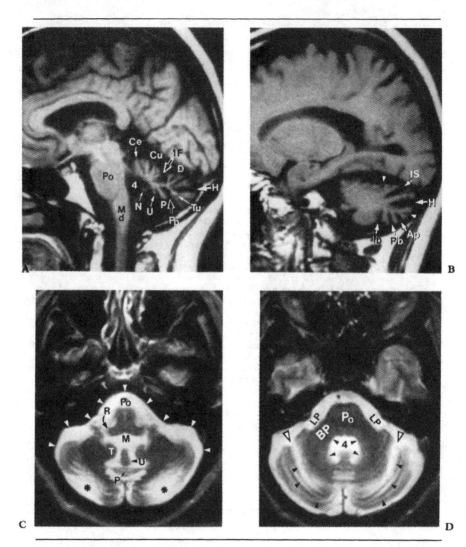

Figure 13-9. Dominant cortical cerebellar degeneration (a variant of cerebello-olivary atrophy?). Female, age 48, affected since age 44. Clinical features included gait and limb ataxia, dysarthria, and oculomotor deficits that reflect midline cerebellar dysfunctions. In addition, the patient's feet are high arched and her lower legs have a decreased vibratory sense. **A**: Sagittal section through the brain stem and cerebellar vermis. Atrophy of the vermis, especially the declive (D), folium, tuber (Tu) and, to a lesser extent, the culment (Cu) is typical. The primary (1F) and great horizontal fissures (H), and the prepyramidal sulcus (Pp) are prominent. The reduced size of the nodulus (N), uvula (U), and pyramid (P) has resulted from a slanted paramedian cut. The central lobule (Ce), medulla oblongata (Md), pons (Po), and fourth ventricle (4) are labeled. **B**: Parasagittal cut through the midportion of a cerebellar hemisphere. There is coarse widening of the great horizontal fissure (H) and sulci of the posterior, lateral, and inferior parts of the cerebellar hemispheres (arrowheads). Due to severe cortical atrophy, much of the normal fine folia pattern has been lost, leaving atrophied white matter of the cerebellar lobules and creating a "coral reef" or "fish bone" deformity. The great horizontal fissure is particularly wide, giving the appearance of a "fish mouth" deformity. The widened prebiventral (Pb), intrasemilunar (IS), ansoparamedian (Ap), and intrabiventral sulci (Ib) are also labeled. **C**: Transverse section through the upper part of the medulla oblongata (M). There is widening of the subarachnoid space around the tonsils (T),

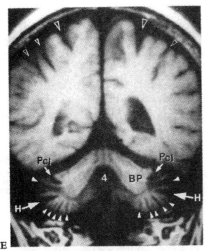
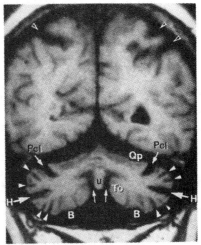
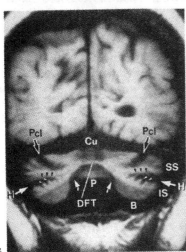

uvula (U), and pyramid (P). The restiform bodies (R) appear small. The basal cisterns (white arrowheads) and great horizontal fissures (*) are widened. The lower end of the pons (Po), seen anterior to the medulla oblongata (M), is also labeled. **D**: Transverse cut through the midportion of the somewhat ballooned (radiating arrowheads) fourth ventricle (4). The postclival sulci and the sulci within the semilunar lobules are markedly widened (empty and black arrowheads). The pons (Po), brachium pontis (BP), and lateral pontine cisternal (LP) are labeled. **E**: Coronal section of the lower part of the fourth ventricle (4), parallel to its floor, cutting primarily through the petrosal aspect (mostly the biventral and inferior semilunar lobules) of the cerebellar hemispheres. The pattern of cerebellar atrophy is that of mixed "fine and coarse comb" (unlabeled small arrowheads). Note the normal brachia pontis (BP). The fourth ventricle (4), postclival sulci (Pcl), great horizontal fissures (H), and parietal sulci (empty arrowheads) are dilated. **F**: Coronal cut through the uvula (U) and tonsils (To), again showing "coarse comb" with some mixture of "fine comb" atrophy. The lateral parts of the cerebellar hemispheres are mainly involved. "Coarse comb" atrophy seems to manifest itself in the widened postclival sulci (Pcl) and great horizontal fissures (H) with nearly burned-out secondary and tertiary folia. The paravermian sulci (unlabeled arrows) are slightly widened. Although the sulci of the biventral lobules (B), the tonsils (To), and the posterior part of the quadrangular lobules (Qp) are scarcely widened; their heights are nevertheless decreased. The parietal sulci (empty arrowheads) are prominent. **G**: Coronal section of the posterior part of the cerebellum showing the connection between the vermis and the hemispheres at the level of the declive, folium, and tuber (DFT). Atrophy of the "coarse comb" type is best demonstrated in this image and in Figure 13-10B. The great horizontal fissures (H) are markedly widened. In some areas, secondary and tertiary folia have dropped out of the original folia, creating a "fish mouth" deformity without teeth (black arrowheads), while in other areas only the tertiary folia have burned out, leaving a "fish mouth" deformity with teeth (tiny white arrowheads correspond to the teeth). The markedly laterally-retracted inferior paravermian sulci, which meet the lateral extension (called the pyramidal band on each side) of the pyramid (P), are also labeled (unlabeled curved arrows). The culmen (Cu) is markedly diminished in size. The superior (SS) and inferior semilunar lobules (IS) and postclival sulci (Pcl) are labeled.

the definite classification of which requires the determination of their primary genetic defects.

Based on the clinical criteria specified above, 11 of 29 patients with dominantly inherited cerebellar degeneration who have been personally evaluated (see Chapter 8) were classified as suffering from Menzel's disease. Their age ranged from 23 to 58 years (mean age 43.6 years), and the duration of their disease ranged from 2 to 30 years (mean duration 13.7 years). In eight of these patients, the onset of the disease occurred during the second decade of life (20–28 years), whereas in the remaining three patients it occurred during the fourth and fifth decades of life. An intrafamilial correlation of age onset was noted in this patient population.

6.1.2. Neuropathologic changes

Atrophy of the cerebellar hemispheres, basis pontis, transverse pontine fibers, middle cerebellar peduncles, inferior olives, and spinal cord have been the main neuropathologic alterations in this disorder [44,45]. The vermis is relatively spared, while the dentate nucleus shows variable neuronal losses. Pontine atrophy, however, varies, even within families [69]. In addition, some patients show considerable loss of motor neurons in the spinal cord, degeneration of lower cranial nerve nuclei, substantia nigra, and striatum [45,69].

6.1.3. Neuroimaging abnormalities

We have previously reported the CT findings of five patients with Schut-Haymaker OPCA [1]. The early abnormalities observed were atrophy of the brachium pontis and the cerebellum [1]. Cerebellar atrophy was most pronounced in the anterior lobe and the superior part of the middle lobe. The pons, midbrain, and frontal cortex also showed atrophic changes. The fourth and third ventricle were enlarged. In advanced disease stages, the brain stem, including the pontine tegmentum, showed severe atrophy. However, these patients lacked the extreme enlargement or "ballooning" of the fourth ventricle, which was characteristic for OPCA patients with slowed saccades [1].

Here we studied the brain MRI of 40-year-old patient from the Schut-Haymaker kindred in an advanced disease stage. A brain CT scan, performed at age 30 in the early stage of the disease, has been previously described [1]. It revealed a moderate to marked atrophy of the brachia pontis and cerebellar cortex. The latter was most pronounced in the anterior cerebellar lobe (quadrangular lobule). In contrast, the vermis was relatively spared, a finding that is consistent with the neuropathologic data mentioned above.

The present MRI study revealed a marked atrophy involving the lower brain stem (Figure 13-10). In the pons the atrophic changes were most pronounced in the rostro-caudal portion in misagittal section (Figure 13-10A). The brachium pontis, the medulla oblongata, and spinal cord were also severely atrophic (Figure 13-10A), whereas the mesencephalon was less involved (Figures 13-10A and 13-10E). In the cerebellum, there was diffuse atrophy involving the hemispheres, particularly the anterior cerebellar lobe. A comparison of the present MRI with the previous CT study, obtained 10 years earlier, revealed a worsening of the degree of atrophy involving the brain stem.

6.2. Dominant OPCA with slowed saccades

6.2.1. Clinical features

Although the association of cerebellar degeneration and slowed eye movements has been known for almost a half century [80,81], it was Wadia and Swami [82] who in 1971 first provided a clear clinical description of slowed saccadic eye movements and linked their occurrence to a form of dominant OPCA associated with peripheral neuropathy. The authors distinguished this from other types of eye-movement abnormalities that can occur in ataxic patients, such as oculomotor nerve palsies resulting from brain-stem or peripheral nerve lesions, progressive external ophthalmoplegia due to ocular myopathy, or pseudo-ophthalmoplegia due to conjugate gaze palsy.

Since Wadia and Swami's account, additional studies describing similar cases have appeared in the literature [83–89]. We have personally examined seven such patients. Five of these were members of three families afflicted by dominantly inherited neurological disorder and two were sporadic occurrences. They all showed progressive cerebellar ataxia, with gait being preodominantly affected, slowed voluntary eye movements, weakness of eye closure, dysarthria, corticospinal deficits, and peripheral neuropathy. Muscle fasciculations and amyotrophy were present in two patients. Although none of these patients had overt visual failure, the optic discs were pale in most of these patients.

The hallmark of the eye-movement abnormality is a marked slowing of the saccadic eye movements. This can be observed clinically and has been confirmed by electro-oculography [8]. All types of saccades, such as those of volitional refixation, as well as the fast saccadic components of caloric and optokinetic nystagmus, are affected. The degree of saccadic slowness generally correlates with the stage of the illness. However, in the same patient saccades are not equally affected in every direction. Thus, in one of two affected brothers of a kindred we have studied [87], the horizontal saccades were more involved than the vertical saccadic eye movements, but the converse was true for his sibling. In contrast, the smooth-pursuit eye movements remain intact. There is no spontaneous or gaze-evoked nystagmus. The eye movements generated by optokinetic and caloric stimuli tended to become pendular, lacking the normally elicited quick components. In advanced stages, the eyes showed a very limited range of motion; however, this could be overcome by the doll's eye maneuver.

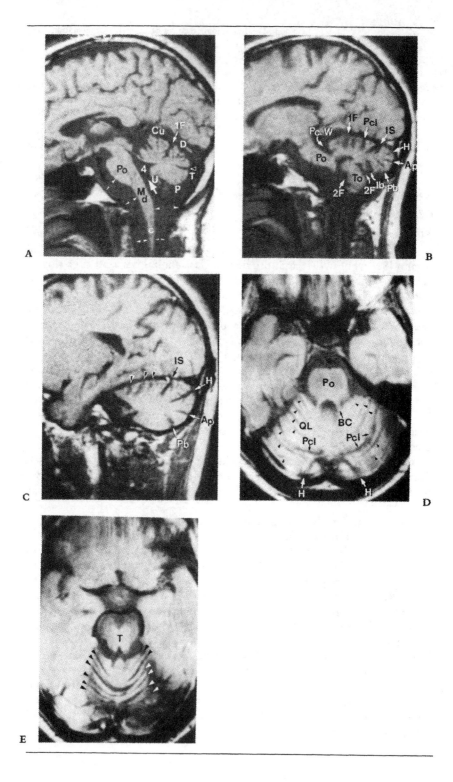

Figure 13-10. Dominant OPCA (Schut-Haymaker family). A brain CT scan was performed on this patient at age 30 and it has been previously reported [1]. This MRI was performed at age 40. **A:** Midsagittal section. The most striking finding is marked atrophy of the medulla oblongata (Md), and to a lesser extent, the pons (Po) and the small atrophic cervical cord (c). This resulted in widening of the medullary and pontine cisterns, the cisterna magna, and the cervical subarachnoid space (unlabeled arrowheads). The superior vermis [particularly the culmen (Cu) and declive (D)] has shrunken. The primary fissure (1F) and sulci within the culmen are wide. In the inferior vermis, a sulcus within the tuber (T) (intratuberal sulcus, empty arrowhead) is prominent. The inferior margin of the uvula (U) is convex upward due to a slightly slanted off-midsagittal cut. The foramen Magendie (large unlabeled arrow) is, however, slightly large. The pyramid (P) is also labeled. **B:** Parasagittal section through the lateral portion of the pons. There is widening of the cerebellar paravermian sulci; the intrasemilunar sulci (IS), sulci within the superior semilunar lobules (esp. superior semilunar lobules) secondary fissure (2F), intrabiventral sulcus (Ib), wing of the precentral cerebellar sulcus (Pc-W), and mildly, the primary fissure (1F) and postclival sulcus (Pcl). The lateral part of the pons (Po), tonsil (To), ansoparamedian sulcus (Ap), great horizontal fissure (H), and prebiventral sulcus (Pb) are also labeled. **C:** Parasagittal section through the lateral part of the paraflocculus. The intrasemilunar sulcus (IS), great horizontal fissure (H), ansoparamedian sulcus (Ap), prebiventral sulcus (Pb), and sulci within the posterior quadrangular lobule (empty arrowheads) are prominent for the patient's age. **D:** Transverse section through the midpons, upper fourth ventricle, and quadrangular lobules (QL) of the cerebellum. The visualized cerebellar folia are uniformly fine and thin, and the in-between sulci are diffusely and finely widened. This constitutes a typical "fine comb" type of atrophy. The postclival sulci (Pcl) and great horizontal fissures (H) are, however, more prominent than others. The pons (Po) and brachium conjunctivum (BC) are labeled. **E:** Transverse section through the mesencephalon and upper cerebellum. "Fine comb" atrophy is best demonstrated in this and previous figures; thin folia alternate with fine and long sulci (arrowheads). The tegmentum (T) of the mesencephalon appears somewhat smaller than usual.

6.2.2. Neuropathologic findings

Koeppen and Hans [84] have studied the brain of one patient with OPCA associated with slowed saccades (case 3 in their series) and have reported marked atrophy of the cerebellar cortex, demyenlination of the cerebellar white matter, and patchy neuronal loss of the dentate nucleus, with involvement of its efferent fibers. The pons was also markedly atrophic, with the basis pontis showing neuronal loss and demyelination of the transverse pontine fibers. The inferior olives were also atrophic, and the dorsal columns of the spinal cord were said to show demyelination. The parietal and occipital lobes showed atrophy of their subcortical white matter. Wadia [89] has also reported histologic evidence consistent with OPCA in his patients.

6.2.3. Neuroimaging abnormalities

We have previously reported detailed CT findings of four patients (from two families) with dominant OPCA-associated slowed saccades [1]. These included pancerebellar atrophy and diffuse involvement of the pons, including atrophy of the pontine tegmentum. Due to excavation of the floor of the fourth ventricle and thining of the brachia pontis and conjuctiva, the

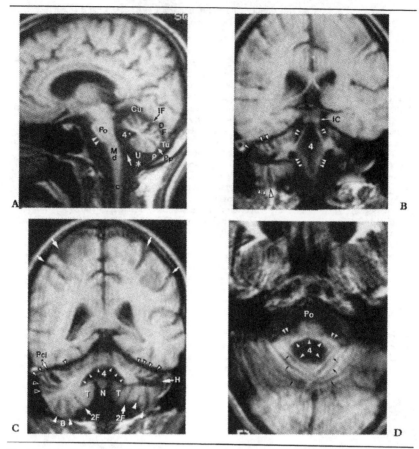

Figure 13-11. Dominant OPCA with slowed saccades. Male, age 24, affected since age 21 by a progressive degenerative disorder, characterized by ataxia, dysarthria, slowed saccadic eye movements, corticospinal deficits, anterior horn cell signs, and peripheral neuropathy. The family history is consistent with an autosomal dominant disorder. **A**: Midsagittal section. The pons, particularly the lower pons (Po) appears markedly flattened (double arrowheads) due to shrinkage and its height has considerably decreased. Additional atrophy of the medulla oblongata (Md) and of the cervical cord (C), has resulted in a widening of the medullary and pontine cisternal and of the cisterna magna (*). The superior vermis is atrophic and shows sulcal widening. The large foramen Magendie (unlabeled large arrow) and the short uvula (U) and nodulus (unlabeled tiny arrowhead) are likely to be due to the large fourth ventricle with the excavated floor, mild atrophy of the uvula and nodulus, and a slightly slanted off-midsagittal cut. The cerebellar vermis, including the culmen (Cu), declive (D), tuber (Tu), folium (F), primary fissure (1F), and prepyramidal sulcus (Pp), are labeled. **B**: Coronal section through the inferior colliculi (IC) and the lower part of the fourth ventricle (parallel to its floor). The fourth ventricle (4) is widened and the walls of its roof—the superior (double arrowheads) and inferior (triple arrowheads) cerebellar peduncles—are thin. The visualized sulci of the anterior surface of the quadrangular, superior, and inferior semilunar and biventral lobules are also widened (empty black arrowheads). **C**: Coronal section through the lower part of the fourth ventricle parallel to its floor. The fourth ventricle (4) is markedly widened transversely (arched row of small white arrowheads). The sulci of the posterior quadrangular and semilunar lobules (empty arrowheads), the postclival sulci (Pcl), the great horizontal fissures (H), and to a much lesser degree, the secondary fissure (2F) and the sulci (arrowheads) of the biventral lobules (B) are considerably widened. The tonsils (T) and the nodulus (N) are slightly atrophic. The parietal sulci are prominent (unlabeled white arrows). **D**: Transverse section through the lower pons (Po) and the midcerebellum showing ballooning (radiating arrowheads) of the fourth ventricle (4). The pons (Po) and brachia pontis (double arrowheads) are markedly atrophic. Finely widened sulci (small black arrows) are also seen in the region of the quadrangular lobules ("fine comb" atrophy).

brain stem showed a "molar tooth" apperance in CT. A greatly ballooned fourth ventricle of ex vacuo type was a characteristic finding [1]. These changes were thought to be specific, indicating the selective atrophic changes in the pontine tegmentum, which could account for the slowed saccades found in this disorder [1].

Here, we studied the brain MRI of a dominant case from an independent family as well as an MRI of a patient with the sporadic form of the disease. As indicated above, the MRI technique provides a higher resolution than CT, particularly for the posterior fossa structures, allowing the study of the gross living neuropathology of this disorder. In the familial case (Figure 13-11), the most striking MRI abnormalities were severe pancerebellar atrophy and "ballooning" of the fourth ventricle. The superior and lateral cerebellar hemispheres were markedly affected, while the vermis was less involved (Figures 13-11A and 13-11C). The brain stem showed a marked atrophy, with the caudal pons being most prominently involved (Figure 13-11). All three pairs of cerebellar peduncles (inferior, middle, and superior) were atrophic (Figures 13-11B and 13-11D). The medulla oblongata was comparably less involved (Figure 13-11A). A slight atrophy of the posterior parietal cortex was also present (Figure 13-11C). These MRI data agree with the CT findings in the four dominant cases previously reported [1].

Similarly, the sporadic case (Figure 13-12) also showed pancerebellar atrophy and "ballooning" of the fourth ventricle. The pons, restiform body, brachium pontis and conjuctivum, and medulla oblongata were all atrophic (Figures 13-12D and 13-12E). The cerebral cortex was diffusely involved, but perhaps more so in the parietal region (Figure 13-12E). The spinal cord was atrophic and showed a decrease in its anteroposterior diameter in both dominant and sporadic cases (Figures 13-11A and 13-12A).

6.3. Dominant OPCA with retinal degeneration, slowed saccades, and other features

6.3.1. Clinical features

Dominant OPCA associated with blindness has been the subject of several previous publications [90–94]. The majority of the cases occurred in black families. The disorder is characterized by a variable age of onset, even within the same family, with some individuals being affected in their adult life and others in their infancy or childhood [90–94]. Generally, the natural history of the disease tends to be shorter in cases that start very early in life. The majority of the infantile cases reported were dead within 1 or 2 years from onset, while adults survived for as long as 40 years after symptoms were first noticed.

Similar to the variation noted in the disease's onset and natural history, the clinical expression of the illness is often pleomorphic. This has been observed within affected families with some individuals presenting with progressive visual loss, whereas others presented, with cerebellar ataxia. In some patients,

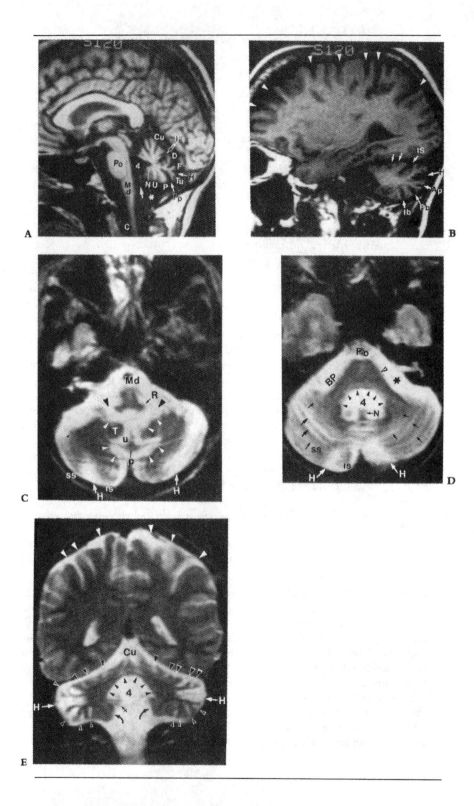

Figure 13-12. Olivopontocerebellar atrophy with slowed saccades (sporadic case). Female, age 38, affected since age 32. Saccades were markedly retarded, especially in the horizontal plane. The clinical features were marked ataxia, dysarthria, paraparesis, and proprioceptive sensory loss. The family history was negative for similar disorders. **A**: Midsagittal section of the posterior fossa. Marked atrophy of the spinal cord (C), medulla oblongata (Md), and pons (Po) resulted in widening of the pontine and medullary cisternal, and the cervical subarachnoid space. Atrophy of the declive (D), folium (F), tuber (Tu), and to a lesser extent, the culmen (Cu) and the widened primary fissure (1F), great horizontal fissure (H), and prepyramidal sulcus (Pp) indicate a considerable volume loss of the vermis. Although the sulci of the pyramid (P) and uvula (U) are not so wide as in the declive, they too are short vertically. Although these changes may partly be caused by a slightly tilted sagittal cut, posterosuperior retraction of the inferior vermis, together with excavation of the floor of the fourth ventricle (4), which is better seen in Figure 13-12E) have caused the fourth ventricle to become large, particularly inferiorly (unlabeled large arrow). The foramen of Magendie, which leads the fourth ventricle into the cisterna magna (*), is usually large. **B**: Parasagittal section of the cerebellar hemisphere. Note marked widening of the primary fissure (1F), the intrasemilunar (IS), prebiventral (Pb), intrabiventral (Ib), and ansoparamedian sulci (Ap), and the sulci within the posterior quadrangular and superior semilunar lobules (unlabeled arrows). The great horizontal fissure (H) is particularly large and the folia within the fissure have shrunken to the extent that a large number of folia have fallen off entirely. Therefore, it has a "fish mouth" or "shark teeth" appearance. This is due to the burned-out secondary or tertiary folia, and atrophy of the adjacent medulla. The sulci over the frontal and parietal convexity are widened (unlabeled large arrowheads). **C**: Transverse section through the upper medulla oblongata (Md), tonsils (T), and uvula (U). The medulla oblongata is decreased in size with the slender restiform bodies (R). The medullary cistern anterior to the brain stem is wide. The markedly widened cerebellomedullary (black arrowheads), vallecular, and hemispheric portions (row of white arrowheads) of the cisterna magna, and the thin lower part of the uvula (U) and pyramid (P) are labeled. The posterior part of the wide great horizontal fissure (H) separates the superior (SS) from the inferior semilunar (IS) lobules. **D**: Transverse section through the midportion of the fourth ventricle. The fourth ventricle (4) is huge and ballooned; its floor is concave (radiating arrowheads), the most characteristic finding in this condition. The pons (Po) is atrophic, with its shrunken and retracted lateral apects (empty arrowhead). The brachia pontis (BP) are thin. As a result, the pons has a triangular rather than a convex belly shape. Note the small pontine tegmentum, wide lateral pontine cisterns (*), and prominent great horizontal fissures (H). The posterior quadrangular and superior semilunar (SS) lobules show slender widened sulci (unlabeled black arrows), and thin and fine atrophic folia—mixed "coarse and fine comb" atrophy. The inferior semilunar lobules (IS) and atrophic nodulus (N) are labeled. **E**: Coronal section of the fourth ventricle parallel to and behind the brain stem. The half-dome shaped (radiating black arrowheads) fourth ventricle (4), atrophic laterally retracted cerebellar tonsils (crossed arrow), widened tonsilar vallecula (curved arrows pointing to the wings of the tonsils and the adjacent biventral lobules), and diffusely widened cerebellar sulci (empty arrowheads, "fine and coarse comb" atrophy) are shown here. The very prominent great horizontal fissures (H) and adjacent sulci indicate the greater involvement of the lateral part of the cerebellar hemispheres, particularly the semilunar lobules. The culmen (Cu) is, however, only minimally atrophic. The dilated parietal sulci are indicated (white arrowheads).

however, cerebellar dysfuction occurs at around the same time as visual loss. Regardless of the presenting symptom, however, most patients will eventually develop a progressive cerebellar syndrome (gait and limb ataxia, dysarthria, and dysdiadochokinesis) associated with decreased visual acuity, slowed eye movements, corticospinal deficits, amyotrophy, and peripheral neuropathy. The visual disturbances are usually progressive and often lead to complete blindness with dilated pupils.

In addition to progressive visual loss, the patients also exhibited slowed eye movements, as indicated above. In personally examined cases, there was a progressive involvement of the fast saccadic eye movements, similar to that described above in patients with dominant OPCA with slowed saccades. In advanced disease stages, voluntary eye movements became extremely slowed and of limited amplitude. In the end stages, the eyes can become totally immobile, and as such, it is not surprising to see that several authors have used the term *ophthalmoplegia* to refer to this type of oculomotor disturbance [91].

Progressive bulbar dysfunction, such as dysarthria and dysphagia, also occurs and can lead to severe malnutrition and aspiration pneumonia. The latter is the most common cause of death in these patients. Corticospinal deficits, such as muscle weakness, hyperreflexia, and positive Babinski signs, are often encountered, usually along with amyotrophy and fasciculations, indicative of additional lower motor neuron involvement. Some authors observed a severe degree of amyotrophy, particularly in patients with end-stage disease. Many patients will also exhibit action myoclonus or even generalized seizures [90–94]. Depending on the age at onset, this may be an early or late manifestation of the illness, or it may be absent in some affected individuals. Mental changes can also occur, including decreased IQ, progressive dementia, or psychosis.

Some patients, particularly those with a short natural history, may exhibit only part of the above-described clinical manifestations. Thus, cases have been encountered with predominat ataxia, bulbar dysfuction, and amyotrophy, but without the visual disturbances. In contrast, others have experienced visual loss for several decades without manifesting additional neurologic deficits.

6.3.2. Neuropathologic findings

Pathologic studies have been carried out in a sizeable number of cases [90–94]. CNS changes consistent with olivopontocerebellar atrophy have been described by several authors [90–93], although atrophy of the basis pontis or cerebellar cortex have not been consistent findings, even within families. In most cases in which the spinal cord has been examined, there was histologic evidence for degeneration of the anterior horn cells, posterior columns, spinocerebellar tracts, pyramidal tracts, and/or Clarke's column. Involvement of the substantia nigra, locus coeruleus, putamen, and parietal and frontal cortex was also described. In infantile and childhood cases, the entire brain may be abnormally small. Histopathologic studies [95] of eyes obtained at autopsy from such patients have shown striking changes in the retinal photoreceptor cell layer, with the macula area being most severely affected. In severe cases, the degenerative process may extend to other retinal layers, including the ganglion cells.

6.3.3. Neuroimaging abnormalities

As described in Chapters 8 and 14, we have been following two affected members of a black kindred in which a dominantly inherited multisystem degenerative disorder, characterized by cerebellar ataxia, retinal degeneration, slowed eye movements, corticospinal deficits, amyotrophy, peripheral neuropathy, and myoclonus, has occurred in several generations. A 13-year-old boy died after a 6-year progressive course. His brain was examined at autopsy. Histologic changes similar to those previously described were found. We have obtained CT scans and MRIs of the brain in a similarly affected 18-year-old brother and the 46-year-old father of this patient.

In the childhood case, marked atrophic changes were present throughout the entire neuraxis, with the relative severity of involvement in the following the order: brain stem > cerebellum > cerebral hemispheres (Figure 13-13). Of the brain-stem structures, the pons and brachium pontis showed the most marked changes (Figures 13-13A and 13-13C). Both the basis and the tegmentum of the pons were involved (Figures 13-13C and 13-13D). Due to the excavation of the floor of the fourth ventricle and the atrophy of the brachium pontis and adjacent cerebellar parenchyma, there was "ballooning" of the fourth ventricle of the ex vacuo type (Figure 13-13C). There was an alteration in the MRI signals within the basis and the posterior part of the tegmentum of the pons, extending into the brachium pontis (Figures 13-13C and 13-13D). This probably indicates degenerative changes. Other brain-stem regions, such as the midbrain and medulla oblongata, were also atrophic. The changes were milder than those seen in the pons (Figure 13-13A). There was pancerebellar atrophy of the "fine comb" type with widened fine sulci and diffusely atrophic folia found in the superior and lateral, and to a lesser extent, infero-medial cerebellar regions (Figures 13-13C, 13-13D, and 13-13F). As in other types of OPCA with slowed saccades, the spinal cord was small and atrophic (Figure 13-13A). As such, these MRI findings are very similar to those observed in patients with OPCA associated with slowed saccades [1].

7. ADULT-ONSET RECESSIVE CEREBELLAR DEGENERATION

7.1. Clinical features

Recessive cerebellar degeneration of adult onset is considered by some [3,96] to be extremely rare. This, however, has not been our experience [8,97,98]. These differences may be due to the fact that we have undertaken vigorous investigations of the family of every patient with cerebellar degeneration. Such investigations, which included neurological examination of close relatives of the patients, not often revealed additional affected family members, some of which had been considered as suffering from unrelated neurological disorders. Of 182 patients with adult-onset cerebellar degenerations, we evaluated (Chapter 8, Table 8-3) 24 (13%) were found to have similarly

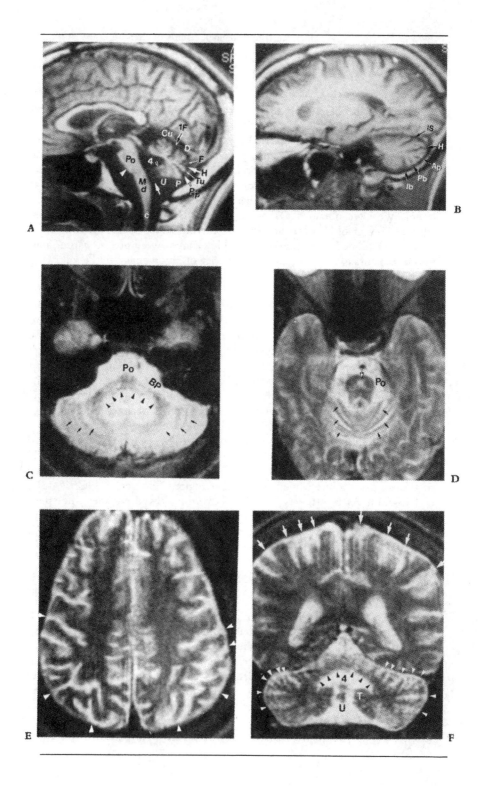

affected siblings. However, neither the parents nor offsprings were neurologically involved, thus suggesting the recessive inheritance of these disorders.

Out of the 24 patients studied, four evidenced a rather pure cerebellar syndrome, justifying their classification under recessive cortical cerebellar degeneration or cerebello-olivary atrophy (Holmes' type). All four patients were females, and their age ranged from 59 to 74 years (mean age 66.5 years). The duration of the disease ranged from 1 to 39 years (mean duration 18 years). They were all ambulatory at the time of their examination. Given the long duration of their disease, it can be concluded that the evolution of the illness was particularly slow in these patients.

The remaining 20 patients showed a combination of progressive cerebellar dysfunction and other neurologic deficits, suggesting the diagnosis of OPCA. In the majority of these cases (13 patients from eight families), the associated clinical feature was corticospinal dysfunction (hyperreflexia,

---

Figure 13-13. OPCA with retinal degeneration, slowed saccades, and other features. Black male, age 15, affected since age 6. Clinical features other than retinal degneration and slowed saccades are progressive cerebellar ataxia, dementia, corticospinal deficits, motor neuron signs, and peripheral neuropathy. His 42-year-old father was similarly affected since age 27. A brother died of a similar disorder at age 15; he was affected since age 7. Hence, the genetic transmission in this family is dominant, with variable age onset. **A**: Midsagittal section. Atrophy of the brain stem is most pronounced; particularly, the anterior aspect of the belly of the pons (Po) has markedly shrunken, especially in its inferior portion (unlabeled white arrowhead). Similarly, the medulla oblongata (Md) and the cervical cord (c) are also atrophied, but the changes are, in comparison, less pronounced. The pontine and medullary cisternal and the cisterna magna behind the brain stem are wide. Changes in the vermis include widening of the primary fissure (1F), and of the prepyramidal (Pp) and postclival sulci. The pyramid (P), uvula (U), and nodulus (empty arrowhead) are less involved but are still somewhat shorter than normal in length. The aqueduct and the inferior part of the fourth ventricle (4), including the foramen Magendie (unlabeled arrow), are enlarged. The atrophic declive (D), folium (F), and tuber (Tu), and the minimally prominent great horizontal fissure (H) and the culmen (Cu), are labeled. **B**: Parasagittal section of a cerebellar hemisphere showing widening of the great horizontal fissure (H), the ansoparamedian (Ap), prebiventral (Pb), intrabiventral (Ib), and intrasemilunar (IS) sulci. **C**: Transverse section of the cerebellum and the pons, showing a marked ballooning of the fourth ventricle with excavation of its floor (arched row of arrowheads), the most characteristic findings in OPCAs with slow saccades, including those with retinal degeneration. The pons (Po) and brachia pontis (BP) are markedly atrophic. Note the presence of hyperintense signals within the pons extending into the brachia pontis, indicative of degenerative changes. The quadrangular lobules show "fine comb" atrophy (black arrows). The pontine and lateral pontine cisternal are widened. **D**: Transverse section through the pons and upper cerebellum showing marked atrophy of the upper pons (Po). There are increased signal intensities within the midline basis pontis, indicative of degenerative changes (large empty arrow). The pontine cistern is wide, and the sulci (black arrows) of the anterior quadrangular lobules are prominent ("fine comb" atrophy). **E**: Transverse section through the centrum semiovale of the cerebral hemispheres showing widened sulci of the parietal and frontal lobes (arrowheads). **F**: Coronal section through the dilated (radiating black arrowheads) fourth ventricle (4). The dilatation is more transverse than superoinferior. Note the finely dilated sulci of the superior and inferior semilunar and posterior quadrangular lobules (white arrowheads). The parietal sulci are also widened (white arrows). The small uvula (U) and tonsils (T) are labeled.

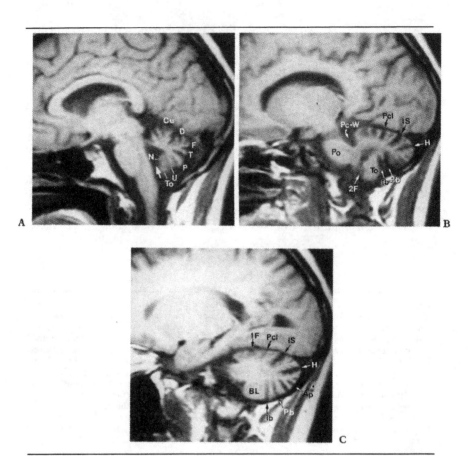

Figure 13-14. Adult-onset recessive cerebellar degeneration. Male, age 23, mildly affected for about 1 year. Neurological features included mild ataxia of gait and limbs, downbeat nystagmus, broken-up pursuit, slight dysarthria, corticospinal deficits (hyperreflexia, extensor plantor responses), and mild sensory neuropathy. A 31-year-old brother is similarly affected (see Figure 13-15). The father (age 56) and mother (age 54) of these patients were found to be neurologically intact. **A**: Midsagittal section. The vermis, especially the culmen (Cu), declive (D), folium (F), tuber (T), and to a much lesser extent, the pyramid (P) and uvula (U), are atrophic. The foramen Magendie (unlabeled large white arrow) is minimally wide. The nodulus (N) and tonsil (To) are labeled. **B**: Parasagittal section through the lateral portion of the pons. The wing of the precentral cerebellar sulcus (Pc-W) is wide. Fissures and sulci of the cerebellar paravermian area, such as the postclival sulcus (Pcl), intrasemilunar sulcus (IS), great horizontal fissure (H), prebiventral sulcus (Pb), intrabiventral sulcus (Ib), and secondary fissure (2F), are widened. The tonsil (To) and the cut surface of the lateral part of the pons (Po) are labeled. Note an early "fish mouth" atrophy of the intrasemilunar sulcus (IS). **C**: Parasagittal section through the brachium pontis. Cerebellar sulci in the paravermian area are diffusely widened. The intrasemilunar sulcus (IS) and the great horizontal fissure (H) are particularly wide and show early "fish mouth" atrophy. The changes involving the primary fissure (1F), postclival sulcus (Pcl), pre- and intra-biventral (Pb and Ib), and ansoparamedian sulci (Ap) are milder. The biventral lobule (BL) are labeled. Cerebral sulci appear unremarkable.

spasticity, and/or extensor plantar responses) with or without peripheral neuropathy. In addition, vertical (downbeat) nystagmus was observed in the affected members of four families. Parkinson's was the second major associated neurologic feature, observed in six patients from four families. One of these patients also showed corticospinal deficits. Finally, progressive ataxia associated only with peripheral neuropathy was observed in two patients. Although the sample size studied here is rather small, there appeared to be an intrafamilial correlation of (1) age of onset and (2) associated clinical features (corticospinal deficits, downbeat nystagmus, and Parkinsonism).

In 7 of the 13 families we studied (see Chapter 8, Table 8-3), the affected members were all males, thus raising the possibility of an X-linked recessive disorder [98]. However, this issue has been confounded by the generally small family size and the difficulties encountered in obtaining adequate information about maternal relatives in our patients. A partial deficiency of glutamate dehydrogenase (GDH) has been shown in several recessive OPCA kindreds [97,98]. In some of these familial forms the possibility of X-linked inheritance has been raised [97]. It is of particular interest, in this regard, that a GDH-specific gene has been assigned to human chromosome X, although its functional significance has not as yet been established (seëbelow).

7.2. Neuroimaging findings

Here we studied the MRIs of four patients from three families. The most striking abnormality was atrophy of the cerebellar hemispheres and the vermis (Figures 13-14 and 13-15). Although the cerebellum was diffusely and markedly affected, the lateral hemispheres (semilunar lobules) showed the worst changes (Figures 13-14C, 13-13C, and 13-13D), while the inferomedial region of the hemispheres (tonsil and biventral lobule) was minimally involved. The great horizontal fissure, the intrasemilunar sulcus, the postclival sulcus, and the prepyramidal fissure were especially enlarged (Figure 13-15C). The degree of involvement of the vermian lobules followed in this order: declive > folium > culmen > tuber (Figure 13-14A). In contrast, the pyramid, uvula, and nodule showed the least involvement. In the brain stem, the anterior aspect of the belley of the pons showed a characteristic flattening (Figure 13-15).

8. LATE-ONSET OPCA (PREDOMINANTLY SPORADIC)

OPCA that starts in late life often occurs sporadically and seems unrelated to genetic factors. However, this often depends on how vigorously the patient's family history has been explored and whether living relatives were evaluated neurologically. Such investigations may uncover additional affected family members and suggest the possibility of a genetically transmitted disease. The frequent onset of the illness in late adult life and declining family size could be the two main factors responsible for the fact that many familial

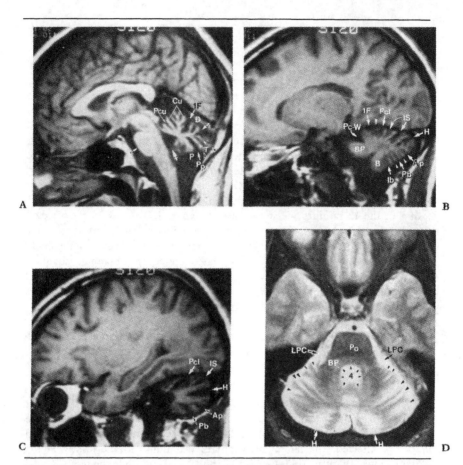

Figure 13-15. Adult-onset recessive cerebellar degeneration. Male, age 31 (a younger brother of the patient in Figure 13-14), affected since age 27. Neurological features included impaired coordination (mild gait and limb ataxia), downbeat nystagmus, impaired smooth-pursuit eye movements, hyperreflexia with ankle clonus, and mild sensory neuropathy. **A:** Midsagittal section. The entire vermis, particularly the declive (D), folium, tuber (T), and culmen (Cu), are atrophic. The primary fissure (1F), and the postclival (double crossed arrow) and prepyramidal sulci (Pp) are widened. The superimposed cerebellar hemisphere-inferior semilunar lobule (empty arrowheads) obscures the posterior border (unlabeled white arrowheads) of the tuber (T). The foramen Magendie (unlabeled crossed arrow) is only minimally widened. The preculminate sulcus (Pcu) and the slightly flattened anterior aspect of the pons (unlabeled arrow) are indicated. **B:** Parasagittal section through a brachium pontis (BP). "Fine comb" atrophy (small arrowheads) located below the ansoparamedian sulcus (Ap) and "coarse comb" atrophy involving the intrasemilunar sulci (IS) and the great horizontal fissure (H) are seen. The wing of the precentral cerebellar sulcus, sulci of the quadrangular lobules, and the postclival sulcus (Pcl) are prominent. The biventral lobules (B) are also labeled. **C:** Parasagittal section through the lateral portion of the cerebellar hemisphere. The postclival sulcus (Pcl), intrasemilunar sulcus (IS), great horizontal fissure (H), ansoparamedian sulcus (Ap), and prebiventral sulcus (Pb) are widened. "Fish mouth"-type atrophy, particularly in the region of the great horizontal fissure, is well demonstrated in this figure. Note the rather wide and deep fissures and sulci, reaching the medullary body of the cerebellar hemisphere. **D:** Transverse section through the pons and the mid quadrangular lobules (inferior orbitomeatal plane). The pons (Po), brachium pontis (BP), and slightly dilated roundish (radiating arrowheads) fourth ventricle (4). The lateral pontine cisternal (LPC) are wide bilaterally. The cerebellar hemispheric sulci (unlabeled black arrowheads) are widened (mixture of "fine and coarse comb" atrophy). The dilated great horizontal fissures (H) are indicated.

disorders (particularly those that are inherited recessively) often appear to occur sporadically.

In view of these considerations, we have undertaken thorough clinical and genetic investigations of all our patients who were diagnosed as suffering from late-onset cerebellar degenerative disorders. Based on data accumulated during the past 10 years, we have separated our patients into two broad groups. The first includes patients with *late-onset OPCA associated with autonomic failure*, with orthostatic hypotention being an obligatory criterion, and the second includes patients with *late-onset OPCA not associated with autonomic failure*. Besides differing in their clinical manifestations and their natural history, these two types of late adult-onset OPCA appear to have different etiopathogenetic factors. Although all cases associated with autonomic failure were sporadic occurrences, a substantial proportion of patients with OPCA not associated with autonomic failure had similarly affected relatives; as such, they are probably of genetic origin.

8.1. Late-Onset OPCA with progressive autonomic failure

8.1.1. Clinical features

Our investigation is based on the study of 30 cases (Table 8-3 in Chapter 8). Although gait ataxia was the most common presentation, in some patients the initial symptoms were autonomic dysfunctions, such as sexual impotence, bladder and bowel dysfunction, or fainting spells due to orthostatic hypotension. The disease progressed rather rapidly in most patients, leading to marked gait disturbances and wheelchair confinement within 2–4 years from the onset. Many patients complained of dizziness on changing position or while standing. With disease progress, symptoms and signs of dysautonomia tended to be more prominent and troublesome. The patients developed difficulty in emptying their bladder, urinary incontinence, constipation, changes in their sweating pattern, and/or intolerance to strong light. Some patients developed lightheadedness or frank faintings on getting up from bed in the morning or after eating meals.

Two thirds of our patients with autonomic failure (20 of 30 cases studied) showed a predominant cerebellar syndrome, characterized by progressive gait and limb ataxia, dysdiadochokinesis, and dysarthria. The oculomotor deficits primarily reflected a midline cerebellar dysfunction and consisted of ocular dysmetria, broken-up pursuit, and inability to suppress the vestibulo-ocular reflex. Interestingly spontaneous or gaze-evoked nystagmus was not a common neurologic manifestation in these patients. The extraocular movements, however, were of full range, and saccades appeared to be of normal velocity. Pupillary changes observed were anisocoria and/or a poor light reflex. The major associated neurological feature in these patients (with predominant ataxia) was corticospinal deficits (hyperreflexia, increased muscle tone, and extensor plantar responses). The age of these patients ranged from

45 to 70 years (mean 57.4 years). Their age at onset ranged from 38 to 63 years (mean 54.3 years), and the disease duration ranged from 1 to 8 years (mean duration 3.7 years).

In one third of our cases with dysautonomia (10 of 30 patients studied), symptoms and signs of Parkinsonism (bradykinesia, tremor, rigidity) were evident. Besides the cerebellar dysfunction, these were often associated with corticospinal deficits and oculomotor abnormalities (multisystem involvement). The age of these patients ranged from 54 to 78 years, and the age at onset of the disease ranged from 50 to 70 years. As with the above patients with predominant ataxia, the duration of the disease was shorter than that of other adult-onset cerebellar degenerations, ranging from 1 to 7 years (mean duration 3.4 years). Pupillary changes, such as anisocoria and/or poorly reactive pupils, were present in some patients. The oculomotor disturbances observed rather consistently were those associated with midline cerebellar dysfunction (broken-up pursuit, inacurate saccades, and inability to suppress the vestibulo-ocular reflex). However, the extraocular movements were of full range and nystagmus was usually absent.

8.1.2. Neuropathologic changes

Following the original neuropathological description by Shy and Drager [99] of a case with autonomic failure, Johnson et al. [100] reported the case of a 54-year-old man who died of progressive neurological disorder of $4\frac{1}{2}$ years duration, characterized by a combination of cerebellar dysfunction and autonomic failure (AF). Pathologically, there was olivopontocerebellar atrophy, along with degeneration of the putamen, substantia nigra, vestibular nuclei, and the intermediolateral columns of the spinal cord. It was estimated that cell loss in the lateral columns was about 75%, and these changes were thought to be responsible for the autonomic failure. Oppenheimer [101] described a similar case in which morphological changes of OPCA were found to be combined with those of nigrostriatal degeneration (SND). He suggested that SND and OPCA were two different forms of the same disease and proposed the term *multiple system atrophy* (MSA).

Oppenheimer [45] recently reviewed the pathology of 46 cases of MSA with AF. Atrophy of the cerebellar cortex was present in 31 cases, pontine nuclei in 30, inferior olives in 34, dorsal vagal nucleus in 29, pyramidal tracts in 19, anterior horns in 23, substantia nigra in 46, locus coeruleus in 33, striatum in 37, and thoracic intermediate columns in 42, occurring in various combinations. Cell counts have been carried out by a number of authors [100,102], including Oppenheimer [102], who compared pateints with MSA with AF to (1) normal controls, (2) patients with Parkinson's disease, and (3) patients with MSA without autonomic failure. He found an obvious correlation between loss of cells in lateral horns and autonomic failure. The cell counts of patients with MSA associated with AF were lower than those found in controls and patients with Parkinson's disease. However, in cases with MSA without AF, there was a 50% depletion in the lateral column cells as compared to the controls. These results do not fully provide an anatomic explanation for the pathogenesis of autonomic dysfunction in OPCA. Moreover, degeneration of Clarke's column has been reported in a variety of cerebellar degenerations, including Friedreich's ataxia and dominant forms of OPCA [45] in which there is no clinical evidence for the presence of autonomic dysfunction of the type seen in patients with MSA with AF.

In these cases there was a 2:1 preponderance of affected males over females. The duration of the illness ranged from 2 to 11 years (mean 5.5 years) and the age at death from 41 to 79 years (mean 58.6 years). As such, these autopsy data confirm our clinical impression of the rather rapid downhill course of this disorder.

8.1.3. Neuroimaging abnormalities

We have previously reported CT findings in four patients with OPCA associated with autonomic failure [1]. They revealed atrophy of the cerebellar hemispheres, vermis, pons, and brachium pontis. The fourth ventricle was enlarged and the frontal cortex showed mild atrophic changes. Here, we studied 12 patients with clinically diagnosed OPCA with autonomic failure. CT scans of the brain were performed in all these patients. In addition, in four of these MRI studies were also obtained. Results revealed (Figures 13-16 and 13-17) a variable degree of pontine atrophy, particularly of the caudal part of the basis pontis, ranging from minimal (Figure 13-17A) to severe (Figures 13-16A, 13-16B, and 13-16C). The brachium pontis was also variably involved. The midbrain and medulla oblongata showed slight changes (Figures 13-16A, 13-16D, and 13-17). The fourth ventricle showed a moderate degree of enlargement in most patients (Figures 13-16C and 13-16D), while the third ventricle was, as a rule, of normal size [1] (not shown here).

The vermis showed the most striking changes in the region of the folium declive and, to a lesser degree, of the culmen and tuber (Figures 13-17 and 13-18). The pyramid, uvula, and nodulus were less involved (Figures 13-17 and 13-18). The cerebellar hemispheres showed atrophic changes, particularly in the region of the superior and inferior semilunar lobules (Figure 13-16E). There was enlargement of the great horizontal fissure and the postclival sulcus. The superior cerebellum was less affected. The restiform body was also atrophic (Figures 13-16A and 13-16E), while the spinal cord was visualized in a few cases and was found to be essentially of normal size. The cerebral cortex showed atrophic changes in the parietal region (Figure 13-16E) and less so in the frontal area (not shown). In some patients, on whom repeated CT or MRI studies were performed, we observed a rapidly advancing vermian and brain-stem atrophy (Figure 13-18).

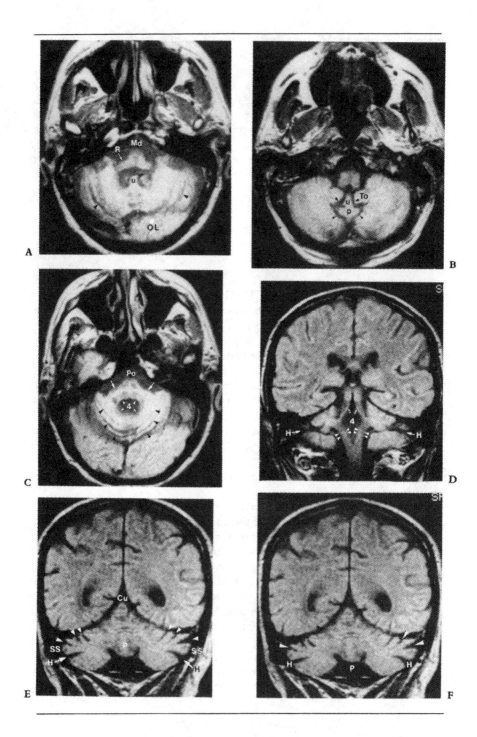

Figure 13-16. OPCA with progressive autonomic failure. Male, age 64, with a 2-year history of progressive gait impairment, sexual impotence, sphincter disturbances, alteration in his sweating pattern, and intolerance to sunlight. Neurologic features included gait and limb ataxia, anisocoria, impaired pursuit eye movements, corticospinal deficits, and orthostatic hypotension. **A:** Transverse section of the medulla oblongata, tonsils (To), pyramid (P), and uvula (u). The inferior paravermian sulci (unlabeled arrows) are prominent. **B:** Transverse section through the medulla oblongata and uvula (inferior orbitomeatal plane). The medulla oblongata (Md) is small. The restiform bodies (R) and uvula (U) are atrophic. The postclival sulci (arrowheads) and the peritonsilar spaces are wide. The left occipital lobe (OL) is labeled. **C:** Transverse section of the pons and the mid quadrangular lobules (inferior orbitomeatal plane). The pons (Po), especially the basis pontis, is markedly atrophic, and its lateral surfaces are retracted (unlabeled white arrows). The fourth ventricle (4) is slightly dilated and roundish (radiating white arrowheads). The sulci (black arrowheads) of the midportion of the quadrangular lobules show widening ("fine comb" atrophy). **E:** Coronal section of the fourth ventricle parallel and close to its floor. The brain stem is small and the fourth ventricle (4) is slightly wide. The visualized restiform bodies (opposing double arrowheads) and brachia conjunctiva (empty arrowheads) are thin. The great horizontal fissures (H) are slightly wide. **F:** Coronal section of the cerebellum through the anterior part of the pyramid (P). The most conspicuous finding is widening of the great horizontal fissures (H) and intrasemilunar sulci (unlabeled arrowheads), indicating that the greatest involvement lies in the superior semilunar lobules. The culmen (Cu) appears unremarkable.

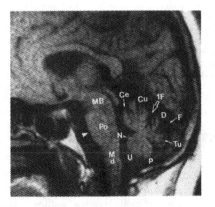

Figure 13-17. Another case of OPCA with progressive autonomic failure. Male, age 51, with poor balance and dizziness since age 48. Neurological features included gait and limb ataxia, impaired smooth-pursuit eye movements, corticospinal deficits, and orthostatic hypotension. **A:** Midsagittal section of the brain stem and cerebellum. The most striking findings are the markedly atrophied declive (D), folium (F), mildly shrunken tuber (Tu), and culmen (Cu) with widened sulci, particularly the primary fissure (1F). The midbrain (MB), pons (Po), and medulla oblongata (Md) appear unremarkable, except for a slight flattening of the anterior aspect of the pons (unlabeled arrowhead). The aqueduct and fourth ventricle including the foramen Magendie are slightly enlarged. The pontine and medullary cisternal are slightly prominent. The pyramid (P), uvula (U), and nodulus (N) appear normal.

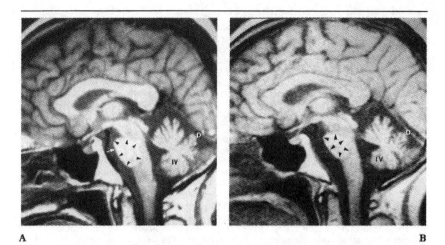

Figure 13-18. A third case of OPCA with progressive autonomic failure. Female, age 57, with a 2-year history of progressive incoordination, speech disturbances, dysarthria, and urination difficulties, including episodes of incontinence.

Neurological findings included impaired smooth-pursuit eye movements, inability to suppress the vestibular occular reflex, dysarthria, gait and limb ataxia, corticospinal deficits and orthostatic hypotension. **A:** MRI (midsagittal section) obtained within a few months from the onset of symptoms (early stage) showing mild atrophic changes involving the brain stem, particularly the anterosuperior portion of the belly of the pons (arrow). The vermis appears to be essentially normal except for a questionable atrophy of the declive (D) and its neighbouring vermis. **B:** MRI of the same patient obtained 2 years later (moderately to severely advanced stage). In comparison with the initial study (Figure 13-18A), the most striking change is atrophy of the pons, the belly of which shows a further flattening. The prepontine cistern is increased in size whereas the anteroposterior diameter of the basis pontis (radiating row of arrowheads) is substantially decreased, losing a normal round contour of the basis pontis.

An increased atrophy has also occurred in the vermis particularly in the region of the declive (D). The cisterna magna has become larger indicating wasting of the inferior vermis (IV).

8.2. Late-Onset OPCA without autonomic failure

8.2.1. Clinical features

The initial neurologic manifestation is gait disturbance—slow and unsteady. The patient usually complains of feeling "dizzy," "out of balance," or "lacking equilibrium." Others will describe a "hesitation in step" or "their legs become weak or stiff after walking for a while." In some patients the presenting symptom may be a difficulty in performing acts that require learned coordination skills, such as dancing. Relatives or friends may sometimes trace back the initial manifestations of the illness to a time when they first noticed his or her "slowed down" or abnormal body posture during walking. In one case, the presenting complaint was difficulty in walking on the sand during a vacation.

As the disease progresses, the gait becomes more affected. The patient develops the well-known broad-based, ataxic gait. This may lead to repeated falls, although some patients may learn compensatory techniques that prevent them from falling. Few patients may, however, present with a combination of cerebellar (broad-based, jerky) and extrapyramidal (small-stepped, shuffling) gait. Gradually, the fine finger coordination (dexterity) becomes affected. At around the same time, disturbances of articulation appear. Speech becomes dysarthric, dysrhythmic, and slowed. Dysphagia usually appears in advanced disease states. Also, a variety of oculomotor disturbances are seen. These include gaze-evoked or primary-gaze nystagmus, ocular dysmetria, abnormal smooth persuit, inability to suppress the vestibulo-ocular reflex, impaired OKNs (particularly in the vertical plane), and upward gaze paresis. Optic atrophy with or without decreased visual acuity may also be found, especially in patients with advanced disease.

Our experience with adult-onset sporadic OPCA without autonomic failure is based on a study of 61 patients (Chapter 8, Table 8-3). Many of such patients had GDH deficiency (see Chapter 14). All evidenced progressive cerebellar dysfunction associated with other neurologic deficits. In 26 out of these 61 patients, the main additional neurologic features observed were corticospinal deficits, such as hyperreflexia, increased muscle tone, and extensor plantar responses. However, pyramidal weakness and frank spasticity were rarely encountered. The age of these patients ranged from 26 to 88 years (mean 54.2 years). The age at onset of symptoms ranged from 20 to 69 years (mean 46.2 years), and the duration of the disease from 1 to 30 years (mean 8.1 years).

Parkinsonism, such as bradykinesia, tremor, rigidity, and/or facial hypomimia with staring, was observed in 17 patients. In some of these patients extrapyramidal manifestations occurred in various combinations with bulbar palsy, corticospinal deficits, amyotrophy, or peripheral neuropathy (multiple system atrophy). The age of these patients ranged from 49 to 96 years (mean 67.0 years). In 13 patients the predominant associated neurologic deficits were signs of motor and/or sensory neuropathy. The age of these patients ranged from 38 to 79 years (mean 61.1 years), and the duration of the disease ranged from 1 to 19 years (mean 10.0 years). Finally, frank dementia, often in association of other deficits, was observed in five patients. Their age ranged from 49 to 79 years (mean 70.4 years); and the duration of the disease ranged from 1 to 11 years (mean 5 years).

8.2.2. Neuropathologic changes

As indicated above, the majority of late-adult onset OPCA cases occur sporadically, and as such, they are classified as Dejerine Thomas type, as proposed by Greenfield [44], although in Greenfield's years autonomic failure with sporadic OPCA had not been described. In his latest account,

Oppenheimer [45] did not separate the pathologic findings of the sporadic form of the disease from those of the dominantly inherited OPCAs. Because much of the neuropathology of the sporadically occurring OPCA became known from original case reports (which differ from author to author), it is difficult to ascertain the frequency of involvement of the various CNS regions, aside from the cerebellum. Moreover, most of these reports dated back to several decades, at which time the association of OPCA with autonomic failure had not been recognized. It is therefore difficult to ascertain whether the cases of Dejerine Thomas disease, studied pathologically, were mixed with MSA associated with autonomic failure.

Sporadic cases are most likely under-represented in the contemporary literature, since isolated cases no longer arouse the interest they did a century ago, and therefore, they do not, as a rule, become a subject of publication. Another reason for under-reporting sporadic OPCA seems to be that neuropathologists are generally reluctant to diagnose OPCA in cases that do not show atrophy of the basis pontis, especially in the absence of a positive family history. This, however, is not true for neuropathologic studies performed in large dominant OPCA kindreds in which disparities in pontine involvement among affected members has been well established and has been attributed to a variable pathologic expressivity of the genetic defect. Given these considerations, our present knowledge of the neuropathology of sporadic OPCA should be regarded as incomplete. More definite clinico-pathological correlations may become evident in the future, with the development of biochemical markers that could allow a more accurate diagnosis and classification of these heterogenous disorders in living patients.

With these considerations in mind, we should now review the pathologic reports of cases classified as Dejerine-Thomas OPCA or sporadic OPCA. According to such descriptions (previously reviewed by Eadie [103]), the cerebellum appears macroscopically to be atrophic, with the cerebellar hemispheres being more affected than the vermis. The ventral pons is reduced in size and the inferior olivary prominences may be less noticeable. Microscopically, there is degeneration of both the cerebellar cortex and cerebellar white matter; the neocerebellum is more severely affected than the paleocerebellum. The dentate nucleus is less frequently affected. In the brain stem, there is involvement of the pontine nuclei, transverse pontine fibers, and middle cerebellar peduncles. In the medulla oblongata, there is atrophy and gliosis of the inferior olives, and there is a considerable loss of the olivocerebellar fibers, with wasting of the inferior cerebellar peduncles. In the midbrain, there is degeneration of the substantia nigra and in some cases involvement of the subthalamic nucleus and red nucleus. The basal ganglia have been reported to be involved rather frequently in sporadic OPCA, while the thalamus is affected less often. Finally, pathologic changes have been noted in the cerebral cortex, particularly in the prefrontal and temporal regions.

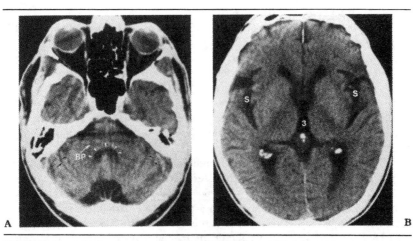

Figure 13-19. Multiple system atrophy of the OPCA type with glutamate dehydrogenase (GDH) deficiency (sporadic case). See text for details of neurological manifestations and autopsy findings. **A**: Transverse section through the fourth ventricle. The visualized fourth ventricle (row of arrowheads) is slightly widened. The sulci (black arrows) of the posterior quadrangular lobules are slightly prominent. The brachia pontis (BP) are slightly decreased in size. **B**: Transverse section through the third ventricle and atria. The third ventricle (3) and sylvian fissures (S) are wide.

8.2.3. Neuroimaging studies

Patients with late-onset sporadic OPCA associated with partial GDH deficiency show variable degrees of cerebellar and brain-stem atrophy. The third ventricle was enlarged in the majority of the cases [1]. Here we studied one patient who came to autopsy. He was a 79-year-old man with a 5-year history of progressive ataxic disorder associated with Parkinsonism, oculomotor abnormalities (anisocoria, difficulty initiating volitional saccades, decreased upward gaze), corticospinal deficits, and a mild to moderate degree of dementia. Neuropathologic findings included atrophy of the cerebellum and brain stem (particularly of the medulla oblongata), as well as diffuse cortical atrophy with dilatation of the lateral ventricles. There was neuronal loss in the inferior olives, nuclei of the pontine tegmentum, dorsal nuclei of the medulla oblongata, substantia nigra, and hippocampus. In the cerebellum, there was focal atrophy of the Purkinje cells, thinning of the internal granular layer, and atrophy of the dentate nuclei with accompanying gliosis.

A CT scan obtained about a year prior to his death revealed some degree of atrophy of the cerebellum and brain stem, and enlargement of the fourth ventricle (Figure 13-19A). In addition there was enlargement of the third ventricle and widening of the Sylvian fissure and the frontal sulci (Figure 13-19B).

9. SUMMARY OF MRI AND CT FINDINGS

A careful analysis of the morphological changes of the brain observed in MRI and CT suggests that the various types of cerebellar degeneration studied here show distinct neuroimaging features, which may be of value in the diagnosis and classification of these disorders. In cerebellar disorders with well-established neuropathologic changes, the MRI and CT findings reported here generally agree with the previously described anatomical changes. In addition, these neuroimaging studies have provided insights into nosological entities, such as OPCA with slowed saccades, for which no adequate pathologic studies are known.

In *Friedreich's ataxia*, the principal finding was atrophy of the posterior half of the spinal cord and, to a much lesser degree, of the superior vermis and lateral cerebellar hemispheres. In *hexosaminidase deficiency*, there was a marked pancerebellar atrophy associated with a profound enlargement of the fourth ventricle; the brain stem appeared rather normal. In addition, there was a variable degree of atrophy involving the frontal and parietal lobes. In *cortical cerebellar degeneration* or *cerebello-olivary atrophy*, the most conspicuous finding was atrophy of the supero-lateral portion of the cerebellar hemispheres and of the middle and upper vermis. In addition, there was a mild parietal atrophy, while the brain stem remained essentially intact.

In *Menzel's-type, HLA-linked dominant OPCA (Schut-Haymaker)*, the most prominent MRI features were atrophy of the spinal cord, medulla oblongata, inferior pons, and anterior cerebellar lobe, associated with a mild frontal cortical atrophy and a moderate enlargement of the third ventricle. In the variant of *dominant OPCA with slowed saccades*, there was pancerebellar atrophy, ballooning of the fourth ventricle, wasting of the tegmentum and basis of the pons, and shrinkage of the spinal cord (Figure 13-11). In the variant of *dominant OPCA with retinal degeneration, slowed saccades, and other features*, the most conspicuous finding was a severe degree of atrophy of the brain stem and brachium pontis, and ballooning of the fourth ventricle. The cerebellum, although markedly atrophic, was comparably less involved than the brain stem. Moreover, cerebellar atrophy was most prominent in the lateral part of the hemispheres and the vermis was spared. In addition, there was atrophy of the cerebral cortex, especially in the parietal region.

In *recessive cerebellar degeneration*, there was conspicuous atrophy of the cerebellum associated with lesser brain-stem changes. The subcortical white matter (medullary body) and its extension into the folia of the cerebellar hemispheres was markedly involved, giving rise to a "fishmouth" type of atrophy. The spinal cord was normal. In the *OPCA with progressive autonomic failure*, there was striking atrophy of the middle vermis (folium and declive) and atrophy of the lateral and superior portions of the cerebellar hemispheres associated with a variable degree of brain-stem atrophy; but the supratentorial structures were intact. Finally, in *GDH-deficient OPCA*, the most conspicuous finding was enlargement of the third ventricle and Sylvian fissures, associated with a variable degree of atrophy involving the brain stem, the lateral portion of the cerebellar hemisphere, and frontal lobe.

10. CLINICOPATHOLOGIC CORRELATIONS

Since ataxia is the cardinal neurologic feature of all cerebellar degenerations, it is not surprising to see that cerebellar atrophy was the most consistent morphologic alteration observed in these afflictions. However, the magnitude and the pattern of this atrophy varied from one clinical syndrome to the other, and even within the same clinical type. Thus, atrophy of the cerebellum was minimally and rather focally expressed in patients with Friedreich's ataxia, while on the other end of the spectrum, cerebellar atrophy was severe and diffuse in patients with forms of OPCA, such as the variants associated with slowed saccades. Despite the mild anatomic changes observed in the cerebellum of patients with FA ataxia, the clinical manifestations of the patients studied here were rather marked. This is consistent with the concept that the functional deficits of this disorder result primarily from deafferentation of the cerebellum due to degeneration of specific spinal-cord afferent pathways terminating in the cerebellar cortex.

In the cortical cerebellar atrophies, degeneration was most pronounced in the upper and midline cerebellar zone. This was associated with a predominant gait ataxia and characteristic oculomotor abnormalities. In recessive cerebellar degeneration, there was a striking discrepancy between the degree of cerebellar atrophy and the severity of clinical symptoms. Thus, despite extensive cerebellar atrophy, demonstrated by MRI or CT, most patients had mild clinical ataxia. This, however, was not the case for the dominantly inherited OPCAs in which cerebellar atrophy grossly correlated with the severity of clinical symptoms. Further analysis of the data suggested that the main difference between these two clinical syndromes lies in the degree of brain-stem atrophy, which was mild in the recessive ataxia, but it was severe in most forms of dominant OPCA. As such, these findings underscore the fact that the brain stem is of particular importance in the clinical expressivity of cerebellar diseases. Moreover, the pattern of brain-stem atrophy appeared to correlate with particular variants of the dominant disease. Thus, in the OPCA variants associated with slowed saccades, the atrophic changes were found to extend into the tegmentum of the pons, the paramedian zone of which has been thought to be responsible for the generation of saccadic eye movements [1].

11. PATHOGENETIC MECHANISMS

The factor(s) determining the selective CNS regional vulnerability and the corresponding neurologic manifestations that characterize the various cerebellar degenerations have been poorly understood. Since most of these disorders are genetically determined, they must depend on mutations or enzymes or other proteins. As such, the regional sensitivity may arise from a differential cellular expression of the mutant gene product and/or modifying factors. In this regard, evidence has emerged in recent years which suggests that the pattern of neurochemical connections of neurons may play a role in rendering them vulnerable to neurodegenerative process [1,104].

One such early indication has been the observation that the metabolism of the neuroexcitotoxic amino acid glutamate is altered in certain forms of adult-onset recessive or sporadic OPCA not associated with autonomic failure. Such patients have low activity of the enzyme glutamate dehydrogenase GHD [97,98], which has been shown to be particularly enriched in glial cells associated with glutamatergic nerve terminals (see Chapter 6). This may cause the defective metabolism of transmitter glutamate at the nerve terminals, leading to the accumulation of excessive amounts of glutamate at the synapses and degeneration of postsynaptic neurons [1].

We have previously reported [1] that the topographic pattern of CNS lesions in patients with GDH-deficient OPCA is rather specific and can be attributed to selective degeneration of structures that receive direct glutamatergic innervation or are sensitive to experimental systemic glutamate toxicity (Figure 13-19). Elegant immunocytochemical investigations by Aoki et al. (see Chapter 6) have supported this hypothesis by showing that GDH immunoreactivity is selectively enriched in CNS regions that receive glutamatergic innervation and that degenerate in OPCA [1].

Although GDH deficiency has not been implicated in most dominant forms of OPCA or many of the sporadic types of the disease, the basic neuropathologic changes of these afflictions are common and may therefore be related to similar but alternate metabolic defects. In this regard, there is evidence that the metabolism of aspartate is altered in non-GDH-deficient disorders, such as the Schut-Haymaker kindred [105].

The present neuroimaging studies showed that atrophy of the *semilunar lobule* of the cerebellar hemisphere is a consistent change observed in all forms of cerebellar degenerative disorders, thus indicating a regional vulnerability common to these afflictions. In this regard, it is of particular interest to note that in rat cerebellum GDH immunoreactivity, studied by Aoki et al. (see Chapter 6, Figure 6-3C), was particularly enriched in the lateral hemispheres. Since GDH may be a marker for glutamatergic terminals, it seems possible that atrophy of the semilunar lobule in human cerebellar disorders is mediated through glutamatergic dysfunction.

Recent reports [106–110] (see Chapter 5) showing significant decreases in glutamate receptors in the cerebellum of patients affected by distinct forms of OPCAS support this concept by suggesting that the degeneration of the glutamatergic postsynaptic neurons is a process common to all forms of the disease [106,107]. The studies of Tsiotos et al. [106] further revealed that the affinities and pharmacologic properties of L-glutamate binding sites of the cerebellum of these patients were not altered. Consequently, it is unlikely that structural abnormalities of glutamate receptors are responsible for the

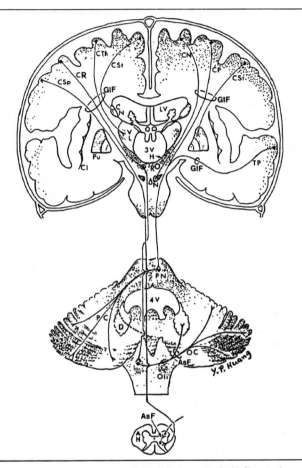

Figure 13-20. Diagram showing putative glutamatergic corticofugal fibers and areas of morphological changes of the brain occurring in the GDH-deficient OPCA in the supra- and infra-tentorial regions and in the spinal cord. Labeled structures include the anterior horn (AH), aspartergic fibers (AsF), claustrum (Cl), caudate nucleus (CN), corticopontine tract (CP), corticorubral tract (CR), corticospinal tract (CSp), corticostriate tract (CSt), corticothalamic tract (CTh), dentate nucleus (D), glutamatergic fibers (GlF), hypothalamus (H), lateral ventricle (LV), substantia nigra (N), olivocerebellar tract (OC), olive (Oli), pontocerebellar tract (PC), pontine nuclei (PN), putamen (Pu), red nucleus (R), temporopontine tract (TP), ventrolateral nucleus (VL) of the thalamus, third ventricle (3V), and fourth ventricle (4V). For more details, see Huang and Plaitakis [1]. This figure is from [112a] with permission.

neurodegenerative processes. Instead, based on the GDH-deficient OPCA model, it can be argued that altered presynaptic glutamatergic mechanisms are involved in the pathogenesis of cerebellar disorders and, perhaps, other primary neurodegenerations [104].

The strategic localization of GDH in glial processes that surround glutamatergic nerve terminals has led to suggestions that the enzyme may be of particular importance in protecting postsynaptic neurons from the neurotoxic effects of transmitter glutamate released during neurotransmission. It can be further argued that the selective vulnerability of certain neuronal systems is determined by the pattern of their chemical connections [104]. More specifically, the premature loss of certain neuronal populations in human cerebellar disorders may be programmed during the development of these cells by their establishment of synapses with glutamatergic nerve terminals.

According to this hypothesis, regional GABAergic and cholinergic neuronal systems, which are thought to receive excitatory innervation by glutamatergic terminals, ought to be selectively vulnerable to the neurodegenerative process. The available data in this regard are indeed consistent with this possibility. Thus, the putative GABAergic Purkinje cells of the cerebellar cortex, innervated by the glutamatergic parallel fibers [1], are markedly affected in most cerebellar disorders. Immunochemical studies by Aoki et al. (see Chapter 6) have shown GDH immunoreactivity to be concentrated within fine processes closely encircling unlabeled Purkinje cells perikarya. The granule cells, which are also affected in cerebellar degenerative disorders, albeit less than the Purkinje cells, are innervated by the mossy fibers, which are also thought to be glutamatergic (see Chapters 2–4).

In addition, cholinergic neurons found in many CNS regions, such as in the striatum, hippocampus, brain stem, and spinal cord, appear to receive excitatory input from glutamatergic tracts. Kish et al. [110] have shown a marked depletion of these cholinergic systems in dominantly inherited OPCA, as evidenced by marked reductions in acetylcholinetransferase activity in the brain of these patients. Additional studies are certainly needed to test these hypotheses and to provide answers to the questions raised by the present studies. Such investigations are expected to advance our understanding of the selective neuronal vulnerability phenomenon that is central to all disorders with primary system degeneration.

ACKNOWLEDGMENTS

This work was supported by NIH grants NS-16871 and RR-71, Division of Reasearch Resources, General Clinical Research Center Branch. We would like to thank Dr. L.J. Schut for providing us with some of the MRI and CT scans used in this studies. We are intebted to many colleagues who made these investigations possible by referring their patients to us. We are also indebted to Mr. J. Huang for his editorial assistance during the preparation of this manuscript.

REFERENCES

1. Huang Y.P., Plaitakis A. (1984). Morphological changes of olivopontocerebellar atrophy in computed tomography and comments on its pathogenesis. Adv. Neurol. 41:39–85.
2. Friedreich N. (1863). Uber degenerative atrophie der spinalen hinterstrange. Virchows

Arch. Pathol. Anat. Physiol. 26:391–419; 433–459.
3. Harding A.E. (1984). The Hereditary Ataxias and Related Disorders. In: Clinical Neurology and Neurosurgery Monographs. London: Churchill Livingstone, pp. 1–226.
4. Bouchand J.P., Barbeau A., Bouchand R., et al. (1979). A cluster of Friedreich's ataxia in Rimouski, Quebec. Can. J. Neurol. Sci. 6:205–208.
5. Geoffroy G., Barbeau A., Breton A., Lemieux B., Aube M., Leger C., Bouchard J.B. (1976). Clinical description and radiologic evaluation of patients with Friedriech's ataxia. Can. J. Neurol. Sci. 3:279–286.
6. Harding A.E. (1981). Friedreich's ataxia: A clinical and genetic study of 90 with an analysis of early diagnostic criteria and intrafamilial clustering of clinical features. Brain 104:589–620.
7. Pasternac A., Krol R., Peticlerc R., et al. (1980). Hypertrophic cardiomyopathy in Friedreich's ataxia: Symmetric or assymmetric. Can. J. Neurol. Sci. 7:379–382.
8. Plaitakis A. (1987). Cerebellar degenerations. Curr Neurol. 7:159–192.
9. Leclercq M., Harmant J., de-Barcy T. (1985). Psychometric studies in Friedreich's ataxia. Acta Neurol. Belg. 85:209–221.
10. Hart R.P., Kwentus J.A., Leshmer R.T., et al. (1985). Information processing speed in Friedreich's ataxia. Ann. Neurol. 17:612–614.
11. Cote M., Davignon A., Slignac A., et al. (1976). Cardiologic signs and symptoms in Friedreich's ataxia. Can. J. Neurol. Sci. 3:319–321.
12. Berg R.A., Kaplan A.M., Jarrett P.B., et al. (1980). Friedreich's ataxia with acute cardiomyopathy. Am. J. Dis. Child. 134:390–393.
13. Smith E.R., Sangalang V.E., Hefferman L.P., et al. (1977). Hypertrophic cardiomyopathy: The heart disease of Friedreich's ataxia. Am. Heart J. 94:428–434.
14. Weiss E., Kreonzon I., Winer H.E., et al. (1981). Echocardiographic observations in patients with Friedreich's ataxia. Am. J. Med. Sci. 282:136–140.
15. Cote M., Bureau M., Leger C., et al. (1979). Evolution of cardiopulmonary involvement in Friedreich's ataxia. Can. J. Neurol. Sci. 6:151–157.
16. Gottdiener J.S., Hawley R.J., Maron B.J., et al. (1982). Characteristics of the cardiac hypertrophy in Friedreich's ataxia. Am. Heart J. 103:525–531.
17. Palagi B., Picossi R., Cassasa F., et al. (1988). Biventricular function in Friedreich's ataxia: A radionuclide angiographic study. Br. Heart J. 59:692–695.
18. Allard P., Duhaime M., Raso J.V., et al. (1980). Pathomechanics and management of scoliosis in Friedreich's ataxia patients: Preliminary report. Can. J. Neurol. Sci. 7:383–388.
19. Allard P., Danseraeu J., Thiry P.S., et al. (1982). Scoliosis in Friedreich's ataxia. Can. J. Neurol. Sci. 9:105–111.
20. Vezina J.G., Bouchard J.P., Bouchard R., et al. (1982). Urodynamic evaluation of patients with hereditary ataxias. Can. J. Neurol. Sci. 9:127–129.
21. Leach G.E., Farsaii A., Kark P., et al. (1982). Urodynamic manifestations of cerebellar ataxia. J. Urol. 128:348–350.
22. Satya-Murty S., Cacae A., Hanson P. (1980). Auditory dysfunction in Friedreich's ataxia: Result of spinal ganglion degeneration. Neurology 30:1047–1053.
23. Shannon E., Himmelfarb M.Z., Gold S. (1981). Auditory function in Friedreich's ataxia. Arch. Otolaryngol. 107:254–256.
24. Taylor M.J., McMenamin J.B., Anderman E., et al. (1982). Electrophysiologic investigation of the auditory system in Friedreich's ataxia. Can. J. Neurol. Sci. 9:131–135.
25. Spoendlin H. (1974). Optic and cochleovestibular degeneration in hereditary ataxias. Brain 97:41–48.
26. Nuwer M.R., Perlman S.L., Packwood J.W., et al. (1983). Evoked potential abnormalities in the various inherited ataxias. Ann. Neurol. 13:20–27.
27. Livingstone I.R., Mastaglia F.L., Edis R., et al. (1981). Visual involvement in Friedreich's ataxia and hereditary spastic ataxia: A clinical and visual evoked response study. Arch. Neurol. 38:75–79.
28. Carroll W.M., Kris A., Baraitser M., et al. (1980). The incidence and nature of visual involvement in Friedreich's ataxia: A clinical and visual evoked potential study of 22 patients. Brain 103:423–434.
29. Kirkham T.H., Coupland S.G. (1981). An electroretinal and visual evoked potential study in Friedreich's ataxia. Can. J. Neurol. Sci. 8:289–294.
30. Pinto F., Amantini A., de Scisciolo G., et al. (1988). Visual involvement in Friedreich's

ataxia: PERG and VEP study. Eur. Neurol. 28:246–251.
31. Dale R.T., Dirby A.W., Jampel R.S. (1978). Square wave jerks in Friedreich's ataxia. Am. J. Ophthalmol. 85:400–406.
32. Furman J.M., Perlman S., Baloh R.W. (1983). Eye movements in Friedreich's ataxia. Arch. Neurol. 40:343–346.
33. Cogan D.G., Chu F.C., Reingold D.B. (1982). Ocular signs of cerebellar disease. Arch Ophthalmol. 100:755–760.
34. Monday L.A., Lemieux B., St. Vincent H., et al. (1978). Clinical and electronystamographic findings in Friedreich's ataxia. Can. J. Neurol. Sci. 5:71–78.
35. Kirkham T.H., Guitton D., Katsarkos A., et al. (1979). Oculomotor abnormalities in Friedreich's ataxia. Can. J. Neurol. Sci. 6:176–172.
36. Bouchard J.P., Barbeau A., Bouchard R., et al. (1979). Electromyography and nerve conduction studies in Friedreich's ataxia and autosomal recessive spatic ataxia of Charlevoix-Saguenay. Can. J. Neurol. Sci. 6:185–189.
37. D'Angelo A., Di Donato S., Negri G., et al. (1980). Friedreich's ataxia in northern Italy: I. Clinical, neurophysiological and in vivo biochemical studies. Can. J. Neurol. Sci. 7:359–362.
38. Jones S.J., Baraister M., Halliday A.M. (1980). Peripheral and somatosensory nerve conduction defects in Friedreich's ataxia. J. Neurol. Neurosurg. Pshchiatry. 43:495–503.
39. Bird T.D., Turner J.L., Sumi S.M., et al. (1978). Abnormal function of endocrine pancreas and anterior pituitary in Friedreich's ataxia. Ann. Intern. Med. 88:478–481.
40. Tolis G. (1980). Friedreich's ataxia and oral glucose tolerance: I. The effect of ingested glucose on serum glucose and insulin values in homozygotes and obligated heterozygotes and potential carriers of the Friedreich's ataxia gene. Can. J. Neurol. Sci. 7:397–400.
41. Campanella G., Filla A., De Falco F., et al. (1980). Friedreich's ataxia in the south of Italy: A clinical and biochemical survey of 23 patients. Can. J. Neurol. Sci. 7:351–357.
42. Finocchiaro G., Balo L., Micossi P. (1988). Glucose metabolism alterations in Friedreich's ataxia. Neurology 38:1292–1296.
43. Schoenle E.J., Boltshauser E.J., Baekkeskov S., et al. (1989). Preclinical and manifest diabetes mellitus in young patients with Friedreich's ataxia: No evidence of immune process behind the islet cell destruction. Diabetologia 32:378–381.
44. Greenfield J.G. (1954). The Spino-Cerebellar Degenerations. Oxford: Blackwell.
45. Oppenheimer D.R. (1984). Diseases of the basal ganglia, cerebellum and motor neurons. In: Greenfield's Neuropathology, J. Hume Adams, J.A.N. Corsellis, L.W. Duchen (eds): New York: John Wiley & Sons, pp. 699–747.
46. Harding A.E., Zilkha K.J. (1981). "Pseudodominant" inheritance in Friedreich's ataxia. J. Med. Genet. 18:285–287.
47. Skre H. (1975). Friedreich's ataxia in Western Norway. Clin. Genet. 7:287–298.
48. Anderman E., Remillard G.M., Goyer G. et al. (1976). Genetic and family studies in Friedreich's ataxia. Can. J. Neurol. Sci. 3:287–303.
49. Romeo G., Menozzi P., Ferlini A., et al. (1983). Incidence of Friedreich's ataxia in Italy estimated from consanguinous marriages. Am. J. Hum. Genet. 35:523–529.
50. Gusella J.F., Wexler N.S., Corneally P.M., et al. (1983). A polymorphic DNA marker genetically linked to Huntington's disease. Nature 306:234–238.
51. Chamberlain S., Shaw J., Rowland A., et al. (1988). Mapping of mutation causing Friedreich's ataxia to human chromosome 9. Nature 334:248–249.
52. Wallis J., Shaw J., Wilkes D., et al. (1989). Prenatal diagnosis of Friedreich ataxia. Am. J. Med. Genet. 34:458–461.
53. Chamberlain S., Shaw J., Wallis J., et al. (1989). Genetic homogeneity at the Friedreich's ataxia locus on chromosome 9. Am. J. Hum. Genet. 44:518–521.
54. Fujida R., Agid Y., Trouilas P., et al. (1989). Confirmation of linkage of Friedreich ataxia to chromosome 9 and identification of a new closely linked marker. Genomics 4:110–111.
55. Fujita R., Hanauer A., Sirugo G., et al. (1990). Additional polymorphisms at marker loci D9S5 and D9S15 generate extended haplotypes in linkage disequilibrium with Friedreich ataxia. Proc. Natl. Acad. Sci. USA 87:1796–1800.
56. Hanauer A., Chery M., Fujita R., et al. (1990). The Friedreich's ataxia gene is assigned to chromosome 9q13–q21 by mapping of tightly linked markers and shows linkage disequilibrium with D9S15. Am. J. Hum. Genet. 46:133–137.

57. Rapin I., Suzuki K., Valsamis M.P. (1976). Adult (chronic) GM2-gangliosidosis-atypical spinocerebellar degeneration in a Jewish sibship. Arch. Neurol. 33:120–130.
58. Johnson W.G., Choutorian A., Miranda A. (1977). A new juvenile hexosaminidase deficiency disease presenting as cerebellar ataxia: Clinical and biochemical studies. Neurology 27:1012–1018.
59. Oonk J.G.W., van der Helm H.J., Martin J.J. (1979). Spinocerebellar degeneration: Hexosaminidase A and B deficiency in two adult sisters. Neurology 29:380–383.
60. Willner J.P., Grabowski G.A., Gordon R.E., et al. (1981). Chronic GM2 gangliosidosis masquerading as atypical Friedreich ataxia: Clinical, morphologic and biochemical studies of nine cases. Neurology, pp. 787–798.
61. Holmes G. (1907). A form of familial degeneration of the cerebellum. Brain 30:466–489.
62. Eadie M.J. (1975). Cerebelo-olivary atrophy (Holmes type). In: P.J. Vinken, G.W. Bruyn (eds): Handbook of Clinical Neurology, Vol. 21. Amsterdam: North Holland Publishing, pp. 403–414.
63. Marie P., Foix C., Alajouanine T. (1922). De l'atrophie cerebelleuse tardive a predominance corticale. Rev. Neurol. 2:849–885.
64. Mancall E.L. (1975). Late (acquired) cortical cerebellar atrophy. In: P.J. Vinken, G.W. Bruyn (eds) Handbook of Clinical Neurology, Vol. 21. Amsterdam: North Holland Publishing, pp. 477–508.
65. Menzel P. (1891). Beitrag zur Kenntniss der hereditaren ataxie und klein-hirnatrophie. Arcv. Psychiatrie Nervenkr. 22:160–190.
66. Thomas D.J. (1900). L'atrophie olivo-ponto-cerebelleuse. Nouvelle Iconographie de la Salpetriere. 13:330–376.
67. Konigsmark B.W., Weiner L.P. (1970). The olivopontocerebellar atrophies: A review. Medicine 49:227–241.
68. Schut J.W. (1950). Hereditary ataxia: Clinical study through six generations. Arch. Neurol. Psychiatry 63:535–568.
69. Schut J.W., Haymaker W. (1951). Hereditary ataxia: A pathologic study of five cases of common ancestry. J. Neuropathol. Clin. Neurol. 1:183–213.
70. Gray R.C., Oliver C.P. (1941). Marie's hereditary cerebellar ataxia (olivopontocerebellar atrophy). Minneso. Med. 24:327–335.
71. Kark R.A.P., Rosenberg R.N., Schut L.J. (1978). The inherited ataxias. Biochemical, viral and pathological studies. Adv. Neurol. 21:107–112.
72. Bennett R.H., Ludvigson P., DeLeon G., et al. (1984). Large-fiber sensory neuropathy in autosomal dominant spinocerebellar degeneration. Arch. Neurol. 41:175–178.
73. Carenini L., Finocchiaro G., DiDonato S., et al. (1984). Electromyography and nerve conduction study in autosomal dominant olivopontocerebellar atrophy. J. Neurol. 231:34–37.
74. Landis D.M., Rosenberg R.N., Landis S.C., et al. (1974). Olivopontocerebellar degeneration. Clinical and ultrastructural abnormalities. Arch. Neurol. 31:295–307.
75. Haines J.L., Schut L.J., Weitkamp L.R., et al. (1984). Spinocerebellar ataxia in a large kindred: Age at onset, reproduction, and genetic linkage studies. Neurology 34:1542–1548.
76. Currier R.D., Glover G., Jackson T.F., et al. (1984). Spinocerebellar ataxia: Study of a large kindered: I. General information and genetics. Neurology 34:1542–1548.
77. Jackson J.F., Currier R.D., Terasaki P.I., et al. (1977). Spinocerebellar ataxia and HLA linkage. N. Engl. J. Med. 296:1138–1141.
78. Nino H.E., Noreen H.T., Dufey D.P., et al. (1980). A family with hereditary ataxia: HLA typing. Neurology 30:12–20.
79. Zoghbi H.Y., Sandkuyl L.A., Ott J., et al. (1989). Assignment of autosomal dominant spinocerebellar ataxia (SCA) centromeric to the HLA region on the short arm of chromosome 6, using multifocus linkage analysis. Am. J. Hum. Genet. 44:255–263.
80. Mass O., Scherer H.J. (1933). Zur klinik und anatomie einiger seltener klinhivner-krankungen. Z. Ges. Neurol. Psychiat. 145:420–444.
81. Garcin R., Man H.X. (1959). Sur la lenteru particuliere des mouvements conjugues des yeun observee frequement das degenerations cerebelleuses et spinocerebelleuses: La "viscoides movements voluntaire." Rev. Neurol. 98:672–673.
82. Wadia N.H., Swami R.K. (1971). A new form of heredofamilial spinocerebellar degeneration with slow eye movements (nine families). Brain 94:359–374.

83. Sears E.S., Hammerberg E.K., Norenberg M.D., et al. (1975). Supranuclear ophthalmoplegia and dementia in olivopontocerebellar atrophy: A clinocopathologic study. Neurology. 25:395.
84. Koeppen A.H., Hans M.B. (1976). Supranuclear ophthalmoplegia in olivopontocerebellar degeneration. Neurology 26:764–768.
85. Zee D.S., Optican L.H., Cook J.D., et al. (1976). Slow saccades in spoinocerebellar degeneration. Arch. Neurol. 33:243–251.
86. Murphy M.J., Goldblatt D. (1977). Slow eye movements with absent saccades in a patients with hereditary ataxia. Arch. Neurol. 34:191–195.
87. Plaitakis A., Huang Y.P., Rudolph S. (1983). Clinical, electrophysiological and CT findings in dominant olivopontocerebellar atrophy with slow saccades. Neurology 33 (Suppl. 2):218.
88. Niakam E., Bertonini T.E., Lemmi H., et al. (1984). Spinocerebellar degeneration and slow saccades in three generations of a kinship: Clinical and electrophysiologic findings. Arq. Neuropsiquiatr. 42:232–241.
89. Wadia N.H. (1977). Heredo-familial spinocerebellar degeneration with slow eye movements: Another variety of olivopontocerebellar degeneration. Neurology (India) 25:147–160.
90. Carpender S., Schumacher G.A. (1966). Familial infantile cerebellar atrophy associated with retinal degeneration. Arch. Neurol. 14:82–94.
91. Jampel R.S., Okazaki H., Bernstein H. (1961). Ophthalmoplegia and retinal degeneration associated with spinocerebellar ataxia. Arch. Ophthalmol. 66:247–259.
92. Weiner L.P., Konigsmark B.W., Stoll J. Jr., Magladery J.W. (1967). Hereditary olivopontocerebellar atrophy with retinal degeneration. Arch. Neurol. 16:364–376.
93. Colan R.V., Snead O.C., Ceballos R. (1981). Olivopontocerebellar atrophy in children: A report of seven cases in two families. Ann. Neurol. 10:355–363.
94. Hussain M.M., Zannis V.I., Plaitakis A. (1989). Characterization of glutamate dehydrogenase isoproteins purified from the cerebellum of normal subjects and patients with degenerative neurological disorders, and from human neoplastic cell lines. J. Biol. Chem. 264:20730–20735.
95. Traboulsi E.I., Maumenee I.H., Green W.R., et al. (1988). Olivopontocerebellar atrophy with retinal degeneration. A clinical and ocular histopathologic study. Arch. Ophthalmol. 106:801–806.
96. Koeppen A.H., Barron K.D. (1984). The neuropathology of olivopontocerebellar atrophy. Adv. Neurol. 41:13–38.
97. Plaitakis A., Nicklas W.J., Desnick R.J. (1980). Glutamate dehydrogenase deficiency in three patients with spinocerebellar syndrome. Ann. Neurol. 7:297–303.
98. Plaitakis A., Berl S., Yahr M.D. (1984). Neurological disorders associated with deficiency of glutamate dehydrogenase. Ann. Neurol. 15:144–153.
99. Shy G.M., Drager G.A. (1960). A neurological syndrome associated with orthostatic hypotension. Arch. Neurol. 2:511–527.
100. Johnson R.H., Lee G. de j, Oppenheimer D.R., Spalding J.M.K. (1966). Autonomic failure with orthostatic hypotension due to intermediolateral column degeneration. Q. J. Med. 35:276–292.
101. Oppenheimer D.R. (1980). Lateral horn cells in progressive autonomic failure. J. Neurol. Sci. 46:393–404.
102. Oppenheimer D.R. (1988). Neuropathology and neurochemistry of autonomic failure. In: R. Banniste (ed): Autonomic Failure. A Textbook of Clinical Disorders of the Autonomic Nervous System. Oxford: Oxford Medical Publishers, pp. 451–464.
103. Edie M.J. (1975). Olivopontocerebellar atrophy (Dejerine-Thomas type). In: P.J. Vinken, G.W. Bruyn (eds): Handbook of Clinical Neurology. Vol. 21. Amsterdam: North Holland Publishing pp. 415–431.
104. Plaitakis A. (1990). Glutamate dysfunction and selective motor neuron degeneration in amyotrophic lateral sclerosis: A hypothesis. Ann. Neurol. 28:3–8.
105. Plaitakis, A., Berl S., Schut L.J., Yahr M.D. (1982). Abnormal aspartate/malate metabolism in dominantly inherited olivopontocerebellar degeneration. Ann. Neurol. 12:79.
106. Tsiotos P., Plaitakis A., Mitsakos A., et al. (1989). L-glutamate binding sites of normal and

atrophic human cerebellum. Brain Res. 481:87–96.
107. Hatziefthimiou A., Mitsakos A., Mitsaki E., et al. (1990). Quantitative autoradiographic study of L-glutamate binding sites in normal and atrophic human cerebellum. J. Neurosci. Res., in press.
108. Albin R.L., Gilman S. (1990). Autoradiographic localization of inhibitory and excitatory amino acid neurotransmitter receptors in human normal and olivopontocerebellar atrophy cerebellum. Brain Res. 522:37–45.
109. Makowiec R.L., Albin R.L., Cha J.J., et al. (1990). Two types of quisqualate receptor are reduced in human OPCA cerebellar cortex. Brain Res. 523:309–312.
110. Kish S.J., Schut L.J., Simmons J., et al. (1988). Brain acetylcholinetransferase activity is markedly reduced in dominantly inherited olivopontocerebellar atrophy. J. Neurol. Neurosurg. Psychiatry 51:544–554.
111. Carlson M., Nakamura Y., Krapcho K., et al. (1987). Isolation and mapping of a polymorphic DNA sequence pMCT112 on chromosome 9 q (D9S15). Nucleic Acids Res. 15:10614.
112. McKusic V.A. (1990). The human gene map. In: S.J. O'Brien (ed): Genetic Maps, 5th ed. Cold Spring Harbor NY: Cold Spring Harbor Laboratory Press, pp. 5.47–5.257.
112a. Plaitakis A. (1992). Olivopontocerebellar atrophy with glutamate dehydrogenase deficiency. In: de Jong (ed): Hereditary Neuropathies and Spinocerebellar Atrophies, Handbook of Clinical Neurology, Amsterdam: Elsevier Science Publishers, 60:551–568.

III. ETIOPATHOGENESIS OF CEREBELLAR DISORDERS

14. GLUTAMATE DEHYDROGENASE DEFICIENCY IN CEREBELLAR DEGENERATIONS

ANDREAS PLAITAKIS AND P. SHASHIDHARAN

1. INTRODUCTION

Glutamate in mammalian tissues is known to play a key role in intermediary metabolism, linked directly with the Kreb's cycle via glutamate dehydrogenase (GDH). The amino acid is of particular importance for ammonia metabolism and the synthesis of many compounds, including the major putative inhibitory transmitter GABA [1]. In the central nervous system (CNS), glutamate is present in concentrations that are 2–4 times greater than that found in other organs. The need for the CNS to maintain such high glutamate levels (highest of any other amino acid) is not fully understood, but it may be related to the many and complex functions the amino acid is thought to serve in the nerve tissue [2]. Of these, the putative neurotransmitter role [1] has attracted much attention lately due to rapidly accumulating evidence that it may be involved in the pathogenesis of neurodegenerative disorders [3,4].

Lucas and Newhouse [5] were the first to observe that the systemic administration of glutamate to infant animals resulted in toxic effects on the retina ganglion cells. Olney et al. [6] subsequently showed that glutamate and other acidic amino acids, when administered to infant animals, can also produce necrosis of certain CNS structures, such as the hypothalamus, which lack a mature blood-brain barrier. Microscopically, the degenerative changes

A. Plaitakis (ed.), CEREBELLAR DEGENERATIONS: CLINICAL NEUROBIOLOGY. Copyright © 1992.
Kluwer Academic Publishers, Boston. All rights reserved.

involved primarily postsynaptic elements, such as dendrites and cell bodies of neurons (dentro-somatic degeneration) [6]. The mechanism of this neurotoxicity was thought to be directly related to the ability of these compounds to produce neuronal excitation ("neuroexcitotoxic hypothesis"). Moreover, the intracerebral injection of glutamate, and particularly its potent analogs kainic, ibotenic, and quinolinic acids, can similarly result in neuronal degeneration, with certain neuronal systems, being especially sensitive to these neurotoxic effects [7,8]. Because the histopathological and biochemical alterations produced by these neuroexcitotoxic compounds were similar to those occurring in certain human neurodegenerative disorders [7,8], it was considered that potent neuroexcitotoxic compounds may accumulate in the brain of patients with resultant neurodegeneration. However, to this date this has not been substantiated. Instead, evidence has emerged in recent years, indicating that defective metabolism or transport of *endogenous* glutamate underlies certain degenerative diseases. Of these, olivopontocerebellar atrophy (OPCA) was the first disorder with primary system degeneration to be associated with a systemic defect in glutamate metabolism [3], resulting from deficiency of the enzyme glutamate dehydrogenase (GDH) [9,10].

2. ANIMAL MODELS FOR OPCA

Clues to the presence of GDH abnormalities in human degenerative disorders were provided by the mode of action of the selective neurotoxin 3-acetylpyridine (3AP), an analog of nicotinamide. The systemic administration of the toxic agent to rats has resulted in selective lesioning of the inferior olives, olivocerebellar fibers, cranial nerve nuclei, and areas of the pons and nigra [11], thus providing an experimental model for olivopontocerebellar atrophy (OPCA) [9,10]. 3AP, acting as a nicotinamide antagonist, is incorporated into the nicotinamide-adenine dinucleotide phosphate (NADP) system, forming substantial amounts of 3APADP in the brain of experimental animals [12]. Inhibition of NADP-dependent oxidoreductases by 3APADP is thought to mediate the neurotoxic effects of 3AP. Neuronal systems that degenerate both in 3AP toxicity and in OPCA may be selectively sensitive to malfunction of one of these oxidoreductases [9,10]. These considerations led us to search for abnormalities of 3AP-inhibited enzymes in patients with OPCA and to find that the activity of glutamate dehydrogenase was selectively decreased in such patients.

3. DETECTION OF GDH DEFICIENCY

Our initial studies [9,10] were carried out in three patients. The first was a 19-year-old male affected since age 11 by a progressive neurological disorder characterized by juvenile Parkinsonism, bulbar palsy, cerebellar ataxia, amyotrophy, and peripheral neuropahty. Both parents of this patient were neurologically normal, and no other family member was known to be affected by a similar disorder.

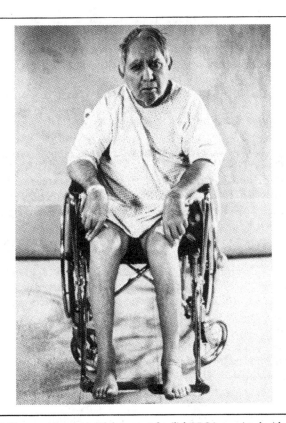

Figure 14-1. A 71-year-old patient with late-onset familial OPCA associated with reduced GDH activity in leukocytes (case 3 in Ref. 10). He has been affected since age 50 by a progressive multisystemic neurological disorder, characterized by spastic paraparesis, Parkinsonism, dysarthria, oculomotor disturbances, ataxia, decreased visual acuity, anterior horn-cell signs, peripheral neuropathy, and a mild dementia. This illustration shows muscle wasting in the lower extremities, Parkinsonian features, and dystonic posturing of the right fingers. A 65-year-old brother has been similarly affected since age 49. The mode of genetic transmission in this family was thought to be either autosomal recessive or X-linked [10]. This figure is from [15] with permission.

The other two patients were siblings. The oldest patient was a 71-year-old male who developed diplopia and spastic paraparesis at around age 50 and thereafter developed a variety of additional neurologic deficits, such as Parkinsonism, dysarthria, dysphagia, cerebellar ataxia, amyotrophy, fasciculations, sensory changes, and a mild dementia (Figure 14-1). His younger brother, a 65-year-old man, also became involved at around age 49 with progressive cerebellar ataxia, head tremor, dysarthria, downbeat nystagmus, anterior horn cell signs, sensory deficits, and behavioral changes. The parents of these patients were reportedly not affected, and no other family members were known to have been afflicted with a similar disorder.

These genetic data were thought to suggest an autosomal recessive or X-linked inheritance [10].

Four NADP-dependent oxidoreductases were estimated in cultured skin fibroblasts obtained from the juvenile patient according to the above-described rationale. Results revealed that GDH activity was selectively reduced (to 22% and 30% of control), while other oxidoreductases, such as glucose-6-phosphate dehydrogenase, isocitrate dehydrogenase, and glutathione reductase, were not significantly altered. [10].

4. GDH ACTIVITY IN VARIOUS TYPES OF OPCA AND OTHER MULTISYSTEM ATROPHIES

Following these initial results, we undertook more extensive investigations, using primarily leukocytes isolated from peripheral blood [13-15]. Eighty-eight patients with various types of degenerative neurological disorders affecting primarily the cerebellum and/or the basal ganglia were studied. Twelve patients with slowly progressive multiple-system atrophic disorders were found to have a partial deficiency of this enzyme (54% of control level) [14]. The age range of the GDH-deficient patients was 43-73 years (mean age 64 years). Although some of these patients showed a combination of cerebellar, bulbar, oculomotor, and extrapyramidal deficits, suggesting, as in the above three patients, the diagnosis of OPCA, others were atypical. Thus, GDH-deficient cases were encountered with predominantly extrapyramidal manifestations (atypical Parkinson's disease), bulbar palsy, motor neuronal degeneration (Figure 14-2), or cerebellar ataxia with peripheral neuropathy. Progressive autonomic failure with orthostatic hypotension was not generally encountered in the GDH-deficient cases. Five of these cases were sporadic, and seven were familial with siblings but not parents or offspring found to be neurologically affected, thus raising the possibility of recessive inheritance [14]. Also, the possibility of dominant transmission with incomplete penetrance has been considered [15].

However, patients affected by dominant OPCA with complete penetrance, particularly the form of the disease that has been linked to the HLA locus on the sixth chromosome, such as the Schut-Haymaker kindred [16,17], had normal GDH activity in their leukocytes. The same was true for patients with cerebello-olivary atrophy and Friedreich's ataxia (FA), although the mean activity of the latter group was lower than that found in controls (80-90% of control). Also, small but significant reductions in leukocytic GDH activity (85-90% of control; $p < 0.05$) were found in a group of patients with atypical Parkinson's disease (PD) and progressive supranuclear palsy [18].

4.1. Evidence for selective deficiency of GDH isoforms in OPCA

In mammalian tissues GDH was thought originally to be localized in the mitochondrial matrix and to be easily released in a soluble form by tissue

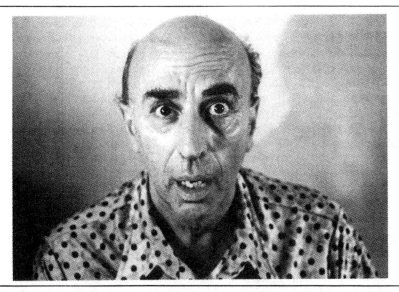

Figure 14-2. A 54-year-old patient with GDH-deficient neurological disorder with predominant motor neuron symptomatology (case 4 in Ref. 14). He developed a "staring" facial appearance at around age 50, and thereafter progressive bulbar dysfunction and leg weakness. There was decreased facial expression with staring, marked dysarthria, dysphagia, oculomotor disturbances, slight ataxia, muscle weakness, and wasting and corticospinal deficits. The patient died at home about 7 months after this photograph was taken due to a rapidly evolving bulbar palsy. Autopsy was not obtained. One 60-year-old paternal cousin has been affected by a slowly progressive multisystem disorder, characterized by spastic paraparesis, dysarthria, slight ataxia, and extrapyramidal features. GDH activity was reduced in leukocytes and glutamate loading was abnormal in both patients.

homogenization. As such, most workers in the field have not used detergents for the estimation of GDH in tissue homogenates. Our earlier studies with leukocyte and fibroblast homogenates were accordingly carried out without the use of detergents. Additional studies, however, revealed that non-ionic detergents, such as Lubrol or Triton X-100 (0.025–0.05%), when used together with the enzyme activators ADP (1 mM) or L-leucine (1–10 mM), produce a marked potentiation of GDH activity (increase by 300–450%), thus raising the possibility that a major fraction of GDH activity in human tissues is bound to membranes.

Additional fractionation studies showed that most GDH present in tissue homogenates sediments with the membrane fraction (100,000 xg for 1 hr) [14]. Studies on leukocyte homogenates isolated from patients and controls revealed that the decrease in GDH activity in the patients was limited to the membrane-bound component of the enzyme [14,15,19], thus raising the possibility that the two GDH activities may be under different genetic controls. Further studies on partially purified leukocytic preparations from control subjects showed that the two enzyme activities differed in their relative resistance to heat denaturation, with the soluble enzyme being thermostable and the particulate GDH thermolabile [14]. Incubation of the two enzyme fractions at 47.5°C in 50 mM Tris HCL, pH 7.40 buffer, showed that the half-life of the soluble enzyme was 10 hr, while that of the membrane-bound GDH was 24 min. Increasing the buffer pH to 8.00 and 8.50 increased the rate of heat inactivation, with the half-life of the soluble enzyme becoming 4 hr and 1 hr, respectively, and that of the particulate GDH 13 min and 4 min, respectively [20].

In view of the above findings, heat-stable and heat-labile forms of GDH were measured in whole homogenates of leukocytes from patients and controls by incubation at 47.5°C in 50 mM Tris HCL, pH 7.40, for 60 min or in 50 mM Tris HCL, pH 8.00, for 30 min. Under these conditions most of the membrane-bound and very little of the soluble GDH become inactivated, thus allowing the differentiation of the two enzyme forms. The study of neurologic patients using these methods has suggested the presence of distinct enzyme abnormalities associated with the following nosologic entities: (1) In some patients with late-onset sporadic or OPCA associated with partial GDH deficiency (total activity reduced to 50–75% of control), there was a selective reduction in the "particulate" and "heat-labile" GDH activities [14,15,21]. GDH activity was also partially deficient (60% of control) in cultured skin fibroblasts from such patients, with the detergent (Triton X-100)-activated fraction of the enzyme showing greater reductions (25% of control) [19]. Hence, these data indicate that the GDH defect in these patients involves a membrane-bound component of the enzyme.

(2) In atypical Parkinsonism and supranuclear palsy, disorders associated with small reductions of total GDH activity (85–90% of control), no alterations were found in either the "particulate" or "soluble" GDH components, although the "heat-labile" enzyme activity was significantly decreased (30–40% of control) [18]. As such, these results suggest an alteration in the thermal stability characteristics of GDH, as tested in crude leukocyte homogenates of these patients [18].

(3) In a family afflicted by a variant of dominant OPCA that is characterized by retinal degeneration, slowed saccades, and other features (see Chapter 13), a selective reduction in the "soluble-thermostable" GDH component has been found without changes in the "particulate-thermolabile" fraction. Studies in brain tissue obtained at autopsy from one patient of this family revealed an almost complete deficiency of GDH isoprotein 1 (see below).

5. STUDIES BY OTHER INVESTIGATORS
Since the initial description of partial GDH deficiency in OPCA patients, studies from 10 different laboratories have been reported in the literature.

Table 14-1.

Reports	Clinical types: N = Number of patients with low GDH	Tissue GDH activity (% of control)
Yamaguchi et al. [23], 1988	Late-onset sporadic OPCA: N = 5; cortical cerebellar atrophy: N = 2	Platelets, total GDH (71-79%)
Duvoisin et al. [24], 1983	Late-onset sporadic OPCA or atypical PD: N = 9	Leukocytes, total GDH (<50%)
Duvoisin et al. [26], 1988	Multiple system atrophy-heterogenous forms: N = 20	Leukocytes, total GDH (80%)
Sorbi et al. [27], 1986	Nondominant OPCA: N = 4 Dominant OPCA: N = 2	Platelets, total GDH (<50%) Only ADP/Triton activated GDH (<50%)
Sorbi et al. [28], 1989	Nondominant OPCA: N = 8 Dominant OPCA: N = 6	Platelets, total GDH (50%); other mitochondrial enzymes (80-90%) Platelets, total GDH (50%); other mitochondrial enzymes normal
Finochiaro et al. [29], 1986	Dominant OPCA: N = 13 Sporadic OPCA: N = 2	Leukocytes, fibroblasts, muscle mitochondria (46-68%)
Konagaya et al. [21], 1986	Late-onset sporadic OPCA: N = 12	Leukocytes, total GDH (78%) heat labile (27%)
Aubby et al. [30], 1988	Ataxic disorders: N = 5; MSA: N = 3; Juvenile PD: N = 1; Dystonia Parkinsonism: N = 1; Parkinson's disease: N = 4	Leukocytes, total GDH (40%); heat stable (36%); heat labile (56%)
Orsi et al. [32], 1988	Ataxic disorders with additional features: N = 8	Leukocytes, total GDH (68%)
Kajiyama et al. [33], 1988	Cerebellar disorders: N = 8; juvenile PD: N = 1	Leukocytes, total GDH (57%) heat labile (43%); decreased GDH protein by radioimmunoassay
Tatsumi et al. [34], 1989	Spinocerebellar degeneraration or atypical Parkinsonism: N = 21	Leukocyte, skin fibroblast; total GDH (46-62%); fibroblasts; increased sensitivity to L-glutamate
Kostic et al. [31], 1989	Sporadic OPCA: N = 7	Platelets: total GDH (<50%)
Iwattsuji et al. [35], 1989	OPCA form of spinocerebebellar degeneration: N = 24	Lymphocytes: total GDH (78%); heat labile (71%); heat stable (80%)
	Parkinson's disease: N = 17	Total GDH (69%); heat stable (65%)
	Alzheimer's disease: N = 17	Total GDH (74%); heat stable (60%)
	Motor neuron disease: N = 37	Total GDH (85%); In controls, total and heat-stable GDH increased with aging

Table 14-1 summarizes the results of these investigation. Decreased GDH activities were found in leukocytes, platelets, lymphocytes, cultured skin fibroblasts, and muscle mitochondria. In accord with our data described above, decreased GDH activity has been correlated primarily with late-onset sporadic OPCA. In addition, some investigators [27-29], but not others [24,30,31], found GDH abnormalities in dominantly inherited forms of OPCA. The heat-labile GDH activity showed greater decreases than the total activity [21,33]; however, other investigators have reported decreases in both heat-labile and heat-stable GDH [26,30]. Aubby et al. [30], in addition, found GDH deficiency in 4 patients with Parkinson's disease.

Small but significant decreases in GDH activity were also found in patients with juvenile Parkinsonism, atypical Parkinsonism, and progressive supranuclear palsy [30,33–35]. Iwatsuji et al. [35] recently showed that in lymphocytes from controls total and heat-stable GDH activity increased with aging but not in lymphocytes from OPCA patients. Patients with Parkinson's and Alzheimer's disease showed small but significant decreases in total and heat stable GDH whereas heat-labile GDH decreased only in OPCA patients. Kajiyama et al. [33] reported decreased GDH protein as determined by radioimmunoassay. The same group [34] also reported that cultured skin fibroblasts from such patients showed an increased sensitivity to L-glutamate toxicity. These results may have importance in the pathogenesis of the disorder (see below).

6. PATHOLOGY

Brain pathological studies have been performed on four patients with reduced leukocytic and fibroblast GDH who came to autopsy [22,25,37,38]. The first patient was a 47-year-old mentally ill man affected by a progressive neurological disorder, characterized by ataxia, amyotrophy, corticospinal deficits, bulbar palsy, slowed eye movements, and peripheral neuropathy. Pathological examination of the brain [36] revealed a marked involvement of the Purkinje and granule cell layers of the cerebellar cortex, inferior olives, and the anterior horns and posterior columns of the spinal cord. Neurons in the pons and locus coeruleus contained lipofuscin, as did neurons in the medulla, midbrain, thalamus, and hypothalamus. Intraneuronal lipofuscin was also present in the cerebral cortex, including the hippocampal formation, which, in addition, showed a moderate number of neurofibrillary tangles.

The second patient was a 72-year-old man who was affected since age 63 by slowly progressive ataxia, dysarthria, Parkinsonism, and amyotrophy [37]. Increasing gait and limb ataxia, dysarthria, dysphagia, oculomotor disturbances, rigidity, hand and foot deformities, and muscle wasting and fasciculations were prominent clinical features at the end stages of this patient's illness. No other families were known to be affected by a similar disorder (sporadic case), GDH activity was partially deficient in leukocytes of this patient, with the "particulate"-"thermolabile" fraction showing greater decreases. Oral loading with monosodium glutamate showed abnormal glutamate clearance [14]. Examination of the brain showed that the cerebellum and the pons were atrophic, the third ventricle was enlarged, and the caudal putamen was shrunken. The cerebellar folia and white matter were atrophic, as were the middle cerebellar peduncles, basis pontis, and inferior olives. Microscopically, the Purkinje cells were markedly depleted in the cerebellar cortex with less granule cell loss. Neuronal loss and gliosis were present in the inferior olives, pontine nuclei, substantia nigra, putamen, and anterior horns. A marked fiber loss was present in the brachium pontis, transverse pontine fibers, and inferior cerebellar peduncle.

The third patient died of aspiration pneumonia at 23 years of age after a 12-year progressive neurological disorder, characterized by juvenile Parkinsonism, oculomotor disturbances, bulbar palsy, cerebellar ataxia, amyotrophy, and peripheral neuropathy. He had been previously studied by us [10, case 1] Pathologically [38] there was barreling of the third ventricle, mild enlargement of the lateral ventricles, marked atrophy of the cerebellum, and depigmentation of the nigra and locus coeruleus. Severe neuronal loss was observed in the Purkinje cell layer of the cerebellar cortex, anterior horns, and posterior columns of the spinal cord, inferior olives, hypoglossal nuclei, substantia nigra, basal nuclei, locus ceruleus, hippocampus, and the peripheral nerves. Eosinophilic intranuclear inclusion bodies of the Cowdry type A and B were present in neurons and glial cells throughout the CNS, as well as in the ganglion cells of the bowel, bladder, and esophagus.

The fourth patient was a 15-year-old black youth who developed blindness at around 7 years of age and thereafter a progressive cerebellar disorder associated with mild dementia, dysarthria, slowed eye movements, macular degeneration, facial hypomimia, corticospinal deficits, myoclonus, amyotrophy, and peripheral neuropathy. He gradually became severely incapacitated. He was followed neurologically at another hospital, where he died of bronchopneumonia complicated by massive upper gastrointestinal hemorrhage. Autopsy and the pathological studies were performed at that hospital.

Pathologically, the brain was atrophic, including the cerebellum and brain stem. The substantia nigra was depigmented. A moderate to marked neuronal loss and gliosis were present in the molecular (Purkinje cell) and granule cell layers of the cerebellar cortex, whereas the dentate nucleus showed fewer changes. Moderate to marked atrophic changes were also present in the anterior horns, inferior olives, cranial nerve nuclei (hypoglossal and oculomotor), and the tegmentum of the medulla, pons, and midbrain. Neuronal loss was also observed in the substantia nigra, locus coeruleus, globus pallidus, anterior putamen, and ventromedial nucleus of the thalamus.

A 19-year-old brother and his 46-year-old father have shown similar neurologic manifestations since age 6 and 27, respectively. HLA typing in members of this family failed to show a linkage of the gene locus to the HLA complex on the sixth chromosome. GDH activities were measured in leukocyte preparations of the last two patients, and the results revealed a partial deficiency of the "soluble," heat-stable component. Thus, the disorder in this family is distinct from other forms of dominant OPCA, both clinically and biochemically. As described in Chapters 8 and 13, this disease constitutes a separate variant of dominant OPCA that is characterized by retinal degeneration, slowed saccades, and other features.

Although the neuropathologic changes observed in these four patients were somewhat variable, atrophy of cerebellar cortex, inferior olives, and pons were common features. In addition, most patients showed degeneration of the anterior horns, hypoglossal nuclei, substantia nigra, locus ceruleus, putamen, and hippocampus, and enlargement of the third ventricle. In some patients, particularly in those with slowed saccadic eye movements, there was atrophy of the tegmentum of the pons, and these findings agree the results of previous [40] and present neuroimaging studies (see Chapter 12). As such, these pathologic alterations confirm indications from previous clinical studies [14] that GDH deficiency is associated with multisystem atrophic disorders.

The significance of the intranuclear inclusions, seen in the patient with the juvenile Parkinsonism [38], remains enigmatic. Similar inclusions have been previously described in patients with multisystem atrophic disorders, which often present as juvenile Parkinsonism and which may appear sporadically or be genetically transmitted [43,44]. As described above, GDH deficiency has been described by several authors in patients with similar clinical and genetic characteristics. Additional studies are, therefore, needed to establish whether this disorder is causally related to GDH deficiency. It is of interest, in this regard, to note that these inclusion bodies have been shown to share antigenic determinants with normal neurofilaments [44] and that abnormal neurofilaments can be induced by glutamate toxicity [45]. Similar considerations also apply to the neurofibrillary tangles observed in two of the above patients.

6.1. Biochemical studies in brain tissue

6.1.1. Studies in crude homogenates

GDH activity in rat and human brain has been shown to exist in distinct "soluble" and "particulate" forms [46], which may have a different subcellular localization [47]. For estimation of the "particulate" enzyme, the use of a detergent (Triton X-100 or Lubrol), together with an allosteric activator (ADP or L-leucine), is required during the assay, as described above. With these considerations in mind, we determined GDH activity (in the presence or absence of Triton X-100) in whole-tissue homogenates, as well as in "soluble" and "particulate" fractions of the cerebellum obtained at autopsy from two OPCA patients described above (second and fourth patients).

In the patient with late-onset sporadic OPCA [37] (second patient described above), Triton X-100-activated GDH activity was markedly reduced (to about 10% of controls) with the "particulate, but not the "soluble" brain GDH being selectively decreased (to 25% of control). In contrast, in the patient with the childhood-onset variant of dominant OPCA, which is characterized by retinal degeneration (fourth patient described above), there was a decrease in the "soluble" cerebellar fraction without a reduction in the "particulate" or the Triton X-100-activated GDH activity, and these results agree with those obtained in leukocytes from the affected brother and father of this patient.

6.1.2. GDH purification

Characterization of human brain GDH has been facilitated by the development of a highly efficient purification method that has permitted the isolation of substantial amounts of the human enzyme in a homogeneous form. Glutamate dehydrogenase has been previously purified from various animal and plant sources using different methods, the yield of which has ranged from 5 to 25% of the original amount of enzyme activity [46,48]. We have recently optimized GDH purification by a careful application of specific fractionation steps tailored to the properties of the human enzyme and were able to achieve an overall recovery of more than 50% of the original GDH activity [22]. The enzyme is extracted from homogenized brain tissue by Triton X-100 (1%) and high concentrations of salt (500 mM NaCl), and the obtained extract is subjected to fractionation by ammonium sulfate. The human brain enzyme precipitates at 30–65% of ammonium sulfate [22]. Hydrophobic interaction chromatography is then applied with the use of Phenyl-Sepharose 4B (Pharmacia) and a double-gradient ethylene glycol-ammonium sulfate elution system. GDH is highly stable under these conditions and, due to its rather high hydrophobicity, is eluted as a rather narrow peak at high ethylene glycol concentrations. We have previously shown [20] that ethylene glycol, L-leucine, ADP, and the salts NaCl and KCl exert protective effects of the denaturation of human GDH. The last step involves affinity chromatography with the use of a GTP-sepharose column. GDH strongly binds to GTP linked to sepharose resin and is efficiently eluted in pure form with a KCl gradient.

6.1.3. Characterization of four isoforms of GDH purified from human brain

Purification of human brain GDH made possible the separation of different isoproteins of the enzyme by electrophoresis [22]. Because human GDH has a rather high isoelectric point (over 6.2), the electrophoretic separation of GDH isoforms is best achieved with the use of a nonequilibrium pH gradient (pH 3.5–10.0) two-dimensional polyacrylamide gel electrophoresis technique [22]. Electrophoretic analysis of GDH, purified from the cerebellum of non-neurologic controls, revealed that the human enzyme consisted of four major isoproteins (designated GDH 1, 2, 3, and 4), which differed in their charge and size (Figure 14-3). The relative abundance and the molecular mass of these isoproteins followed the order $1 > 2 > 3 > 4$. The GDH isoforms were not affected by treatment with neuraminidase and alkaline phosphatase, thus indicating that they did not derive from post-translational sialation or phosphorylation.

Using the above techniques, GDH was purified to homogeneity from whole cerebellar tissue of two of the above OPCA patients (cases 2 and 4) who came to autopsy and controls, and was analyzed by the high-resolution electrophoretic technique indicated above [22]. These studies showed that the cerebellar enzyme of the patient with the childhood-onset variant of OPCA associated with retinal degeneration was distinctly different from that of

380 III. Etiopathogenesis of cerebellar disorders

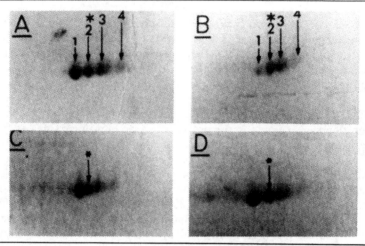

Figure 14-3. Glutamate dehydrogenase isoproteins purified from cerebellar autopsy tissue and analysed by two-dimensional nonequilibrium pH gradient electrophoresis, as described by Hussain [22]. **A**: Neurologically normal subject who died of an acute myocardial infarction. **B**: Patient with childhood-onset variant of dominant OPCA, characterized by macular degeneration, slowed saccades, myoclonus, amyotrophy, and peripheral neuropathy (see text for details). **C**: Patients from the Schut-Haymaker kindred (non-GDH-deficient). **D**: Patient with ALS. Asterisk indicates ^{35}S-labeled GDH synthesized by Hep G2 cells. The labeled enzyme was immunoprecipitated from cell lysates and mixed (in trace amounts) with the purified human brain GDH. Its position, in relation to the four brain GDH isoproteins (comigration with the GDH isoprotein 2) was revealed by autoradiography of the slab gel. As seen here, GDH isoprotein 1, which in normal and disease controls is more abundant than other isoproteins, was markedly reduced in the patient with the childhood-onset variant of dominant OPCA. The data are from Ref. 22.

neurologically normal controls, as well as patients with Parkinson's disease, ALS, and the Schut-Haymaker OPCA (non-GDH deficient) (Figure 14-3). Thus, GDH isoprotein 1, which in non-neurologic and neurologic controls was found to be more abundant than the other GDH isoproteins, was markedly reduced in this patient (Figure 14-3). GDH was also purified from the "soluble" brain fraction of this patient and controls, and its kinetic properties were studied. Results revealed that the Km for the enzyme's substrates—α-ketoglutarate, glutamate, NADH, and NADPH—were significantly increased (by three- to fivefold), thus indicating that the altered electrophoretic pattern of this patient's enzyme is associated with abnormal catalytic properties and function [22]. Electrophoretic analysis of GDH, purified form the cerebellum of the patient with the late-adult onset sporadic OPCA, showed a relative enrichment in GDH isoprotein 1, an electrophoretic pattern that is distinct from that of the above childhood case [49]. The kinetic properties of this patient's brain GDHs were not significantly altered as compared to controls.

Glutamate levels in cerebellar cortical tissue were decreased both in the childhood-onset dominant OPCA (to 25% of control) as well as in the late-onset sporadic disease (to 40% of control). The levels of other amino acids were not significantly altered, except for the level of aspartate, which was decreased to 40% of control in the dominant case [22].

7. PATHOGENESIS

Because GDH deficiency occurs in heterogenous forms of OPCA and other multi-system atrophies and because the enzymatic defect is partial, it is presently unclear whether these biochemical abnormalities result from primary genetic mutations or reflect non-specific mitochondrial alterations (see Chapter 15). As indicated below, multiple GDH-specific mRNAs and genes have been detected in human tissues and these could be the basis for a genetic heterogeneity of GDH deficient disorders. Also, abnormalities in the metabolism of glutamate, characteristic topographic brain changes and glutamate receptor losses have been detected in such patients (see below), and these lend further support to this possibility.

7.1. Glutamate metabolism in GDH-deficient OPCA

To determine whether the above-described GDH abnormalities are associated with altered metabolism of the enzyme's substrates glutamate, α-ketoglutarate and ammonia, we measured these compounds in the fasting plasma of GDH-deficient patients and controls. Results revealed significant increases in plasma glutamate (by 137–150%) and decreases in α-ketoglutarate (by 21–22%) [3,13], thus indicating that a partial metabolic block at the oxidation of glutamate to α-ketoglutarate exists in the patients with low enzyme activity. This possibility was further tested by performing glutamate loading tests (60 mg/kg body weight taken orally after overnight fasting). These studies showed that glutamate levels increased excessively in the plasma of the patients as compared to controls, and this was associated with proportional increases in the plasma aspartate levels, thus indicating an intact transamination pathway. Similar observations have been made by other investigators in patients with dominant and sporadic OPCA associated with reduced GDH activity [29,34].

7.2. Pathophysiology of brain topographic lesions

Because GDH is an enzyme that is directly involved in the metabolism of the neuroexcitatory amino acid glutamate, detection of its deficiency in some forms of OPCA has uncovered a direct link between a genetic molecular defect and neuroexcitotoxic mechanisms of neurodegeneration. Thus, the study of GDH-deficient OPCA, a disorder characterized by classic system degeneration, has permitted us to gain insights into the mechanisms involved in the phenomenon of neuronal degeneration that is of fundamental im-

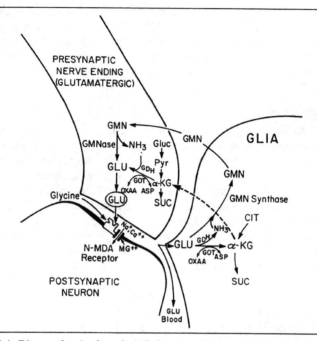

Figure 14-4. Diagram showing hypothetical glutamatergic nerve-ending synapse with hypothetical postsynaptic neuron. Glutamate (GLU) released from glutamatergic terminals during neurotransmission is thought to act on three different types of postsynaptic receptors, one of which, the N-methyl-D-asparatate (N-MDA) receptor, is shown here schematically. This receptor appears to mediate excitatory transmission by activating a Na^+ channel, which is also permeable to Ca^{2+}. The N-MDA receptor has been shown to be regulated by glycine, which acts on strychinine-insensitive allosteric site of the receptor. The action of synaptic glutamate may be terminated by uptake, primarily into the surrounding astrocytes (GLIA), where it is thought to be metabolized to α-ketoglutarate (α-KG) via glutamate dehydrogenase (GDH) or via glutamate oxaloacetate transaminase (GOT), or it is converted to glutamaine (GMN) via glutamine synthase. The glial cell, in turn, supplies the nerve terminal by glutamine and/or α-KG, which can both serve as precursors of transmitter glutamate [53]. The hypothesized malfunction of synaptic glutamate in these disorders may be due to a defect in its transport (uptake) or metabolism (by glial cells) or to its release from the nerve terminals, causing enhanced synaptic (excitatory) action. This figure is from [66] with permission.

portance in all human disorders with primary system atrophy. Moreover, these investigations have provided clues to the factors determining the topographically characteristic brain lesions, and as such, these observations appear to have implications that extend beyond OPCA.

Immunocytochemical investigations by Aoki et al. [50] (see Chapter 6) have shown that, under conditions expected to reveal the membrane-bound GDH, enzyme immunoreactivity was markedly enriched in CNS regions that receive putative glutamatergic innervation. In these regions GDH seems to be localized primarily in astrocytic processes associated with glutamatergic terminals. These astrocytes are thought to be primarily responsible for removing synaptic glutamate by an energy-dependent high-affinity uptake system [51]. Glutamate taken up by these cells seems to be catabolized mainly via GDH [52] to α-ketoglutarate, which may be transported back to the nerve terminal to serve as a precursor of transmitter glutamate [53] (Figure 14-4). As such, the enzyme may be of particular importance in glutamatergic transmission mechanisms.

A defect in GDH may thus impair the ability of the glial cells to metabolize glutamate, leading to excessive accumulation of the transmitter at the synaptic cleft and neuroexcitotoxic degeneration of postsynaptic neurons (Figure 14-4). Moreover, impaired glutamate degradation by synaptic astrocytes may disrupt the recycling of the transmitter (due to a decreased supply of precursors to the nerve terminals), with a resultant depletion of the intracellular glutamate content.

The study of brain morphological changes in living patients with the use of neuro-imaging techniques [40] (see Chapter 13), as well as examination of the brain at autopsy (see above), have indeed revealed a characteristic topography of brain lesions that can be attributed to the degeneration of CNS structures that receive putative glutamatergic innervation and that are rich in GDH activity, as indicated above. Moreover, glutamate content in nerve tissues of patients with OPCA associated with GDH abnormalities has been found to be markedly reduced [22]. Since most free glutamate in nerve tissue is intracellular, these data are consistent with a depletion of the intraneuronal pools of this amino acid, probably resulting from impaired recycling of the transmitter at the nerve endings according to above-described model.

It should, however, be indicated that the exact function of GDH in the nervous tissue has not been fully understood, with some authors suggesting that GDH in neurons may be primarily involved in the synthesis of glutamate from glucose [54], rather than in the breakdown of the amino acid. Such a hypothesis will predict that the marked decreases found in the content of glutamate in the CNS result from a metabolic block affecting the synthesis of the amino acid from glucose. This interpretation, however, cannot readily account for a neuroexcitotoxic cell loss, since it would be expected to lead to a decrease rather than an excess of synaptic glutamate. Moreover, most recent evidence points not to GDH, but rather to glutamate-oxaloacetate transaminase as being the primary enzyme involved in the synthesis of transmitter glutamate [53].

Although glutamate receptors have been thought primarily to mediate neuroexcitotoxicity [6], there are recent intriguing observations indicating that stimulation of these receptors may in fact have a neurotrophic function, promoting the normal growth and development of cerebellar cells [55,56] (see Chapter 3). As such, an alternative possibility that needs to be considered is that the decreased neuronal content of glutamate may lead to premature degeneration by affecting a neurotrophic support essential for these cells.

Further studies are certainly needed to address these exciting issues, which may have implications for understanding the human neurodegenerations in general.

8. L-GLUTAMATE RECEPTORS IN OPCA

If the above neuroexcitotoxic hypothesis for the pathogenesis of OPCA is correct, one would expect to find in this disorder a selective disappearance of glutamatergic postsynaptic neurons bearing glutamate receptors. To test this hypothesis, L-glutamate binding was measured in the cerebellar cortical tissue of four OPCA patients. Two of these were associated with a GDH deficiency, as described above, whereas the other two were members of the Schut-Haymaker dominant OPCA kindred (non-GDH deficient). Results revealed the L-glutamate binding was significantly decreased to 30% of control in all these patients without a change in the affinities and the pharmacologic properties of the binding sites [57]. Autoradiographic studies [58] further showed significant decreases in quisqualate-sensitive L-glutamate binding in the molecular layer of the cerebellum. Also, in some patients, particularly those associated with GDH abnormalities, L-glutamate binding was decreased in the granule layer as well (see Chapter 5). Similar results have also been obtained in these patients by Albin [59] et al., who reported significant reductions in quisqualate receptor and non-N-methyl-D-aspartate/nonquisqualate glutamate binding in the molecular layer and in the N-methyl-D-aspartate receptor in the granule layer. Moreover, these authors [60] also showed that the ionotropic and metabotropic quisqualate receptors were significantly decreased in the OPCA cerebellum.

9. GENETIC CONSIDERATIONS

9.1. Genetics of GDH-deficienct disorders

In studies that evaluated a rather large number of patients suffering from degenerative neurological disorders of various types, GDH deficiency was found in 1/2 to 1/5 of cases in which a clinical diagnosis of OPCA appeared tenable [14,24,30,31]. At present, there are no reliable clinical criteria that can distingish between the GDH-deficient and the GDH-normal cases, although the frequent association of neuropathy [25] in the former, and of progressive autonomic failure with orthostatic hypotension in the latter [14,31], have been suggested.

Genetic data have been reported in a total number of 72 patients with GDH deficiency. Forty-two of these cases were sporadic and 30 were familial. Seventeen of the latter cases had similarly affected siblings, but no parents nor offspring were neurologically affected, thus suggesting a possible autosomal recessive inheritance pattern in these cases. However, in some pedigrees an autosomal dominant transmission with incomplete penetrance, or even an X-linked inheritance, has been considered [10,15].

In the remaining 13 familial cases, the mode of inheritance was clearly dominant. Omitting our case of a childhood-onset variant of dominant OPCA, almost all other dominant GDH-deficient cases occurred in Italian patients. In these, as in our dominant cases, the onset of the disease was often in childhood. Thus, in the series of Orsi et al. [32], two dominant OPCA cases were included, which began at 5 and 9 years of age. Also, all but 1 of the 4 cases reported by Sorbi et al. [27] had the onset of their disease in childhood or adolescence. In this regard, Finocchiaro et al. [29], who also described decreased GDH activity in dominant OPCA cases, did not indicate in their report the age at disease onset of their patients.

However, the majority of dominant OPCA kindreds studied by us (16 dominant OPCA cases from five families of our series), particularly those linked to the HLA locus on the sixth chromosome [42], have been shown to have normal GDH activity [14,15]. Aubby et al. [30] also found no GDH-deficient cases among 19 patients from 15 families with autosomal dominant cerebellar ataxia. Similar conclusions were also drawn by Kostic et al. [31], who reported normal GDH activity in all five patients (from three families) with dominant OPCA.

9.2. Molecular genetics of human GDH

Substantial progress has been recently made in understanding the genetics of human GDH following the cloning of cDNAs encoding for human GDH [61,62]. This has permitted the detection of four different-sized GDH-specific mRNAs in human tissues and has provided evidence for the presence of a multigene family (probably three to five genes) [62]. Recent studies on genomic DNA have indeed identified the presence of structurally distinct GDH-specific genes in the human [65, Papamatheakis, personal communication]. In addition, two chromosomal loci have been demonstrated for human GDH: One mapping on the long arm of chromosome 10:q22.3–23 [63,64] and another, lacking introns and thought to represent a pseudogene, on the long arm of chromosome X:q22–23 [64] or q24 [63] (Figure 14-5). An additional DNA band has also been detected, showing no concordance with either chromosome 10 or X [64]; its chromosomal localization is currently under investigation.

Whether the four different-sized GDH-specific mRNAs represent distinct species or differential processing of RNA from a single functional gene has not as yet been established. Over 20 cDNA clones from brain, liver, and fibroblasts have been sequenced and have been found to be identical. Because, as described above, retinal degeneration is the feature of a form of OPCA associated with GDH abnormalities, it is possible that a particular species of GDH-specific mRNA is expressed in human retina. To test for this possibility, we screened a human retinal cDNA library and isolated several GDH-positive clones. One of these clones was sequenced and found to be distinct, showing a 97.7% homology at the nucleotide level and a 96%

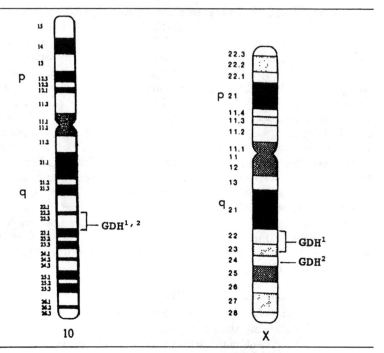

Figure 14-5. Physical mapping of GDH genes to chromosomes 10 and X (according to Jung et al. [63][1] and Anagnou et al. [64][2]), as described in the text.

homology at the amino acid level, as compared to the published human liver cDNA [62]. Interestingly, work done independently at the Molecular Biology Institute of Crete [Papamatheakis, personal communication] using genomic clones led to the characterization of an intronless GDH-specific gene, the sequence of which was found to be identical to our retinal cDNA. Thus these findings taken together indicate that this intronless gene is transcribed and, as such, it may be functional. Moreover, if this intronless gene proves to be the one that has been mapped to the X chromosome (see above), it may be involved in the X-linked forms of OPCA. This remains to be further tested.

To establish whether GDH deficiency constitutes the primary genetic defect of the above neurodegenerations, the demonstration of abnormal DNA sequence is required. If such studies show that gene mutations are present in the patients with reduced enzymatic activity, it will then be important to determine whether distinct gene mutations account for the clinical and genetic heterogeneity encountered. The use of highly efficient and sensitive protein chemistry techniques will permit the characterization of

GDH at the primary structure level. Mutations that can be uncovered by these techniques are those that alter either the quantity (Figure 14-3) or the electrophoretic mobility of the different GDH isoproteins. At present this method is suitable for studying brain autopsy material. The application of these techniques in more readily available non-neural tissues, such as leukocytes, platelets, or fibroblasts, remains problematic because these cells are usually not available in amounts sufficient for the purification and analysis of GDH. However, the use of the Western blotting technique in conjunction with the electrophoretic analysis method may permit the visualization of the different GDH isoproteins in crude tissue extracts, thus providing a suitable method of testing for GDH polymorphism in neurologic patients.

The ultimate proof, however, for a gene mutation will be the demonstration of an abnormal DNA sequence. Although GDH-specific cDNAs have been recently cloned, the structure, organization, and function of the various GDH-specific genes have not as yet been fully elucidated. Additional studies are, therefore, needed as a prerequisite for the analysis of the GDH-deficient disorders at the genomic DNA level. Meanwhile, the currently available cDNA probes will permit the analysis of these disorders at the mRNA level. The sequencing of cDNAs obtained from patients with GDH protein alterations will be required to establish whether DNA mutations are indeed responsible for these protein changes. Once DNA mutations have been identified, mutant oligonucleotides will be synthesized and used for screening large number of patients and for performing family studies to test whether particular GDH gene mutations correlate with specific clinico-pathologic entities.

ACKNOWLEDGMENTS

This work was supported by NIH grants NS-16871 and RR-71, Division of Research Resources, General Clinical Research Center Branch. We are indebted to Drs. L.J. Schut and G.E. Solomon for providing us with some of the autopsy tissue, as well as the clinical and pathological summaries of their patients. We thank Drs. D.P. Perl and J. Smith for performing pathological examinations in the patients described here. We are also thankful to many colleages who made these investigations possible by referring their patients to us.

REFERENCES

1. Fonnum F. (1984). Glutamate: A neurotransmitter in mammalian brain. J. Neurochem. 42:1–11.
2. Berl S., Takagaki G., Clarke D.D., Waelsch H. (1962). Metabolic compartments in vivo: Ammonia and glutamic acid metabolism in brain and liver. J. Biol. Chem. 237:2562–2569.
3. Plaitakis A., Berl S., Yahr M.D. (1982). Abnormal glutamate metabolism in adult-onset degenerative neurological disorder. Science 216:193–196.
4. Plaitakis A. (1990). Glutamate dysfunction and selective neuronal degeneration in amyotrophic lateral sclerosis. A hypothesis. Ann. Neurol. 28:3–8.

5. Lucas D.R., Newhouse J.P. (1957). The toxic effect of sodium L-glutamate on the inner layers of the retina. Arch. Ophthalmol. 58:193–201.
6. Olney J.W. (1989). Excitatory amino acids and neuropsychiatric disorders. Biol. Psychiatry 26:505–525.
7. Olney J.W., Ho O.L., Rhee V. (1971). Cytotoxic effects of acidic and sulfur containing amino acids of the infant mouse central nervous system. Exp. Brain. Res. 14:61–76.
8. McGeer E.G., McGeer P.L. (1976). Duplication of biochemical changes of Hunting-ton's chorea by intrastriatal injection of glutamic and kainic acids. Nature 263:517–519.
9. Plaitakis A., Nicklas W.J., Desnick R.J. (1979). Glutamate dehydrogenase deficiency in three patients with spinocerebellar ataxia: A new enzymatic defect? Trans. Am. Neurol. Assoc. 104:54–57.
10. Plaitakis A., Nicklas W.J., Desnick R.J. (1980)). Glutamate dehydrogenase deficiency in three patients with spinocerebellar syndrome. Ann. Neurol. 7: 297–303.
11. Desclin J.C., Escubi J. (1974). Effects of 3-acetylpyridine on the central nrvous system of the rat as demonstrated by silver methods. Brain Res. 77:349–364.
12. Herken H. (1968). Functional disorders of the brain induced by synthesis of nucleotides containing 3-acetylpyridine. Z. Klin. Chem. 6:635–367.
13. Plaitakis A., Berl S. (1983). Oral glutamate loading in disorders with spinocerebellar and extrapyramidal involvement: Effect on plasma glutamate, aspartate and taurine. J. Neural Transmission, Suppl. 19: 65–74.
14. Plaitakis A., Berl S., Yahr M.D. (1984). Neurological disorders associated with deficiency of glutamate dehydrogenase. Ann. Neurol. 15:144–153.
15. Plaitakis A. (1984). Abnormal glutamate metabolism of neuroexcitatory amino acids in olivopontocerebellar atrophy. In: R.C. Duvoisin, A. Plaitakis (eds): The Olivopontocerebellar Atrophies. Advances in Neurology, New York: Raven Press, 41:225–243.
16. Schut J.W., Haymaker W. (1951). Hereditary ataxia: A pathologic study of five cases of common ancestry. J. Neuropathol. Clin. Neurol. 1:183–213.
17. Haines J.L., Schut L.J., Weikamp L.R., et al. (1984). Spinocerebellar ataxia in a large kinded: Age at onset, reproduction, and genetic linkage studies. Neurology 34:1542–1548.
18. Plaitakis A., Yahr M.D. (1986). Abnormalities in GDH and glutamate metabolism in atypical Parkinson's disease and progressive supranuclear palsy. Neurology 35(Suppl. 1):110.
19. Plaitakis A. (1984). Biochemistry of recessive olivopontocerebellar atrophy. Ital. J. Neurol. Sci. Suppl. 4:65–73.
20. Plaitakis A., Berl S. (1988). Pathology of glutamate dehydrogenase. In: E. Kvamme (ed): Glutamine and Glutamate in the Mammals, Vol. 2. Boca Raton, FL: CRC Press, pp. 127–142.
21. Konagaya Y., Konagaya M., Takayanagi T. (1986). Glutamate dehydrogenase and its isozyme activity in olivopontocerebellar atrophy. J. Neurol. Sci. 74:231–236.
22. Hussain M.M., Zannis V., Plaitakis A. (1989). Characterization of glutamate dehydrogenase isoproteins purified from the cerebellum of normal subjects and patients with degenerative neurological disorders, and from human neoplastic cell lines. J. Biol. Chem. 264:20730–20735.
23. Yamaguchi T., Hayashi K., Murakami H., et al. (1982). Glutamate dehydrogenase deficiency in spinocerebellar degenerations. Neurochem. Res. 7:627–636.
24. Duvoisin R.C., Chokroverty S., Lepore F., Nicklas W.J. (1983). Glutamate dehydrogenase deficiency in patients with olivopontocerebellar atrophy. Neurology 33:1322–1326.
25. Chokroverty S., Duvoisin R.C., Sacheo R., et al. (1985). Neurophysiologic study of olivopontocerebellar atrophy with or without glutamate dehydrogenase deficiency. Neurology 35:652–659.
26. Duvoisin R.C., Nicklas W.J., Ritchie V., et al. (1988). Low leukocyte glutamate dehydrogenase activity does not correlate with a particular type of multiple system atrophy. J. Neurol. Neurosurg. Psychiatry. 51:1508–1511.
27. Sorbi S., Tonini S., Giannini E., Piacetini S., et al. (1986). Abnormal platelet glutamate dehydrogenase activity and activation in dominant and non-dominant olivopontocerebellar atrophy. Ann. Neurol. 19:239–245.
28. Sorbi S., Piacetini S., Fani C., et al. (1989). Abnormalities of mitochondrial enzymes in hereditary ataxias. Acta Neurol. Scand. 80:103–110.

29. Finocchiaro G., Taroni F., Di Donato S. (1986). Glutamate dehydrogenase in olivopontocerebellar atrophies: Leukocytes, fibroblasts, and muscle mitochondria. Neurology 36: 550-553.
30. Aubby D., Saggu H.K., Jenner P., et al. (1986). Leukocyte glutamate dehydrogenase activity in patients with degenerative neurological disorders. J. Neurol. Neurosurg. Psychiatry 51:893-902.
31. Kostic V.S., Mojsilivic L.J., Stojanovic M. (1989). Degenerative neurological disorders associated with deficiency of glutamate dehydrogenase. J. Neurol. 236:111-114.
32. Orsi L., Bertolotto A., Bringolio F., et al. (1988). Glutamate dehydrogenase (GDH) deficiency in different types of progressive hereditary cerebellar ataxia. Acta Neurol. Scand. 78, 394-400.
33. Kajiyama K., Ueno S., Tatsumi T., et al. (1988). Decreased glutamate dehydrogenase protein in spinocerebellar degeneration. J. Neurol. Neurosurg. Psychiatry 51:1078-1080.
34. Tatsumi C., Yorifuji S., Takahashi M., Tarui S. (1989). Decreased viability of skin fibroblasts from patients with glutamate dehydrogenase deficiency. Neurology 39:451-452.
35. Iwattsuji, K., Nakamura S., Kameyama M. (1989). Lymphocyte glutamate dehydrogenase activity in normal aging and neurological diseases. Gerontology 35:218-224.
36. Chokroverty S., Khedekar R., Derby B., et al. (1984). Pathology of olivopontocerebellar atophy with glutamate dehydrogenase deficiency. Neurology 34:1451-1455.
37. Plaitakis A., Smith J. (1986). Biochemical and morphological changes of brain in a patient dying of GDH deficient olivopontocerebellar atropy. Ann. Neurol. 20:152-153.
38. Parker J.C. Jr., Dyer M.L., Paulsen W.A. Neuronal intranuclear hyaline inclusion disease associated with premature coronary atherosclerosis. J. Clin. Neuro-Ophthalmol. 7:244-249.
39. Plaitakis A., Berl S., Yahr M.D. (1983). The treatment of GDH-deficient olivopontocerebellar atrophy with branched chaim amino acids. Neurology 33(Suppl. 2):78.
40. Huang Y.P., Plaitakis A. (1984). Morphological changes in olivopontocerebellar atrophy in computed tomography and comments on its pathogenesis. In: Duvoisin R.C., Plaitakis A. (eds): The Olivopontocerebellar Atrophies. Advances in Neurology, Vol. 41. New York: Raven Press, pp. 39-81.
41. Jackson J.F., Currier R.D., Terasaki P.I., et al. (1977). Spinocerebellar ataxia and HLA linkage. N. Engl. J. Med. 236:138-141.
42. Rich S.S., Wilkie P.J., Schut L.J., et al. (1987). Spinocerebellar ataxia: Localization of an autosomal dominant locus between two markers on human chromosome 6. Am. J. Hum. Genet. 41:524-531.
43. Haltia M., Sommer H., Palo J., Johnson W.G. (1984). Neuronal intranuclear inclusion disease in identical twins. Ann. Neurol. 15:316-321.
44. Palo J., Haltia M., Carpender S., et al. (1984). Neurofilament subunit-related protein in neuronal intranuclear inclusions. Ann. Neurol. 15:322-328.
45. De Boni V., Crapper-McLachlan D.R. (1985). Controlled induction of paired helical filamants of the Alzheimer's type in cultured human neurons by glutamate and aspartate. J. Neurol. Sci. 68:105-118.
46. Colon A., Plaitakis A., Perakis A., et al. Purification and characterization of a soluble and a particulate glutamate dehydrogenase from rat brain. J. Neurochem. 46:1811-1819.
47. Plaitakis A., Berl S., Clarke D.D. (1986). Glutamate dehydrogenase in rat brain mitochondria and synaptosomes. Trans. Am. Soc. Neurochem. 17:218.
48. Smith E.E., Austen B.M., Blumenthal K.M., Nyc J.F. (1975). Glutamate dehydrogenases. In: Boyer (ed), The Enzymes, New York: Academic Press, pp. 293-367.
49. Plaitakis A., Shashidharan P. Unpublished data.
50. Aoki C., Milner T.A., Rex Sheu K.-F.R., et al. (1987). Regional distribution of astrocytes with intense immunoreactivity for glutamate dehydrogenase in rat brain: Implications for neuronglia interactions in glutamate transmission. J. Neurosci. 7:2214-2231.
51. Balcar V.J., Borg J., Mandel P. (1977). High affinity uptake of L-glutamate and L-aspartate by glial cells. J. Neurochem. 28:27-28.
52. Hertz L., Schousboe A. (1988). Metabolism of glutamate and glutamine in neurons and astrocytes in primary cultures. In: E. Kvamme (ed): Glutamine and Glutamate in Mammals, Boca Raton, FL: CRC Press, 2:39-55.
53. Shank R.P., Aprison M.H. (1988). Glutamate as neurotransmitter. In: E. Kvamme (ed): Glutamine and Glutamate in Mammals, Vol. II. CRC Press, Boca Raton, FL: pp. 3-19.

54. Chee P.Y., Dahl L.J., Fahien A. (1979). The purification and properties of rat brain glutamate dehydrogenase. J. Neurochem. 33:53–60.
55. Pearce I.A., Cambray-Deakin M.A., Burgyne R.D. (1987). Glutamate acting on NMDA receptors stimulates neurite outgrowth from cerebellar granule cells. FEBS Lett. 233:143–147.
56. Balazs R., Hack N., Jorgensen O.S. (1988). Stimulation of the N-methyl-D-aspartate receptor has a trophic effect on differentiating cerebellar granule cells. Neurosci. Lett. 87:80–86.
57. Tsiotos P., Plaitakis A., Mitsakos A., et al. (1989). L-glutamate binding sites of normal and atrophic human cerebellum. Brain Res. 481:87–96.
58. Hatziefthimiou A., Mitsakos A., Mitsaki E., et al. (1990). Quantitative autoradiographic study of L-glutamate binding studies in normal and atrophic human cerebellum. J. Neurosci. Res. 28:361–375.
59. Albin R.L., Gilman S. (1990). Autoradiographic localization of inhibitory and excitatory amino acid transmitter receptors in human normal and olivopontocerebellar atrophy cerebellum. Brain Res. 522:37–45.
60. Macowiec R.L., Albin R.L., Cha J.J., et al. (1990). Two types of quisqualate receptor are reduced in human OPCA cerebellar cortext. Brain Res. 552:37–45.
61. Banner C., Silverman S., Thomas J.W., et al. (1987). Isolation of a human cDNA clone for glutamate dehydrogenase. J. Neurochem. 49:246–252.
62. Maurothalassitis G., Tzimagiorgis G. Mitsialis A., Zannis V., Plaitakis A., Papamatheakis J., Moschonas N. (1988). Isolation and characterization of cDNA clones encoding human liver glutamate dehydrogenase: Evidence for a small gene family. Proc. Natl. Acad. Sci. USA 85:3494–3498.
63. Jung K.Y., Warter S., Rumpler Y. (1989). Assignment of the GDH loci to human chromosomes 10q23 and Xq24 by in situ hybridization. Annales Genetique 32:109–110.
64. Anagnou N.P., Senanez H., Modi S.J., et al. (1992). Chromosomal mapping of the human glutamate dehydrogenase (GLUD) genes to chromosomes 10 q22.3–23 and X q22–23, submitted.
65. Amuro N., Goto Y., Okazaki T. (1990). Isolation and characterization of two distinct genes for human glutamate dehydrogenase. Biochem. Biophys. Acta 1049:216–218.
66. Plaitakis A. (1990). Altered glutamate metabolism in amyotrophic lateral sclerosis and treatment with branched chain amino acids. In: G. Lubec and G.A. Rosenthal (eds): Amino Acids: Chemistry, Biology and Medicine. Leiden: ES COM Science Publishers BV, pp. 379–385.

15. MITOCHONDRIAL ABNORMALITIES IN HEREDITARY ATAXIAS

SANDRO SORBI, AND JOHN P. BLASS

1. INTRODUCTION

The hereditary ataxias or *spinocerebellar degenerations* are a group of hereditary neurodegenerative disorders that have been the subject of intensive investigations for over 100 years, since Friedreich wrote a landmark paper distinguishing the hereditary ataxic syndrome that bears his name from neurosyphilis and other conditions [1]. A forest of erudition has resulted, as illustrated by many reviews of this subject [2], including this volume. Although individual syndromes, such as Friedreich's ataxia, can be clearly delineated, there is much overlap in clinical presentation, and even in neuropathology, among these conditions. Variant and intermediate forms of these syndromes are common. Patients change over time, sometimes leading to changes in the diagnostic category in which they are classified. Patients within the same family, who presumably inherit the same gene defect, can present with syndromes so different that they would be classified as having different disorders were they not from the same kindred [3]. These disorders thus provide an example of pleiotropism, a phenomenon well documented in neurology for both genetic diseases [4] and nongenetic conditions, such as syphilis and stroke. As usual [4], phenotypic variation tends to be more marked among patients with a pattern of dominant inheritance (i.e., a single abnormal gene leading to clinical signs and symptoms) than in those with a recessive pattern (i.e., two abnormal genes necessary for phenotypic

A. Plaitakis (ed.), CEREBELLAR DEGENERATIONS: CLINICAL NEUROBIOLOGY. Copyright © 1992.
Kluwer Academic Publishers, Boston. All rights reserved.

expression). The distinction between "dominant" and "recessive" inheritance is, however, often less than clear cut. For instance, if an abnormally high arch is observed in an otherwise asymptomatic parent of a child with Friedreich's ataxia, is the inheritance in this family dominant or recessive? Classifications of this group of disorders are less like trees than like Ven diagrams.

A series of molecular abnormalities have been associated with hereditary ataxic syndromes [5]. They include abnormalities in lipid, protein, and carbohydrate metabolism, and in oxygen utilization. These observations have led to the suggestion that the hereditary ataxias result from a family of inborn errors of metabolism [5]. Thus, the group of hereditary ataxias, as a whole, demonstrate genetic heterogeneity.

There is less agreement about whether or not genetic heterogeneity exists within single syndromes identified among these overlapping diagnostic categories [6]. If genetic heterogeneity exists within a single syndrome, such as Friedreich's ataxia, then the discovery of an enzyme deficiency or another molecular abnormality in a subgroup of patients with this syndrome may be of etiological significance. The abnormality might even represent the product of the defective gene. If genetic heterogeneity does not exist within a single syndrome, then a molecular abnormality that is found in only a portion of the affected patients is not likely to represent the primary genetic defect. Implicit assumptions about the degree of genetic heterogeneity *within defined syndromes* underlie some of the controversies about the significance of reported abnormalities in enzymes such as mitochondrial malate dehydrogenase [7-9] and lipoamide dehydrogenase [10-12] in Friedreich's ataxia [13]. Chamberlain and coworkers [14] studied kindreds in which the patients conformed strictly to the diagnostic criteria developed by Geoffroy and coworkers [15], including recessive inheritance. They found the clinical disorder closely linked to a pericentromeric site on chromosome 9 in studies of 553 individuals, including 202 patients from 80 families of diverse ethnic origins in Europe and North America, with a cumulative LOD score above 25. These findings support the existence of a genetically homogeneous "core" syndrome of Friedreich's ataxia. DiMauro and DeVivo [12] have proposed that patients with pyruvate dehydrogenase deficiency who have the clinical and pathological findings of Friedreich's ataxia represent the overlap of these two syndromes. The overlap of clinical and pathological findings in patients with different inborn errors can also be called *genetic heterogeneity*. A review of British cases led Harding [13] to propose that there is genetic heterogeneity in Friedreich's ataxia, with a mutation at a single locus accounting for the largest proportion of patients. The published data on Friedreich's ataxia conform to Harding's hypothesis.

Further correlations of clinical and molecular genetic studies, including studies of "variant" forms of ataxic syndromes, are likely to lead to important advances in the nosology of these conditions [4]. Identification of the product(s) of the affected gene(s) can also be expected to provide clues to the pathophysiological mechanisms involved.

From the viewpoint of disease mechanisms, it is notable that abnormalities in mitochondrial constituents have been reported frequently [5,16–18] in association with hereditary ataxic syndromes of both the spinal ("Friedreich") type and the multiple-system ("olivo-ponto-cerebellar") type. In the rest of this chapter, we examine the hypothesis that hereditary ataxic syndromes often represent the clinical expression of mitochondrial defects.

2. METABOLIC ABNORMALITIES IN VIVO

It is a common observation that patients with hereditary ataxias have a high incidence of abnormal glucose tolerance curves [16,19–21]. The abnormality has often been attributed to a smaller metabolic sink for taking up glucose in these muscle-wasted patients. On the other hand, the impaired ability to clear glucose from the blood is also consistent with a metabolic defect in the ability to metabolize glucose via oxidative pathways in mitochondria.

Barbeau and coworkers [21] demonstrated a decreased ability to clear pyruvate from the blood after a glucose load in a subgroup of patients with Friedreich's ataxia. The impaired pyruvate clearance appeared unrelated to the degree of muscle wasting. Indeed, the clinically unaffected, non-muscle-wasted, presumed heterozygote, parents of these patients showed a decreased ability to clear pyruvate after a similar glucose load. The values for the presumed heterozygotes fell between those for the patients and those for controls, as expected for an abnormality closely linked to the primary gene abnormality. Dijkstra and coworkers [20] have reported that clearance of a pyruvate load is characteristically slowed in patients with Friedreich's ataxia, apparently unrelated to the degree of muscle wasting.

The interpretation of in vivo metabolic studies is always difficult. Patients may suffer from a variety of secondary abnormalities related to the consequences of the disease process, including nutrition, medication, activity, and muscle wasting. The data summarized above, however, are consistent with the existence of mitochondrial abnormalities affecting glucose oxidation in a significant proportion of patients with hereditary ataxic syndromes.

3. MITOCHONDRIAL ENZYME ACTIVITIES IN VITRO

Abnormalities in a number of mitochondrial enzymes have been associated with ataxic syndromes; they include the pyruvate dehydrogenase complex and its components, mitochondrial malate enzyme, glutamate dehydrogenase, and components of the electron transport chain (see Blass et al. [16] for review). These studies have classically been done in non-neural tissues from living patients suffering from hereditary ataxic syndromes, although some studies of autopsy tissue have been possible.

A hereditary deficiency in the pyruvate dehydrogenase complex (PDHC) is reasonably well established as one cause of the syndrome of intermittent ataxia of childhood [22]. It is not, however, the only hereditary abnormality that can cause this syndrome, i.e., the syndrome shows genetic heterogeneity [22,23]. The chemistry of PDHC is complex and has been reviewed elsewhere [22]. PDHC consists of a precise geometric array of three proteins: E1, which contains alpha and beta subunits; E2, the core protein of the complex; and E3, also known as lipoamide dehydrogenase (LAD), which is a component that is common to both PDHC and the ketoglutarate (KGDHC) complex. When the syndrome of intermittent ataxia is associated with PDHC deficiency, the abnormality is most often in the alpha E1 peptide. This conclusion was first reached on the basis of studies of enzyme activities [22,23] and has been confirmed by molecular biological techniques using cDNA probes for this gene [24]. It is possible that the *alphE1* gene, which is located on the X chromosome, is a relatively frequent site of mutation, i.e., a genetic "hot spot" [24]. Apparently a variety of abnormalities can affect this locus and lead to neurological abnormalities, including ataxia [24].

Attempts to associate deficiencies in specific mitochondrial enzymes with other, more common mitochondrial syndromes have not led to clear-cut results. For instance, deficiencies of PDHC have been found by some workers, but not by others, to be associated with Friedreich's ataxia [10, 11,17–19,25,26]. Deficiencies of the activity of mitochondrial malic enzyme have also been reported in this condition but have not been confirmed by other researchers in other patients [7–9]. The occurrence of glutamate dehydrogenase deficiency in some but not all patients with forms of multiple system degenerations is discussed in detail elsewhere in this volume. Patients with defects in electron transport often, although not invariably, suffer from ataxia [27,28]. Ataxia is characteristically part of the syndrome of "myoclonus epilepsy with ragged red fibers" (MERFF), which exhibits maternal inheritance and appears to be associated with hereditary abnormalities of mitochondrial DNA [27]. The controversies about the association of specific enzyme defects with specific ataxic syndromes may reflect partly technical factors and partly genetic heterogeneity within single syndromes. It is also possible that the common factor in these patients—even among patients suffering from a clearly defined entity such as Friedreich's ataxia—is damage to mitochondria with secondary and variable damage to a number of mitochondrial enzymes.

To examine directly the hypothesis that mitochondrial damage is a frequent finding among patients with hereditary ataxias, a series of mitochondrial enzymes were assayed in platelets from such patients and controls studied at the Burke Rehabilitation Genter in White Plains, NY [16–18]. The reductions in mean activity for the group of ataxics, compared to the group of controls, were statistically significant for both PDHC and GDH, but were quantitatively small (Table 15-1). The biological significance of these small changes may be the evidence they provide for the existence of mitochondrial abnormalities in these groups of patients, rather than for specific deficiencies

Table 15-1. Mitochondrial enzyme activities in platelets from patients with hereditary ataxias (White Plains series)

	Pyruvate dehydrogenase	Glutamate dehydrogenase	Fumarase
Controls	101 ± 3 (42)	98 ± 4 (38)	100 ± 3 (37)
Ataxics	77 ± 6 (23)[1]	84 ± 7 (20)	89 ± 7 (14)
Spinal ataxias	85 ± 9 (10)[3]	87 ± 9 (8)	93 ± 2 (4)[3]
Multiple system Degenerations	71 ± 8 (13)[2]	84 ± 7 (12)[3]	87 ± 9 (10)

Values are percent of simultaneous control ± SEM (number of subjects studied). Absolute values in nmole/min/mg protein ± SD were pyruvate dehydrogenase, 1.58 ± 0.25; glutamate dehydrogenase, 10.3 ± 2.3; fumarase, 49.7 ± 10.3.
[1] $p < 0.001$;
[2] $p < 0.01$;
[3] $p < 0.05$ (one-tailed t test; comparisons to control group; one-tailed t test used because the hypothesis being tested was of reductions in mitochondrial enzyme activities).
Data are from Sheu et al. [18] (which describes the methods used) and Cedarbaum and Blass [17]. For a discussion of classification, see the text, Blass et al. [16], and Cedarbaum and Blass [17].

of the restricted number of mitochondrial enzymes assayed. Abnormalities in one or another of the three enzymes studied (PDHC, GDH, and fumarase) occurred in over 30% of the patients studied, as evidenced by the activities of more than two standard deviations between the means for controls [17]. No relationship was evident between the enzyme found to have deficient activity and the clinical syndrome.

More extensive studies have been carried out in the University Medical School in Florence, Italy [25], and these are summarized in Tables 15-2 and 15-3. Of the 24 patients studied, 20 (83%) had activities of one or more mitochondrial enzymes below the lowest values for any of the 33 controls (Table 15-2). Ten patients showed deficiencies in only one mitochondrial enzyme, two had deficiencies in two enzymes, one had deficiencies in three enzymes, and one had deficiencies in four enzymes. GDH was deficient in 14 patients, valine dehydrogenase in eight, PDHC in seven, and succinate dehydrogenase in four. In this study, no patients were found deficient in citrate synthase, malate dehydrogenase, or fumarase. Again, no clear relationship emerged between the mitochondrial enzyme(s) found deficient and the clinical syndrome. The net reductions for the groups of ataxics as a whole were relatively mild, except for GDH, where a decrease of over 50% was found for patients with either "Friedreich's ataxia" or variants of olivo-ponto-cerebellar ataxias (Table 15-3).

The studies summarized in Tables 15-1 to 15-3 indicate that damage to mitochondria, at least at the enzyme level, is a frequent finding in hereditary ataxiias. It should be noted that these abnormalities have been described primarily in *clinically and histologically normal* tissues, both platelets [16–18] and cells cultured from the skin [16,25,26]. These findings indicate that the damage to mitochondria is unlikely to be secondary to histological damage,

Table 15-2. Distribution of mitochondrial enzyme deficiencies in hereditary ataxias

	PDHC	GDH	FUM	CS	SDH	MDH	VDH
FRIEDREICH'S ATAXIA							
1	+	+	−	−	+	−	+
2	+	−	−	−	+	−	−
3	+	−	−	−	−	−	−
4	+	−	−	−	−	−	−
5	−	+	−	−	−	−	+
6	−	+	−	−	−	−	+
7	−	+	−	−	−	−	+
8	−	+	−	−	−	−	−
9	−	−	−	−	−	−	−
10	−	−	−	−	−	−	−
OLIVOPONTO-CEREBELLAR ATROPHY (NON-DOMINANT FORM)							
1	+	−	−	−	+	−	−
2	+	+	−	−	−	−	+
3	+	+	−	−	−	−	−
4	−	+	−	−	−	−	+
5	−	+	−	−	−	−	−
6	−	+	−	−	−	−	−
7	−	−	−	+	−	−	+
8	−	−	−	−	−	−	+
OLIVOPONTO-CEREBELLAR ATROPHY (DOMINANT FORM)							
1	−	+	−	−	−	−	−
2	−	+	−	−	−	−	−
3	−	+	−	−	−	−	−
4	−	+	−	−	−	−	−
5	−	−	−	−	−	−	−
6	−	−	−	−	−	−	−

+ = activity lower than the lowest control; − = activity within the control range. PDHC = pyruvate dehydrogenase complex; GDH = glutamate dehydrogenase; FUM = fumarase; CS = citrate synthase; SDH = succinic dehydrogenase; MDH = malate dehydrogenase; VDH = valine dehydrogenase. For original data and methods, see Sorbi et al. [25].

as might be the case for brain or muscle. The persistence of abnormalities in cultured cells argues that they are in some way related to an abnormality in an information molecule that is accurately copied as the cells proliferate; other materials would be diluted out. Such an information molecule might be viral in origin, but the most straightforward assumption in these hereditary disorders is that the abnormal information is in the abnormal gene responsible for the disorder. The abnormalities in mitochondrial enzymes may, however, themselves be the consequence of other genetically determined abnormalities that lead to secondary damage to mitochondria.

4. POSSIBLE PATHOPHYSIOLOGICAL MECHANISMS

If mitochondrial damage is important in the development of hereditary ataxic syndromes, rather than being an epiphenomenon in these disorders, it should be possible to link mitochondrial damage to the pathological characteristics of these disorders using plausible mechanisms. The hereditary ataxias are

Table 15-3. Mitochondrial enzyme activities in platelets from patients with hereditary ataxias (Florentine series)

	Controls (33)	Friedreich Ataxia (10)	Olivoponto-cerebellar atrophy (dominantly inherited) (6)	Olivoponto-cerebellar atrophy (nondominantly inherited) (8)
Pyruvate DHC	3.2 ± 0.4	2.4 ± 0.8[1]	3.0 ± 0.3	2.6 ± 0.4[1]
Glutamate DH	32.1 ± 9.8	15.4 ± 6.7[1]	15.9 ± 15.1[2]	16.6 ± 7.3[1]
Fumarase	30.2 ± 4.5	29.1 ± 4.0	31.0 ± 2.8	28.9 ± 1.4
Citrate synthase	57.1 ± 6.2	58.9 ± 6.8	52.8 ± 4.1	52.9 ± 2.9[2]
Succinate DH	44.1 ± 3.4	39.4 ± 6.0[2]	44.2 ± 1.4	37.7 ± 5.5[2]
Malate DH	14.1 ± 2.2	14.5 ± 2.4	13.5 ± 1.3	15.9 ± 1.7
Valine DH	26.0 ± 2.9	22.3 ± 4.3[2]	26.7 ± 2.6	22.4 ± 4.5[2]

Values are mean nmol/min/mg protein ± SD. The number of subjects studied is in parentheses.
[1] $p < 0.01$; [2] $p < 0.05$.
Comparison to control group by Mann-Whitney test. See Sorbi et al. [25] for original data and methods.

classical examples of "system degenerations," in which certain systems in the brain are much more vulnerable than others, with relatively less evident damage to other parts of the nervous system or to other tissues. Proposing a link between generalized mitochondrial dysfunctions, which can be demonstrated in extraneural tissues, and the hereditary ataxias requires dealing with several questions. Why is the nervous system particularly affected? How can a generalized metabolic abnormality particularly affect certain populations of neurons? Why are cells with long axons particularly affected [16,17] in Friedreich's ataxia and other "spinal ataxias"? Why are the particular systems affected in olivopontocerebellar atrophies and other "multiple system atrophies" particularly susceptible to mitochondrial disorders?

It is reasonable that the nervous system is particularly affected in disorders that impair oxidative metabolism, since the nervous system is more dependent on continuous oxidative metabolism than is any other organ system, even the heart [29]. A few seconds of anoxia lead to the loss of consciousness; more than 4–10 minutes of total anoxia-ischemia (for instance, in cardiac arrest) lead to irreversible brain damage. Although there are mitochondrial disorders in which clinical and mitochondrial abnormalities have been demonstrated only in muscle, there is a large and increasing group of "encephalomyopathies" in which brain damage is prominent as well [27]. As discussed elsewhere in this volume, in many of the "encephalomyopathies" ataxia is a prominent symptom [16,27]. They include Kearns-Sayre syndrome (KSS) and myoclonus epilepsy with ragged red fibers (MERRF).

Studies of both humans who have undergone hypoxic or ischemic episodes [30] and experimental animal models, such as the four-vessel occluded rat [31], indicate that impairments of oxidative metabolism can lead to damage that is relatively restricted to certain populations of neurons. For instance, the CA1 neurons of the hippocampus appear to be particularly vulnerable to hypoxia/ischemia compared to CA2 cells [30,31]. The mechanisms are not yet well defined but have been proposed to involve excitotoxic effects on cells receiving glutamatergic innervation, mediated by NMDA receptors and the resultant excess influx of calcium into the cells [32]. In this regard, it is notable that Hirsch and Gibson [33] have shown that impairments of oxidative metabolism lead to *excess* release of glutamate (and dopamine) and *reduced* release of acetylcholine. Although the mechanisms of selective vulnerability have not been worked out in detail for hypoxic/ischemic insults or for inborn errors of metabolism, there are ample clinical and experimental precedents for selective vulnerability in disorders of oxidative metabolism.

Neurons with long axons have been proposed to be particularly susceptible to disorders of mitochondria [16,17]. These cells carry out axoplasmic transport over distances as long as a meter or more. Fast axoplasmic transport requires energy generated in mitochondria clustered at the paranodal regions of the axon; it is exquisitely sensitive to the disruption of oxidative metabolism and less so to the disruption of glycolysis [34]. Spencer and coworkers [35] have pointed out that toxic agents that lead to dying-back neuropathies typically disrupt oxidative metabolism. This is the type of neuropathy typically seen pathologically in Friedreich's ataxia and related variants [16,36]. In a proposed classification of hereditary ataxic syndromes based on cellular mechanisms of degeneration, those ataxias in which the deterioration of cells with long axons is particularly prominent and in which these mechanisms may have played a part have been called *spinal ataxias* or *axonal ataxias* [16,17].

Mechanisms that lead to relatively selective vulnerability of neurons affected in multiple-system degenerations are discussed in detail elsewhere in this volume. They may involve cells that receive glutamatergic innervation, and may therefore be linked to mitochondrial abnormalities and resultant impaired oxidative metabolism through excess glutamate release, as discussed above. It is notable that the mitochondrial enzyme glutamate dehydrogenase (GDH) appears to be concentrated in glial cells surrounding structures that in the rat are analogous to structures particularly susceptible to damage in GDH deficiency in humans [37,38]. This observation supports the hypothesis that impaired removal of glutamate from the synaptic cleft contributes to the premature demise of these cells by excitotoxic mechanisms [39]. Another potential mechanism is impaired transport of trophic substances by fast axoplasmic transport [34], which might underlie the phenomenon of "chain degeneration" that has been hypothesized to underlie the degeneration of related neural systems in multiple-system atrophies [16,36,40]. These mechanisms are not, of course, mutually exclusive. In the mechanistic classification discussed above, the term *multiple systems disorders* has been used as a category to described the hereditary ataxias in which a number of such neuronal groups deteriorate, including the olivo-ponto-cerebellar atrophies [16,17].

5. IMPLICATIONS FOR FUTURE RESEARCH

Extensive studies over more than a decade have indicated that hereditary ataxic syndromes can be associated with mitochondrial damage. They have led to the hypothesis that a significant proportion—perhaps the majority—of hereditary ataxias are mitochondrial disorders [16,17].

Most of the studies implicating mitochondrial abnormalities in hereditary ataxic syndromes have been done at the level of enzyme assays of mitochondrial components. Current molecular biological techniques allow the demonstration of mitochondrial abnormalities at the level of the mitochondrial or genomic DNA [27]. cDNA probes are now available for some of the enzymes that have been proposed to be associated with hereditary ataxic syndromes and can be considered as "candidate genes" for linkage (RFLP) analyses of informative kindreds with these syndromes. For instance, the gene for lipoamide dehydrogenase (LAD) has been localized to chromosome 7 [41], suggesting that an abnormality of this gene is **not** responsible for the common, chromosome 9-linked form of Friedreich's ataxia studied by Chamberlain et al. [15]. Newer techniques to study mitochondria, including the use of fluorescent probes, should facilitate the examination of these intracellular organelles for possible dysfunction in cultured cells from patients with hereditary ataxic syndromes. Applications of these newer molecular and cellular biological techniques will allow more definitive testing of the hypothesis that has underlain the studies reported in Tables 15-1 to 15-3: *"hereditary ataxias are often mitochondrial disorders."*

REFERENCES

1. Adams R.D. (1977). Spinocerebellar degenerations. Introduction I. In: D.A. Rottenberg, F.H. Hochberg (eds): Neurological Classics in Modern Translation. New York: MacMillan, pp. 203–204.
2. Refsum S., Skre H. (1978). Nosology, genetics, and epidemiology of hereditary ataxias, with particular reference to the etiology of these disorders in Western Norway. Adv. Neurol. 19:497–508.
3. Konigsmark B.W., Wiener L.P. (1970). The olivopontocerebellar atrophies: A review. Medicine 49:227–241.
4. Rosenberg, R.N. (1986). Neurogenetics: Principles and Practice. New York: Raven Press.
5. Blass J.P. (1981). Hereditary ataxias. Curr. Neurol. 3:66–91.
6. Barbeau A. (1982). A tentative classification of recessively inherited ataxias. Can. J. Neurol. Sci. 9:95–98.
7. Stumpf D.A., Parks J.K., Parker W.D. (1983). Friedreich's disease IV. Reduced mitochondrial malic enzyme activity in heterozygotes. Ann. Neurol. 33:780–783.
8. Bottacchi E., DiDonato S. (1983). Skeletal muscle NAD(P) and NADP dependent malic enzyme in Friedreich's ataxia. Neurology 33:712–716.
9. Chamberlain S., Lewis P.D. (1983). Normal mitochondrial malic enzyme levels in Friedreich's ataxia fibroblasts. J. Neurol. Neurosurg. Psychiat. 47:1050–1051.
10. Kark R.A.P., Budelli M.M., Becker D.M., Wiener L.P., Forsythe A.B. (1981). Lipoamide dehydrogenase: Rapid heat inactivation in platelets of patients with recessively inherited ataxia. Neurology 31:199–201.
11. Stumpf D.A., Parks J.K., Parker W.D. (1979). Friedreich ataxia, II. Normal kinetics of lipoamide dehydrogenase. Neurology 29:802–806.
12. DiMauro S., De Vivo D.C. (1989) Diseases of carbohydrate, fatty acid, and mitochondrial metabolism. In: D.A. Siegel, B. Agranoff, R.W. Albers, P. Molinoff (eds): Basic

Neurochemistry: Molecular, Cellular, and Medical Aspects, 4th ed. New York: Raven Press, pp. 647-670.
13. Harding A.E. (1981). Friedreich's ataxia: A clinical and genetic study of 90 families with an analysis of early diagnostic criteria and intrafamilial clustering of clinical features. Brain 104:589-620.
14. Chamberlain S., Shaw J., Wallis J., Rowland A., Chow L., Farrel M., Keats B., Richter A., Roy M., Melancon S., Deufel T., Berciano J., Williamson R. (1989). Genetic homogeneity at the Friedreich ataxia locus on chromosome 9. Am. J. Hum. Gen. 44:518-521.
15. Geoffroy G., Barbeau G., Breton B., Lemieux B., Aube M., Leger C., Bouchard J.B. (1976). Clinical description and roentgenologic evaluation of patients with Friedreich's ataxia. Can. J. Neurol. Sci. 3:279-287.
16. Blass J.P., Sheu R.K.F., Cedarbaum J.M. (1988). Energy metabolism in disorders of the nervous system. Rev. Neurol. (Paris) 144:543-563.
17. Cedarbaum J.M., Blass J.P. (1986). Mitochondrial dysfunction and spinocerebellar degenerations. Neurochem. Pathol. 4:43-63.
18. Sheu K.F.R., Blass J.P., Cedarbaum J.M., Kim Y.T., Harding B.J., DeCicco J. (1988). Mitochondrial enzymes in hereditary ataxias. Metab. Brain Dis. 3:151-160.
19. Barbeau A. (1982). Friedreich's disease 1982: Etiologic hypotheses—a personal analysis. Can. J. Neurol. Sci. 9:243-256.
20. Dijkstra U., Gabreels F., Joosten E., Wevers R., Lamers K. (1984). Friedreich's ataxia: Intravenous pyruvate load to demonstrate a defect in pyruvate metabolism. Neurology 34:1493-1497.
21. Barbeau A., Butterworth R.F., Ngo T., Breton G., Melancon S., Shapcott D., Geoffrey G., Lemieux B. (1976). Pyruvate metabolism in Friedreich's ataxia. Can. J. Neurol. Sci. 3:379-388.
22. Blass J.P. (1983). Inborn errors of pyruvate metabolism. In: J.B. Stanbury, J.B. Wyngaarden, D.S. Fredrickson, J.L. Goldstein, M.S. Brown (eds). The Metabolic Basis of Inherited Disease, 5th ed. New York: McGraw-Hill, pp. 193-203.
23. Blass J.P. (1979). Disorders of pyruvate metabolism. Neurology 29:280-286.
24. Roche T.E., Patel M.S. (eds). Alpha-keto acid dehydrogenase complexes. Ann NY Acad Sci, in press.
25. Sorbi S., Piacentini S., Fani C., Tonini S., Marini P., Amaducci L. (1989). Abnormalities of mitochondrial enzymes in hereditary ataxias. Acta Neurol. Scand. 80:103-110.
26. Blass J.P., Kark R.A.P., Menon N.K., Harris S. (1976). Low activities of the pyruvate and oxoglutarate dehydrogenase complexes in five patients with Friedreich's ataxia. N. Engl J. Med. 295:62-67.
27. DiMauro S., Bonila E., Zeviani M., Nakagawa M., DeVivo D.C. (1985). Mitochondrial myopathies. Ann. Neurol. 17:521-538.
28. Rivner M.H., Shamsnia M., Swift T.R., Trefz J., Roesel R.A., Carter A.L., Yamamura W., Hommes F.A. (1989). Kearns-Sayre syndrome and complex II deficiency. Neurology 39:693-696.
29. Gibson G.E., Pulsinelli W., Blass J.P., Duffy T.E. (1981). Brain dysfunction in mild to moderate hypoxia. Am. J. Med. 70:1247-1254.
30. Volpe B.T., Petito C.K. (1985). Dementia with bilateral medial temporal lobe ischemia. Neurology 35:1793-1797.
31. Volpe B.T., Pulsinelli W.A., Tribuna J., Davis H.P. (1984). Behavioral performance of rats following transient forebrain ischemia. Stroke 15:558-562.
32. Rothman S. (1984). Synaptic release of excitatory amino acid neurotransmitter mediates anoxic neuronal death. J. Neurosci. 4:1884-1891.
33. Hirsch J.A., Gibson G.E. (1984). Selective alterations of neurotransmitter release by low oxygen in vitro. Neurochem. Res. 9:1039-1049.
34. Grafstein B., Forman D.S. (1980). Intracellular transport in the nervous system. Physiol. Rev. 60:1167-1283.
35. Spencer P.S., Sabri M.I., Shaumburg H.H., Moore C.O. (1979). Does a defect of energy metabolism in the neurofibre underlie axonal degeneration in polyneuropathy? Ann. Neurol. 5:501-507.
36. Holmes G. (1907). A form of familial degeneration of the cerebellum. Brain 30:466-488.
37. Aoki C., Milner T.A., Sheu K.F.R., Blass J.P., Pickel V.M. (1987). Regional distribution of

astrocytes with intense immunoreactivity for glutamate dehydrogenase in rat brain: Implications for neuron-glia interactions in glutamate transmission. J. Neurosci. 7:2214–2231.
38. Aoki G., Milner T.A., Berger S.B., Sheu K.F.R., Blass J.P., Pickel V.M. (1987). Glial glutamate dehydrogenase: Ultrastructural localization and regional distribution in relation to the mitochondrial enzyme, cytochrome oxidase. J. Neurosci. Res. 18:305–318.
39. Plaitakis A., Berl S., Yahr M.D. (1984). Neurological disorders associated with deficiency of glutamate dehydrogenase. Ann. Neurol. 15:144–153.
40. Menzel P. (1891). Beitrage zur kenntnis der hereditaren ataxie und kleinhirn atrophie. Arch. Psychiat. Nervenkr. 22:160–190.
41. Pons G., Raefsky-Estrin C., Carothers D.J., Pepin R.A., Javed A.A., Jesse B.W., Ganapathi M.K., Sands D., Patel M.S. (1988). Cloning and cDNA sequence of the dehydrolipoamide dehydrogenase component of human alpha-ketoacid dehydrogenase complexes. Proc. Natl. Acad. Sci. USA 85:1422.

16. CEREBELLAR DISORDER IN THE HEXOSAMINIDASE DEFICIENCIES

WILLIAM G. JOHNSON

Hexosaminidase deficiency causes a spectrum of neurological diseases that are astonishing in their variability. The hexosaminidase deficiency diseases have been reviewed previously [1-9]. One important phenotype of hexosaminidase deficiency diseases is cerebellar disorder. The focus of this discussion is the nature of the cerebellar disorder and its relation to the other phenotypes of hexosaminidase deficiency disease. An ancillary question is whether the different hexosaminidase deficiency phenotypes are quantitatively or qualitatively different; that is, are the hexosaminidase deficiency phenotypes simply the result of increasingly severe deficiencies of enzyme activity, or do the different mutations cause different biochemical abnormalities, which then affect different brain areas to cause the different phenotypes?

1. THE PHENOTYPES OF HEXOSAMINIDASE DEFICIENCY DISEASE

Genetically determined deficiency of activity of the lysosomal hydrolase enzyme hexosaminidase (HEX) results in a number of progressive neurological disease phenotypes. These range from infantile encephalopathy, with onset at age 5-6 months and rapid progression, to disorders presenting in the third decade or later as very slowly progressive motor neuron disease. In between are late-infantile and juvenile encephalopathy, and syndromes presenting as cerebellar ataxia and dystonia. Mixed syndromes are common: Patients presenting with cerebellar disorder may develop motor neuron

A. Plaitakis (ed.), CEREBELLAR DEGENERATIONS: CLINICAL NEUROBIOLOGY. Copyright © 1992. Kluwer Academic Publishers, Boston. All rights reserved.

disease; patients presenting with encephalopathy may develop cerebellar disorder and motor neuron disease. Psychosis may occur as a feature of a more complex neurological disorder. These disorders are classified (1) according to the phenotype and the gene locus (alpha, beta, or activator locus), and if possible, the allele. Some, and probably most, of these later onset phenotypes are genetic compounds.

1.1. Infantile encephalopathy

1.1.1. Alpha-locus disorders

Classical infantile Tay-Sachs disease is the most severe HEX deficiency phenotype and also the first to be described. Warren Tay in 1881 and Bernard Sachs in 1887 described a disorder with onset at age 5–6 months in a previously healthy child. The initial features [10] are slowing of psychomotor development, diminished visual acuity, and myoclonic seizures. Auditory evoked myoclonus, an exaggerated startle reflex to loud sounds, may be prominent. A whitish deposit surrounds the normally red maculae of the ocular fundus, giving rise to the appearance of the macular cherry-red spot. The infant loses milestones, enters a vegetative state by the second year of life, and dies in the third or fourth year of life.

Infants never walk and do not develop enough voluntary limb movement to allow adequate clinical examination for cerebellar function. However, hypotonia is an early sign, often present by age 4 months. Likewise, infants do not develop early weakness or wasting, and are not regarded as having spinal muscular atrophy. Nonetheless, the limb wasting of the late stages of the disease, ascribed to inanition, occurs in spite of adequate diet by tube feeding, and probably, in fact, results from spinal muscular atrophy. Infants often have spasticity before age 6 months, in the presence of hypotonia and hyperactive deep-tendon reflexes.

Some patients with the B1 variant, an alpha-locus defect [11,12], may have the phenotype of infantile encephalopathy. These patients may be erroneously diagnosed as having the AB variant, a defect of the HEX A activator protein.

1.1.2. Beta-locus disorders

The clinical phenotype of classical infantile Sandhoff disease [13–17] is quite similar to that of classical infantile Tay-Sachs disease. The patients may have mild liver enlargement or may have cardiac involvement with congestive heart failure [18]. This disorder may occur in either homozygotes or genetic compounds for defects at the HEX beta locus. These patients, like the alpha-locus patients, cannot be readily examined clinically for cerebellar defects.

1.1.3. Activator-locus disorders

These patients have a clinical phenotype [19] which resembles that of classical infantile Tay-Sachs and Sandhoff disease. However, this disorder may progress more slowly.

1.2. Late-infantile and juvenile encephalopathy

1.2.1. Alpha-locus disorders

This is a heterogeneous group of disorders with onset from late in the first year of life to the first few years of life. In general, the same clinical features are present as in classical infantile Tay-Sachs disease, but the onset is later, the clinical course is slower, and the presenting features are more variable [20–33].

These patients develop dementia, seizures, and spasticity, which are progressive. Usually, these patients have begun to walk before the onset of symptoms, and they are observed to have gait ataxia. Muscle wasting due to anterior horn cell disease is often a feature. Macular cherry-red spot occasionally occurs. In at least one case, the presence of prominent upper and lower motor neuron findings led to the clinical diagnosis of Fazio-Londe disease.

In general, the earlier the onset, the more rapid the progression and the shorter the survival. These patients usually die in the first decade of life. Some survive into the second decade.

1.2.2. Beta-locus disorders

These patients are a heterogeneous group with clinical disorders that present after the first year of life, in the infantile or juvenile period. The patients closely resemble those with alpha-locus disorders [32,34–37]. Dementia, spasticity, ataxia, seizures, and muscle wasting due to anterior horn cell disease are the clinical features, and their mode and order of presentation are somewhat variable. Dementia, spasticity, and cerebellar ataxia tend to be earlier features, while seizures, dystonic posturing, and muscle wasting are later features. Macular cherry-red spot is a variable feature.

These patients have relentlessly progressive disease with death in the first or second decade.

1.3. Adult-onset encephalopathy

1.3.1. Activator-locus disorder

1.3.1.1. AB VARIANT WITH ADULT-ONSET ENCEPHALOPATHY. This phenotype is documented [38] in a single, incompletely studied case and therefore must be regarded as tentative. The patient had onset at age 18 years of progressive encephalopathy with dementia and seizures. Normal pressure hydrocephalus was also present. On autopsy after a 4-year course, light microcopy showed PAS-positive material in cells of the brain and spleen. Ultrastructural study showed multilamellar cytoplasmic inclusions in cells of the brain, spleen, and arachnoid granulations, thus explaining the normal pressure hydrocephalus.

The conclusion that this was an example of the AB variant rests solely on the observation that G_{M2} ganglioside was increased in cerebral cortex and that HEX A and B were elevated. It would be of interest to restudy this case using PCR methods.

1.4. Cerebellar ataxia

1.4.1. Alpha-locus disorders

A group of patients present with atypical spinocerebellar ataxia in the first decade of life or early adolescence [33,39,40]. These patients present with dysarthria, or ataxia of the limbs or gait. The disorder is slowly progressive. Although the clinical features may be those of spinocerebellar ataxia initially, with time a mixed syndrome develops. Upper motor neuron features develop or may be noted at the outset. These include hyper-reflexia, spasticity, and less frequently Babinski signs. Lower motor neuron signs, weakness and wasting, may absent or may be missed early in the course. The patient may simply appear to have thin arms and legs. However, testing strength by the ability to rise from a squatting position with support for balance may elicit unsuspected weakness. Dystonic postures may be noticed.

Dementia may be present or absent in this group of patients. This variation may occur even in patients in the same sibship. In patients with early-onset and severe dysarthria, mild dementia may be difficult to detect. Psychosis may occur and when it does is usually a precursor of dementia. Seizures may occur, usually as a late feature.

These patients do not develop visual loss with macular cherry-red spot, sensory loss, or bladder involvement.

1.4.2. Beta-locus disorders

1.4.2.1. JUVENILE CEREBELLAR ATAXIA. In patients presenting after the second year of life, cerebellar ataxia of gait and limbs may be the presenting feature [41]. The limb ataxia may be striking with brisk intention tremor of the arms, giving the appearance of the Ramsey Hunt syndrome, dyssynergia cerebellaris progressiva. The macular cherry-red spot, when present, may be atypical and difficult to see. Ocular fundus photography gives a clearer picture than simple funduscopic observation through the undilated pupil.

The initial presenting motor features may be isolated and slowly progressive for several years, but eventually dementia, seizures, spasticity, and muscle wasting supervene. Death is in the first or second decade.

1.4.2.2. ADULT-ONSET SPINOCEREBELLAR DEGENERATION. Onset after age 20 years has been reported [42,43] of a progressive spinocerebellar syndrome with gait ataxia and limb ataxia. Additional features were limb chorea, hyperreflexia, and facial grimacing. Decreased vibratory sense and slightly decreased joint position sense, unusual features in hexosaminidase deficiency disease, have been noticed in this syndrome. Macular cherry-red spots were absent.

1.5. Dystonia

1.5.1. Alpha-locus disorder

1.5.1.1. PROGRESSIVE JUVENILE DYSTONIA. This disorder resembles idiopathic dystonia musculorum deformans, except for the presence of additional features, especially dementia [44,45]. These patients become symptomatic in the early juvenile period with signs of dystonia, which include abnormal postures of the face, trunk, and limbs; abnormal gait due to dystonic posture of the foot; and speech abnormality with stuttering and dysarthria. Voluntary movement is clumsy but without dysmetria. However, other features are atypical for dystonia musculorum deformans. Speech delay, low IQ, behavioral disturbances, and learning disability indicate a mental defect, which becomes progressive. Hyperreflexia and action tremor may be noted, along with nystagmus.

The additional clinical features place this disorder in the group of secondary or symptomatic dystonias. Interestingly, dystonia has also been reported in a form of G_{M1} gangliosidosis [46–48].

1.6. Motor neuron disease

1.6.1. Alpha-locus disorders

A group of late-onset hexosaminidase deficiencies present with the clinical features of progressive spinal muscular atrophy or may have in addition signs of upper motor neuron dysfunction [5,49–67].

1.6.1.1. SPINAL MUSCULAR ATROPHY. This disorder presents as juvenile spinal muscular atrophy (Kugelberg-Welander syndrome) or adult spinal muscular atrophy (Aran-Duchenne syndrome). Patients present with signs of lower limb weakness, such as gait disturbance, stumbling, trouble climbing stairs, difficulty depressing an automobile brake pedal, or trouble getting up out of a chair. Weakness, wasting and fasciculations, and decreased or absent deep-tendon reflexes are found on examination, without clear signs of upper motor neuron involvement. The disorder begins in the legs and progresses to the arms late or not at all.

Onset of the disorder is in the late teens, the second or third decade, or later. Precise onset may be difficult to date because patients often remember being unathletic or clumsy as children. Macular cherry-red spot, cerebellar signs, and seizures are absent. Dementia may be present or absent. Patients may continue for decades as intelligent, active, and gainfully employed. Psychosis may occur, usually as a sign of early dementia. Psychosis may resemble schizophrenia, depression, or manic-depressive disorder.

The disorder is progressive. Patients are usually wheelchair bound by age 40, though their life span may be normal.

1.6.1.2. "ALS" PHENOTYPE. These patients closely resemble the group described as having spinal muscular atrophy, except that they have upper motor neuron signs. Usually, this means hyperactive deep-tendon reflexes in the legs and often in the arms as well. Babinski signs occur less commonly. Weakness occurs first in the lower limbs and later, or not at all, in the arms. Hyper-reflexia and clumsiness may occur in the upper limbs without weakness. Patients with spinal muscular atrophy and hyperreflexia are more common than those with spinal muscular atrophy alone. Patients with spinal muscular atrophy alone may develop hyperreflexia.

The resemblance of this disorder to amyotrophic lateral sclerosis (ALS) is not close, except for the neurological examination. Patients with ALS usually have onset of symptoms after age 40 years, usually have rapidly progressive disease, and are of no particular ethnic predilection. Patients with motor neuron disease resulting from hexosaminidase alpha-locus deficiency usually have the onset of symptoms before age 40 years, have slowly progressive disease, and are usually of Ashkenazi Jewish background.

1.6.2. Beta-locus disorders

Defects at the hexosaminidase beta locus can produce a very similar clinical picture to the two disorders just described [42,43,58,68–71].

1.7. Hexosaminidase deficiency in asymptomatic adults

Asymptomatic adults have been described with the biochemical findings of either Tay-Sachs disease or Sandhoff disease. In almost all cases, these individuals have later become symptomatic with one of the adult-onset phenotypes of hexosaminidase deficiency disease. Thus, these asymptomatic individuals are better regarded as presymptomatic.

2. PATHOLOGY

The pathology of Tay-Sachs disease, Sandhoff disease, and some other forms of hexosaminidase deficiencies has been described in detail. However, certain features are of special interest here.

2.1. Classical infantile Tay-Sachs disease

In the case of classical infantile Tay-Sachs disease [10,72–74], the cerebellum has been noted to be remarkably small in comparison with other parts of the brain. After age 2 years, the weight of the cerebral hemispheres progressively increases in comparison to unaffected controls. This increase occurs despite extensive demyelination in the hemispheres. However, after age 2 years, the weight of the cerebellum progressively decreases in comparison with unaffected controls [10], reaching less than 50% of control weight by age 5 years.

The same observation has been made by computerized tomography: The cerebellum and brain stem have been found to be small [74].

On gross examination of the brain in patients who have survived more than 2 years, the cerebellar folia are atrophic and thinned with prominent sulci. The inferior portions of the cerebellum, especially the inferior vermis, tonsils, and biventral lobes are shrunken [10].

The intensity of the histologic changes is remarkably variable, although neurons in the entire central nervous system are involved. Neurons in the substantia nigra, hippocampal cortex, pons, and globus pallidus are relatively uninvolved [10]. However, extensive degenerative changes occur in the cerebellum. Purkinje cells and cells of the internal granular layer show extensive destruction and loss. Cells in the molecular layer are lost. Cells in the dentate nucleus and other deep nuclei show marked ballooning due to lipid storage. Diffuse demyelination and axonal loss is seen in the cerebellar white matter [10,74]. Similar changes are widespread, though less pronounced, in the cerebral hemispheres. The anterior horn cells of the spinal cord are especially severely affected.

The same pattern of changes in seen in the fetus where ultrastructural changes are seen early and frequently in the spinal cord but very rarely in the cerebral cortex [73].

2.2. Infantile sandhoff disease

Cerebellar pathology is similar in infantile Sandhoff disease. Cerebellum has been reported [13-17] to show atrophy, often marked, of the folia with variable atrophy of the white matter. On microscopic examination, severe parenchymatous degeneration has been noted with a marked decrease of Purkinje cells and severely decreased granular cells. Interestingly, the cerebral cortex may be less severely affected than the cerebellum [15]. The brain stem, pyramidal tracts, and spinal cord (anterior horn cells and white matter tracts) are also involved.

2.3. Other hexosaminidase deficiency diseases

In a case [21] of juvenile G_{M2} gangliosidosis with hexosaminidase A deficiency, autopsy showed severe cerebellar involvement. There was marked atrophy of the cerebellum with slight atrophy of the cerebellar white matter. Cerebral cortical atrophy was less marked than that of the cerebellar cortex. On microscopic examination, there was severe cerebellar parenchymatous degeneration, with a marked decrease in Purkinje cells and a severe decrease in granular cells. Interestingly, the degree of cellular involvement was less severe in the cerebral cortex than in some other areas, including the brain stem and anterior horn cells of the spinal cord. Interestingly, there was a normal number and distribution of cells in all retinal layers, although some cellular pathology was noted. Cerebellar atrophy has also been noticed on CT scan in patients in this group [29].

In one case of adult G_{M2} gangliosidosis with hexosaminidase A deficiency, autopsy findings are available [39]. This patient survived until age 16 and two similarly affected sibs are alive in the fifth decade. However, this was not an adult-onset case, since onset was at about age 2 years, but is a case of chronic childhood-onset disease. The clinical picture was that of atypical spinocerebellar atrophy. There was involvement of cerebellar and pyramidal systems, along with spinal muscular atrophy and mild dementia without macular cherry-red spot. At autopsy the cerebellum was very small, with extreme atrophy of the folia in the vermis and hemispheres. Cerebral hemispheres showed only mild atrophy. The brain stem and spinal cord appeared normal. On microscopic examination, there was loss of granular cells and Purkinje cells. Many of the remaining cells contained storage material. Neurons throughout the nervous system showed evidence of storage. Swollen neurons were prominent in the spinal-cord grey matter. Under the electron microscope, intracytoplasmic lamellar inclusions were most prominent in the anterior horn cells and Purkinje cells.

Cerebellar atrophy has also been noted by CT scan in patients in this group [56,57].

3. BIOCHEMICAL ABNORMALITIES IN HEXOSAMINIDASE DEFICIENCY DISEASES

HEX substrates [3] are complex polymeric glycans, characterized by a terminal, beta-linked, N-acetylated galactosaminide or glucosaminide moiety. Most of these are membrane-related compounds, that is, they consist of a hydrophobic portion embedded in the membrane and a hydrophilic portion extending into the extracellular fluid.

Glycosaminoglycans include gangliosides, glycolipids, oligosaccharides, mucopolysaccharides, and glycoproteins. Storage of these compounds occurs in various combinations in a large number of lysosomal diseases. In the hexosaminidase deficiencies, storage of gangliosides and glycolipids occurs in the alpha-locus disorders [75]. Additional storage of oligosaccharides occurs in the beta-locus disorders [76]. G_{M2} ganglioside, but not the glycolipids or oligosaccharides, requires HEX A and activator protein for catabolism. Sulfated mucopolysaccharide is also reported to require HEX A for catabolism [77].

4. ENZYMATIC ABNORMALITIES IN HEXOSAMINIDASE DEFICIENCY DISEASES

Three major HEX isoenzymes make up the HEX ABS system in humans [7]. Hex A is composed of alpha and beta subunits [78]; HEX S and HEX B are homopolymers of the alpha-and beta-subunit, respectively. Only HEX A is active against G_{M2}-ganglioside [75,79] and sulfated hexosaminides [77]. HEX A and B are active against glycolipids, most oligosaccharides [80], and glycoproteins. For full HEX A activity, a protein activator [81–86] is required. HEX enzyme subunits, synthesized as prepropeptides [78,87,88], are processed in the cell to mature peptides of HEX A and HEX B [88,89], and are then glycosylated and phosphorylated [90]. A leader sequence appears to exist [87].

5. MOLECULAR BASIS OF HEXOSAMINIDASE DEFICIENCIES

The nucleotide sequences of the HEX alpha-locus coding region [91,92], of the HEX beta-locus coding region [92,93], and in part, the gene organization [94,95], have been determined. Various defects of HEX beta-subunit mRNA have been found in Sandhoff disease [96,97]. Defects have been found in alpha-locus disorders affecting exon 11 [98], intron 12 [99–101], exon 1 [102], exon 13 [103–104] (two defects), exon 5 [105–107], and exon 7 [108–109]. cDNA for the hexosaminidase activator protein has been cloned [102]. The HEX developmental regulation has been studied for evidence of temporal gene alleles thus far in a mouse [111] and an inbred mouse [112] system.

6. RELATIONSHIP OF THE PHENOTYPE TO THE ENZYMATIC AND MOLECULAR DEFECTS

An unanswered question that arises from this discussion is how deficiency of activity of a single enzyme system, HEX, can result in so many different clinical disorders where the disability is (1) confined to the nervous system and (2) referable in each to different groups of neurons.

The neuronal dysfunction in the hexosaminidase deficiencies is commonly believed to be caused by storage of G_{M2} ganglioside [2]. In part this is because gangliosides were the first compounds whose storage was recognized in the hexosaminidase deficiency diseases. Also, gangliosides and other lipids are easier to study than hexosaminidase substrates, such as oligosaccharides or glycoproteins. HEX A is deficient in all of the hexosaminidase deficiency diseases and HEX A, with its activator protein, is required to cleave G_{M2} ganglioside. Membranous cytoplasmic bodies (MCBs), which contain G_{M2} ganglioside, are the most visible ultrastructural abnormality in the hexosaminidase deficiencies.

However, morphologic abnormalities of dendrites and their appendages are not explained entirely by the MCBs. These MCBs are thought to cause disease by filling up the central cytoplasmic space and interfering with cell metabolism. Glycosaminoglycans are membrane components and are found on the plasma membrane as well. The cell plasma membrane is a crucial element in neuronal function [113–120]. It is possible that membrane components, including gangliosides or other hexosaminidase substrates may cause the functional disorder of neurons in other ways than by neuronal storage in MCBs; for example, by changing the structure and function of the plasma membrane. Therefore, it is not clear that neuronal dysfunction in the hexosaminidase deficiencies results entirely from the storage of ganglioside in lysosomes. It is possible that glycosaminoglycans at the cell surface contribute to the neuronal dysfunction.

The developmental program for hexosaminidase also remains unclear. Hexosaminidase substrates include gangliosides, which show significant developmental change [121–124]. G_{M2} ganglioside is suspected of having a developmental role in cell adhesion and recognition in development of the retino-tectal system [125]. Many glycoproteins are also developmentally regulated. The developmental program of hexosaminidase has been studied in mice [111] and inbred mice [112] only in whole organs, including the brain. The developmental program in neurons and in specific neuronal populations is unknown.

7. THE CEREBELLAR PHENOTYPE OF HEXOSAMINIDASE DEFICIENCY DISEASE

As the previous discussion has shown, cerebellar disorder is a feature of several different clinical phenotypes of hexosaminidase deficiency diseases. In infants who have never walked, gait ataxia cannot be detected. Limb ataxia is difficult to appreciate in infants who are just beginning to reach for objects and to transfer. However, hypotonia may be a manifestation of cerebellar disease. Also, as discussed earlier, both neuroimaging procedures and neuropathological examination give clear evidence of cerebellar disease in the infantile encephalopathy phenotype.

In the late-infantile and juvenile encephalopathy phenotype, dementia is the most common feature, but pyramidal system involvement and cerebellar system involvement are nearly as common. In addition, any of these may be the presenting feature. Gait abnormality as a presenting feature may be due either to pyramidal involvement, cerebellar involvement, or both. However, no matter how the disorder presents, the disorder usually evolves to include dementia, pyramidal involvement, and cerebellar involvement. Macular cherry-red spot is present in about one third of the cases. Interestingly cherry-red spot is usually absent in patients with an onset of symptoms after age 18-24 months and is usually present in patients with an onset of symptoms before that. Seizures are noted in about half the cases. Muscle wasting has been noted in about one third of these cases, most often as a late manifestation. Probably it is present in most patients, but late in the course it is overlooked as a specific manifestation, because it is attributed to "inanition" rather than anterior horn cell involvement.

In the phenotype of atypical dystonia, cerebellar involvement was noted in one case but not in the other. Pyramidal involvement and mental defect were noted in both.

In the adult or chronic phenotypes, cerebellar involvement may be present or absent. The age of onset and severity of the disease seem to be important determining factors. In one group of patients with the so-called adult form of hexosaminidase deficiency disease, the patient is asymptomatic until adulthood—the third decade or later. These patients, then, have adult-onset disease. However, if a careful history is taken, it is usually noted that the patient was clumsy as a child, not an able athlete, and was not able to keep up with the other children. Of course, the same is true of a large number of individuals who do not have hexosaminidase deficiency. Indeed, most such people do not have any disease at all. However, in retrospect, in hexosaminidase-deficient patients who later turn out to become symptomatic as adults, clumsiness as a child is probably an early manifestation of their disease. If they had been carefully examined as a child, mild neurological abnormalities would probably have been detected. In families with multiple cases of adult-onset hexosaminidase deficiency disease, it is not uncommon to find that an apparently asymptomatic sib has neurological findings, usually evidence of anterior horn cell disease and upper-limb brisk reflexes.

A second group of patients with the adult phenotype have onset of symptoms in childhood, but the disease is only slowly progressive and the patients survive into adulthood with chronic progressive disability. These patients, have childhood-onset chronic disease with survival into adulthood.

In general, this second group of adult patients, those with childhood-onset chronic disease, have more severe disease and are more likely to have cerebellar manifestations.

What, then, can be said about the existence of a cerebellar phenotype of hexosaminidase deficiency disease? Does such a phenotype exist or not? Clearly, some patients with onset in the late-infantile period or later do present with cerebellar syndromes, most commonly ataxia, tremor (sustention, intention or both), or dysmetria.

Consequently, it is important to recognize hexosaminidase deficiency disease as a cause of cerebellar syndromes and to carry out diagnostic testing for hexosaminidase deficiency in this group of patients. The most important tests at present are measurements of hexosaminidases in serum, leukocytes, or cultured skin fibroblasts. Rectal biopsy for ganglion cell ultrastructure is an important adjunct test to confirm the causative relation of an enzyme defect to a clinical syndrome and to detect variants that may be missed by the commonly used tests. DNA-based testing is now becoming possible for some forms of hexosaminidase deficiency and will become more important in the future.

On careful neurological examination or during follow-up examinations, patients with hexosaminidase deficiency disease presenting with a cerebellar syndrome will almost always have involvement of other systems, most commonly evidence of pyramidal involvement, dementia, or evidence of anterior horn cell disease. Anterior horn cell involvement is particularly easy to overlook, even by experienced examiners. Testing the ability to rise from a squatting position and to walk on toes and on the heels is an efficient way to avoid missing this.

8. THE BASIS OF SELECTIVE SYSTEM INVOLVEMENT IN HEXOSAMINIDASE DEFICIENCY DISEASE

As already discussed, the hexosaminidase deficiencies comprise a varied group of clinical disorders with several clinical phenotypes. What is the basis for these different phenotypes? Are the different hexosaminidase deficiency phenotypes quantitatively or qualitatively different; that is, are the hexosaminidase deficiency phenotypes simply the result of increasingly severe deficiencies of enzyme activity, or do the different mutations cause different biochemical abnormalities, which then affect different brain areas to cause the different phenotypes?

8.1. Phenotype specificity from multiple substrates

One way in which different mutations can cause different biochemical abnormalities is for a mutation to affect differentially the action of the enzyme protein on its various substrates. O'Brien [126] has suggested that this might explain why some of G_{M1} gangliosidosis affect the brain and the skeleton to different degrees. This type of explanation could be relevant for the hexosaminidase deficiencies because there are a variety of different classes of hexosaminidase substrates. As discussed earlier, hexosaminidase A and B differ in their ability to cleave G_{M2} ganglioside and other charged substrates. Unfortunately for this hypothesis, the alpha-locus and beta-locus defects differ little in their clinical picture, and their differences relate more to peripheral involvement than to nervous system involvement. The chief substrate associated with nervous system involvement seems, at present, to be G_{M2} ganglioside. Perhaps other substrates, such as minor gangliosides, may play a role in pathogenesis, or other classes of substrates, such as glycoproteins, could play a role. However, there is little support at present for the hypothesis that the specificity for selective system involvement in the nervous system depends upon the differential storage of multiple substrates.

8.2. Phenotype specificity from residual enzyme levels

Another hypothesis for explaining multiple clinical phenotypes is that the differences are only quantitative; that is, the different mutations determine the amount of residual G_{M2} ganglioside-cleaving activity, and this is the sole determinant of the clinical picture. Studies of residual G_{M2} ganglioside-cleaving activity [119,120] document that the amount of residual activity correlates well with the age of onset and the severity of the clinical phenotype. Patients with infantile encephalopathy have undetectable G_{M2} ganglioside-cleaving activity; patients with late-infantile or juvenile encephalopathy have higher levels; patients with the adult syndromes have still more.

8.2.1. A hierarchy of neuronal system involvement

Even if the correlation of clinical phenotype and residual G_{M2}-cleaving hexosaminidase activity is correct, how could it explain the occurrence of a cerebellar syndrome in one patient, a cherry-red spot in another, and spinal muscular atrophy in a third? In Table 16-1, the various clinical syndromes of hexosaminidase deficiency are ordered from top to bottom by decreasing severity and increasing age of onset. Across the top are the six major neuronal systems involved. All six are involved in infantile encephalopathy.

Table 16-1.

Phenotype	Macular cherryred spot	Seizures	Cerebellar disorder	Dementia	Pyramidal dysorder (UMN)	Spinal muscular atrophy (LMN)
Infantile encephalopathy	+	+	(+)	+	+	+ (late)
Late infantile-juvenile encephalopathy	1/3	1/2	+	+	+	+ (late)
Juvenile cerebellar ataxia	±	+ (late)	+	+ (late)	+ (late)	+ (late)
Juvenile dystonia	−	−	±	+	+	±
Chronic childhood-onset form	−	±	±	+	+	+
Adult motor neuron disease	−	−	−	±	+	+
Adult spinal muscular atrophy	−	−	−	±	late	+

Fewer and fewer neuronal systems are clinically involved as the severity decreases, the age of onset increases, and the amount of residual G_{M2}-cleaving activity increases. There appears to be a hierarchy of neuronal system involvement with respect to the severity of disease required for their clinical involvement. Neuronal systems at the right of Table 16-1 are even involved with mild disease of adult onset and relatively high residual G_{M2}-cleaving hexosaminidase activity. For clinical involvement of the systems toward the left of the table, progressively more severe disease with earlier onset and less residual G_{M2}-cleaving hexosaminidase activity is required.

For example, clinical involvement of retinal ganglion cells (giving decreased visual acuity and macular cherry-red spot) is regularly seen only in infantile encephalopathy, the most severe phenotype. Retinal ganglion cells are involved in only a minority (one third) of cases with late-infantile and juvenile encephalopathy; and even in this group, the macular cherry-red spot is seen almost solely in patients with an onset of symptoms before 18–24 months. Macular cherry-red spot and decreased visual acuity are not a feature of the adult-onset phenotypes.

On the other hand, spinal muscular atrophy is seen with infantile encephalopathy and with all of the other phenotypes. Spinal muscular atrophy is under-reported for the reasons given earlier. Spinal muscular atrophy may be the presenting feature, or even the only feature, in patients with adult-onset disease. These patients may have only progressive spinal muscular atrophy for a long time. If an additional system is involved or becomes involved, this is usually the pyramidal system, when brisk tendon reflexes in the upper limbs are noticed. If yet another system becomes involved, psychiatric disorder, often the precursor of dementia, is usually the next to appear. Many patients with adult-onset disease presenting with spinal muscular atrophy never develop mental involvement or dementia. Likewise, cerebellar disorder appears less frequently in these patients; seizures are uncommon, and macular cherry-red spot is not a feature.

Retinal involvement and spinal muscular atrophy are at opposite poles of this hierarchy. Cerebellar disorder, dementia, and pyramidal involvement are intermediate and are less clearly distinguished in their order of appearance. The juvenile dystonia phenotype appears to be a partial exception to this hierarchy; however, these patients have not been followed for very long. There are also occasional other exceptional cases.

8.2.2. Neuronal susceptibility as the basis for phenotypic hierarchy

The hierarchy of neuronal system involvement and its relation to clinical severity, age of onset, and the level of residual enzyme activity suggest that the specificity for the clinical phenotype may reside in differential susceptibility of cells in different neuronal systems to a deficiency of G_{M2}-ganglioside-cleaving hexosaminidase activity. This hypothesis suggests that anterior horn cells are the most susceptible to hexosaminidase deficiency, while retinal ganglion cells are the least susceptible. It is possible that G_{M2} ganglioside is for some reason especially toxic to anterior horn cells, compared with retinal ganglion cells, or that anterior horn cells normally have less hexosaminidase activity than retinal ganglion cells. Thus this is a testable hypothesis.

8.3. Developmental regulation

An additional consideration is a possible role of development regulation. Developmental changes in enzyme activity are found in many systems and are a prominent feature of most lysosomal hydrolase enzymes that have been studied.

Regulation of lysosomal hydrolase enzymes, especially beta-glucuronidase, has been extensively studied in inbred mice, chiefly by Paigen and coworkers [129–132]. The HEX developmental regulation has been studied for evidence of temporal gene alleles thus far in a mouse [111] and inbred mouse [112] system.

The subject has been reviewed [130–131]. The mechanism that has emerged from this work [130–131] is that at least five classes of genes are active in the genetic control of lysosomal acid hydrolases: structural genes, processing genes, systemic regulator genes, inducibility genes, and temporal genes.

The picture that has emerged from these and many other studies of acid hydrolase genes in inbred mice and other organisms is that for each enzyme, a gene complex exists whose components are closely linked to or in some way are part of the structural gene. Temporal genes may be part of this complex or may be located elsewhere, usually on another chromosome. Still obscure are the mechanisms of action of temporal regulators. Those that act *trans* presumably produce a diffusible product, possibly a non-histone chromatin protein [132]. Those that act *cis* may involve receptors of a diffusable product. A related (but nontemporal) gene in the beta-glucuronidase complex, which regulates the magnitude of enzyme induction by androgen, appears to act at a pretranslational level [133-134]. Interestingly hexosaminidase is one of several enzymes in mice for which the urinary enzyme (ultimately renal in origin) shows androgen inducibility [135]. Also obscure are the mechanisms by which multiple enzymes show coordinated but tissue-specific temporal developmental programs.

Hexosaminidase in humans is likely to be subject to these various levels of regulation, including temporal regulation. It is possible that this could be of clinical significance if mutation altered the developmental program in such a way that the level of hexosaminidase activity was insufficient to prevent lysosomal storage of hexosaminidase substrates in a group of neurons at a particular developmental stage. If the brain hexosaminidase level normally rises and falls rapidly during early development and falls further slowly later in life, as is the case in mice, a mutation in a temporal regulator gene could affect this developmental program and cause hexosaminidase deficiency disease.

At present the quantitative hypothesis, that levels of residual enzyme determine in a hierarchical fashion the selective system involvement in hexosaminidase deficiency disease, is the simpliest and most plausible. However, as the molecular basis for various kinds of developmental regulation become better understood [136], it is possible that these too may play a role.

ACKNOWLEDGMENTS

Support is gratefully acknowledged from the National Institutes of Health (NS-15281 and NS-11766), the Muscular Dystrophy Association, the March of Dimes Birth Defects Foundation, and the Alexander Rapaport Foundation.

REFERENCES

1. Sandhoff K., Conzelmann E., Neufeld E.F., Kaback M.M., Suzuki K. (1989). The G_{M2} gangliosidoses. In: C.R. Soriver, A.L. Beaudet, W.S. Sly, D. Valle (eds): The Metabolic Basis of Inherited Disease, 6th ed. New York: McGraw-Hill, pp. 1807-1839.
2. Neufeld E.F. (1989). Natural history and inherited disorders of a lysosomal enzyme, beta-hexosaminidase. J. Biol. Chem. 264:10927-10930.
3. Johnson W.G. (1987). Neurological disorders with hexosaminidase deficiency. In: A.J. Moss (ed): Pediatrics Update. New York: Elsevier, pp. 91-104.
4. Suzuki K. (1984). Gangliosides and disease: A review. Adv. Exp. Med. Biol. 174:407-418.
5. Argov Z., Navon R. (1984). Clinical and genetic variations in the syndrome of adult G_{M2} gangliosidosis resulting from hexosaminidase A deficiency. Ann. Neurol. 16:14-20.
6. Johnson W.G. (1982). Hexosaminidase deficiency: A cause of recessively inherited motor neuron diseases. Adv. Neurol. 36:159-164.
7. Johnson W.G. (1983). Genetic heterogeneity of hexosaminidase deficiency diseases. ARNMD 60:215-237.
8. Johnson W.G. (1981). The clinical spectrum of hexosaminidase deficiency disorders. Neurology 31:1453-1456.

9. Beutler E. (1979). The biochemical genetics of the hexosaminidase system in man. Am. J. Hum. Genet. 31:95–105.
10. Volk B.W., Schneck L., Adachi M. (1970). Clinic, pathology and biochemistry of Tay-Sachs disease. In: P.J. Vinken G.W. Bruyn (eds): Leucodystrophies and Poliodystrophies. Handbook of Clinical Neurology, Vol. 10. New York: American Elsevier, pp. 385–426.
11. Gordon B.A., Gordon K.E., Hinton G.G., et al. (1988). Tay-Sachs disease: B1 variant. Pediatr. Neurol. 4:54–57.
12. Tanaka A., Ohno K., Sandhoff K., et al. (1990). G_{M2}-gangliosidosis B1 variant: Analysis of β-hexosaminidase α gene abnormalities in seven patients. Am. J. Hum. Genet. 46:329–339.
13. Pilz H., Muller D., Sandhoff K., ter Meulen V. (1968). Tay-Sachssche krankheit mit hexosaminidase-defect. Deutsch Med. Wochenschr. 39:1833–1839.
14. Suzuki Y., Jacob J.C., Suzuki K., Kutty K.M. (1971). G_{M2}-gangliosidosis with total hexosaminidase deficiency. Neurology 21:313–328.
15. Bain A.D., Tateson R., Anderson J.M. (1972). Sandhoff's disease (G_{M2} gangliosidosis, type 2) in a Scottish family. J. Ment. Defic. Res. 16:119–127.
16. Dolman C.L., Chang E., Duke R.J. (1973). Pathologic findings in Sandhoff disease. Arch. Pathol. 96:272–275.
17. Tatematsu M., Imaida K., Ito N., Togari H., Suzuki Y., Ogiu T. (1981). Sandhoff disease. Acta. Pathol. Jpn. 31:503–512.
18. Blieden L.C., Desnick R.J., Carter J.B., Krivit W., Moller J.H., Sharp H.L. (1974). Cardiac involvement in Sandhoff's disease. Am. J. Cardiol. 34:83–88.
19. De Baecque C.M., Suzuki K., Rapin I., Johnson A.B., Wethers D.L. (1975). G_{M2}-gangliosidosis, AB variant. Acta Neuropathol. (Berlin) 33:207–226.
20. Suzuki Y., Suzuki K. (1970). Partial deficiency of hexosaminidase component A in juvenile G_{M2}-gangliosidosis. Neurology 20:848–851.
21. Suzuki K., Rapin I., Suzuki Y., Ishii N. (1970). Juvenile G_{M2}-gangliosidosis. Clinical variant of Tay-Sachs disease or a new disease. Neurology 20:190–204.
22. Schneck L., Friedland J., Pourfar M., Saifer A., Volk B.W. (1970). Hexosaminidase activities in a case of systemic G_{M2} gangliosidosis of late infantile type. Proc. Soc. Exp. Biol. Med. 133:997–998.
23. Okada S., Veath M.L., O'Brien J.S. (1970). Juvenile G_{M2} gangliosidosis: Partial deficiency of hexosaminidase. A. J. Pediatr. 77:1063–1065.
24. Menkes J.H., O'Brien J.S., Okada S., Grippo J., Andrews J.M., Cancilla P.A. (1971). Juvenile G_{M2} gangliosidosis. Biochemical and ultrastructural studies on a new variant of Tay-Sachs disease. Arch. Neurol. 25:14–22.
25. Borri P.L., Bugiani O., Lauro G., Palladini G., Ravera G. (1971). Juvenile G_{M2}-gangliosidosis. A morphological and chemical study of a cerebral biopsy. Acta Neurol. Belg. 71:309–318.
26. Buxton P., Cumings J.N., Ellis R.B., et al. (1972). A case of G_{M2} gangliosidosis of late onset. J. Neurol. Neurosurg. Psychiatry 35:685–692.
27. Brett E.M., Ellis R.B., Haas L., et al. (1973). Late onset G_{M2}-gangliosidosis. Clinical, pathological, and biochemical studies on 8 patients. Arch. Dis. Child. 48:775–785.
28. Ben Yoseph Y., Baylerian M.S., Momoi T., Nadler H.L. (1983). Thermal activation of hexosaminidase A in a genetic compound with Tay-Sachs disease. J. Inherited Metab. Dis. 6:95–100.
29. Mantovani J.F., Vidgoff J., Cass M. (1985). Brain dysfunction in an adolescent with the neuromuscular form of hexosaminidase deficiency. Dev. Med. Child. Neurol. 27:664–667.
30. Charrow J., Inui K., Wenger D.A. (1985). Late onset G_{M2} gangliosidosis: An alpha-locus genetic compound with near normal hexosaminidase activity. Clin. Genet. 27:78–84.
31. Besley G.T., Broadhead D.M., Young J.A., (1987). G_{M2}-gangliosidosis variant with altered substrate specificity: Evidence for alpha-locus genetic compound. J. Inherited Metab. Dis. 10:403–404.
32. Adams C., Green S. (1986). Late-onset hexosaminidase A and hexosaminidase A and B deficiency: Family study and review. Dev. Med. Child. Neurol. 28:236–243.
33. Oates C.E., Bosch E.P., Hart M.N. (1986). Movement disorders associated with chronic G_{M2} gangliosidosis. Case report and review of the literature. Eur. Neurol. 25:154–159.
34. Goldie W.D., Holtzman D., Suzuki K. (1977). Chronic hexosaminidase A and B

deficiency. Ann. Neurol. 2:156–158.
35. MacLeod P.M., Wood S., Jan J.E., Applegarth D.A., Dolman C. (1977). Progressive cerebellar ataxia, spasticity, psychomotor retardation, and hexosaminidase deficiency in a 10-year-old child: Juvenile Sandhoff disease. Neurology 27:571–573.
36. Spence M.W., Ripley B.A., Embil J.A., Tribbles J.A.R. (1974). A new variant of Sandhoff's disease. Pediatr. Res. 8:628–637.
37. Van Hoof G., Evrard P., Hers H.G. (1972). An unusual case of G_{M2}-gangliosidosis with deficiency of hexosaminidase A and B. In: B.W. Volk, S.M. Aronson (eds): Sphingolipids, Sphingolipidoses and Allied Disorders. New York: Plenum, pp. 343–350.
38. O'Neill B., Butler A.B., Young E., Falk P.M., Bass N.H. (1978). Adult-onset G_{M2}-gangliosidosis. Neurology 28:1117–1123.
39. Rapin I., Suzuki K., Valsamis M.P. (1976). Adult (chronic) G_{M2} gangliosidosis. Arch. Neurol. 33:120–130.
40. Willner J.P., Grabowski G.A., Gordon R.E., Bender A.N., Desnick R.J. (1981). Chronic G_{M2} gangliosidosis masquerading as atypical Friedreich ataxia: Clinical morphologic, and biochemical studies of nine cases. Neurology 31:787–798.
41. Johnson W.G., Chutorial A., Miranda A. (1977). A new juvenile hexosaminidase deficiency disease presenting as cerebellar ataxia—clinical and biochemical studies. Neurology 27:1012–1018.
42. Oonk J.G.W., Van der Helm H.J., Martin J.J. (1979). Spinocerebellar degeneration: Hexosaminidase A and B deficiency in two adult sisters. Neurology 29:380–384.
43. Bolhuis P.A., Oonk J.G., Kamp P.E., et al. (1987). Ganglioside storage, hexosaminidase lability, and urinary oligosaccharides in adult Sandhoff's disease. Neurology 37:75–81.
44. Hardie R.J., Young E.P., Morgan Hughes J.A. (1988). Hexosaminidase A deficiency presenting as juvenile progressive dystonia (letter). J. Neurol. Neurosurg. Psychiatry 51:446–447.
45. Meek D., Wolfe L.S., Andermann E., Andermann F. (1984). Juvenile progressive dystonia: A new phenotype of G_{M2} gangliosidosis. Ann. Neurol. 15:348–352.
46. Goldman J.E., Katz D., Rapin I., Purpura D.P., Suzuki K. (1981). Chronic G_{M1} gangliosidosis presenting as dystonia: I. Clinical and pathological features. Ann. Neurol. 9:465–475.
47. Kobayashi T., Suzuki K. (1981). Chronic G_{M1} gangliosidosis presenting as dystonia: II. Biochemical studies. Ann. Neurol. 9:476–483.
48. Nakano T., Ikeda S., Kondo K., Yanagisawa N., Tsuji S. (1985). Adult G_{M1}-gangliosidosis: Clinical patterns and rectal biopsy. Neurology 35:875–880.
49. Kaback M., Miles J., Yaffe M., et al. (1978). Hexosaminidase-A (HEX A) deficiency in early adulthood: A new type of G_{M2} gangliosidosis. Am. J. Hum. Genet. 30:31A–31A.
50. Johnson W.G., Wigger H.J., Karp H.R., Glaubiger L.M., Rowland L.P. (1982). Juvenile spinal muscular atrophy—a new hexosaminidase deficiency phenotype. Ann. Neurol. 11:11–16.
51. Jellinger K., Anzil A.P., Seemann D., Bernheimer H. (1982). Adult G_{M2} gangliosidosis masquerading as slowly progressive muscular atrophy: Motor neuron disease phenotype. Clin. Neuropathol. 1:31–44.
52. Kolodny E.H., Lyerla T., Raghavan S.S., Seashore G., Fogelson H., Pope H.G. (1982). Significance of hexosaminidase A deficiency in adults. Neurology 32:A81–A82.
53. Johnson W.G., Hogan E., Hanson P.A., DeVivo D.C., Rowland L.P. (1983). Prognosis of late-onset hexosaminidase deficiency with spinal muscular atrophy. Neurology 33(Suppl. 2):155–155.
54. Dale A.J.D., Engel A.G., Rudd N.L. (1983). Familial hexosaminidase A deficiency with Kugelberg-Welander phenotype and mental change. Ann. Neurol. 14:109–109.
55. Parnes S., Karpati G., Carpenter S., Ng Ying Kin N.M.K., Wolfe L.S., Suranyi L. (1985). Hexosaminidase-A deficiency presenting as atypical juvenile-onset spinal muscular atrophy. Arch. Neurol. 42:1176–1180.
56. Mitsumoto H., Sliman R.J., Schafer I.A., et al. (1985). Motor neuron disease and adult hexosaminidase A deficiency in two families: Evidence for multisystem degeneration. Ann. Neurol. 17:378–385.
57. Harding A.E., Young E.P., Schon F. (1987). Adult-onset supranuclear ophthalmoplegia, cerebellar ataxia, and neurogenic proximal muscle weakness in a brother and sister:

Another hexosaminidase A deficiency syndrome. J. Neurol. Neurosurg. Psychiatry 50:687–690.
58. Federico A. (1987). G_{M2} gangliosidosis with a motor neuron disease phenotype: Clinical heterogeneity of hexosaminidase deficiency disease. Adv. Exp. Med. Biol. 209:19–23.
59. Karni A., Navon R., Sadeh M. (1988). Hexosaminidase A deficiency manifesting as spinal muscular atrophy of late onset. Ann. Neurol. 24:451–453.
60. Navon R., Sandbank U., Frisch A., Baram D., Adam A. (1986). Adult-onset G_{M2} gangliosidosis diagnosed in a fetus. Prenat. Diagn. 6:169–176.
61. Navon R., Argov Z., Frisch A. (1986). Hexosaminidase A deficiency in adults. Am. J. Med. Genet. 24:179–196.
62. Navon R., Argov Z., Brand N., Sandbank U. (1981). Adult G_{M2} gangliosidosis in association with Tay-Sachs disease: A new phenotype. Neurology 31:1397–1401.
63. Navon R., Brand N., Sandbank U. (1980). Adult (G_{M2}) gangliosidosis: Neurologic and biochemical findings in an apparently new type. Neurology 30:449–450.
64. Sliman R.J., Mitsumoto H., Schafer I.A., Horwitz S.J. (1983). A study of hexosaminidase-A deficiency in a patient with 'atypical amyotrophic lateral sclerosis.' Ann. Neurol. 14:148–149.
65. Willner J.P., Bender A.N., Strauss L., Yahr M., Desnick R.J. (1979). Total beta-hexosaminidase A deficiency in two adult Ashkenazi Jewish siblings: Report of a new clinical variant. Am. J. Hum. Genet. 31:86A–86A.
66. Yaffe M.G., Kaback M., Goldberg M., et al. (1979). An amyotrophic lateral sclerosis-like syndrome with hexosaminidase-A deficiency: A new type of G_{M2} gangliosidosis. Neurology 29:611–611.
67. Kaback M., Miles J., Yaffe M., et al. (1979). Type VI G_{M2} gangliosidosis: An amyotrophic lateral sclerosis phenocopy. West. Pedi. Neurosci. 121A–121A.
68. Rubin M., Karpati G., Wolfe L.S., Carpenter S., Klavins M.H., Mahuran D.J. (1988). Adult onset motor neuronopathy in the juvenile type of hexosaminidase A and B deficiency. J. Neurol. Sci. 87:103–119.
69. Federico A., Ciacci G., D'Amore I., et al. (1986). Late onset G_{M2} gangliosidosis with atypical motor neuron disease phenotype and hexosaminidase A and B deficiency. Neurology 36(Suppl. 1):301–301.
70. Cashman N.R., Antel J.P., Hancock L.W., et al. (1986). N-acetyl-beta-hexosaminidase beta locus defect and juvenile motor neuron disease: A case study. Ann. Neurol. 19:568–572.
71. Hancock L.W., Horwitz A.L., Cashman N.R., Antel J.P., Dawson G. (1985). N-acetyl-beta-hexosaminidase B deficiency in cultured fibroblasts from a patient with progressive motor neuron disease. Biochem. Biophys. Res. Commun. 130:1185–1192.
72. Schmitt H.P., Berlet H., Volk B. (1979). Peripheral intraaxonal storage in Tay-Sachs' disease (G_{M2}-gangliosidosis type 1). J. Neurol. Sci. 44:115–124.
73. Yamada E., Matsumoto M., Hazama T., Momoi T., Sudo M. (1981). Two siblings, including a fetus, with Tay-Sachs disease. Acta. Pathol. Jpn. 31:1053–1061.
74. Watanabe K., Mukawa A., Muto K., Nishikawa J., Takahashi S. (1985). Tay-Sachs disease with conspicuous cranial computerized tomographic appearances. Acta Pathol. Jpn. 35:1521–1532.
75. Sandhoff K., Harzer K., Wassle W., Jatzkewitz H. (1971). Enzyme alterations and lipid storage in three variants of Tay-Sachs disease. J. Neurochem. 18:2469–2489.
76. Warner T.G., deKremer R.D., Sjoberg E.R., Mock A.K. (1985). Characterization and analysis of branched-chain N-acetylglucosaminyl oligosaccharides accumulating in Sandhoff disease tissue. Evidence that biantennary bisected oligosaccharide side chains of glycoproteins are abundant substrates for lysosomes. J. Biol. Chem. 260:6194–6199.
77. Thompson J.N., Stoolmiller A.C., Matalon R., Dorfman A. (1973). N-acetyl-beta-hexosaminidase: Role in the degradation of glycosaminoglycans. Science 181:866–867.
78. Hasilik A., Neufeld E.F. (1980). Biosynthesis of lysosomal enzymes in fibroblasts. Synthesis as precursors of higher molecular weight. J. Biol. Chem. 255:4937–4945.
79. Bach G., Suzuki K. (1975). Heterogeneity of human hepatic N-acetyl-beta-D-hexosaminidase A activity toward natural glycosphingolipid substrates. J. Biol. Chem. 250:1328–1332.
80. Seyama Y., Yamakawa T. (1974). Multiple components of beta-N-acetylhexosaminidase from equine kidney. J. Biol. Chem. 75:495–507.

81. Conzelmann E., Sandhoff K., Nehrkorn H., Geiger G., Arnon R. (1978). Purification, biochemical and immunological characterization of hexosaminidase A from variant AB of infantile G_{M2} gangliosidosis. Eur. J. Biochem. 84:27–33.
82. Conzelmann E., Sandhoff K. (1978). AB variant of infantile G_{M2} gangliosidosis: Deficiency of a factor necessary for stimulation of hexosaminidase A-catalyzed degradation of ganglioside G_{M2} and glycolipid G_{A2}. Proc. Natl. Acad. Sci. USA 75:3979–3983.
83. Li Y.T., Mazzotta M.Y., Wan C.C., Orth R., Li S.C. (1973). Hydrolysis of Tay-Sachs ganglioside by beta-hexosaminidase A of human liver and urine. J. Biol. Chem. 248:7512–7515.
84. Li S.C., Nakamura T., Ogamo A., Li Y.T. (1979). Evidence for the presence of two separate protein activators for the enzymatic hydrolysis of G_{M1} and G_{M2} gangliosides. J. Biol. Chem. 254:10592–10595.
85. Hechtman P., LeBlanc D. (1977). Purification and properties of the hexosaminidase A-activating protein from human liver. Biochem. J. 167:693–701.
86. Hechtman P., Kachra Z. (1989). Interaction of activating protein and surfactants with human liver hexosaminidase A and G_{M2} ganglioside. Biochem. J. 185:583–591.
87. Proia R.L., Neufeld E.F. (1982). Synthesis of beta-hexosaminidase in cell-free translation and in intact fibroblasts: An insoluble precursor alpha chain in a rare form of Tay-Sachs disease. Proc. Natl. Acad. Sci. USA 79:6360–6364.
88. dAzzo A., Proia RL., Kolodny E.H., Kaback M.M., Neufeld E.F. (1984). Faulty association of alpha- and beta-subunits in some forms of beta-hexosaminidase A deficiency. J. Biol. Chem. 259:11070–11074.
89. Little L.E., Lau M.M., Quon D.V., Fowler A.V., Neufeld E.F. (1988). Proteolytic processing of the alpha-chain of the lysosomal enzyme, beta-hexosaminidase, in normal human fibroblasts. J. Biol. Chem. 263:4288–4292.
90. Sonderfeld-Fresko S., Proia R.L. (1989). Analysis of the glycosylation and phosphorylation of the lysosomal enzyme, beta-hexosaminidase B, by site-directed mutagenesis. J. Biol. Chem. 264:7692–7697.
91. Myerowitz R., Piekarz R., Neufeld E.F., Shows T.B., Suzuki K. (1985). Human beta-hexosaminidase alpha chain: Coding sequence and homology with the beta chain. Proc. Natl. Acad. Sci. USA 82:7830–7834.
92. Korneluk R.G., Mahuran D.J., Neote K., et al. (1986). Isolation of cDNA clones coding for the alpha-subunit of human beta-hexosaminidase. Extensive homology between the alpha- and beta-subunits and studies on Tay-Sachs disease. J. Biol. Chem. 261:8407–8413.
93. ODowd B.F., Quan F., Willard H.F., et al. (1985). Isolation of cDNA clones coding for the beta subunit of human beta-hexosaminidase. Proc. Natl. Acad. Sci. USA 82:1184–1188.
94. Proia R.L., Soravia E. (1987). Organization of the gene encoding the human beta-hexosaminidase alpha-chain [published erratum appears in J. Biol. Chem. 1987; 262(31):15322]; 262:5677–5681.
95. Proia R.L. (1988). Gene encoding the human beta-hexosaminidase beta chain: Extensive homology of intron placement in the alpha- and beta-chain genes. Proc. Natl. Acad. Sci. USA 85:1883–1887.
96. ODowd B.F., Klavins M.H., Willard H.F., Gravel R., Lowden J.A., Mahuran D.J. (1986). Molecular heterogeneity in the infantile and juvenile forms of Sandhoff disease (O-variant G_{M2} gangliosidosis). J. Biol. Chem. 261:12680–12685.
97. Nakano T., Suzuki K. (1989). Genetic cause of a juvenile form of Sandhoff disease. Abnormal splicing of beta-hexosaminidase beta chain gene transcript due to a point mutation within intron 12. J. Biol. Chem. 264:5155–5158.
98. Myerowitz R., Costigan F.C. (1988). The major defect in Ashkenazi Jews with Tay-Sachs disease is an insertion in the gene for the alpha-chain of beta-hexosaminidase. J. Biol. Chem. 263:18587–18589.
99. Myerowitz R. (1988). Splice junction mutation in some Ashkenazi Jews with Tay-Sachs disease: Evidence against a single defect within this ethnic group. Proc. Natl. Acad. Sci. USA 85:3955–3959.
100. Arpaia E., Dumbrille-Ross A., Maler T., et al. (1988). Identification of an altered splice site in Ashkenazi Tay-Sachs disease. Nature 333:85–86.
101. Ohno K., Suzuki K. (1988). A splicing defect due to an exon-intron junctional mutation results in abnormal beta-hexosaminidase alpha chain mRNAs in Ashkenazi Jewish patients

with Tay-Sachs disease. Biochem. Biophys. Res. Commun. 153:463–469.
102. Myerowitz R., Hogikyan N.D. (1986). Different mutations in Ashkenazi Jewish and non-Jewish French Canadians with Tay-Sachs disease. Science 232:1646–1648.
103. Nakano T., Muscillo M., Ohno K., Hoffman A.J., Suzuki K. (1988). A point mutation in the coding sequence of the beta-hexosaminidase alpha gene results in defective processing of the enzyme protein in an unusual G_{M2}-gangliosidosis variant. J. Neurochem. 51:984–987.
104. Lau M.M., Neufeld E.F. (1989). A frameshift mutation in a patient with Tay-Sachs disease causes premature termination and defective intracellular transport of the alpha-subunit of beta-hexosaminidase. J. Biol Chem. 264:21376–21380.
105. Ohno K., Suzuki K. (1988). Mutation in G_{M2}-gangliosidosis B1 variant. J. Neurochem. 50:316–318.
106. Tanaka A., Ohno K., Suzuki K. (1988). G_{M2}-gangliosidosis B1 variant: A wide geographic and ethnic distribution of the specific beta-hexosaminidase alpha chain mutation originally identified in a Puerto Rican patient. Biochem. Biophys. Res. Commun. 156:1015–1019.
107. Nakano T., Nanba E., Tanaka A., Ohno K., Suzuki Y., Suzuki K. (1990). A new point mutation within exon 5 of beta-hexosaminidase alpha gene in a Japanese infant with Tay-Sachs disease. Ann. Neurol 27:465–473.
108. Navon R., Proia R.L. (1989). The mutations in Ashkenazi Jews with adult G_{M2} gangliosidosis, the adult form of Tay-Sachs disease. Science 243:1471–1474.
109. Paw B.H., Kaback M.M., Neufeld E.F. (1989). Molecular basis of adult-onset and chronic G_{M2} gangliosidoses in patients of Ashkenazi Jewish origin: Substitution of serine for glycine at position 269 of the alpha-subunit of betah-hexosaminidase. Proc. Natl. Acad. Sci. USA 86:2413–2417.
110. Schroeder M., Klima H., Nakano T., et al. (1989). Isolation of a cDNA encoding the human G_{M2} activator protein. FEBS Lett. 251:197–200.
111. Johnson W.G., Hong J.L. (1984). Organ-specific developmental regulation for 10 lysosomal hydrolase enzymes. Trans. Am. Soc. Neurochem. 15:249 (abstract).
112. Salviati A., Johnson W.G., Hong J.L., Wu P.M. (1986). A new system for study of beta-hexosaminidase (HEX) regulation. Neurology 36(Suppl. 1):301 (abstract).
113. Dodd J., Solter D., Jessell T.M. (1984). Monoclonal antibodies against carbohydrate differentiation antigens identify subsets of primary sensory neurones. Nature 311:469–472.
114. Rutishauser U., Jessell T.M. (1988). Cell adhesion molecules in vertebrate neural development. Physiol. Rev. 68:819–857.
115. Regan L.J., Dodd J., Barondes S.H., Jessell T.M. (1986). Selective expression of endogenous lactose-binding lectins and lactoseries glycoconjugates in subsets of rat sensory neurons. Proc. Natl. Acad. Sci. USA 83:2248–2252.
116. Tanaka H., Obata K. (1984). Developmental changes in unique cell surface antigens of chick embryo spinal motoneurons and ganglion cells. Dev. Biol. 106:26–37.
117. Chou D.K., Dodd J., Jessell T.M., Costello C.E., Jungalwala F.B. (1989). Identification of alpha-galactose (alpha-fucose)-asialo-G_{M1} glycolipid expressed by subsets of rat dorsal root ganglion neurons. J. Biol. Chem. 264:3409–3415.
118. Jessell T.M., Dodd J. (1985). Structure and expression of differentiation antigens on functional subclasses of primary sensory neurons. Phil. Trans. R. Soc. Lond. (Biol.) 308:271–281.
119. Dodd J., Jessell T.M. (1986). Cell surface glycoconjugates and carbohydrate-binding proteins: Possible recognition signals in sensory neurone development. J. Exp. Biol. 124:225–238.
120. Denburg J.L., Caldwell R.T., Marner J.M. (1987). Differences in surface molecules of motor axon terminals correlated with cell-cell recognition. J. Neurobiol 18:407–416.
121. Svennerholm L., Bostroem K., Fredman P., Maansson J.-E., Rosengren B., Rynmark B.-M. (1989). Human brain gangliosides: Developmental changes from early fetal stage to advanced age. Biochim. Biophys. Acta 1005:109–117.
122. Nudelman E.D., Mandel U., Levery S.B., Kaizu T., Hakomori S. (1989). A series of disialogangliosides with binary 2→3 sialosyllactosamine structure, defined by monoclonal antibody NUH2, are oncodevelopmentally regulated antigens. J. Biol. Chem. 264:18719–18725.

123. Hirabayashi Y., Hirota M., Suzuki Y., Matsumoto M., Obata K., Ando S. (1989). Developmentally expressed O-acetyl ganglioside GT3 in fetal rat cerebral cortex. Neurosci. Lett. 106:193–198.
124. Hirabayashi Y., Hyogo A., Nakao T., et al. (1990). Isolation and characterization of extremely minor gangliosides, G_{M1b} and G_{D1a}, in adult bovine brains as developmentally regulated antigens. J. Biol. Chem. 265:8144–8151.
125. Roth S., Pierce J.M. (1982). A possible role for the ganglioside G_{M2} in the development of the avian visual projections. Progr. Clin. Biol. Res. 97:165–172.
126. O'Brien J.S. (1975). Molecular genetics of G_{M1} beta-galactosidase. Clin. Genet. 8:303.
127. Conzelmann E., Kytzia H.J., Navon R., Sandhoff K. (1983). Ganglioside G_{M2} N-acetyl-beta-D-galactosaminidase activity in cultured fibroblasts of late-infantile and adult G_{M2} gangliosidosis patients and of healthy probands with low hexosaminidase level. Am. J. Hum. Genet. 35:900–913.
128. Kolodny E.H., Raghavan S.S. (1983). G_{M2}-gangliosidosis. Hexosaminidase mutations not of the Tay-Sachs type produce unusual clinical variants. Trends Neurosci. 6:16–20.
129. Paigen K. (1961). The genetic control of enzyme activity during differentiation. Proc. Natl. Acad. Sci. USA 47:1641–1649.
130. Paigen K. (1979). Acid hydrolases as models of genetic control. Annu. Rev. Genet. 13:417–466.
131. Paigen K. (1980). Temporal genes and other developmental regulators in mammals. In: T. Leighton, W.F. Loomis (eds): The Molecular Genetics of Development. New York: Academic Press, pp. 419–470.
132. Lusis A.J., Chapman V.M., Wangenstein R.W., Paigen K. (1983). Trans-acting temporal locus within the beta-glucuronidase gene complex. Proc. Natl. Acad. Sci. USA 80:4398–4402.
133. Pfister K., Watson G., Chapman V., Paigen K. (1984). Kinetics of beta-glucuronidase induction by androgen. J. Biol. Chem. 259:5816–5820.
134. Paigen K., Jakubowski A.F. (1982). Progressive induction of beta-glucuronidase in individual kidney epithelial cells. Biochem. Genet. 20:875–881.
135. Swank R.T., Novak E., Brandt E.J., Skudlarek M. (1978). Genetics of lysosomal functions. In: D. Doyle (ed): Protein Turnover and Lysosome Function. New York: Academic Press, pp. 251–271.
136. Funkenstein B., Leary S.L., Stein J.C., Catteral J.F. (1988). Genomic organization and sequence of the Gus-sa allele of the murine beta-glucuronidase gene. Mol. Cell Biol. 8:1160–1168.

17. DOMINANT OLIVOPONTOCEREBELLAR ATROPHY MAPPING TO HUMAN CHROMOSOME 6p

MARTHA A. NANCE AND LAWRENCE J. SCHUT

1. INTRODUCTION

The adult-onset form of autosomal dominant spinocerebellar ataxia genetically linked to chromosome 6p is the subject of this chapter. Although the first genetic linkage studies of spinocerebellar ataxia were performed in the 1950s using blood markers, it was not until the 1970s that a biochemical marker genetically linked to the disorder was identified. This marker was human leukocyte antigen (HLA), a highly variable cell-surface protein marker for which individuals can easily be typed. More precise mapping of the gene locus for the HLA-linked form of spinocerebellar ataxia, now known as spinocerebellar ataxia type 1 (SCA 1), required the sophisticated statistical methods of computerized linkage analysis that were developed in the 1970s and 1980s. In 1990, a new DNA marker, called D6S89, was found to be very closely linked to SCA 1; using this marker, ataxia families can now be screened rapidly for SCA 1. After a brief description of the clinical features of SCA 1, we review below the genetic linkage studies that resulted in the localization of the SCA 1 gene to chromosome 6p.

2. CLINICAL ASPECTS OF SCA 1

Broadly speaking, the hereditary ataxias can be classified according to their inheritance pattern into autosomal dominant, autosomal recessive, and X-

A. Plaitakis (ed.), CEREBELLAR DEGENERATIONS: CLINICAL NEUROBIOLOGY. Copyright © 1992.
Kluwer Academic Publishers, Boston. All rights reserved.

linked forms. X-linked ataxia is rare and will not be discussed here [1]. Autosomal recessive forms of hereditary ataxia include Friedreich's ataxia, ataxia telangiectasia, and those due to various disorders of metabolism that result in chronic or episodic cerebellar dysfunction. Most of these disorders are recognizable by their clinical or laboratory findings, and some are discussed elsewhere in this volume. The autosomal dominant ataxias are a group of related neurodegenerative disorders, primarily of adult onset, whose common feature is degeneration of the cerebellum or its afferent and efferent connections; the SCA 1 gene is the first dominant ataxia gene to be localized to a specific chromosome. The SCA 1 gene is now known to be located on chromosome 6p, distal to the HLA group of genes and proximal to the gene that encodes the coagulation factor subunit called Factor 13a.

Although SCA 1 within a given family can now be distinguished genetically from other ataxias by virtue of its linkage to markers derived from chromosome 6p, the clinical distinction between SCA 1 and other ataxias is less certain. Over the years, a number of authors have attempted to classify the adult-onset, autosomal dominant ataxias according to their clinical or pathologic features, but none of these classifications has proven entirely satisfactory [2–6]. Although a "pure" cerebellar degeneration, usually of later onset, was described by Holmes [7], most classification schemes have been based primarily on the frequency of associated involvement of noncerebellar systems. The hereditary ataxias with pancerebellar, olivary, and pontine degeneration are known as the olivopontocerebellar atrophies (OPCA). These are often referred to as multiple-system atrophies because of associated pyramidal tract involvement, degeneration of brain-stem nuclei, and less frequently, involvement of extrapyramidal tracts, amyotrophy, and dementia. The form of ataxia present in SCA 1 can be characterized as an OPCA; other terms that have been used to describe the condition include Marie's ataxia and Menzel's ataxia, and Schut-Swier or Haymaker-Schut ataxia. Machado-Joseph disease, spinopontine atrophy, and hereditary spastic paraplegia with ataxia are also autosomal dominant neurodegenerative disorders, which may include cerebellar dysfunction, but these are usually pathologically and clinically distinguishable from SCA 1 [8,9].

One of the great obstacles to the development of a scientifically meaningful and clinically useful classification of the dominant ataxias has been the variability of the symptoms and signs within a single kindred. J.W. Schut, for instance, separated members of the Minnesota kindred into three groups based on the age of onset and the degree of sensory, cerebellar, and corticospinal tract involvement [10]; Konigsmark and Weiner unwittingly classified some members of this same kindred under OPCA I and others under OPCA IV [4]. The basis for clinical variability within a single ataxia kindred remains unknown, but has a parallel in the phenotypic variability of other dominant neurologic diseases, such as Huntington disease (HD), myotonic dystrophy, and neurofibromatosis. In the case of HD, the parental origin of the HD mutation appears to be responsible for some of the variability in the age of onset of disease symptoms (juvenile-onset HD is much more frequently due to a paternally derived HD gene than to a maternally derived HD gene). The mechanism of this effect is uncertain, but genetic imprinting has been suggested as a possible mechanism [11]. Some evidence suggests that paternal inheritance of SCA 1 leads to an earlier onset of symptoms [12], but this remains controversial [13,14].

Clinical categorization of the ataxias is also hampered by variability in presentation among different kindreds. This variation could reflect mutations in two different genes, different mutations in the same gene, or additional hereditary or environmental factors remote from the primary gene defect. The genetic linkage studies described below suggest that at least two different genes can be responsible for clinically indistinguishable forms of hereditary ataxia, but whether the variation among families is due in part to different mutations in the same gene remains unknown. In the future, a greater understanding of the genetic lesions underlying hereditary ataxia will allow a new classification based on genetic linkage groups or underlying genetic etiology, rather than on the often overlapping or variable clinical and pathologic end products.

Table 17-1 shows the clinical features of affected patients in all reported SCA 1 families (Table 17-1a); for comparison, the clinical features of non-SCA 1 families are also shown (Table 17-1b). The affected members of SCA 1 families generally have an illness that could be called Marie's or Menzel's ataxia, OPCA I or OPCA IV, in that involvement of extrapyramidal tracts, higher cortical functions, peripheral nervous system, or sensory systems (retina) are minor features. The most prominent signs are the onset in adulthood of progressive gait ataxia, followed by dysarthria, dysphagia, and later by truncal and limb ataxia. Corticospinal tract involvement may be seen in the form of spasticity or hyperreflexia. Other neurologic findings are variable and are much less common, but include muscle fasciculations, optic nerve atrophy, amyotrophy, and dementia. The age of disease onset has shown variation within and among kindreds; additional (noncerebellar) signs are more common in individuals with a younger age of onset, and the rate of progression is faster and the average age at death younger in families with a lower mean age of onset [15]. Ophthalmoplegia or slow eye movements have been a prominent finding in two families proven not to have a chromosome 6p-linked disorder; and the dystonia, rigidity, and staring eyes that are prominent features in Machado-Joseph kindreds are not seen in SCA 1.

A number of smaller families of uncertain genetic linkage status have been reported [65–68,71–72]; these families, too, show cerebellar and bulbar dysfunction. Eye-movement abnormalities [65,66,72], dementia [65,71], and extrapyramidal signs [67,72] were each prominent in certain families but not others, and some families had additional features of tremor [67,68,72], skeletal deformity [67,71,72], and urinary dysfunction [67]. Now that it is possible to categorize families as having SCA 1 or non-SCA 1 by linkage

Table 17-1. Clinical findings in dominant hereditary ataxia

KINDREDS SHOWING PROBABLE OR DEFINITE LINKAGE TO HLA OR D6S89

Kindred	Mississippi	Michigan	Minnesota	Houston	Italian	Louisiana	Denmark	Denmark
References	46, 47, 50	51, 52, 56	10, 14, 40	13	58, 59	53	64, 65	66
Clinical finding								
1. Age of onset (mean)	20–40	26–59 (39)	15–35 (26)	6–60 (35)	20–54	15–35	18–58 (35)	20–50 (34)
2. Bulbar dysfunction	Frequent	Frequent	Frequent	Frequent	Frequent	Frequent	Frequent	Frequent
3. Nystagmus	Frequent	Frequent	Frequent	ND	Frequent	ND	Frequent	Frequent
4. Ocular motor palsy	Rare	ND	Frequent	Frequent	Frequent	ND	Rare	ND
5. Corticospinal deficits	Frequent	Frequent	Frequent	Occasional	Frequent	Frequent	Occasional	Occasional
6. Lower motor neuron signs	ND	ND	Frequent	Occasional	Frequent	Frequent	ND	ND
7. Sensory deficits	Frequent	Occasional	Occasional	Occasional	Absent	Frequent	Frequent	Frequent
8. Extrapyramidal signs	Occasional	ND	Occasional	ND	Absent	ND	ND	ND
9. Dementia	Rare	Rare	Occasional	ND	Absent	ND	Rare	Absent

KINDREDS UNLINKED TO HLA OR D6S89

Kindred	Cuban	Nebraska
References	54, 63	57
Clinical finding		
1. Age of onset (mean)	2–65 (32)	16–50 (30s)
2. Bulbar dysfunction	Frequent	Frequent
3. Nystagmus	Rare	Frequent
4. Ocular motor palsy	Frequent	Occasional
5. Corticospinal deficits	Occasional	Occasional
6. Lower motor neuron signs	ND	Occasional
7. Sensory deficits	Frequent	Frequent
8. Extrapyramidal signs	Rare	Occasional
9. Dementia	Rare	Absent

ND = Not described.

analysis, careful studies will be needed to determine which clinical, biochemical, radiologic, pathologic, or electrophysiologic features can distinguish these disorders in the clinical setting, which features are related to the severity or duration of the condition, and which features reflect true individual variability in the expression of the same genetic disorder.

3. GENETIC LINKAGE STUDIES

3.1. Rationale

The classical approach to the pathophysiology of genetic diseases begins with the identification of a protein, enzyme, or other identifiable biochemical marker of the disease. From this marker, the protein defect responsible for the disease can be identified (for instance, the hemoglobin S molecule of sickle-cell disease or the deficiency of the enzyme hexosaminidase A in Tay-Sachs disease). After the identity and mechanism of action of the normal protein are determined, the relationship between the altered protein and the disease can be elucidated. Modern molecular genetic techniques make it possible to identify the gene that encodes a protein, as well as the mutations that result in a nonfunctioning or dysfunctioning protein. Finally, the factors responsible for the normal regulation of the gene can be determined.

Unfortunately for the spinocerebellar ataxias, as for many other neurogenetic disorders, no clear biochemical marker or enzyme deficiency has been identified. Thus, a different approach is needed to identify the genetic abnormality or abnormalities that underlie these disorders. Genetic linkage studies provide the first step in the search for the genetic cause of a disease of unknown protein defect. Through linkage analysis, a disease gene can be localized to within a few million DNA base pairs on a chromosome, without any understanding of the identity, function, or protein product of the gene. Once the general location of the gene is known, molecular genetic methods can be used to identify and isolate the gene itself. When the disease gene is isolated, its protein product can be identified and studied, and the relationship between the abnormal protein and the resulting disease can be delineated. Using this "reverse-genetics" approach, then, the genetic basis of a disease of unknown biochemical cause can be established. A reverse-genetics approach has led to the identification of the proteins that are altered in Duchenne and Becker muscular dystrophies, neurofibromatosis I, and cystic fibrosis [16–18]. Linkage analysis has resulted in the approximate localization of many other genes, including those whose mutations result in Friedreich's ataxia, ataxia telangiectasia, Huntington disease, neurofibromatosis II, Charcot-Marie-Tooth disease, myotonic dystrophy, facioscapulohumeral dystrophy, and spinal muscular atrophy [19–30].

The purpose of a linkage study is to identify a biochemical marker that, not necessarily because of any causal relationship to the disease process, but because of its genetic location on a chromosome near the disease gene, identifies individuals in a family who have a genetic disease. Such a marker must have variable, distinguishable forms (polymorphisms). The most commonly used polymorphic markers in the past were enzymes or other proteins for which normal electrophoretic or immunologic variants exist (such as the ABO blood group). More recent technology permits the identification of normal variations in the DNA sequence itself; these DNA polymorphisms can also be used as markers, and they have several advantages over protein markers. They are more numerous than protein markers, and they may involve segments of DNA between genes or within a gene of unknown protein product (regions from which analogous protein markers could not be developed).

In order to distinguish between individuals in a family, the marker must show variability within the particular family being studied. If all individuals in a family have type AA blood, for instance, then ABO blood typing will not be helpful in distinguishing affected family members from unaffected family members. The ABO protein is uninformative in that particular family. A marker with few polymorphic forms, or a marker whose frequency of polymorphism is low, is more likely to be uninformative than a highly polymorphic marker. Highly variable markers, such as HLA or a special class of DNA markers, called variable number of tandem repeat markers (VNTRs), are likely to be informative in most families. The D6S89 marker, which has revolutionized linkage analysis in autosomal dominant ataxia, is a type of VNTR known as a GT-repeat marker.

Finally, the linkage study must demonstrate, to a statistically valid level of certainty, that a marker polymorphism passes from one generation to the next in association with the disease, such that affected members of a family show one polymorphism, while unaffected individuals have an alternate form of the marker. If a marker polymorphism passes with the disease frequently enough, the marker or the gene that encodes it must be located on the same chromosome as the disease gene, close to the disease gene.

The statistical analysis of genetic linkage data results in (1) an estimate of the genetic distance between the two loci, which can, in turn, be translated into a rough estimate of the physical distance between the loci, measured in DNA base pairs; and (2) a measure of the statistical likelihood of the genetic distance estimate. At best, a linkage study can place a disease gene to within about 1–2 million base pairs of a marker locus. As each human chromosome contains several thousand genes, each of which is composed of a sequence of a thousand to several million DNA base pairs, with variable lengths of DNA interspersed between them, isolating a gene is no trivial matter, even once it has been localized by genetic linkage. For example, although the first report of linkage between Huntington disease and a polymorphic DNA marker was published in 1983, and despite intensive eforts, by mid-1991 the HD gene has yet to be identified.

In addition to providing information useful to researchers attempting to discover the gene that causes a disease, a genetic linkage study can have immediate clinical benefits. If one form of the marker is found to be associated with the family disease, there is great likelihood that a family member with this marker form will also have the disease gene. Linkage studies have been used clinically in families for prenatal testing of fetuses at risk for Duchenne muscular dystrophy, Huntington disease, and myotonic dystrophy; and for predictive diagnosis of adult-onset disorders such as HD, adult polycystic kidney disease, and multiple endocrine neoplasia. Predictive testing of a child or adult, before the onset of clinical symptoms of a disorder, may have benefits for presymptomatic therapy, heightened medical vigilance, or for family, financial, or career planning.

3.2. Methods

The reader is referred to other sources for a thorough discussion of the genetic and mathematical concepts important to genetic linkage analysis [31–34]; only a brief description will be given here.

Two genes or DNA segments are said to be genetically linked if they pass through meiosis together in a nonrandom fashion. Meiosis is the special chromosome division that results in the passage of only one member of each chromosome pair into the developing egg or sperm. A normal feature of meiosis is the interchange, or crossing over, of genetic material between homologous chromosomes. This recombination of genetic material occurs frequently, at many points along the chromosome, and is, to a first approximation, a random event. Recombination between a marker and a disease gene on a chromosome in a developing egg or sperm cell causes the marker form previously located on the homologous chromosome no longer to accompany the disease gene in the progeny formed from that gamete. The chance that two genes or DNA segments far apart on a single chromosome will be separated during meiosis is 50%, depending on whether an odd or even number of recombination events have taken place between them. DNA segments that lie close to one another on the same chromosome, however, because of their physical proximity, have a smaller chance of being separated by recombination. The chance that these two nearby, or genetically linked, DNA segments will be separated by recombination varies from 0%, for segments that are identical or overlapping, up to 50%, the likelihood of recombination for unlinked loci. The frequency of recombination between two loci is called *theta*; theta can vary from 0.0 to 0.5.

For short distances, theta is proportional to the genetic distance, or map distance, between the loci. The map distance is measured in Morgans, where one centiMorgan (one 100th of a Morgan; cM) is roughly equal to a 1% recombination frequency. For the average chromosome region, one cM corresponds to approximately one million DNA base pairs. Thus, a theta value of 0.10 corresponds to a recombination frequency of 10%, a genetic distance of 10 cM, or about 10 million DNA base pairs. The correlations between recombination frequency, map distance, and physical distance on the chromosome are most accurate for closely linked loci. Recombination frequencies tend to be higher in females than in males, for unknown reasons. In addition, certain structural features of a chromosome region or DNA sequence may alter the relationship between the map distance and physical distance by an order of magnitude in either direction by creating recombination "hot spots" or areas particularly resistant to recombination.

In a genetic linkage study, each family member is scored for the presence or absence of the disease, as determined by physical examination, laboratory evaluation, or historical information, and for the marker or markers to which linkage is being sought. The body of data from the entire family or kindred are then assessed statistically for evidence of linkage between the marker and the disease. If one polymorphic form of the marker is always seen in association with the disease, then the marker locus and the disease gene locus are likely to be genetically linked, although it always remains possible that the observed association is a chance occurence. The odds that the data obtained for the family are the result of linkage between the disease locus and the marker locus (theta < 0.5) are compared with the odds that the data could have been obtained by chance, using a statistical formula that may also take into account the frequencies of the marker polymorphisms in the population and the age of onset of the disease symptoms in the family, among other variables. The logarithm of the resulting probability ratio is determined; this value is called the *lod score* (logarithm of the *od*ds). A lod score of three indicates 1000:1 odds in favor of linkage between the loci at that theta value, while a lod score of −2 indicates 100:1 odds against linkage. Lod scores above 3 or below −2 are generally accepted as statistically significant. Lod scores between 0 and 3 are suggestive of linkage but are not statistically conclusive; similarly, lod scores between 0 and −2 do not statistically exclude the possibility of linkage. The analysis is repeated several times using several possible theta values, such as 0.05, 0.1, 0.2, 0.3, and 0.4, and the results are often reported in a table or graph of lod scores for these assigned theta values. The theta value that corresponds to the highest lod score can be determined by a maximum likelihood function; this theta value represents the most likely distance between the loci. For example, a maximum lod score of 4.2 at theta = 0.12 indicates significant (>10,000:1) odds in favor of linkage between the two loci with a recombination frequency of 12%, corresponding to a map distance of 12 cM. For a typical chromosome region, this would represent on the order of 12 million bases of DNA.

Because the lod score is a logarithmic function, lod scores obtained in independent studies can be combined by simple addition. Thus, it may be possible to obtain statistically significant results by combining lod scores obtained from the analysis of several small families that individually lack statistical power.

Two-point linkage analysis works well for closely linked loci, but for loosely linked loci other kinds of analysis may be necessary. In particular, it may be difficult to determine whether a disease gene is on the centromeric side (toward the middle of the chromosome) or on the telomeric side (toward the end of the chromosome) of a marker. Multipoint linkage analysis allows the order of several disease genes or markers along a chromosome to be determined; in a fashion analogous to the two-point analysis, multipoint analysis provides a statistical estimate of the relative likelihoods of different proposed orders, called a *location score*. Multipoint analysis was very important in localizing the SCA 1 gene relative to HLA, Factor 13a, and other chromosome 6p markers, to which it was only loosely linked.

For any but the simplest pedigrees, the calculation of the probability functions, lod scores, the maximum likelihood function, and location scores is quite complex. Several computer programs have been written to perform these calculations, the most popular of which for two-point linkage analysis is LIPED [35], and for multilocus analysis, LINKAGE [36].

Clinical factors that complicate the interpretation of or invalidate linkage studies include (1) incorrect diagnosis (a major problem for studies of adult-onset disorders, such as the ataxias or dementias, in which the disease may not yet have become manifest in young or middle-aged adults; and for such disorders as the epilepsies, in which the clinical and electroencephalographic diagnoses may differ); (2) nonpaternity, which in large surveys has been shown to occur about 5% of the time; (3) incomplete or small families, whose linkage results often do not reach statistical significance; and (4) genetic heterogeneity for the disorder, which makes the interpretation and attempts to combine conflicting results from studies of different families difficult. Clinical and genetic heterogeneity have presented a major problem for the interpretation of genetic studies of the ataxias.

3.3. Linkage studies in spinocerebellar ataxia

Although Schut and his coworkers performed the first genetic linkage studies of spinocerebellar ataxia in the 1950s and 1960s [10,37,38], Yakura deserves the historical credit for being the first to identify an association between HLA and autosomal dominant ataxia [39]. His brief letter in 1974 described a nuclear family with Marie's ataxia; all affected children in the family had inherited the same HLA-(A-B) haplotype from their affected father, while all the unaffected children had inherited his other HLA haplotype.

Since 1974, linkage analysis with HLA has been performed in at least 35 ataxia families of varying ethnic origins. Table 17-2 summarizes the results of these studies, as well as the more recent studies using the DNA locus D6S89 (see below). Only four large kindreds of around 100 or more individuals have individually shown significantly positive linkage with HLA: the Minnesota kindred (originally from Holland) [10,14,37,38,40–44], the Houston kindred (a black American family) [13,45,59], the Mississippi kindred (of British ancestry) [46–50], and the Michigan kindred (of Prussian

Table 17-2. Linkage with HLA and D6S89 in autosomal dominant ataxia

Kindred name	Reference	Marker(s)*	Max lod score (Z max)	Theta value at Z max	Comments
A. KINDREDS SHOWING PROBABLE OR DEFINITE LINKAGE					
Yakura	39	HLA-A	1.2	0	Analysis in [47]
Mississippi	50	HLA-A	2.33	0.21	Individual lod scores not reported
3 Danish families	64, 65	HLA-(A-B-C)	1.67	0.01	
Denmark	66	HLA-(A-B-C)	1.37	0.05	
Houston	13, 45	HLA-(A-B)	3.62	0.14	
	59	D6S89	5.86	0.05	
Minnesota	44	HLA-A	8.11	0.18	Combined in [44]
	44, 57	D6S89	11.37	0	
Michigan	44	HLA-A	4.34	0.22	Combined in [44]
	44, 57	D6S89	9.78	0.03	
Louisiana	53	HLA	0	0.5	
	53	D6S89	4.9	0	
5 Italian families	58, 59	D6S89	16.35	0.01	
7 Japanese families	60	D6S89	4.95	0	
B. FAMILIES WITH INCONCLUSIVE RESULTS					
M I	67	HLA-(A-B)	0.54	0	Analysis in [49]
M II	67	HLA-(A-B-D)	0.22	0.21	Analysis in [49]
M II	67	HLA-(A-B)	0	0.5	Analysis in [49]
FC	68	HLA-(A-B)	0.02	0.37	Analysis in [49]
A	50	HLA-A	0	0.5	
O	50	HLA-A	0	0.5	
CA6	64, 65	HLA-(A-B-C)	0	0.5	Linkage excluded to theta ≈0.1
5 families	69, 70	HLA-(A-B)	0	0.5	Individual lod scores not reported
A	71	HLA-(A-B-C-Dr)	0.3	0.4	
B	71	HLA-(A-B-C-Dr)	0.5	0.3	
A	66	HLA-(A-B-C)	0	0.5	Linkage excluded to theta ≈0.1
C	66	HLA-(A-B-C)	0.19	0.1	
I	72	HLA-(A-B-C-Dr)	0.02	0.5	Linkage excluded to theta ≈0.1
II	72	HLA-(A-B-C-Dr)	0.26	0.45	Linkage excluded to theta ≈0.1
III	72	HLA-(A-B-C-Dr)	0	0.05	
IV	72	HLA-(A-B-C-Dr)	0	0.5	
V	72	HLA-(A-B-C-Dr)	0.56	0.05	
C. KINDREDS SHOWING PROBABLE OR DEFINITE ABSENCE OF LINKAGE					
Cuban	54, 55	HLA-Dq	0.05	0.4	Linkage excluded to theta ≈0.18
Nebraska	57	HLA-A	0	0.5	Linkage excluded to theta =0.08
	57	D6S89	0	0.5	Linkage excluded to theta =0.11
Rhode Island	73	D6S89			Linkage excluded to theta =0.06
8 Japanese families	60	D6S89			Reported to have negative lod scores

*Most studies used HLA serotyping, though some used DNA polymorphisms in the HLA region

origin) [44,51,52]. These families clearly demonstrated loose linkage between SCA 1 and HLA, at a distance of 14–22 cM. Although a number of authors have studied smaller families in an attempt to confirm or deny this linkage, all these studies have lacked the statistical power to exclude linkage at this distance (Table 17-2b). In fact, one family that showed negative lod scores at all theta values with HLA was subsequently shown to have SCA 1 by linkage analysis with D6S89 [53]. Thus, HLA linkage studies claiming to exclude SCA 1 as the genetic basis of ataxia in a family should be interpreted with caution until the studies are repeated with more closely linked markers, such as D6S89.

At least two large ataxia kindreds have been shown with reasonable certainty to be unlinked to the region of chromosome 6p that contains SCA 1. These are the Cuban kindred [54,55,63], and the Nebraska kindred of German origin [56,57] (Table 17-2c). In addition, several American and Japanese ataxia families, a number of Machado-Joseph kindreds, and at least two kindreds with pathologically proven spinopontine atrophy have shown no linkage to HLA or D6S89 [8,60–62,73]. Thus, the non-SCA 1 ataxias remain a heterogeneous groups of disorders, some clinically or pathologically distinct from SCA 1 (Machado-Joseph disease, spinopontine atrophy; see Chapter 8), but not yet distinguished clinicopathologically or biochemically from SCA 1 (Japanese, Cuban, and Nebraska kindreds). One author has suggested that SCA 1 presents a clinical phenotype intermediate between that of Machado-Joseph disease and the non-SCA 1 forms of ataxia, with the more frequent presence of spasticity, amyotrophy, lingual fasciculations, and external ophthalmoparesis in the SCA 1 kindreds [60]. Clearly, more careful clinical and pathologic characterization of both SCA 1 and non-SCA 1 ataxias will be necessary in the future.

Attempts to link spinocerebellar ataxia with other protein markers other than HLA whose genes are known to reside on chromosome 6p, including glyoxalase I enzyme (GLO) and Factor XIIIa (F13A), have been reserved for the larger kindreds, as these markers are not polymorphic enough to provide informative results in smaller families. Even in the larger kindreds, these markers have been more useful in multipoint linkage studies than in two-point analyses. Because HLA is only loosely linked to the SCA 1 gene, at a distance of approximately 14–20 cM, multilocus analyses were important in helping to determine on which side of the HLA locus the search for SCA 1 should focus. Multilocus studies in the Minnesota and Michigan kindreds, utilizing HLA-A, GLO, and F13A as well as DNA markers 1.10.2 and D6S7, showed conclusively that the SCA 1 gene in these families was located telomeric to HLA-A and centromeric to F13A [41–44], and it is in this region that the search for more closely linked DNA markers has focused.

Within the last 2 years, several advances in molecular genetics have greatly enhanced the search for the SCA 1 gene. A new class of polymorphic DNA markers, the GT-repeat markers, has been characterized. These repetitive sequences are interspersed throughout the genome at frequent intervals, and the number of repeats in a sequence is highly variable among individuals. In addition, these small variations in repetitive sequence length can be detected quickly in the laboratory by means of a reaction called the polymerase chain reaction (PCR), in which a small segment of DNA (containing the repetitive sequence) is recopied ("amplified") manyfold; this amplified segment of DNA can then be detected directly on a gel following electrophoresis, and segments differing in length by as little as two DNA base pairs can be distinguished. Standard polymorphic DNA markers (restriction fragment length polymorphisms, or RFLPs) are usually detected by laborious methods involving gel transfer and marker detection by autoradiography using radioactively labelled DNA probes, and the level of resolution is less using these methods than with the PCR-based markers.

A GT-repeat marker localizing to chromosome 6p, called D6S89 [74], has now been analyzed in over 20 ataxia kindreds. This marker appears to be extremely closely linked to the SCA 1 gene, with a maximum theta value of 0.0–0.03 [44,53,58–60]. Because this highly polymorphic marker is so closely linked to SCA 1, it can be used much more efficiently in smaller families than HLA. Reported linkage studies using D6S89 are shown in Table 17-2. Roughly half of the families studied to date show definitive evidence of linkage with D6S89, while the other half appear to be unlinked. Reanalysis of the families that yielded inconclusive results with HLA (Table 17-2b) should allow the definitive assignment of SCA 1 or non-SCA 1 in most cases.

As shown in Figure 17-1, linkage analysis with SCA 1, D6S89, and other nearby markers (D6S88 and D6S109) suggests that SCA 1 is proximal to D6S89, but the results are not conclusive [44,58,59]. Although the linkage studies performed to date would suggest that D6S89 may be as little as 2–3 cM from SCA 1, these studies too must be interpreted with caution, as few genetic recombinations have been identified with this marker. Independent evidence that D6S89 may be extremely close to SCA 1 was presented by Frontali et al. [58], who found significant linkage disequilibrium between SCA 1 and D6S89 in a series of Italian families. Linkage disequilibrium is the linkage of a disorder to a single marker polymorphism in unrelated families; its presence suggests that particular polymorphism was present on the ancestral chromosome on which the SCA 1 mutation occurred, and that recombination between the marker and the disease gene has been extremely rare over many generations. Based on their observation, these authors felt that the SCA 1 gene may be within 1000 kb of D6S89.

In summary, after 40 years of research using progressively more sophisticated methods of laboratory and statistical analysis, one gene responsible for adult-onset autosomal dominant ataxia has been localized within a small region of chromosome 6p. This genetic disorder, SCA 1, remains at this time clinically indistinguishable from other forms of dominant ataxia, which are apparently caused by a mutation or mutations at some other genetic locus.

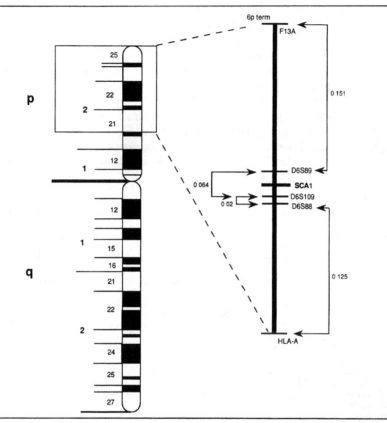

Figure 17-1. Current map of chromosome 6p. Distances are given as recombination frequencies between adjacent markers in a large reference panel of families. The most likely location of SCA 1 is at $\varnothing = 0.027$ proximal to D6S89.

SCA 1 appears to be genetically, pathologically, and clinically distinct from other neurodegenerative disorders, such as Machado-Joseph disease, spinopontine atrophy, Huntington disease, familial Alzheimer's disease, and familial amyotrophic lateral sclerosis.

The future will bring a number of developments in SCA 1 research and clinical management. The D6S89 marker can be used in both clinical and research settings to screen ataxia families for SCA 1. Once appropriate clinical protocols and ethical guidelines are established, this marker and others like it could be used for prenatal or presymptomatic testing for SCA 1. For research purposes, this marker can be used to identify patients with SCA 1 (as opposed to other forms of hereditary ataxia), which will facilitate clinical, biochemical, pathologic, radiologic, and electrophysiologic characterization of this disorder and comparison with non-SCA 1 forms of ataxia. Families previously felt to have non-HLA-linked forms of ataxia should be recategorized on the basis of linkage with this much more powerful marker; the search may then begin for the other gene or genes that result in disorders clinically similar but genetically distinct from SCA 1. Finally, localization of the SCA 1 gene to within 0–6 million DNA base pairs allows researchers to consider using physical molecular genetic means to isolate the SCA 1 gene itself. While there is no promise that the gene will be isolated quickly, these approaches could not have been considered 2 years ago, and there is reason to be optimistic that the SCA 1 gene will be identified and isolated in the future.

REFERENCES

1. McKusick V.A. (1990). Mendelian Inheritance in Man, 9th ed. Baltimore: Johns Hopkins University Press, p. 1567.
2. Greenfield J.G. (1954). The Spino-Cerebellar Degenerations. Springfield, IL: Charles C. Thomas, pp. 19–20.
3. Pratt R.T.C. (1967). The Genetics of Neurologic Disorders. London: Oxford University Press, pp. 31–44.
4. Konigsmark B.W., Weiner L.P. (1970). The olivopontocerebellar atrophies: A review. Medicine 49:227–241.
5. Currier R.D. (1984). A classification for ataxia. In: R.C. Duvoisin, A. Plaitakis (eds). The Olivopontocerebellar Atrophies. New York: Raven Press, pp. 1–4.
6. Harding A.E. (1984). The Hereditary Ataxias and Related Disorders. Edinburgh: Churchill Livingstone, pp. 174–204.
7. Holmes G. (1907). A form of familial degeneration of the cerebellum. Brain 30:466–488.
8. Bale A.E., Bale S.J., Schlesinger S.L., McFarland H.F. (1987). Linkage analysis in spinopontine atrophy: Correlation of HLA linkage with phenotypic findings in hereditary ataxia. Am. J. Med. Genet. 27:595–601.
9. Rosenberg R.N. (1986). Neurogenetics: Principles and Practice. New York: Raven Press, pp. 124–129.
10. Schut J.W. (1950). Hereditary ataxia: Clinical study through six generations. Arch. Neurol. Psychiatry 63:535–568.
11. Reik W. (1988). Genomic imprinting: A possible mechanism for the parental origin effect in Huntington's chorea. J. Med. Genet. 25:805–808.
12. Harding A.E. (1988). Genetic aspects of autosomal dominant late onset cerebellar ataxia. J. Med. Genet. 18:436–441.
13. Zoghbi H.Y., Pollack M.S., Lyons L.A., et al. (1988). Spinocerebellar ataxia: Variable age of onset and linkage to human leukocyte antigen in a large kindred. Ann. Neurol. 23:580–584.
14. Haines J.L., Schut L.J., Weitkamp L.R., et al. (1984). Spinocerebellar ataxia in a large kindred: Age at onset, reproduction, and genetic linkage studies. Neurology 34:1542–1548.
15. Currier R.D., Jackson J.F., Maydrech E.F. (1982). Progression rate and age at onset are related in autosomal dominant neurologic diseases. Neurology 32:907–909.
16. Hoffman E.P., Brown R.H., Kunkel L.M. (1987). Dystrophin: The protein product of the Duchenne muscular dystrophy locus. Cell 51:919–928.
17. Riordan J.R., Rommens J.M., Kerem B.-S., et al. (1989). Identification of the cystic fibrosis gene: Cloning and characterization of the complementary DNA. Science 245:1066–1073.
18. Xu G.F., et al. (1990). The neurofibromatosis gene encodes a protein related to GAP. Cell 62:599–608.
19. Gusella J.F., Wexler N.S., Conneally P.M., et al. (1983). A polymorphic DNA marker genetically linked to Huntington's disease. Nature 306:234–238.
20. Bates G.P., MacDonald M.E., Baxendale S., et al. (1991). Defined physical limits of the

Huntington disease gene candidate region. Am. J. Hum. Genet. 49:7-16.
21. Rouleau G.A., Wertelecki W., Haines J.L., et al. (1987). Genetic linkage of bilateral acoustic neurofibromatosis to a DNA marker on chromosome 22. Nature 329:246-248.
22. Rouleau G.A., Seizinger B.R., Wertelecki W., et al. (1990). Flanking markers bracket the neurofibromatosis type 2 (NF 2) gene on chromosome 22. Am. J. Hum. Genet. 46:323-328.
23. Middleton-Price H.R., Harding A.E., Monteiro C., et al. (1990). Linkage of hereditary motor and sensory neuropathy type I to the pericentromeric region of chromosome 17. Am. J. Hum. Genet. 46:92-94.
24. Yamaoka L.H., Pericak-Vance M.A., Speer M.C., et al. (1990). Tight linkage of creatine kinase (CKMM) to myotonic dystrophy on chromosome 19. Neurology 40:222-226.
25. Harley H.G., Brook J.D., Floyd J., et al. (1991). Detection of linkage disequilibrium between the myotonic dystrophy locus and a new polymorphic DNA marker. Am. J. Hum. Genet. 49:68-75.
26. Sanal O., Wei S., Foroud T., et al. (1990). Further mapping of an ataxia telangiectasia locus to the chromosome 11q23 region. Am. J. Hum. Genet. 47:860-866.
27. Shaw J., Lichter P., Driesel A.J., et al. (1990). Regional localization of the Friedreich ataxia locus to human chromosome 9q13-q21.1. Cytogenet. Cell Genet. 53:221-224.
28. Wijmenga C., Frants R.R., Brouwer O.F., et al. (1990). Location of the facioscapulohumeral dystrophy gene on chromosome 4. Lancet 336:651-653.
29. Melki J., Abdelhak S., Sheth P., et al. (1990). Gene for proximal spinal muscular atrophies maps to chromosome 5q. Nature 344:767-768.
30. Brzustowicz L.M., Lehner T., Castilla T., et al. (1990). Genetic mapping of chronic childhood spinal muscular atrophy to chromosome 5q11.2-13.3. Nature 344:540-541.
31. Haldane J.B.S., Smith C.A.B. (1947). A new estimate of linkage between the genes for colorblindness and hemophilia in man. Ann. Eugen. 14:10-31.
32. Morton N.E. (1955). Sequential tests for the detection of linkage. Am. J. Hum. Genet. 7:277-318.
33. Conneally P.M., Rivas M.L. (1980). Linkage analysis in man. In: H. Harris, K. Hirschhorn (eds): Advances in Human Genetics, Vol. 10. New York: Plenum, pp. 209-266.
34. Conneally P.M., Edwards J.H., Kidd K.K., et al. (1985). Human gene mapping 8: Report of the committee on methods of linkage analysis and reporting. Cytogenet. Cell. Genet. 40:356-359.
35. Ott J. (1974). Estimation of the recombination fraction in human pedigrees: Efficient computation of the likelihood for human linkage studies. Am. J. Hum. Genet. 26:588-597.
36. Lathrop G.M., Lalouel J.M., Julier C., Ott J. (1984). Strategies for multilocus linkage analysis in humans. Proc. Natl. Acad. Sci. USA 80:4808-4812.
37. Schut J.W. (1951). Hereditary ataxia: A survey of certain clinical, pathologic, and genetic features with linkage data on five additional hereditary factors. Am. J. Hum. Genet. 3(2):93-110.
38. Matson G.A., Schut J.W., Swanson J. (1961). Hereditary ataxia: Linkage studies in hereditary ataxia. Ann. Hum. Genet. 25:7-23.
39. Yakura H., Wakisaka A., Fujimoto S., Itakura K. (1974). Hereditary ataxia and HL-A genotypes. N. Engl. J. Med. 291:154-155.
40. Landis D.M.D., Rosenberg R., Landis S.C., et al. (1974). Olivopontocerebellar degeneration: Clinical and structural abnormalities. Arch. Neurol. 31:295-307.
41. Haines J.L., Trofatter J.A. (1986). Multipoint linkage analysis of spinocerebellar ataxia and markers on chromosome 6. Genet. Epidemiol 3:399-405.
42. Rich S.S., Wilkie P.J., Schut L.J., et al. (1987). Spinocerebellar ataxia: Localization of an autosomal dominant locus between two markers on human chromosome 6. Am. J. Hum. Genet. 41:524-531.
43. Wilkie P.J., Schut L.J., Orr H., Rich S.S. (1991). Spinocerebellar ataxia: Multipoint linkage analysis of genes associated with the disease locus. Hum Genet, in press.
44. Ranum L.P.W., Duvick L.A., Rich S.S., et al. (1991). Localization of the autosomal dominant HLA-linked spinocerebellar ataxia (SCA 1) locus, in two kindreds, within an 8-cM subregion of chromosome 6p. Am. J. Hum. Genet. 49:31-41.
45. Zoghbi H.Y., Sandkuyl L.A., Ott J., et al. (1989). Assignment of autosomal dominant spinocerebellar ataxia (SCA 1) centromeric to the HLA region on the short arm of chromosome 6, using multilocus linkage analysis. Am. J. Hum. Genet. 44:255-263.

46. Currier R.D., Glover G., Jackson J.F., Tipton A.C. (1972). Spinocerebellar ataxia: Study of a large kindred. I. General information and genetics. Neurology 22:1040–1043.
47. Jackson J.F., Currier R.D., Terasaki P.I., Morton N.E. (1977) Spinocerebellar ataxia and HLA linkage: Risk prediction by HLA typing. N. Engl. J. Med. 296:1138–1141.
48. Jackson J.F., Whittington J.E., Currier R.D., et al. (1978). Genetic linkage and spinocerebellar ataxia. In: R.A.P. Kark, R.N. Rosenberg, L.J. Schut (eds): Advances in Neurology, Vol. 21. New York: Raven Press, pp. 315–318.
49. Morton N.E., Lalouel J.-M., Jackson J.F., et al. (1980). Linkage studies in spinocerebellar ataxia (SCA). Am. J. Med. Genet. 6:251–257.
50. Whittington J.E., Keats B.J.B., Jackson J.F., et al. (1980). Linkage studies on glyoxalase I (GLO), pepsinogen (PG), spinocerebellar ataxia (SCA 1), and HLA. Cytogenet. Cell Genet. 28:145–150.
51. Nino H.E., Noreen H.J., Dubey D.P., et al. (1980). A family with hereditary ataxia: HLA typing. Neurology 30:12–20.
52. Noreen H.J., Nino H.E., Dubey D.P., et al. (1979). Genetic linkage with HLA in spinocerebellar ataxia. Transplant. Proc. 11:1729–1731.
53. Keats B.J.B., Pollack M.S., McCall A., et al. (1991). Tight linkage of the gene for spinocerebellar ataxia to D6S89 on the short arm of chromosome 6 in a kindred for which close linkage to both HLA and Factor 13A-1 is excluded. Am. J. Hum. Genet. 49:972–977.
54. Orozco Diaz G., Nodarse Fleites A., Cordoves Sagaz R., Auburger G. (1990). Autosomal dominant cerebellar ataxia: Clinical analysis of 263 patients from a homogeneous population in Holguin, Cuba. Neurology 40:1369–1375.
55. Auburger G., Orozco Diaz G., Ferreira Capote R., et al. (1990). Autosomal dominant ataxia: Genetic evidence for locus heterogeneity from a Cuban founder-effect population. Am. J. Hum. Genet. 46:1163–1173.
56. Aita J.F. (1978). Cranial computerized tomography and Marie's ataxia. Arch. Neurol. 35:55–56.
57. Ranum L.P.W., Rich S.S., Nance M.A., et al. Autosomal dominant spinocerebellar ataxia: Locus heterogeneity in a Nebraska kindred. Neurology, in press.
58. Frontali M., Jodice C., Lulli P., et al. (1991). Spinocerebellar ataxia (SCA 1) in two large Italian kindreds: Evidence in favour of a locus position distal to GLO1 and the HLA cluster. Ann. Hum. Genet. 55:7–15.
59. Zoghbi H.Y., Jodice C., Sandkuijl L.A., et al. (1991). The gene for autosomal dominant ataxia (SCA 1) maps telomeric to the HLA complex and is closely linked to the D6S89 locus in three large kindreds. Am. J. Hum. Genet. 49:23–30.
60. Sasaki H., Wakisaka A., Takada A., et al. (1991). Linkage study of dominant OPCA and Machado-Joseph disease in Japan (abstract). International Symposium on Ataxia, Boston, MA, April 1991.
61. Carson W.J., MacLeod P.L., Vincent D., et al. (1991). Linkage studies in Machado-Joseph disease (abstract). International Symposium on Ataxia, Boston MA, April 1991.
62. MacLeod P.M., Rouleau G., Simpson N.E., et al. (1991). Machado-Joseph disease: Linkage analysis between the loci for the disease and a number of conventional and DNA markers (abstract). International Symposium on Ataxia, Boston MA, April 1991.
63. Orozco G., Estrada R., Perry T.L., et al. (1989). Dominantly inherited olivopontocerebellar atrophy from eastern Cuba. Clinical, neuropathological, and biochemical findings. J. Neurol. Sci. 93:37–50.
64. Pedersen L. (1980). Hereditary ataxia in a large Danish pedigree. Clin. Genet. 17:385–393.
65. Pedersen L., Platz P., Ryder L.P., et al. (1980). A linkage study of hereditary ataxias and related disorders: Evidence of heterogeneity of dominant cerebellar ataxia. Hum. Genet. 54:371–383.
66. Werdelin L., Platz P., Lamm L.U. (1984). Linkage between late-onset, dominant spinocerebellar ataxia and HLA. Hum. Genet. 66:85–89.
67. Moller E., Hindfelt B., Olsson J.-E. (1978). HLA-determination in families with hereditary ataxia. Tissue Antigens 12:357–366.
68. Wastiaux J.P., Lamoureux G., Bouchard J.P., et al. (1978). HLA and complement typing in olivo-ponto-cerebellar atrophy. J. Can. Sci. Neurol. 5:75–81.
69. Koeppen A.H., Goedde H.W., Hirth L., Benkmann H.-G. (1980). Genetic linkage in hereditary ataxia. Lancet 1:92–93.

70. Koeppen A.H., Goedde H.W., Hiller C., et al. (1981). Hereditary ataxia and the sixth chromosome. Arch. Neurol. 348:158–164.
71. Van Rossum J., Veenema H., Went L.N. (1981). Linkage investigations in two families with hereditary ataxia. J. Neurol. Neurosurg. Psychiatr. 44:516–522.
72. Kumar D., Blank C.E., Gelsthorpe C. (1986). Hereditary cerebellar ataxia and genetic linkage with HLA. Hum. Genet. 72:327–332.
73. Lazzarini A., Zimmerman T.R., Johnson W., et al. (1991). A new family with autosomal dominant spinocerebellar ataxia linkage mapping studies. International Symposium on Ataxia, Boston MA, April 1991.
74. Litt M., Luty J.A. (1990). A TG microsatellite VNTR detected by PCR is located on 6p (HGM10 No. D6S89). Nucleic Acids Res. 18:4301.

18. POSITRON EMISSION TOMOGRAPHY STUDIES OF CEREBELLAR DEGENERATION

SID GILMAN

1. INTRODUCTION

The cerebellar degenerations consist of a group of progressive neurological diseases, usually characterized by ataxia of speech, limb movements, and gait, some of which are hereditary and some of which are sporadic. Many cerebellar disorders, particularly those encountered in infancy and childhood, are associated with specific metabolic disturbances, such as storage diseases, disorders of amino acid or lipid metabolism, leukodystrophies, dysproteinemias, and disorders of vitamin function [1]. Others are associated with alcoholism and malnutrition, exposure to toxins such as heavy metals, and neoplasms [2]. Many cerebellar degenerations, however, particularly those occurring in adulthood, are of unknown cause. Even in the cerebellar degenerations with a clear hereditary transmission and an established chromosomal location for the responsible gene [3,4], the biochemical pathogenesis for the disease is unknown. In some of the degenerations, such as olivopontocerebellar atrophy, biochemical analysis of postmortem tissue has disclosed disorders of amino acid neurotransmitter levels [5]. Only limited numbers of cases have been studied in this way, and several different abnormalities have been found [6-12].

Positron emission tomography (PET) offers an opportunity to study cerebral metabolic rates, cerebral blood flow, the density of neurotransmitter

A. Plaitakis (ed.), CEREBELLAR DEGENERATIONS: CLINICAL NEUROBIOLOGY. Copyright © 1992.
Kluwer Academic Publishers, Boston. All rights reserved.

presynaptic uptake sites, and the density of neurotransmitter receptors in living patients with neurological disorders. These studies are noninvasive and may be used safely in large numbers of patients. Thus, PET may provide an opportunity to understand more fully the cerebral metabolic characteristics and biochemical abnormalities of the cerebellar degenerations.

Recently, PET has been used to study cerebral glucose metablic activity in patients with two types of cerebellar degenerations: Friedreich's ataxia [13] and olivopontocerebellar atrophy [14–16]. The findings of these studies will be reviewed in this chapter and the prospects for examination of the biochemical disturbances in the cerebellar degenerations will be discussed.

2. FRIEDREICH'S ATAXIA

The cerebellar degenerations are notoriously difficult to classify, largely because the basic biochemical processes underlying most of them are unknown and because many of them are associated with diverse symptoms and a variable course. Friedreich's ataxia is one of the cerebellar degenerations that tends to display similar symptoms and signs from case to case and to run a similar course in most patients. It is a chronic progressive disease that is inherited as an autosomal recessive trait. The site of the abnormality is on chromosome 9 [3], but the biochemical abnormalities leading to the expression of the disease are unknown.

The disease is characterized by degeneration of peripheral nerves, dorsal columns, spinocerebellar tracts, and corticospinal pathways [2,17–20]. The symptoms usually begin before puberty, when progressive muscle weakness and ataxia appear. As the disease progresses, other signs and symptoms are added, including limb ataxia; muscular weakness; pes cavus; scoliosis; loss of position, vibration, and light touch sensation of the limbs; loss of the deep tendon reflexes; and the development of extensor plantar responses. Cardiomyopathy appears consistently in Friedreich's ataxia and usually is responsible for death by middle age [21]. Dysarthria is a regular feature of the disease. Muscle atrophy and degeneration of the optic nerves occur late in the disease. Diabetes mellitus occurs in a substantial number of patients [19].

Neuropathological examination of patients with Friedreich's ataxia reveals extensive changes throughout the nervous system. The peripheral sensory fibers are degenerated, including the posterior roots and ganglion cells, particularly those in the lower spinal segments [22]. Within the spinal cord, there is degeneration of the dorsal columns, lateral corticospinal tracts, Clarke's columns, and the dorsal and ventral spinocerebellar tracts [23–25]. The cerebellar cortex usually shows no change, but decreased numbers of neurons are seen in the dentate nuclei [23,25]. Degeneration frequently occurs in the vestibular and cochlear nuclei and in the optic tracts.

Many biochemical abnormalities have been found in fibroblasts, leukocytes, and muscle tissue in patients with Friedreich's ataxia, but some of the findings have been controversial. Deficiencies in the activity of the pyruvate dehydrogenase complex and alpha-ketoglutarate dehydrogenase complex have been found [26,27]. Other abnormalities include deficiency of mitochrondial malic enzyme activity [28,29] and of serum lipoprotein lipase [30], abnormalities of erythrocyte membrane phospholipids, increased plasma catecholamines, and decreased leukocyte glutamate dehydrogenase activity [19].

Few neurochemical investigations of postmortem tissues from patients with Friedreich's ataxia have been reported. Robinson found decreased glutamate and aspartate, and increased glucose-6-phosphate dehydrogenase in the spinal cord [31–34]. Butterworth and Giguère [35,36] reported reduced concentrations of glutamate and glutamine but normal levels of aspartate, and increased levels of taurine, in the spinal cord. Huxtable et al. [37,38] found that glutamate, aspartate, phenylalanine, and GABA concentrations were selectively decreased in the cerebellar hemispheres and vermis. Taurine was increased in the cerebellar cortex.

Only one study has been reported of neurotransmitter receptor binding in the central nervous system of patients with Friedreich's ataxia. Reisine et al. [39] found that [^{3}H]flunitrazepam binding was decreased in the inferior olive and dentate nucleus but was unchanged in the cerebellar cortex. The binding of [^{3}H]QNB to muscarinic cholinergic receptors was increased in the inferior olives and cerebellar vermis but was unchanged in the dentate nucleus and cerebellar hemispheres. The binding of [^{3}H]dihydroalprenolol to alpha-adrenergic sites was increased in the inferior olive and unchanged in the dentate nucleus and cerebellar cortex.

Anatomical imaging studies, such as computed tomography and magnetic resonance, are often not helpful in the diagnosis of Friedreich's ataxia. A large number of patients, even those with longstanding disease, may have a normal cerebellum and brain stem on CT [40]. Cerebellar and sometimes cerebral atrophy has been reported in Friedreich's ataxia, but no consistent pattern has been described [40–42]. Often the CT scan will show atrophy in the rostral parts of the vermis, with relative sparing of the cerebellar hemispheres. In general, as the clinical disorder becomes more severe, the degree of cerebellar atrophy on CT worsens. MR scanning is helpful since it allows visualization noninvasively of both the cerebellum and the cervical spinal cord. Spinal-cord atrophy occurs with Friedreich's ataxia, and although only marked degrees of atrophy can be diagnosed with MR, useful conclusions can be provided by MR study [40].

My colleagues and I have performed PET scanning with [^{18}F]2-fluoro-2-deoxy-D-glucose ([^{18}F]FDG) in 22 patients with Friedreich's ataxia [13]. The studies were performed in the University of Michigan Cyclotron/PET facility with the subjects lying awake, supine, and blindfolded in a quiet room. Scans were performed 30–75 min after the injection of 5–10 mCi of [^{18}F]FDG. The scans were taken with a TCC-PCT 4600A tomograph having an in-plane resolution of 11 mm full width at half maximum (FWHM) and a z-axis resolution of 9.5 mm FWHM. Five planes with 11.5 mm center-to-

Table 18-1. Average ages of the subjects studied

	Control subjects		Patients with Friedreich's ataxia	
	n	Age (yr) (± SD)	n	Age (yr) (±SD)
Male	13	28 ± 5	9	30 ± 7
Female	10	34 ± 9	13	31 ± 7
All subjects	23	30 ± 8	22	30 ± 7

Reproduced with permission from Gilman et al. [13].

center separation were imaged simultaneously. Four sets of scans were taken per patient, including interleaved sets through the lower brain levels and interleaved sets through the higher levels for a total of 20 slices, each separated by 5.75 mm.

Blood samples were collected from the radial artery. The local cerebral metabolic rate for glucose (LCMRG) was calculated with a three-compartment model and single scan approximation, as described by Phelps et al. [43], with gray-matter kinetic constants derived from normals [44]. Regions of interest (ROIs) were studied in the cerebral cortex, basal ganglia, thalamus, cerebellar hemispheres, cerebellar vermis, mesencephalon, and pons. Data were collected from the cerebral cortex, consisting of the cortical rim from six consecutive slices, beginning with the lowest slice containing the basal ganglia. The ROIs for the basal ganglia consisted of an 11 × 11 mm square on each side of the caudate nucleus and an 11 × 15 mm parallelogram on each side of the lenticular nucleus. The other ROIs consisted of an 11 × 11 mm square over each thalamus, a 22 × 11 mm parallelogram on each cerebellar hemisphere, and an 11 × 18 mm rectangle on the cerebellar vermis. The ROIs for the mesencephalon consisted of midsagittal 11 × 11 × 7.5 mm right parallelepiped, while those for the pons consisted of 15 × 11 × 15 mm right parallelepiped. Each ROI was centered over a local peak in LCMRG.

Table 18-1 provides a comparison of the age and sex distribution of the patients with Friedreich's ataxia and with the normal control subjects. As shown in Table 18-2, LCMRG was significantly *increased* in the cerebral cortex and basal ganglia of the patients with Friedreich's ataxia in comparison with the normal control subjects. To determine whether changes in LCMRG might be related to the severity of the disease, we divided the patients with Friedreich's ataxia into ambulatory and nonambulatory groups (Table 18-3). As shown in Table 18-4, LCMRG in the ambulatory group was significantly increased over the levels in the normal control group in all structures studied except the pons. LCMRG in the nonambulatory group was significantly increased over the levels in the normal control group in the basal ganglia, but not in the other structures studied (Table 18-4). LCMRG was significantly

Table 18-2. Local cerebral metabolic rate for glucose (in mg/100 gm/min) in all subjects[a]

Structure	Control subjects (n = 23)	Patients with Friedreich's ataxia (n = 22)	p value
Cerebral cortex	5.80 ± 1.07	6.64 ± 1.30	0.02
Caudate nucleus	6.20 ± 1.32	7.91 ± 1.56	0.0003
Lenticular nucleus	6.82 ± 1.44	8.10 ± 1.64	0.008
Thalamus	6.81 ± 1.46	7.29 ± 1.87	NS
Cerebellar vermis	5.10 ± 1.05	5.55 ± 1.33	NS
Left cerebellar hemisphere	5.59 ± 1.20	6.04 ± 1.61	NS
Right cerebellar hemisphere	5.47 ± 1.18	6.10 ± 1.66	NS
Mesencephalon	5.06 ± 1.02	5.54 ± 1.21	NS
Pons	4.10 ± 0.79	4.46 ± 1.05	NS

[a] Values given are the mean ± SD; statistical test, Student's t test.
NS = not significant; p > 0.10.
Reproduced with permission from Gilman et al. [13].

Table 18-3. Comparison of ambulatory and nonambulatory patients with Friedreich's ataxia[a]

	Ambulatory	Nonambulatory	p value
Age[b]	30 ± 7	30 ± 7	NS
Age of onset (yr)[c]	17 ± 4	10 ± 4	0.007
Duration (yr)[c]	15 ± 5	21 ± 5	0.08

[a] Values given are the mean ± SD; statistical test, Student's t test.
[b] n = 8 ambulatory, 14 nonambulatory patients.
[c] n = 5 ambulatory, 11 nonambulatory patients (only first patients affected in a family are included).
NS = not significant; p = 0.95.
Reproduced with permission from Gilman et al. [13].

increased in the ambulatory group in comparison with the nonambulatory group in all structures studied, except for the cerebral cortex and the basal ganglia (Table 18-4).

The finding of glucose hypermetabolism in the brains of patients with Friedreich's ataxia appears to represent a fundamental biochemical abnormality in this disease. Our data suggest that the glucose metabolic rate is increased broadly in the central nervous system early in the course of the disease, and then decreases as the disease progresses. The process is not uniform among structures; the basal ganglia retain a high metabolic rate longer than the other structures studied. It is conceivable that the declining metabolic rate over time may result from a progressive decrease in the number of active synaptic terminals and neurons.

Cerebral glucose hypermetabolism may result from abnormalities of carbohydrate metabolism [26,27]. The precise defect has been the subject of considerable controversy, but increasing evidence points to a mitochondrial

Table 18-4. Local cerebral metabolic rate for glucose (in mg/100 g/min) in control subjects and ambulatory and nonambulatory patients with Friedreich's ataxia (FA)[a]

Structure	Control subjects (n = 23)	Ambulatory patients with FA (n = 8)	p value[b]	Nonambulatory patients with FA (n = 14)	p value[c]	p value[d]
Cerebral cortex	5.80 ± 1.07	7.24 ± 0.80	0.01	6.29 ± 1.42	NS	0.07
Caudate nucleus	6.20 ± 1.32	8.30 ± 1.23	0.002	7.68 ± 1.72	0.005	NS
Lenticular nucleus	6.82 ± 1.44	8.51 ± 1.53	0.03	7.86 ± 1.71	0.03	NS
Thalamus	6.81 ± 1.46	8.58 ± 1.43	0.007	6.55 ± 1.71	NS	0.01
Cerebellar vermis	5.10 ± 1.05	6.28 ± 0.79	0.04	5.13 ± 1.41	NS	0.03
Left cerebellar hemisphere	5.59 ± 1.20	7.09 ± 1.03	0.007	5.45 ± 1.61	NS	0.02
Right cerebellar hemisphere	5.47 ± 1.18	7.17 ± 1.02	0.008	5.49 ± 1.67	NS	0.006
Mesencephalon	5.06 ± 1.02	6.22 ± 0.93	0.03	5.15 ± 1.20	NS	0.03
Pons	4.10 ± 0.79	5.09 ± 0.67	0.07	4.10 ± 1.07	NS	0.03

[a] Values given are the mean ± SD; statistical test, Newman-Keuls.
[b] Comparison of normal control subjects with ambulatory patients with FA.
[c] Comparison of normal control subjects with nonambulatory patients with FA.
[d] Comparison of ambulatory with nonambulatory patients with FA.
NS = not significant; $p > 0.20$.
Reproduced with permission from Gilman et al. [13].

abnormality [13]. Glucose hypermetabolism could result from difficulty in the oxidation of pyruvate, resulting in limited access to the Krebs tricarboxylic acid cycle. Further studies are needed to explore this possibility, including combined glucose and oxygen metabolic studies utilizing PET.

In the future, PET studies with specific ligands to label certain synaptic endings in the cerebellum may be a fruitful means of gaining further insight into the pathophysiology of Friedreich's ataxia. As noted above, one of the principal pathological changes in this disease is degeneration of spinocerebellar afferents, which terminate as mossy fibers within the granule cell layer of the cerebellar cortex. These fibers are thought to be excitatory and glutamatergic [45]. Their terminal fields are located principally within the cerebellar vermis, and they send collaterals into the fastigial and interposed nuclei. Since the spinocerebellar glutamatergic afferents degenerate in Friedreich's ataxia, the cerebellar cortex should be deficient in glutamate levels. As noted above, only two studies of amino acid levels within the cerebellum have been reported, and they show decreased glutamate in the cerebellar cortex in three patients [37,38]. A ligand that would bind to these terminal has not been identified as yet, but will be a valuable diagnostic tool when it is developed.

PET studies of neurotransmitter receptor density in the cerebellum of patients with Friedreich's ataxia are also promising avenues of future research. Degeneration of glutamatergic endings in the cerebellum may lead to compensatory upregulation of glutamate receptors in the granule cell layer and in the fastigial and interposed nuclei. PET studies of this abnormality will depend upon the development of ligands capable of specifically binding the glutamate receptors in the cerebellar cortex.

3. OLIVOPONTOCEREBELLAR ATROPHY

Olivopontocerebellar atrophy (OPCA) is a progressive neurological disease characterized neuropathologically by degeneration of neurons in the inferior olives, pons, and cerebellum [2,46,47,48]. Other sites within the central nervous system may be degenerated as well, and in some cases OPCA is associated with multisystem degeneration [49,50]. The disease occurs sporadically and with hereditary transmission. When hereditary, it usually is transmitted as an autosomal dominant trait, although many cases of autosomal recessive transmission have been documented [2,46,49]. The site of the genetic abnormality in dominantly inherited OPCA is on chromosome 6 [4], but the biochemical abnormalities leading to expression of the disease are unknown.

Two adult-onset degenerative diseases affecting the cerebellum may present in similar fashion to OPCA: hereditary cerebellar olivary degeneration of Holmes [51–53] and parenchymatous cerebellar cortical atrophy (also known as cortical cerebellar atrophy) [46,52,54]. Both of these diseases present in adult life with a progressive cerebellar ataxia. It is difficult clinically to distinguish these types of cerebellar degeneration from each other and from OPCA.

Clinically, OPCA usually begins with a progressive cerebellar ataxia in middle age. Typically, a disorder of gait and an ataxic dysarthria are the presenting complaints. With time, incoordination progressively affects all of the limbs. In most patients the dominating clinical feature is cerebellar ataxia, though some may have additional symptoms and signs of spasticity. Many other features have been described, including ophthalmoplegia, optic atrophy, retinal degeneration, dementia, rigidity, chorea, athetosis, and amyotrophy of the limbs or tongue [2]. In some patients a progressive multisystem disease occurs, consisting of a combination of OPCA with the Shy-Drager syndrome. This leads to a mixture of cerebellar ataxia, extrapyramidal symptomatology, and autonomic insufficiency.

Most patients with OPCA on neurological examination have ocular dysmetria, ataxic speech, and ataxia of limb movements and gait. The ataxia of speech usually is accompanied by variable degrees of spasticity of speech [15]. Spasticity of the limbs, hyperactive deep-tendon reflexes, and extensor plantar responses can be seen but are relatively unusual [16]. Dementia has been reported, but the frequency of dementia in OPCA is unclear. In Berciano's review of the literature, dementia was found in 22% of familial cases and 11% of sporadic cases [49]. Harding has found that as many as 40% may have significant intellectual impairment [46], and some authors state that chronic progressive dementia occurs consistently in OPCA [55]. In a recently published study [56], minor intellectual defects were found in OPCA, but dementia was not a prominent feature. In a new cross-sectional study of OPCA patients, dementia was found to be uncommon, and in many of the patients motor disorders accounted for the apparently deficient intellectual performance on IQ testing [57].

As mentioned above, neuropathological features of OPCA consist of gross shrinkage of the ventral portion of the pons, the inferior olives, and the cerebellar cortex [25,58]. There is decreased tissue volume and loss of neurons in the pons and inferior olives, with loss of the pontocerebellar projections and the olivocerebellar fibers [59,60]. The numbers of Purkinje cells and granule cells are markedly reduced. Basket cells, Golgi cells, and dentate nucleus cells tend to remain preserved. The severity of the neuropathological changes in OPCA are highly variable between cases. In some patients the long tracts of the spinal cord are degenerated, and there may be degenerative changes in the striatum and substantia nigra. With multisystem disease, there may be extensive central nervous system degeneration, including the inferior olives, pons, cerebellum, striatum, and the intermediolateral column of the spinal cord.

A limited number of biochemical studies of postmortem brain tissue have been performed in cases of OPCA. These studies have demonstrated changes in the levels of putative neurotransmitters and neurotransmitter receptors [6,7,10,11,61,62]. Several different biochemical profiles have been described [5,12]. One group of patients shows decreased glutamate, aspartate, and GABA levels in the cerebellar cortex, decreased GABA levels in the dentate nucleus, and increased taurine levels in the cerebellar cortex. A second group displays decreased glutamate, aspartate, and GABA levels in the cerebellar cortex and decreased GABA levels in the dentate nucleus, but normal taurine levels in the cerebellar cortex. In a third group, glutamate, aspartate, and GABA are within normal limits in the cerebellar cortex, but the GABA content of the dentate nucleus is reduced. In a fourth group, glutamate, aspartate, and GABA levels are decreased in the cerebellar cortex, GABA levels are decreased in the dentate nucleus, and glutamate is decreased in other brain regions as well, particularly the cerebral cortex and striatum.

The groups described above may or may not represent distinct biochemical subtypes of OPCA. These biochemical changes may reflect only differences in the degree of degeneration of particular types of neurons in different cases of OPCA. Evidence supporting this notion comes from studies correlating neurotransmitters and neuronal cell densities in OPCA. One study, for example, demonstrated that the concentration of glutamate and aspartate in the cerebellar cortex varied considerably from case to case [61]. Glutamate concentrations in the anterior vermis were correlated with the density of granule cells there, and aspartate concentrations in the anterior vermis were related to the density of neurons in the inferior olives. GABA concentrations in the dentate nucleus were decreased in all cases of OPCA and were correlated with the degree of loss of Purkinje cells.

Three studies of neurotransmitter receptor binding in OPCA have been reported [6,7,62]. Using [^{3}H]GABA as a ligand, Kish et al. [6] found a marked increase of GABA receptor binding in the cerebellar cortex at postmortem in patients with a dominantly inherited form of OPCA. GABA levels were markedly decreased in the dentate nucleus, probably due to a considerable loss of Purkinje cells. In a later study using methyl [^{3}H]flunitrazepam as a ligand, Kish et al. [7] found normal or slightly elevated benzodiazepine receptor binding in the cerebellar cortex. Receptor binding in the deep nuclei was not reported. In another study using [^{3}H]flunitrazepam as a ligand for autoradiography, Whitehouse et al. [62] found that benzodiazepine receptor binding was unchanged in the cerebellar cortex but increased in the dentate nucleus.

We have studied benzodiazepine binding in four cases of dominantly inherited OPCA using postmortem tissue kindly supplied by Dr. Andreas Plaitakis [63]. Receptor autoradiography was used to determine the distribution of excitatory and inhibitory amino acid neurotransmitter receptors in the cerebellar cortex of the OPCA cases in comparison with four control specimens. In control specimens, the molecular layer had a high density of benzodiazepine, GABA$_B$, and the quisqualate subtype of glutamate receptors. The granular layer of controls had a high density of benzodiazepine, GABA$_A$, and the N-methyl-D-aspartate (NMDA) subtype of glutamate receptors. All OPCA specimens were characterized by loss of Purkinje cells, marked thinning of the molecular layer, and diminution of the granular layer. Accompanying the thinning of the molecular layer was a substantial decrease in the number of benzodiazepine, GABA$_B$, and quisqualate receptors. The density of benzodiazepine, GABA$_A$, and NMDA receptors in the granular layer was also reduced markedly. Our findings are different from those reported by Kish et al. [17] and Whitehouse et al. [62], probably because of differences between the cases studied. OPCA is a heterogeneous disorder, and as such, it is not surprising to find substantial biochemical differences between cases.

Several other neurochemical abnormalities have been found in postmortem tissues of OPCA patients, including a reduction of noradrenalin in the cerebellar cortex [8], decreased levels of choline acetyltransferase in the cerebral cortex and hippocampus [10], and changes in the concentration of immunoreactive thyrotropin-releasing hormone in the cerebellar cortex, dentate nuclei, and olivary nuclei [64].

The diagnosis of OPCA requires the exclusion of other disorders that may cause a progressive ataxia in middle life. These disorders include demyelinative disease, neoplasms, paraneoplastic processes, vascular disease, toxic/metabolic disorders, other degenerative disease, malformations, and infectious disease. A positive family history can be helpful in the diagnosis,

Table 18-5. Local cerebral metabolic rate for glucose (in mg/100 g/min) in control subjects compared to patients with olivopontocerebellar atrophy

	Controls (n = 35)	OPCA (n = 41)
Cerebellar vermis	5.35 ± 1.05	3.73 ± 1.02[a]
Left cerebellar hemisphere	5.96 ± 1.27	4.16 ± 1.18[a]
Right cerebellar hemisphere	5.84 ± 1.24	4.19 ± 1.17[a]
Brain stem	4.36 ± 0.81	3.40 ± 0.78[a]
Thalamus	6.93 ± 1.14	6.62 ± 1.33
Cerebral cortex	5.73 ± 0.98	5.69 ± 1.13

[a] $p < 0.005$.
Controls average age = 51.3 ± 13.1.
OPCA average age = 53.5 ± 13.1

and anatomical imaging studies are helpful. CSF protein may be somewhat elevated and the EEG may show mild abnormalities [2]. Auditory evoked potentials are abnormal in some patients [65] and impaired sensory nerve conduction has been reported [66].

Imaging studies in OPCA involve CT and MR scans. CT usually is sufficient, but because of beam hardening artifact in the posterior fossa, MR studies often are helpful. MR studies also show pontine atrophy more effectively than CT. Usually there is atrophy of the cerebellar folia and smallness of the brain stem with enlargement of the fourth ventricle and the cisterna magna. Often there is atrophy of the basis pontis in OPCA, though this area may be spared in cerebellar cortical atrophy and in the olivocerebellar atrophy of Holmes. In OPCA there is much more atrophy of the cerebellum than of the cerebrum [41,42,67,68].

Positron emission tomography offers an opportunity to investigate the neurochemical changes in OPCA patients while still living and at various stages of their disease. PET also offers an opportunity to study larger numbers of patients than can be examined with postmortem studies. Positron emission tomography scanning with [^{18}F]FDG has been used to study the pattern of glucose metabolic activity in patients with OPCA. Thus far, 41 patients with OPCA and 35 age-matched and sex-matched normal controls have been studied (Tables 18-5, 18-6 and 18-7). The results in 30 of these cases have been published [14–16].

The normal controls were volunteers with no history of neurological disease, and on general physical and neurological examination, no abnormalities were found. These subjects were taking no medication known to affect the central nervous system. The patients with OPCA were evaluated with a physical examination, neurological examination, and a battery of laboratory tests of blood and urine to exclude other diseases. The finding of cerebellar atrophy on CT was helpful diagnostically, although a few of them did have essentially normal scans. None of the patients had a history of chronic alcoholism, and none had exposure to medications,

Table 18-6. Local cerebral metabolic rate for glucose normalized to the cerebral cortex in control subjects compared to patients with olivopontocerebellar atrophy

	Controls (n = 35)	OPCA (n = 41)
Cerebellar vermis	0.93 ± 0.08	0.66 ± 0.15[a]
Left cerebellar hemisphere	1.04 ± 0.11	0.73 ± 0.15[a]
Right cerebellar hemisphere	1.02 ± 0.11	0.74 ± 0.16[a]
Brain stem	0.76 ± 0.05	0.60 ± 0.09[a]
Thalamus	1.22 ± 0.10	1.17 ± 0.10

[a] $p < 0.005$.

Table 18-7. Local cerebral metabolic rate for glucose normalized to the cerebral cortex in control subjects compared to patients with olivopontocerebellar atrophy

	OPCA males (n = 18)	OPCA females (n = 23)
Cerebellar vermis	0.65 ± 0.16	0.67 ± 0.14
Left cerebellar hemisphere	0.70 ± 0.15	0.75 ± 0.16
Right cerebellar hemisphere	0.71 ± 0.16	0.76 ± 0.16
Brain stem	0.60 ± 0.09	0.60 ± 0.09
Thalamus	1.17 ± 0.11	1.16 ± 0.10
	Control males (n = 18)	OPCA males (n = 18)
Cerebellar vermis	0.92 ± 0.09	0.65 ± 0.16[a]
Left cerebellar hemisphere	1.02 ± 0.13	0.70 ± 0.15[a]
Right cerebellar hemisphere	1.00 ± 0.12	0.71 ± 0.16[a]
Brain stem	0.75 ± 0.06	0.60 ± 0.09[a]
Thalamus	1.21 ± 0.11	1.17 ± 0.11
	Control females (n = 17)	OPCA females (n = 23)
Cerebellar vermis	0.95 ± 0.07	0.67 ± 0.14[a]
Left cerebellar hemisphere	1.05 ± 0.08	0.75 ± 0.16[a]
Right cerebellar hemisphere	1.04 ± 0.08	0.76 ± 0.16[a]
Brain stem	0.77 ± 0.05	0.60 ± 0.09[a]
Thalamus	1.22 ± 0.09	0.16 ± 0.10

[a] $p < 0.005$

such as phenytoin, that might cause cerebellar ataxia. In no patient was there a disorder of sensory function adequate to cause ataxia of gait or limb movement.

The studies were performed in the University of Michigan Cyclotron/PET facility, as described above in the section concerning Friedreich's ataxia. The data analysis for OPCA patients was also similar to that used in studies of Friedreich's ataxia.

The PET scans of OPCA patients in comparison with normal controls showed an obvious decrease of LCMRG in the cerebellar vermis, cerebellar hemispheres, and brain stem. No abnormality was seen in the cerebral cortex, basal ganglia, or thalamus. Statistical testing of the data revealed significant differences in LCMRG between OPCA patients and controls in the cerebellar vermis, cerebellar hemispheres, and brain stem (Table 18-5).

Table 18-8. Local cerebral metabolic rate for glucose (in mg/100 g/min) in patients with sporadic as compared to genetic olivopontocerebellar atrophy

	OPCA sporadic (n = 23)	OPCA genetic (n = 18)
Cerebellar vermis	3.42 ± 0.76	4.12 ± 1.19[b]
Left cerebellar hemisphere	3.80 ± 0.92	4.62 ± 1.34[b]
Right cerebellar hemisphere	3.90 ± 0.94	4.56 ± 1.34[a]
Brain stem	3.23 ± 0.63	3.61 ± 0.91
Thalamus	6.56 ± 1.32	6.70 ± 1.38
Cerebral cortex	5.47 ± 1.01	5.97 ± 1.24

[a] $p < 0.05$.
[b] $p < 0.01$.
OPCA sporadic average age = 57.4 ± 11.1.
OPCA genetic average age = 48.4 ± 13.9.

With the data normalized to the cerebral cortex, the results remained significant (Table 18-6). The data from males as compared to females with OPCA showed no differences from each other, but there were significant differences from the sex-matched controls (Table 18-7).

In our studies, approximately half the patients had a family history of OPCA and the other half apparently had sporadic forms of the disease. The familial cases all had a history compatible with autosomal dominant inheritance, except for two patients in whom the history was compatible with autosomal recessive inheritance. Comparison of these groups revealed that LCMRG of structures in the posterior fossa was lower in the sporadic than in the familial cases (Table 18-8). These difference reached statistical significance for the cerebellar vermis and cerebellar hemispheres, but not for the brain stem. The only other difference found between groups was the age of the patients; the familial cases were on the average younger than the sporadic cases (Table 18-4). These differences may be based on an ascertainment bias; the familial group included many patients whom we recruited into the study after evaluating an afflicted relative. In contrast, the sporadic cases came to us for a diagnostic opinion and thus may have represented individuals with a more advanced illness. We found no differences between groups in the duration of illness and no differences in clinical signs. There was no correlation of the clinical disorder with age.

Many of the patients studied had appreciable atrophy of the cerebellum and brain stem, as detected by CT scans. This can influence the measurement of tissue metabolic rates through partial volume averaging. To study this factor, we quantitated the degree of atrophy in the cerebellum and brain stem as determined by CT with rating scales developed by a neuroradiologist [14, 16]. The resulting data were plotted against metabolic rates. We found a significant relationship between metabolic rate and CT measures of atrophy. Nevertheless, several patients with OPCA had marked hypometabolism but only minor degrees of atrophy, and some had normal metabolic rates despite moderate atrophy.

We correlated the LCMRG of the cerebellum and brain stem with the degree of impairment in clinical neurological function. These studies included assessments of speech and of cranial and somatic motor function [15,16]. Evaluation of speech was with oral motor assessment and perceptual speech analysis. A quantitative rating scale was developed that depended upon the use of the deviant speech dimensions of Darley et al. [69]. We found a significant inverse correlation between the degree of ataxia of speech and the absolute level of metabolic rate within the cerebellar vermis, both cerebellar hemispheres, the brain stem, and the cerebral cortex, but not the thalamus. For studies of the relationship of cranial and somatic motor dysfunction to LCMRG in the cerebellum and brain stem, we devised a rating scale to quantitate the severity of the clinical motor disorders in OPCA [16]. All the signs selected are important in the diagnosis of OPCA. A significant relationship was found between the overall severity of the clinical disturbance and the degree of decrease in LCMRG, normalized to the cerebral cortex, for the cerebellar vermis, left and right cerebellar hemispheres, and the brain stem of OPCA patients.

We also evaluated the degree of tissue atrophy with respect to the severity of the motor disorders [16]. We found a significant relationship for the cerebellar vermis and both cerebellar hemispheres. The correlations were slightly less strong than those found between LCMRG and the clinical score. The degree of tissue atrophy as determined by CT showed a significant correlation with the decline in LCMRG in the cerebellum and brain stem. The clinical score also correlated significantly with both LCMRG and atrophy on CT. Since these measures all correlated with each other, the partial correlations method was used to determine the relationship of PET and CT to clinical score independently of each other. This method revealed that when the effects due to variations in LCMRG were taken into account, the degree of tissue atrophy no longer correlated significantly with the severity of the motor disorder. When the effects due to variations in atrophy were accounted for, however, the degree of hypometabolism still correlated significantly with clinical severity.

Recently, we have begun to undertake PET studies of some of the neurochemical changes in the cerebellar cortex and deep nuclei in OPCA using a ligand for GABA receptors. As mentioned above, we have adduced evidence that GABA/benzodiazepine receptor binding is markedly decreased in the cerebellar cortex in OPCA. We have begun to study the density of GABA/benzodiazepine receptors in the cerebellar cortex and deep nuclei using [^{11}C]flumazenil and PET. Preliminary studies with this agent have been made in six patients with OPCA in comparison with six normal control subjects. In all six patients with OPCA, we have found diminished uptake and retention in the cerebellum as compared with the cerebral cortex, basal ganglia, and thalamus. We are currently investigating additional patients and controls with this agent. Similar studies have been reported in preliminary form from Japan [70].

4. SUMMARY

This chapter concerns the clinical, neuropathological, and neurochemical features of Friedreich's ataxia and olivopontocerebellar atrophy, two frequently encountered progressive cerebellar degenerative diseases. In both diseases, positron emission tomography with [^{18}F]2-fluoro-2-deoxy-D-glucose has shown abnormalities in glucose metabolic rate. In Friedreich's ataxia, widespread hypermetabolism was found in the central nervous system in patients remaining ambulatory. In patients no longer ambulatory, only the basal ganglia remained hypermetabolic. In olivopontocerebellar atrophy there is hypometabolism in the cerebellar vermis, cerebellar hemispheres, and brain stem. The degree of hypometabolism is directly related to the severity of the ataxia of speech and limb motor function. Future studies with positron emission tomography employing new tracers to examine neurotransmitter specific synaptic endings and receptors may make it possible to characterize more fully the neurochemical abnormalities in this group of diseases.

ACKNOWLEDGMENTS

I am indebted to my colleagues Roger Albin, Stanley Berent, Stephen S. Gebarski, Larry Junck, Karen Kluin, Robert A. Koeppe, Dorene Markel, Guy Rosenthal, and Anne B. Young for their participation in these studies and to Diane Vecellio for typing the manuscript. This work was supported in part by NIH grant NS 15655.

REFERENCES

1. Gilman S. (1987). Inherited ataxia. In: R.T. Johnson (ed): Current Therapy in Neurologic Disease, 2nd ed. Toronto: Decker, pp. 224–232.
2. Gilman S., Bloedel J.R., Lechtenberg R. (1981). Disorders of the Cerebellum. Philadelphia: Davis.
3. Chamberlain S., Shaw J., Rowland A., Wallis J., South S., Nakamura Y., von Gabin A., Farrall M., Williamson R. (1988). Mapping of mutation causing Freidreich's ataxia to human chromosome 9. Nature 334:248–250.
4. Rich S.S., Wilkie P., Schut L., Vance G., Orr H.T. (1987). Spinocerebellar ataxia: Localization of an autosomal dominant locus between two markers on human chromosome 6. Am. J. Hum. Genet. 41:524–531.
5. Perry T.L. (1984). Four biochemically different types of dominantly inherited olivopontocerebellar atrophy. In: R.C. Duvoisin, A. Plaitakis (eds): The Olivopontocerebellar Atrophies. New York: Raven Press, pp. 205–216.
6. Kish S.J., Perry T.L., Hornykiewicz O. (1983). Increased GABA receptor binding in dominantly inherited cerebellar ataxias. Brain Res. 269:370–373.
7. Kish S.J., Perry T.L., Hornykiewicz O. (1984). Benzodiazepine receptor binding in cerebellar cortex: Observations in olivopontocerebellar atrophy. J. Neurochem. 42:466–469.
8. Kish S.J., Shannak K.S., Hornykiewicz O. (1984). Reduction of noradrenaline in cerebellum of patients with olivopontocerebellar atrophy. J. Neurochem. 42:1476–1478.
9. Kish S.J., Currier R.D., Schut L., Perry T.L., Morito C.L. (1987). Brain choline

acetyltransferase reduction in dominantly inherited olivopontocerebellar atrophy. Ann. Neurol. 22:272–275.
10. Perry T.L., Hansen S., Currier R.D., Berry K. (1978). Abnormalities in neurotransmitter amino acids in dominantly inherited cerebellar disorders. In: R.A.P. Kark, R.N. Rosenberg, L.J. Shut (eds): Advances in Neurology, Vol. 21. New York: Raven Press, pp. 303–314.
11. Perry T.L., Kish S.J., Hansen S., Currier R.D. (1981). Neurotransmitter amino acids in dominantly inherited cerebellar disorders. Neurology 31:237–242.
12. Perry T.L. (1984). Neurotransmitter abnormalities in dominantly inherited olivopontocerebellar atrophies. Ital. J. Neurol. Sci. Suppl. 4:79–89.
13. Gilman S., Junck L., Markel D.S., Koeppe R.A., Kluin K.J. (1990). Cerebral glucose hypermetabolism in Friedreich's ataxia detected with positron emission tomography. Ann. Neurol. 28:750–757.
14. Gilman S., Markel D.S., Koeppe R.A., et al. (1988). Cerebellar and brain stem hypometabolism in olivopontocerebellar atrophy studied with PET. Ann. Neurol. 213:223–230.
15. Kluin K.J., Gilman S., Markel D.S., Koeppe R.A., Rosenthal G., Junck L. (1988). Speech disorders in olivopontocerebellar atrophy correlate with positron emission tomography findings. Ann. Neurol. 23:547–554.
16. Rosenthal G., Gilman S., Koeppe R.A., Kluin K.J., Markel D.S., Junck L., Gebarski S.S. (1988). Motor dysfunction in olivopontocerebellar atrophy is related to cerebral metabolic rate studied with positron emission tomography. Ann. Neurol. 24:414–419.
17. Geoffroy G., Barbeau A., Breton G. (1979). Clinical description and roentgenologic evaluation of patients with Friedreich's ataxia. Can. J. Neurol. Sci. 3:279–280.
18. Harding A.E. (1981). Friedreich's ataxia: An analysis of 90 families with an analysis of early diagnostic clustering of clinical features. Brain 104:589–620.
19. Stumpf D.A. (1985). The inherited ataxias. Neurol. Clin. 2:47–57.
20. Tyrer J.H. (1975). Friedreich's ataxia. In: Handbook of Clinical Neurology, Vol. 21, pp. 319–364.
21. Heiver R.L. (1969). The heart in Friedreich's ataxia. Br. Heart J. 31:5–14.
22. Hughes J.T., Brownell B., Heiver R.L. (1968). The peripheral sensory pathway in Friedreich's ataxia. Brain 91:803–818.
23. Lamarche J.B., Lemieux B., Lieu H.B. (1984). The neuropathology of "typical" Friedreich's ataxia in Quebec. Can. J. Neurol. Sci. 11:592–600.
24. Oppenheimer D.R. (1979). Brain lesions in Friedreich's ataxia. Can J. Neurol. Sci. 6:173–176.
25. Oppenehimer D.R. (1984). Diseases of the basal ganglia, cerebellum and motor neurons. In: J.H. Adams, J.A.N. Corsellis, L.W. Duchen (eds): Greenfield's Neuropathology, 4th ed. New York: John Wiley Sons, Chap. 15.
26. Blass J.P., Kark R.A.P., Menon N.K. (1976). Low activities of the pyruvate and oxoglutarate dehydrogenase complexes in five patients with Friedreich's ataxia. N. Engl. J. Med. 295:62–67.
27. Kark R.A.P., Blass J.P., Engel W.K. (1974). Pyruvate oxidation in neuromuscular diseases: Evidence of a genetic defect in two families with the clinical syndrome of Friedreich's ataxia. Neurology 24:964–971.
28. Stumpf D.A., Parks J.K., Eguran L.A., et al. (1982). Friedreich ataxia. II. Mitochondrial malic enzyme deficiency. Neurology 32:221–227.
29. Bottachi E., DiDonato S. (1983). Skeletal muscle NAD^+ (P) and $NADP^+$-dependent malic enzyme in Friedreich's ataxia. Neurology 33:712–716.
30. Blache D., Bouthillier D., Barbeau A., Davignon J. (1982). Plasma lipoprotein lipase and hepatic lipase activities in Friedreich's ataxia. Can. J. Neurol. Sci. 9:191–194.
31. Robinson N. (1966). Friedreich's ataxia: A histochemical study. Part 1. Enzymes of carbohydrate metabolism. Acta Neuropath. (Berlin). 6:25–34.
32. Robinson N. (1966). Friedreich's ataxia: A histochemical study. Part 2. Hydrolytic enzymes. Acta Neuropath. (Berlin) 6:35–45.
33. Robinson N. (1966). A histochemical study of motor neuron disease. Acta Neuropath. (Berlin) 7:101–110.
34. Robinson N. (1968). Chemical changes in the spinal cord in Friedreich's ataxia and motor neuron disease. J. Neurol. Neurosurg. Psych. 31:330–333.

35. Butterworth R.F., Giguère J.-F. (1984). Amino acids in autopsied human spinal cord. Biochem. Pathol. 2:7–17.
36. Butterworth R.F., Giguère J.-F. (1982). Glutamic acid in spinal-cord gray matter in Friedreich's ataxia. N. Engl. J. Med. 307:897.
37. Huxtable R. Azari J., Reisine T., Johnson P., Yamamura H., Barbeau A. (1979). Regional distribution of amino acids in Friedreich's ataxia brains. Can. J. Neurol. Sci. 6:255–258.
38. Huxtable R.J., Johnson P., Lippincott S.E. (1984). Free amino acids and calcium, magnesium and zinc levels in Friedreich's ataxia. Can. J. Neurol. Sci. 11:616–619.
39. Reisine T.D., Azari J., Johnson P.C., Barbeau A., Huxtable R., Yamamura H.I. (1979). Brain neurotransmitter receptors in Friedreich's ataxia. Can. J. Neurol. Sci. 6:259–262.
40. Ramos A., Quintana F., Diez C., Leno C., Berciano J. (1987). CT findings in spinocerebellar degeneration. AJNR 8:635–640.
41. Claus D., Aschoff J.C. (1981). Cranial computerized tomography in spinocerebellar atrophies. NYAS 374:831–838.
42. Diener H.C., Muller A., Thron A., Poremba M., Dichgans J., Rapp H. (1986). Correlation of clinical signs with CT findings in patients with cerebellar disease. J. Neurol. 233:5–12.
43. Phelps M.E., Huang S.C., Hoffman E.J., et al. (1979). Tomographic measurement of local cerebral glucose metabolic rate in humans with (F-18) 2-fluoro-2-deoxy-D-glucose: Validation of method. Ann. Neurol. 6:371–388.
44. Hawkins R.A., Mazziotta J.C., Phelps M.E., et al. (1983). Cerebral glucose metabolism as a function of age in man: Influence of the rate constants in the fluorodeoxyglucose method. J. Cereb. Blood Flow Metab. 3:250–253.
45. Olson J.M.M., Greenamyre J.T., Penney J.B., Young A.B. (1987). Autoradiographic localization of cerebellar excitatory amino acid binding sites in the mouse. Neuroscience 22:913–923.
46. Harding A.E. (1984). The Hereditary Ataxias and Related Disorders. London: Churchill Livingstone.
47. Eadie M.J. (1975). Olivo-ponto-cerebellar atrophy (Dejerine-Thomas type). In: P.J. Vinken, G.W. Bruyn (eds): Handbook of Clinical Neurology, Amsterdam: North Holland, pp. 415–431.
48. Eadie M.J. (1975). Olivo-ponto-cerebellar atrophy (Dejerine-Thomas type). In: P.J. Vinken, G.W. Bruyn (eds): Handbook of Clinical Neurology, Vol. 21. Amsterdam: North Holland, pp. 433–449.
49. Berciano J. (1982). Olivopontocerebellar atrophy. A review of 117 cases. J. Neurol. Sci. 53:253–272.
50. Duvoisin R.C. (1984). An apology and an introduction to the olivopontocerebellar atrophies. In: R.C. Duvoisin, A. Plaitakis (eds): The Olivopontocerebellar Atrophies. New York: Raven Press, pp. 5–12.
51. Holmes G. (1907). A form of familial degeneration of the cerebellum. Brain 30:466–489.
52. Greenfield J.G. (1954). The Spino-Cerebellar Degenerations. Oxford: Blackwell.
53. Weber F.P., Greenfield J.G. (1942). Cerebello-olivary degeneration: An example of heredofamilial incidence. Brain 65:220–231.
54. Marie P., Foix C., Alajouanine T. (1922). De l'atrophie cérébelleuse tardive à prédominance corticale. Rev. Neurol. 38:849–1082.
55. Cummings J.L., Benson D.F. (1983). Dementia: A Clinical Approach. Boston: Butterworths.
56. Kish S.J., El-Awar M., Schut L., Leach L., Oscar-Berman M., Freedman M. (1988). Cognitive deficits in olivopontocerebellar atrophy: Implications for the cholinergic hypothesis of Alzheimer's dementia. Ann. Neurol. 24:200–206.
57. Berent S., Giordani B., Gilman S., Junck L., Lehtinen S., Markel D.S., Boivin M., Kluin K.J., Parks R., Koeppe R.A. (1990). Neuropsychological changes in olivopontocerebellar atrophy. Arch. Neurol. 47:997–1001.
58. Koeppen A.H., Barron K.D. (1984). The neuropathology of olivopontocerebellar atrophy. In: R.C. Duvoisin, A. Plaitakis (eds): The Olivopontocerebellar Atrophies. New York: Raven Press, pp. 13–38.
59. Landis D.M.D., Rosenberg R.N., Landis S.C., Schut L., Nyhan W.L. (1974). Olivopontocerebellar degeneration. Clinical and ultrastructural abnormalities. Arch. Neurol. 31:295–307.

60. Petito C.K., Hart M.N., Porro R.S., Earle K.M. (1973). Ultrastructural studies of olivopontocerebellar atrophy. J. Neuropathol. Exp. Neurol. 32:503–522.
61. Kanazawal I., Kwak S., Sasaki H., Mizusawa H., et al. (1985). Studies on neurotransmitter receptors in olivopontocerebellar atrophy: An autoradiographic study. Neurology 36:193–197.
62. Whitehouse P.J., Muramoto O., Troncoso J.C., Kanazawa I. (1986). Neurotrasmitter receptors in olivopontocerebellar atrophy: An autoradiographic study. Neurology 36:193–197.
63. Albin R.L., Gilman S. (1990). Autoradiographic localization of inhibitory and excitatory amino acid neurotransmitter receptors in human normal and olivopontocerebellar atrophy cerebellar cortex. Brain Res. 522:37–45.
64. Mitsuma T., Nogimori T., Adachi K., Mukoyama M., Ando K., Sobue I. (1985). Concentrations of immunoreactive thyrotropin-releasing hormone in the brain of patients with olivopontocerebellar atrophy. J. Neurol. Sci. 71:369–375.
65. Sinatra M.G., Baldini S.M., Baiocco F., Carenini L. (1988). Auditory brainstem response patterns in familial and sporadic olivopontocerebellar atrophy. Eur. Neurol. 28:288–290.
66. Rossi A., Ciacci G., Federico A., Modelli M., Rizzuto N. (1986). Sensory and motor peripheral neuropathy in olivopontocerebellar atrophy. Acta Neurol. Scand. 73:363–371.
67. Yamamoto H., Yasuhiko A., Taktoshi W., Yoshitaka H., Yasushi M., Sobue I. (1986). Evalution of supra- and infratentorial brain atrophy by computerized tomography in spinocerebellar degeneration. Jpn. J. Med. 25:238–245.
68. Bianco F., Bozzao L., Colonnese C., Fantozzi L. (1983). The value of computerized tomography in the diagnosis of cerebellar atrophy. Ital. J. Neurol. Sci. 1:65–68.
69. Darley F.L., Aronson A.E., Brown J.R. (1975). Motor Speech Disorders. Philadelphia: Saunders.
70. Shinotoh H., Hirayama K., Iyo M., Inoue O., Suzuki K., Yamasaki T., Ikehira H., Tateno Y. (1987). Benzodiazepine receptors in olivo-ponto-cerebellar atrophy studied with positron emission tomography and carbon-11 labeled RO 15-1788. Jpn. J. Nucl. Med. 24:688.

19. ATAXIA-TELANGIECTASIA: A HUMAN MODEL OF NEUROIMMUNE DEGENERATION

MASSIMO FIORILLI, MAURIZIO CARBONARI, MICHELA CHERCHI, AND CARLO GAETANO

1. INTRODUCTION

Ataxia-telangiectasia (AT) is an autosomal recessive disorder, characterized phenotypically by early cerebellar ataxia, variable immunodeficiency, and the inability to repair radiation-induced damage to the DNA. Other cardinal features of the disease include oculocutaneous telangiectasia, which usually has a later onset than ataxia; progeric changes of the skin; somatic growth retardation; and hypogonadism in female patients [1–3]. Patients usually have increased levels of alpha-1 fetoprotein [4].

Cytogenetic abnormalities are a typical marker of AT cells and represent an important diagnostic tool. They consist of increased chromosome breakage, occurring both spontaneously and upon exposure to DNA-damaging agents, such as ionizing radiation or bleomycin [5–7]. Chromosome instability is most likely the cause of the high frequency of neoplasia in AT: lymphoid tumors, particularly T-cell leukemias and lymphomas, predominate in younger patients, and epithelial tumors in patients beyond 15 years [8,9].

A unique feature of AT cells is that they are unable to inhibit DNA synthesis following exposure to x-rays [10]. The phenomenon of radioresistant DNA synthesis can also be exploited for diagnostic purposes [11].

Genetically, AT is a heterogeneous disease, with four known major

A. Plaitakis (ed.), CEREBELLAR DEGENERATIONS: CLINICAL NEUROBIOLOGY. Copyright © 1992 Kluwer Academic Publishers, Boston All rights reserved.

complementation groups and some variants [12–14]. The "Nijmegen breakage syndrome" seems to represent a genetically distinct variant of AT [12], presenting with microcephaly instead of cerebellar degeneration [15]. The estimated frequency of homozygosity for AT is about 1 per 40,000 live births [16], while the frequency of heterozygotes has been estimated around 2.8% [17].

The long-lasting interest of many researchers in AT stems from several reasons. This disease represents a cellular and molecular link between cancer, neuropathology, and immune deficiency. The AT mutation brings into focus a relationship between DNA repair mechanisms and the differentiation of lymphocytes and other cell systems. The eventual identification of the AT genes would have not only a theoretical impact on cancer research, but also a practical one on cancer prevention. In fact, it is known that AT heterozygotes have a risk of developing tumors, particularly for breast cancer, which is six times more frequent than in the general population [18]. Finally, AT represents a fascinating model of genetically determined "neuroimmune defect." A growing body of evidence suggests a molecular, developmental, and functional relationship between the nervous and the immune systems [19,20]. Thus, this mutation offers a way of untangling this relationship at the genetic level. We shall review the available information on the pathogenetic mechanisms underlying AT and on the possible links between the immunological and neuropathological abnormalities of this disease.

2. IMMUNE DYSFUNCTIONS IN ATAXIA-TELANGIECTASIA

2.1. Immunological abnormalities

A hallmark of AT is immunodeficiency, involving both humoral and T-cell-mediated immunity. As a consequence, AT patients have unusually frequent infections, expecially involving the respiratory tract, so that pulmonary disease is their major cause of death [21].

The severity of the immune defect is variable [22–26], and also some immunopathogenetic mechanisms appear to be heterogeneous [25,27]. The most common abnormalities of humoral immunity are increased serum IgM, and diminished or absent IgA, IgE, and IgG2 [22,25,27–29]. Despite the lack of some immunoglobulins, B cells bearing surface immunoglobulins of the corresponding isotypes are often present in significant numbers [25,27]. The fact that these B cells appear phenotypically immature (i.e., they carry both IgM and IgG or IgA on the surface) [30,31] indicates a block of differentiation.

T lymphocytes can be impaired both functionally and numerically [23–25]. T-cell responses to mitogens are decreased in some patients [23,24], and helper and suppressor T cells appear to work improperly [25,27]. The proportions of T lymphocyes in the peripheral blood have been reported to be either reduced [25–27] or normal [23,24]. Such discrepancies are difficult to interpret, given the relatively small numbers of patients in most series, the influence of factors such as recent infections, and the intrinsic heterogeneity of AT.

Recently, we have reported that most AT patients have a relative increase of the subtype of T cells expressing the gamma/delta receptor [32]. This finding, and the selective deficiency of some immunoglobulin isotypes, suggests, as will be discussed below, that a problem involving the process of recombination of B- and T-cell receptor genes may cause the immunodeficiency of AT patients.

2.1. Defective DNA recombination as a possible cause of immunodeficiency in AT

B- and T-lymphocytes use surface receptors for antigen, represented by immunoglobulins and by heterodimeric structures called T-cell receptors (TCRs), respectively. An enormous amount of information on the molecular mechanisms controlling the expression of these genes has been obtained over the past years. Immunoglobulins and TCRs share several features, including an immunoglobulinlike-structure and the need for somatic DNA rearrangements in order to generate functional receptors [33].

Immunoglobulin molecules are constituted by 1 of 9 types of heavy chains (mu, delta, gamma-1,2,3,4, alpha-1,2, or epsilon) and by light chains of either the kappa or lambda type. The immunoglobulin heavy (IgH) chain locus, located at band q32 of chromosome 14 [34], is a highly complex region: It contains clusters of variable (V), diversity (D), and joining (J) genes, which together encode for the variable part of the immunoglobulin, and genes for each isotype-specific constant (C) portion of the immunoglobulin [35]. During B-cell differentiation these genes undergo a series of somatic rearrangements. Initially, one D and one J gene are joined by cutting the intervening DNA and ligating the free DNA ends; subsequent rearrangements juxtapose, by an analogous mechanisms, one V gene to DJ and the VDJ unit to C-mu (the gene for the constant portion of IgM molecules) [35]. To switch from one isotype to another, B cells undergo "switch rearrangements," which involve the deletion of C-mu, C-delta, and all the other C genes located between VDJ and the C gene to be juxtaposed to VDJ [35,36]. Thus, the size of the DNA segment to be spliced out for switching to a given isotype increases with the distance of the corresponding gene from the VDJ unit. The order of IgH genes on chromosome 14 is as follows: V, D, J, C-mu, C-delta, C-gamma3, C-gamma1, C-alpha1, C-gamma2, C-gamma4, C-epsilon, and C-alpha2 [36]. Thus, the finding in AT patients of a deficiency of IgG2, IgA, and IgE, with accumulation of IgM, suggests a difficulty of B cells in performing switch rearrangements to the most downstream isotypes.

TCR genes are organized in a similar way. They are also composed of V, D, J, and C genes, which, during differentiation, undergo somatic rearrangements analogous to those of immunolobulin genes [33,37]. There exist four TCR genes, indicated as alpha, beta, gamma, and delta. The delta locus is endowed within the alpha locus, between the J-alpha and V-alpha segments [38] at region q11 of chromosome 14. The beta and gamma loci, both on chromosome 7, contain two C genes each [39,40], recalling the multiplicity of IgH C genes.

The majority of T lymphocytes circulating in the adult peripheral blood express alpha/beta TCR chains, while gamma/delta T cells represent the large majority of immature thymocytes and of intraepithelial lymphocytes in the adult [33,37,41]. The developmental relationship between alpha/beta and gamma/delta T cells is unclear: They probably represent two distinct lineages arising from the same progenitor, but the point at which they diverge is not known. A possible model is the following [33,42]: Early in ontogeny, rearrangements of gamma/delta genes occur in the absence of alpha rearrangements and lead to the exclusive generation of gamma/delta T cells. Rearrangements of the alpha gene start later in ontogeny, involve the deletion of the delta gene, and commit most T cells to the alpha/beta lineage. Thus, the generation of alpha/beta T cells requires rearrangements that are substantially more complex, due to the enormous size of the alpha locus and to the presence of the delta locus within it. By contrast, rearrangements of the delta and gamma genes appear much easier due to the relative simplicity of these two loci [39,40].

The complexity of the alpha locus appears to make of it a "hot spot" for aberrant rearrangements. This is indicated by the fact that chromosomal rearrangements involving band 14q11 (thus, presumably, the alpha locus itself) are the commonest chromosomal abnormality of normal human lymphocytes [43] and that chromosome translocations involving the genes for the alpha chain of the T-cell receptor are common in T-cell leukemias developed by AT patients [44].

The problem of AT lymphocytes consists, most likely, of an inability to rejoin the strand breaks generated during the process of rearrangement of the immunoglobulin and TCR genes. This view is strongly suggested by the spontaneous occurrence in AT lymphocytes of chromosome breaks that involve selectively those regions where the immunoglobulin and TCR genes are located (TCR gamma chain at chromosome 7p12, beta at 7q35, alpha/delta at 14q11, and IgH at 14q32) [38,45,46]. Chromosomal breaks are, instead, randomly distributed in AT fibroblasts [47].

The failure of AT patients to produce the immunoglobulin isotypes whose genes are located most distantly from the VDJ region (i.e., IgG2, IgA, and IgE) suggests that their B cells are unable to perform the rearrangements needed for isotype switching when these involve large DNA regions. By analogy, one should predict that also within the T-cell compartment, complex rearrangements (i.e., those at the alpha locus) are performed less efficiently than simpler ones (i.e., gamma and delta rearrangements). We tested this hypothesis by examining the distribution of alpha/beta and gamma/delta T cells in the peripheral blood of 10 patients with AT. Using monoclonal antibodies specific for the TCR chains, we found that, indeed, alpha/beta T cells were selectively decreased in comparison to gamma/delta cells [32]. Furthermore, we found a peculiar distribution within the gamma/delta subset: Cells expressing the product of the C-gamma-1 gene markedly predominated over those expressing C-gamma-2 [32]. Again, this supports the view that AT cells are unable to perform complex DNA rearrangements. In fact, since the C-gamma-1 gene is located between V-gamma and C-gamma-2, rearrangements at the latter gene must involve the deletion of a large DNA segment encompassing C-gamma-1 [40].

Taken overall, these data suggest that AT lymphocytes are unable to handle those somatic rearrangements of immunoglobulin and TCR genes that invovle the splicing of large DNA segments. This recombinational defect may be the cause of at least the immunological abnormalities of AT patients.

2.2. Immunotherapy of ataxia-telangiectasia

The consequences of immunodeficiency are the major cause of death of patients with AT [21]. Therefore, immunotherapeutic maneuvers might have great importance in the management of these patients. A variety of approaches have been used in an attempt to reconstitute T-cell immunity in AT patients, but the results have been overall, disappointing. Several trials were based on the concept that the thymic hypoplasia, found in virtually all AT cases, is the cause of the deficiency of T cells. However, despite sporadic claims of efficacy [48,49], neither thymus transplantation nor replacement therapy with thymic hormones have resulted in consistent benefits [50–52]. An early trial with transfer factor gave promising results in a few patients [53], but the results were not confirmed in larger studies [54,55]. Bone marrow transplantation has no practical role in AT, and no reports have been published. The failure to reconstitute T-cell immunity using a variety of approaches is consistent with the interpretation that the inefficient production of mature T lymphocytes in AT depends on an intrinsic inability to generate functional T-cell receptors.

Immunoglobulin deficiency plays a major role in the susceptibility of AT patients to infections, and therefore, its correction is of paramount importance in their management. Isolated IgG2 subclass deficiency, with normal levels of total serum IgG, can predispose to severe bacterial infections [56]. Thus, AT patients, who usually have IgG2 deficiency, presenting with recurrent sinopulmonary infections, should, in our opinion, be treated with gammaglobulin replacement therapy, even if they have normal values of total IgG. Gammaglobulin preparations for intravenous use should be administered at an average dosage of 300–400 mg/kg every 3 weeks [57]. Over the past 4 years, we have used this treatment schedule in nine AT patients, and we (unpublished results) have observed a significant reduction of the number and severity of sinopulmonary infections, resulting in a remarkable improvement in the quality of life.

3. NERVOUS SYSTEM ABNORMALITIES IN ATAXIA-TELANGIECTASIA

Clinical and pathologic findings in the central nervous system of AT patients have been extensively described in the literature [for reviews, see 3,16,21]. Although other clinical signs may be present, ataxia is the first symptom of AT, and it is always present. It typically becomes apparent as soon as the child begins to walk, and it has a steadily progressive course. Some other prominent neurological features are choreoathetosis (91%), ocular apraxia (84%), diminished deep reflexes (89%), dysarthric speech (100%), negative Romberg sign (78%), and intact deep and superficial sensation (98%) [3,21].

The neuropathological hallmark of the disease is diffuse cortical cerebellar degeneration, which mainly involves Purkinje and granular cells. Basket cells, on the contrary, are relatively preserved, suggesting that Purkinje cells develop but subsequently degenerate [58]. The presence of ectopic Purkinje cells suggests, furthermore, that the AT mutation directly affects the differentiation of these neurones and that this occurs during the first half of gestation [59].

How the AT mutation affects Purkinje cells is, at present, a mere matter for speculation. It has still to be established, for example, whether the changes in the nervous system are secondary to the immune disorder or if neuronal degeneration is directly determined by the AT gene. In the latter case, many possibilities could be explored, including the role of a defect of DNA recombination or the significance of some molecular markers common to neurons and lymphocytes. The immune abnormalities might, on the other hand, affect the nervous system, for example, through the action of common trophic factors. These possibilities will be discussed in the following sections.

4. ATAXIA-TELANGIECTASIA: A MODEL OF A DEVELOPMENTAL NEUROIMMUNE DEFECT

Over the past years, many observations have made clear that the nervous and the immune systems share several molecular and trophic features and that their disorders are linked in a variety of diseases [60]. An evolutionary relationship between neurons and lymphoid cells [61,62] might provide a developmental basis for those genetic diseases primarily involving the nervous and the immune systems. Besides ataxia-telangiectasia, other examples exist, such as the syndrome of agenesis of the corpus callosum associated with immunodeficiency [63], the 18q− syndrome [64], the Chediack-Higashi syndrome [65], and some cases of the Di George syndrome [66]. We shall discuss the possible significance of the known molecular and trophic links between the nervous and immune systems in the pathogenesis of the "neuroimmune disorder" of AT.

4.1. The nervous and the immune systems share the expression of members of the immunoglobulin superfamily

The so-called immunoglobulin superfamily includes several cell-surface adhesion and recognition molecules, some of which are selectively expressed on lymphoid cells and neurons [61,62]. The fact that several immunoglobulin superfamily genes lie in a restricted portion of human chromosome 11 suggests that this might be a region of extensive genetic duplication of an ancestral immunoglobulin-like receptor structure [62]. The recent mapping of the major AT gene to this region raises important considerations.

The gene responsible for the complementation group AB of AT (about 50% of AT cases) has been localized to region q22–23 of chromosome 11 [67]. Within the same chromosomal region are the following genes belonging to the immunoglobulin superfamily: Thy-1, three chains (gamma, delta, and epsilon) of the CD3 complex, and the neural-cell adhesion molecule (N-CAM) [68]. Interestingly, the expression of some of these genes is shared by cells of the nervous and of the immune systems.

Thy-1 is present on T lymphocytes and on neurons, and it appears to be involved in the activation of both types of cells [69]. It has been the first molecule, besides histocompatibility antigens, to be recognized as a member of the immunoglobulin superfamily [70].

N-CAM is a cell-surface glycoprotein that mediates cell-cell interactions through a homophilic binding [71]. N-CAM is expressed on several cell types during embryonic development, but in the adult individual is basically restricted to neural and muscle cells [72]. Surprisingly, it has been shown that fetal thymocytes and adult natural killer (NK) lymphocytes express N-CAM [73–75]. N-CAM appears to be functionally involved in the process of target recognition and destruction by NK cells [76].

CD3 is a glycoprotein complex associated with the T-cell antigen receptor [77]. Monoclonal antibodies to human CD3 have been reported to react with Purkinje cells of many animal species [78], but whether this reflects the presence of CD3 in these cells is questionable. The CD3 complex is composed of at least three molecules, indicated as gamma, delta, and epsilon [77,79]. Only two anti-CD3 antibodies (Leu4 and UCHT1) stain Purkinje cells [178], and both of them react with the epsilon chain [80]. To our knowledge, it has not been established whether these two antibodies identify the same epitope. Several considerations suggest that the reativity of Leu4/UCHT1 with Purkinje neurons does not reflect the expression of CD3-epsilon. First, this reactivity is seen with Purkinje cells from many animal species, including those where the genuine CD3 of T cells is not detectable by Leu4. Furthermore, we (unpublished data) could not detect any staining of murine Purkinje cells with other antibodies to human CD3-epsilon (monoclonal antibodies SP-6 ans SP-10, [81], which were kindly provided by Silvana Pessano). Finally, the observation that Leu4 (but not other anti-CD3 antibodies) reacts with tubular cells of the human kidney [82] strengthens the suggestion of a "cross-reactivity" of this antibody with structures distinct from CD3. Our attempts to immunoprecipitate the Leu4-reactive material from mouse cerebellum have been unsuccessful thus for.

In addition to those listed above, two other lymphocyte-specific members of the immunoglobulin superfamily are expressed in brain. CD4, which is also the receptor for human immunodeficiency virus, is expressed in the central nervous system as a truncated mRNA [83]. Immunohistochemical analysis suggests that some neurons express the CD4 protein [84], but definitive diochemical evidence is still lacking. OX-2, a surface glycoprotein of lymphocytes, is also present on neurons [85].

Taken overall, these data are consistent with the hypothesis of an evolutionary relationship between the nervous and immune systems [61,62]. Williams [62] suggests that the vertebrate immune system may have evolved from the nervous system: Cytotoxic cells involved in programmed neural cell death might have provided a precursor for a primitive cytotoxic lymphocyte (i.e., the NK cell). As an example, in *Caenorhabditis elegans* some neurons seem to have the capacity to kill other neurons [86]. The recent demonstration that N-CAM may serve as a ligand in NK cell-mediated cytotoxicity provides an important support to this fascinating hypothesis.

4.2. Cytokines and the nervous system

Another important link between the nervous and the immune systems is the reciprocal influence exerted through some cytokines. Nerve growth factor (NGF), interleukin-1 (IL-1), and interleukin-2 (IL-2) appear to be involved in such a network. On one side, NGF acts on lymphoid cells by inducing high-affinity IL-2 receptors on T lymphocytes and by inhibiting the proliferation of a subset of thymocytes bearing NGF receptors [87]. On the other hand, IL-1 and IL-2 are active on cells of the nervous system.

IL-1 is elaborated by macrophages and is primarily involved in the activation of T cells [88]. In addition, IL-1 exerts important effects on the nervous system: It determines drowsiness and synchronization of the electroencephalogram [89], when injected in brain it causes fever and the secretion of adrenal corticosteroids [90], and it stimulates astroglial proliferation after brain injury [91]. The finding of neural elements containing IL-1 within the hypothalamus suggests that this cytokine might act as an intrinsic neuromodulator [92].

IL-2, formerly called T-cell growth factor, is a lymphokine produced by T cells that induces the proliferation of activated T cells expressing high-affinity IL-2 receptors [93]. At high concentrations, IL-2 activates the low-affinity receptors of NK cells and induces them to become highly cytotoxic (lymphokine activated killer [LAK] cells) [94]. Patients treated with high doses of IL-2 show sometimes neurological symptoms [95], suggesting that IL-2 may have effects on the central nervous system. Indeed, it has been shown that cultured oligodendrocytes respond to IL-2 by either increasing or reducing their rate of proliferation [96,97]. IL-2 appears to act also on cells of the neuronal lineage, as we found that, at high concentrations, it inhibited the proliferation of human neuroblastoma cells in vitro [98].

It is presently unknown whether this cytokine network plays a role in the development and functional regulation of the nervous system under normal and pathological conditions. In this regard, it is of great interest a recent finding of Ishida et al. [99], who showed that mice made transgenic for both the IL-2 and IL-2 receptor genes undergo a selective and severe loss of Purkinje cells in the cerebellum by 2 weeks of age. Although no clear explanation can be provided so far, it is possible that the loss of Purkinje cells might depend on the toxicity of IL-2 on these cells or to their extreme sensitivity to cytotoxic lymphocytes activated in vivo by IL-2.

5. CONCLUSIONS

It is fair to say that, at present, there is no reasonable way to relate the cerebellar degeneration to the other features of ataxia-telangiectasia. Some possibilities can, however, be listed.

At patients might have neurochemical abnormalities [100] that affect neuronal metabolism.

The exquisite sensitivity of Purkinje neurons to a variety of noxious agents might render them susceptible to toxic factor(s) somehow generated by the AT mutation [101], which might be of a metabolic or immunological nature. Concerning immunological factors, the only consistent clue has been provided, so far, by the data of Ishida et al. [99] in IL-2 transgenic mice (see above). Some considerations, however, make it difficult to correlate this experimental model with AT. First, while in transgenic mice it is an overproduction of IL-2 that appears to determine the degeneration of Purkinje cells, the production of this interleukin is usually defective in AT patients [102]. Second, it is difficult to understand why Purkinje cells are unaffected in none of the so many other congenital immune defects characterized by a deficiency of IL-2.

Lymphocyte and neuronal degeneration in AT might be the common consequence of a defect in genetic recombination [103,104]. Recently, two independent lines of evidence have provided strong support to the hypothesis that DNA rearrangements analogous to those of lymphocytes occur in the nervous system [105,106]: Chun et al. [105] have shown that the Recombination Activating Gene-1 (RAG-1), which is necessary for the activation of VDJ recombination in lymphocytes, is expressed in the brain. Also, Matsuoka et al. [106] directly demonstrated somatic recombination in the brain of mice transgenic for a DNA construct containing the recombination signal sequences of immunoglobulin genes.

The most fascinating aspect of ataxia-telangiectasia is that the mutation of a single gene brings about the degeneration of the nervous and immune systems, under the common denominator of a defect in DNA recombination. Untangling this disease at the molecular level will help to highlight the many links between lymphocytes and neurons.

ACKNOWLEDGMENTS

The authors are indebted to Drs. Giulio Cossu, Luciana Chessa, Giandomenico Russo, and Marco Crescenzi for many helpful discussions. This work was supported by grants from the Italian Ministry of Education.

REFERENCES

1. Bridges B.A., Harnden D.G. (eds). (1982). Ataxia-Telangiectasia: A Cellular and Molecular Link Between Cancer, Neuropathology, and Immune Deficiency. Chichester, England: John Wiley & Sons.
2. Gatti R.A., Swift M. (eds). (1985). Ataxia-Telangiectasia: Genetics, Neuropathology and Immunology of a Degenerative Disease of Childhood. New York: Alan R. Liss.
3. Boder E., Sedgwick R.P. (1967). Ataxia-telangiectasia: A review of 101 cases. In: G. Walsh (ed): Little Clubs Clinic in Developmental Medicine, No. 8. London: Heinemann Medical Books, pp. 110–118.
4. Waldmann T.A., McIntire K.R. (1972). Serum-alpha-fetoprotein levels in patients with ataxia-telangiectasia. Lancet 2:112.
5. Taylor A.M.R., Oxford J.M., Metcalfe J.A. (1981). Spontaneous cytogenetic abnormalities in lymphocytes from thirteen patients with ataxia-telangiectasia. Int. J. Cancer 27:311–319.
6. Taylor A.M.R., Metcalfe J.A., Oxford J.M., Harnden D.G. (1976). Is chromatid type damage in ataxia-telangiectasia at Go a consequence of defective repair? Nature 260:441–443.
7. Taylor A.M.R., Rosney C.M., Campbell J.B. (1979). Unusual sensitivity of ataxia-telangiectasia cells to bleomycin. Cancer Res. 39:1046–1050.
8. Spector B.D., Filipovich A.H., Perry G.S., Kersey J.H. (1982). Epidemiology of cancer in ataxia-telangiectasia. In: B.A. Bridges and, D.G. Harnden (eds): Ataxia-Telangiectasia. New York: John Wiley & Sons, pp. 103–138.
9. Yoshitomi F., Zaitsu Y., Tanaka K. (1980). Ataxia-telangiectasia with renal cell carcinoma and hepatoma. Virchows Arch. Pathol. Anat. Histol. 389:119–124.
10. Painter R.B., Young B.R. (1980). Radiosensitivity in ataxia-telangiectasia: A new explanation. Proc. Natl. Acad. Sci. USA 77:7315–7317.
11. Young B.R., Painter R.B. (1989). Radioresistant DNA synthesis and human genetic diseases. Hum. Genet. 82:113–120.
12. Jaspers N.G.J., Gatti R.A., Baan C., Linssen P.C.M.L., Bootsma D. (1988). Genetic complementation analysis of ataxia-telangiectasia and Nijmegen breakage syndrome: A survey of 50 patients. Cytogenet. Cell. Genet. 49:259–263.
13. Fiorilli M., Antonelli A., Russo G., Crescenzi M., Carbonari M., Petrinelli P. (1985). Variant of ataxia-telangiectasia with low radiosensitivity. Hum. Genet. 70:274–277.
14. Taylor A.M.R., Flude E., Laher B., Stacey M., McKay E., Watt J., Green S.H., Harding A.E. (1987). Variant forms of ataxia-telangiectasia. J. Med. Genet. 24:669–677.
15. Taalman R.D.F.M., Jaspers N.G.J., Scheres J.M.J.C., De Wit J., Hustinx T.W.J. (1983). Hypersensitivity to ionizing radiation, in vitro, in a new chromosomal breakage disorder, the Nijmegen breakage syndrome. Mutat. Res. 112:23–32.
16. Sedgwick R.P., Boder E. (1972). Ataxia-telangiectsia. In: Handbook of Clinical Neurology, Vol. 14 P.J. Vinken, G.W. Bruyn (eds): Amsterdam: North Holland, pp. 267–339.
17. Swift M., Morrell D., Cromartie E., Chamberlin A.R., Skolnick M.H., Bishop D. (1986). The incidence and gene frequency of ataxia-telangiectasia in the United States. Am. J. Hum. Genet. 39:573–580.
18. Swift M., Reitnauer P.J., Morrell D., Chase C.L. (1987). Breast and other cancers in families with ataxia-telangiectasia. N. Engl. J. Med. 316:1289–1293.
19. Edelman G. (1988). CAMs and Igs: Cell adhesion and the evolutionary origin of immunity. Immunol. Rev. 100:11–64.
20. Williams A.F. (1987). A year in the life of the immunoglobulin superfamily. Immunol. Today 8:298–303.
21. Boder E. (1985). Ataxia-telangiectasia: An overview. In: R. Gatti, M. Swift (eds): Ataxia-Telangiectasia: Genetics, Neuropathology, and Immunology of a Degenerative Disease of Childhood. New York: Alan R. Liss, pp. 1–61.
22. Mc Farlin D.D., Strober S., Waldmann T.A. (1972). Ataxia-telangiectasia. Medicine 51:281–314.
23. Roifman C.M., Gelfand E.W. (1985). Heterogeneity of the immunological deficiency in ataxia-telangiectasia: Absence of a clinical-pathological correlation. In: R.A. Gatti, M. Swift (eds): Ataxia-Telangiectasia: Genetics, Neuropathology, and Immunology of a Degenerative Disease of Childhood. New York: Alan R. Liss, pp. 273–285.

24. Gatti R.A., Bick M., Tam C.F., Medici M.A., Oxelius V.A., Holand M., Goldstein A.L., Boder E. (1982). Ataxia-telangiectasia: A multiparamether analysis of eight families. Clin. Immunol. Immunopathol. 23:501–516.
25. Fiorilli M., Businco L., Pandolfi F., Paganelli R., Russo G., Aiuti F. (1983). Heterogeneity of immunological abnormalities in ataxia-telangiectasia. J. Clin. Immunol. 3:135–141.
26. Waldmann T.A., Misiti J., Nelson D.L., Kraemer K.H. (1983). Ataxia-telangiectasia: A multisystem hereditary disease with immunodeficiency, impaired organ maturation, x-ray hypersensitivity, and a high incidence of neoplasia. Ann. Intern. Med. 99:367–379.
27. Waldmann T.A., Broder S., Goldman C.K., Frost K., Korsmeyer S.J., Medici M.A. (1983). Disorders of B cells and helper T cells in the pathogenesis of the immunoglobulin deficiency of patients with ataxia-telangiectasia. J. Clin. Invest. 71:282–295.
28. Oxelius V.A., Berkel A.I., Hanson L.A. (1982). IgG2 deficiency in ataxia-telangiectasia. N. Engl. J. Med. 306:515–517.
29. Ammann A.J., Cain W.A., Ishizaka K., Hong R., Good R.A. (1969). Immunoglobulin E deficiency in ataxia-telangiectasia. N. Engl. J. Med. 281:469–474.
30. Conley M.E., Cooper M.D. (1981). Immature IgA B cells in patients with IgA deficiency. N. Engl. J. Med. 305:495–497.
31. Fiorilli M., Crescenzi M., Carbonari M., Tedesco L., Russo G., Gaetano G., Aiuti F. (1986). Phenotypically immature IgG-bearing B cells in patients with hypogammaglobulinemia. J. Clin. Immunol. 6:21–25.
32. Carbonari M., Cherchi M., Paganelli R., Giannini G., Galli E., Gaetano C., Papetti C., Fiorilli M. (1990). Relative increase of T cells expressing the gamma/delta rather than the alpha/beta receptor in ataxia-telangiectasia. N. Engl. J. Med. 322:73–76.
33. Strominger J.L. (1989). Developmental biology of T cell receptors. Science 244:943–950.
34. Kirsh I.R., Morton C.C., Nakahara K., Leder P. (1982). Human immunoglobulin heavy chain genes map to a region of translocations in malignant B lymphocytes. Science 216:301–303.
35. Tonegawa S. (1983). Somatic generation of antibody diversity. 302:575–580.
36. Honjo T. (1983). Immunoglobulin genes. Annu. Rev. Immunol. 1:499–528.
37. Allison J.P., Lanier L.L. (1987). Structure, function and serology of the T-cell antigen recptor complex. Annu. Rev. Immunol. 5:503–540.
38. Isobe M., Russo G., Haluska F.G., Croce C.M. (1988). Cloning of the gene encoding the delta subunit of the human T-cell receptor reveals its physical organization within the alpha-subunit locus and its involvement in chromosome translocations in T cell malignancy. Proc. Natl. Acad. Sci. USA 85:3933–3937.
39. Quertermous T., Murre C., Dialynas D., et al. (1986). Human T cell gamma chain genes: Diversity and rearrangement. Science 231:252–255.
40. Sims J.E., Tunnacliffe A., Smith W.J., Rabbitts T.H. (1984). Complexity of the human T-cell antigen receptor beta-chain constant- and variable-region genes. Nature 312:541–545.
41. Janeway C.A., Jones B., Hayday A. (1988). Specificity and function of T cells bearing gamma/delta receptors. Immunol. Today 9:73–77.
42. Winoto A., Baltimore D. (1989). Separate lineages of T cells expressing the alpha/beta and gamma/delta receptors. 338:430–432.
43. Aurias A., Couturier J., Dutrillaux A.M., et al. (1985). Inversion (14)(q12qter) or (14)(q11.2q32.3): The most frequently acquired rearrangements in lymphocytes. Hum. Genet. 71:19–21.
44. Russo G., Isobe M., Gatti R.A., Finan J., Batuman D., Huebner K., Nowell P.C., Croce C.M. (1988). Molecular analysis of a t(14;14) translocation in leukemic T cells of an ataxia-telangiectasia patient: A model of T-cell leukemogenesis. Proc. Natl. Acad. Sci. USA 86:602–606.
45. Fiorilli M., Carbonari M., Crescenzi M., Russo G., Aiuti F. (1985). T cell receptor genes and ataxia-telangiectasia. Nature 313:186.
46. Aurias A., Dutrillaux B. (1986). Probable involvement of immunoglobulin superfamily genes in most recurrent chromosomal rearrangements from ataxia-telangiectasia. Hum. Genet. 72:210–214.
47. Kojis T.L., Schreck R.R., Gatti R.A., Sparkes R.S. (1989). Tissue specificity of chromosomal rearrangements in ataxia-telangiectasia. Hum. Genet. 83:347–352.
48. Bordigoni P., Bene M.C., Bach J.F., Faure G., Dardenne M., Duheille J., Olive D. (1982).

Improvement of cellular immunity and IgA production in immunodeficient children after treatment with synthetic serum thymic factor 4FTS). Lancet 2:293–297.
49. Lopukhin Y., Morosov Y., Petrov R. (1973). Transplantation of neonate thymus-sternum complex in ataxia-telangiectasia. Transplant. Proc. 5:823.
50. Ammann A.J., Wara D.W., Doyle N.E., Golbus M.S. (1975). Thymus transplantation in patients with thymic hypoplasia and abnormal immunolobulin synthesis. Transplantation 20:457–466.
51. Wara D.W., Ammann A.J. (1978). Thymosin treatment of children with primary immunodeficiency diseases. Transplant Proc. 10:203–212.
52. Aiuti F., Businco L., Fiorilli M., et al. (1983). Thymopoietin pentapeptide treatment of primary immunodeficiencies. Lancet 1:551–555.
53. Levin A.S., Spitler L., Fudenberg H.H. (1973). Transfer factor therapy in immunodeficient states. Ann. Rev. Med. 24:175–186.
54. Berkel A.I., Ersoy F., Epstein L.B., Spitler L.E. (1977). Transfer factor therapy in ataxia-telangiectasia. Clin. Exp. Immunol. 29:376–384.
55. Pascual-Pascual S.I., Pascual-Castrovejo I., Fontan G., Lopez-Martin V. (1981). Brain Dev. 3:289–296.
56. Umetsu D.T., Ambrosino D.M., Quinti I., Siber G.R., Geha R.S. (1985). Recurrent sinopulmonary infection and impaired antibody response to bacterial capsular polysaccharide antigen in children with selective IgG-subclass deficiency. N. Engl. J. Med. 313:1247–51.
57. Eibl M., Wedgwood R.J. (1989). Intravenous Immunoglobulins: A review. Immunodefic. Rev. 1(Suppl.):1–96.
58. Gatti R.A., Vinters H.V. (1985). Cerebellar pathology in ataxia-telangiectasia. The significance of basket cells. In: R.A. Gatti, M. Swift (eds): Ataxia-Telangiectasia: Genetics, Neuropathology, and Immunology of a Degenerative Disease of Childhood. New York: Alan R. Liss, pp. 225–232.
59. Vinters H.V., Gatti R.A., Racik P. (1985). Sequence of cellular events in cerebellar ontogeny relevant to expression of neuronal abnormalities in ataxia-telangiectasia. In: R.A. Gatti, M. Swift (eds): Ataxia-Telangiectasia: Genetics, Neuropathology, and Immunology of a Degenerative Disease of Childhood. New York: Alan R. Liss, pp. 233–256.
60. Barnes D. (1986). Nervous and immune system disorders linked in a variety of diseases. Science 232:160.
61. Edelman G. (1988). CAMs and Igs: Cell adhesion and the evolutionary origin of immunity. Immunol. Rev. 100:11–64.
62. Williams A.F. (1987). A year in the life of immunoglobulin superfamily. Immunol. Today 8:298–303.
63. Dionisi Vici C., Sabetta G., Gambarara M., Vigevano F., Bertini E., Boldrini R., Parisi S.G., Quinti I., Aiuti F., Fiorilli M. (1988). Agenesis of the corpus callosum, combined immunodeficiency, bilateral cataract, and hypopigmentation in two brothers. Am. J. Med. Genet. 29:1–8.
64. Schinzel A., Hayashi K., Schmid W. (1975). Structural aberrations of chromosome 18. Humangenetik 26:123–132.
65. Blume R.S., Wolf S.M. (1972). The Chediack-Higashi syndrome: Studies in four patients and a review of the literature. Medicine (Baltimore) 51:247–280.
66. Conley M.E., Beckwitt J.B., Mancer J.F., Tenckhoff L. (1979). The spectrum of Di George syndrome. J. Pediatr. 94:883–890.
67. Gatti R.A., Berkel I., Boder E., et al. (1988). Localization of an ataxia-telangiectasia gene to chromosome 11q22–23. Nature 336:577–580.
68. Evans G.A., Lewis K.A., Lawless G.M. (1988). Molecular organization of the human CD3 gene family on chromosome 11q23. Immunogenetics 28:365–373.
69. Saleh M., Lang R.J., Bartiett P.F. (1988). Thy-1-mediated regulation of a low-threshold transient calcium current in cultured sensory neurons. Proc. Natl. Acad. Sci. USA 85:4543–7.
70. Williams A.F., Gagnon J. (1982). Neuronal cell Thy-1 glycoprotein: Homology with immunoglobulin. Science 216:696–703.
71. Edelman G.M. (1985). Cell adhesion and the molecular process of morphogenesis. Annu. Rev. Biochem. 54:135–169.

72. Couvault J., Sanes J.R. (1985). Neural cell adhesion molecule (N-CAM) accumulates in denervated and paralyzed skeletal muscle. Proc. Natl. Acad. Sci. USA 82:4544–4548.
73. Brunet J.F., Hirsch M.R., Naquet P., et al. (1989). Developmentally regulated expression of the neural cell adhesion meolcule (N-CAM) by mouse thymocytes. Eur. J. Immunol. 19:837–841.
74. Lanier L.L., Testi R., Bindl J., Phillips J.H., (1989). Identity of Leu-19 (CD56) leukocyte differentiation antigen and neural cell adhesion molecule. J. Exp. Med. 169:2233–2237.
75. Carbonari M., Gaetano C., Scarpa S., Criniti A., Cherchi M., Modesti A., Fiorilli M. (1989). Expression of the natural killer cell antigen Leu19/NKH-1 by human neural tumor cell lines. J. Immunol. Res. 1:163–167.
76. Nitta T., Yagita H., Sato K., Okumura K. (1989). Involvement of CD56 (NKH-1/Leu19 antigen) as an adhesion molecule in natural killer-target cell interaction. J. Exp. Med. 170:1757–1761.
77. Kronenberg M., Siu G., Hood L.E., Shastri N. (1986). The molecular genetics of the T-cell antigen receptor and T-cell antigen recognition. 4:529–572.
78. Garson J.A., Beverly P.C.L., Coakham H.B., Harper E.I. (1982). Monoclonal antibodies against human thymus-derived lymphocytes label Purkinje neurons from many species. Nature 298:375–378.
79. Borst J., Coligan J.E., Oettgen H., Pessano S., Malin R., Terhost C. (1984). The delta- and epsilon chains of the human T3/T-cell receptor complex are distinct polypeptides. Nature 312:455–457.
80. Van Dongen J.J.M., Krissansen G.W., Wolvers-Tettero I.L.M., Comans-Bitter W.M., Adriaansen H.J., Hooijkaas H., van Wering E.R., Terhost C. (1988). Cytoplasmic expression of CD3 antigen as a diagnostic marker for immature T cell malignancies. Blood 71:603.
81. Pessano S., Oettgen H., Bhan A.K., Terhost C. (1985). The T3/T cell receptor complex: Antigenic distinction between two 20-Kd T3 (T3-delta and T3-epsilon) subunits. Embo J. 4:337–344.
82. Karlsson-Parra A., Dimeny E., Juhlin C., Fellstrom B., Klareskog L. (1989). The anti-Leu4 (CD3) monoclonal antibody reacts with proximal tubular cells of the human kidney. Scand. J. Immunol. 30:719–722.
83. Maddon P.J., Dalgleish A.G., McDougal J.S., Clapham P.R., Weiss R.A., Axel R. (1986). The T4 gene encodes the AIDS virus receptor and is expressed in the immune system and in brain. Cell 47:333–344.
84. Funke I., Hahn A., Rieber E.P., Weiss E., Riethmuller G. (1987). The cellular receptor (CD4) of the human immunodeficiency virus is expressed on neurons and glial cells in human brain. J. Exp. Med. 165:1230–1235.
85. Webb M., Barclay A.N. (1984). Localization of the MRC OX-2 glycoprotein on the surface of neurons. J. Neurochem. 43:1061.
86. Hedgecock E.M., Sulston J.E., Thomson J.N. (1983). Mutations affecting programmed cell deaths in the nematode *Caenorhabditis elegans*. Science 220:1277–1279.
87. Thorpe L.W., Werrbach-Perez K., Perez-Polo J.R. (1987). Ann. N.Y. Acad. Sci. 496:310–311.
88. Dinarello C.A. (1988). Biology of interleukin-1. FASEB J. 2:108–115.
89. Krueger J.M., Walter J., Dinarello C.A., Wolff S., Chedid L. (1984). Am. J. Physiol. 246:R994.
90. Besedovsky H., del Rey A., Sorkin E., Dinarello C.A. (1986). Immunoregulatory feedback between interleukin-1 and glucocorticoid hormones. Science 233:652–654.
91. Muller H.W., Gebicke-Harter P.J., Hangen D.H., Shooter E.M. (1985). Interleukin-1 stimulation of astroglial proliferation after brain injury. Science 228:497–499.
92. Breder C.D., Dinarello C.A., Saper C.B. (1988). Interleukin-1 immunoreactive innervation of the human hypothalamus. Science 240:321–324.
93. Smith K.A., (1984). Interleukin-2. Annu. Rev. Immunol. 2:319–338.
94. Grimm E.A., Rosenberg S.A. (1984). The human lymphokine activated killer cell phenomenon. In: E. Pick, (ed): Lymphokines New York: Academic Press, 9:279–293.
95. Lotze M.T., Chang A.E., Seipp C.A., Simpson C., Vetto J.T., Rosenberg S.A. (1986). High-dose recombinant interleukin 2 in the treatment of patients with disseminated cancer. JAMA 256:3117–3121.

96. Benveniste E.N., Merril J.E. (1986). Stimulation of oligodendroglial proliferation and maturation by interleukin-2. Nature 321:610–613.
97. Saneto R.P., Altman A., Knobler R.L., Johnson H.M., de Vellis J. (1986). Interleukin 2 mediates the inhibition of oligodendrocyte progenitor cell proliferation in vitro. Proc. Natl. Acad. Sci. USA 83:9221–9225.
98. Gaetano C., Carbonari M., Scarpa S., Modesti A., Giannini G., Fiorilli M. (1988). Interleukin-2 inhibits the proliferation of human neuroblastoma cell lines in vitro. In: F. Aiuti, L. Bonomo, G. Danieli (eds): Topics in Immunology: Proceedings of the Ninth European Immunology Meeting. Rome: Il Pensiero Scientifico, pp. 336–340.
99. Ishida Y., Nishi M., Taguchi O., Inaba K., Hattori M., Minato N., Kawaichi M., Honjo T. (1989). Expansion of natural killer cells but not T cells in human interleukin-2/interleukin-2 receptor (Tac) transgenic mice. J. Exp. Med. 170:1103–1115.
100. Perry T.L., Kish S.J., Hinton D., Hansen S., Becker L.E., Gelfand E.W. (1984). Neurochemical abnormalities in a patient with ataxia-telangiectasia. Neurology 34:187.
101. Herndon R.M. (1985). Selective vulnerability in the nervous system. In: R.A. Gatti, M. Swift (eds). Ataxia-Telangiectasia: Genetics, Neuropathology, and Immunology of a Degenerative Disease of Childhood. New York: Alan R. Liss, pp. 257–268.
102. Paganelli R., Capobianchi M.R., Ensoli B. D'Offizi G.P., Facchini J., Dianzani F., Aiuti F. (1988). Evidence that gamma interferon production in primary immunodeficiencies is due to intrinsic incompetence of lymphocytes. Clin. Exp. Immunol. 72:124–129.
103. Breakefield X.O., Hansen C.R. (1983). Do DNA rearrangements occur in neuronal development? Clues from an inherited human disease. Trends Neurosci. 6:444.
104. Peterson R.D.A., Frunkhouser J.D. (1989). Speculations on ataxia-telangiectasia: Defective regulation of the immunoglobulin gene superfamily. Immunol. Today 10:313.
105. Chun J.M., Schatz D.G., Oettinger M.A., et al. (1989). The Recombination Activating Gene-1 (RAG-1) transcript is present in the murine central nervous system. Cell, 59: 189–200.
106. Matsuoka M., Nagawa F., Okazaki K., et al. (1991). Detection of somatic recombination in the transgenic mouse brain. Science 245:81–86.

20. PARANEOPLASTIC CEREBELLAR DEGENERATION

J.E. HAMMACK AND J.B. POSNER

1. INTRODUCTION

Paraneoplastic cerebellar degeneration (PCD) is a term applied to cerebellar dysfunction occurring in patients with identifiable or occult systemic cancer in whom direct cancer invasion or other defined causes of a cerebellar disorder (i.e., toxic, metabolic, vascular, infection, etc.) have been excluded. PCD is rare: Fewer than 200 cases have been reported (Table 20-1). Nevertheless, PCD is one of the more common and best recognized "remote effects" of cancer on the nervous system [1]. Recent advances (see Pathogenesis) in our understanding of this uncommon disorder have permitted earlier diagnosis of PCD, often leading to the identification of an occult and potentially curable underlying malignancy. These same advances have also given support to the hypothesis that PCD is an immune disorder.

PCD is a pancerebellar syndrome, often with associated signs of brain stem, cerebral, and spinal cord dysfunction. The disorder can begin abruptly or gradually but generally progresses rapidly over weeks or a few months. Progression ceases, usually when the patient is significantly disabled by severe cerebellar dysfunction. Noncerebellar signs, when present, are usually mild. Pathologically, the most prominent histologic feature is severe, diffuse loss of cerebellar Purkinje cells, with or without inflammation (see Pathology).

A. Plaitakis (ed.), CEREBELLAR DEGENERATIONS: CLINICAL NEUROBIOLOGY. Copyright © 1992.
Kluwer Academic Publishers, Boston. All rights reserved.

Table 20-1. Malignancies associated with PCD

Ovarian	50	Lymphoma	32
Papillary	3	Hodgkins	27
Serous	3	NHL	1
Unspecified	43	Unspecified	1
Epithelial	1	T cell	1
Lung	56	Lymphosarcoma	2
Small cell	36	Stomach	2
Squamous	4	Larynx	1
Large cell	4	Prostate	3
Adenocarcinoma	5	Thyroid	1
Unspecified	7	Rectum	1
Mixed	1	Bronchus and rectum	1
Uterus	6	Colon	6
Fallopian tube	4	Ca maxillary antrum	1
Breast	17	Tonsil	1
Adenocarcinoma		Renal cell	1
Unknown primary	9	Chondrosarcoma	1
Uterine and ACA colon	1	AML	1
Ovary and breast	1	Monoclonal gammopathy	1
Polymorphic sarcoma	1		
Total: 199 cases			

PCD has been associated with several systemic malignancies, although ovarian, lung, breast, and lymphoma are the most common (Table 20-1). Recently, a number of antibodies reactive with cerebellar Purkinje cell antigens have been detected in the serum and cerebrospinal fluid (CSF) of some but not all patients with PCD (see Pathogenesis). One specific antibody (Anti-Yo) has been found consistently in the serum and CSF of PCD patients with gynecologic cancers and currently is the focus of attention of a number of investigators.

Both the clinical and pathologic definition of *paraneoplastic* cerebellar degeneration are arbitrary. Neither the clinical findings of a rapidly developing pancerebellar disorder nor the pathological evidence of Purkinje cell loss distinguishes paraneoplastic cerebellar degeneration from some other degenerative cerebellar disorders (see Differential Diagnosis). Only in the presence of relatively specific autoantibodies can a direct causal relationship be established. In the absence of such a marker, we have made a clinical diagnosis of PCD in patients with a pancerebellar disorder in whom no other cause of cerebellar disease can be found and who have cancer within a year preceding or following the onset of cerebellar dysfunction [2].

2. HISTORY

The first reported case of PCD was probably that of Brouwer [3], who in 1919 described a subacute pancerebellar syndrome in 60-year-old woman with "polymorphic sarcoma of the pelvis." Reports of PCD continued to appear [4–10] and by 1982 Henson and Urich [11] were able to find 50 pathologically confirmed cases. There are now 199 reported patients in whom the diagnosis, established either pathologically or by clinical investigation, seems certain.

3. EPIDEMIOLOGY

While ovarian cancer, lung cancer, and lymphoma are the commonest malignancies in patients with PCD, many different tumors are associated with this disorder (Table 20-1) [12–16].

4. CLINICAL MANIFESTATIONS

Unlike most degenerative cerebellar diseases that begin gradually, PCD is usually characterized by an acute or subacute onset of a pancerebellar syndrome. An abrupt onset of vertigo, nausea, and vomiting followed by rapidly progressive truncal, gait, and limb ataxia, as well as ataxic dysarthria, is common. In some patients a flulike illness with headache and myalgias may precede the neurologic symptoms, although no viral illness has been linked etiologically to PCD.

Cerebellar signs may be asymmetric, especially early in the clinical course. Nystagmus, especially downbeating, is common [2,13]. Diplopia secondary to apparent third, fourth or sixth cranial nerve palsies or skew deviation have been reported. Other extracerebellar signs and symptoms, such as ptosis, sensorineural hearing loss, dysphagia, hyperreflexia, upper and lower motorneuron, and extrapyramidal dysfunction have been noted [2,13,16]. Dementia and other abnormalities of mental status have been commonly reported in PCD [16], despite a paucity of pathologic findings outside the cerebellum in most cases. A recent study, using formal cognitive testing, found that dementia was not common when controlling for impaired motor and language skills, and that perceived clinical changes in intellectual performance may be more apparent than real [17].

Recently, the clinical features of Anti-Yo seropositive and seronegative patients have been compared (Table 20-2) [2,12,13]: All seropositive patients studied have been women with gynecologic (including breast) cancer, with the two possible exceptions noted above. The onset is usually abrupt, and symptoms of PCD were more likely to precede the diagnosis of cancer in Anti-Yo positive patients. Abnormalities of affect, mentation, and ocular motility (particularly downbeat nystagmus) were more common in seropositive patients. Clinical impairment was judged as more severe in seropositive patients [2], although no difference in overall clinical impairment has been noted by another investigator [12,13]. LEMS has been seen in association with PCD in 11 reported patients [2,12,18–21], most of whom had small-cell carcinoma of the lung; of those patients tested, all were seronegative for Anti-Yo antibody.

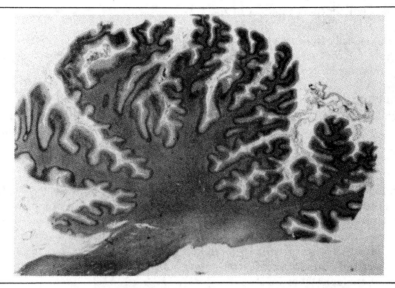

Figure 20-1. A midsagittal section of cerebellum from a patient who died of Hodgkin's disease 6 years after he became incapacitated by PCD. The tumor was in remission when he developed a cerebellar degenerative disorder evolving over several weeks, leaving him unable to walk. The neurological disorder remained static. He died of recurrent Hodgkin's disease 6 years later. The H and E section enlarged four times shows an atrophic cerebellum with relatively well-preserved granule cells. No Purkinje cells were identified on representative microscopic sections. (Compare with Figures 20-6A and 20-6B, MRIs of a patient early and late in the course of the disease.)

5. PATHOLOGY

On gross examination, the cerebellum may appear either normal or atrophic (Figure 20-1). The remainder of the neuraxis usually shows no obvious abnormalities, although generalized cerebral atrophy and brainstem atrophy have been described [16].

Histologically, the hallmark of PCD is a severe and often complete loss of Purkinje cells of the cerebellar cortex (Figure 20-2a). Degenerating Purkinje cells may have swellings, called *torpedos*, along the course of their axons.

Additional pathologic features include thinning of the molecular and granular layers without marked cell loss. "Empty" basket cells, reflecting Purkinje cell loss, are seen, as is a proliferation of Bergmann astrocytes (see Figure 20-2b). The deep cerebellar nuclei are generally well preserved, although there may be some rarefaction of the white matter surrounding these nuclei, corresponding to the loss of Purkinje cell axons.

Pathological changes outside the cerebellum are often absent, but may include dorsal column and pyramidal tract degeneration of the spinal cord [13,15]. Degeneration of the basal ganglia, specifically the pallidum, has been

Table 20-2. Clinical features in anti-Yo positive and anti-Yo negative patients

	Anti-Yo positive	Anti-Yo negative
Sex	All female	Male ≥ female
Primary cancer	Gynecologic tumors Breast cancer	Lung, lymphoma others
Mode of onset	Abrupt or rapid	Rapid or slow
Clinical features	Pancerebellar syndrome Downbeat nystagmus Often severe dysarthria	Pancerebellar syndrome Milder dysarthria and nystagmus Additional paraneoplastic syndromes (i.e., LEMS)
Clinical severity	Usually severe disability	?Less severe disability
Relationship to diagnosis of Ca	Cerebellar syndrome usually precedes diagnosis of Ca	Cancer usually precedes cerebellar syndrome
CSF abnormalities		
Increased protein	50%	50%
Pleocytosis	50%	50%
Incr. IgG index	80%	10%
Oligoclonal bands	70%	20%
MRI	Cerebellar atrophy or normal	Cerebellar atrophy or normal
CT scan	Cerebellar atrophy or normal	Cerebellar atrophy or normal
Response to treatment: Plasmapheresis, immunosuppressants	Usually no response	Usually no response (occasional spontaneous remission)

reported rarely, as has an associated peripheral neuropathy [16,22]. Although dementia is often a clinical feature of PCD, pathologic changes in the cerebral cortex are not common [11]. It is not known if there are differences in the distribution of extra-Purkinje cell pathological changes between anti-Yo positive and anti-Yo negative patients. Insufficient numbers of patients have been examined pathologically since the antibodies were identified.

Inflammation is minimal or absent in most cases of PCD. A subset of patients, however, have an associated encephalomyelitis with a prominent inflammatory reaction [23]. Lymphocytic infiltration may be seen in the meninges, deep cerebellar or brainstem nuclei, dorsal root ganglia, as well as other parts of the neuraxis. Inflammatory infiltrates surrounding Purkinje cells have not been described. The inflammatory and noninflammatory cases may represent variants of a single disease process, perhaps at different points in evolution, although this is controversial. It has not been shown that these patients differ in terms of their clinical disease or associated malignancies, although lung cancer was slightly more common in the "inflammatory" group [11]. The incidence of gynecologic cancers was approximately the same in both the inflammatory or noninflammatory group.

Purkinje cell loss is in no way specific for PCD and may be seen in degenerative, toxic, metabolic, infectious, and vascular insults to the cerebellum, as well as in normal aging. Diffuse, often severe, Purkinje cell loss has also been found postmortem in cancer patients without the symptoms

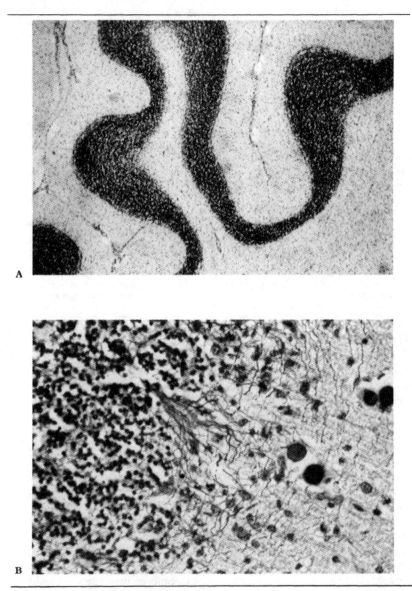

Figure 20-2. A: A low-power photomicrograph of the cerebellum from a patient with PCD. The granule layer and molecular layer are well preserved, although no Purkinje cells are identified in the entire section. Bergmann astrocytes are increased in number. **B**: A higher power photomicrograph from another patient with PCD. This Bodian stain demonstrates the absence of Purkinje cells. Basket cell fibers (arrow) were present, but the Purkinje cell that they should be surrounding is absent. Bergmann astrocytes are increased in number.

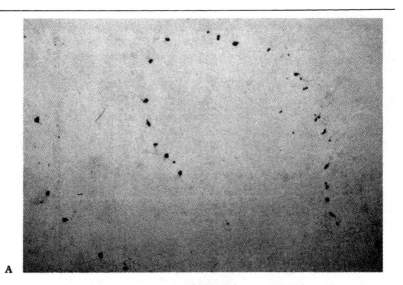
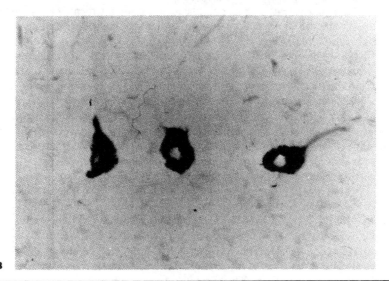

Figure 20-3. Immunohistochemistry of anti-Yo antibody. Biotinylated IgG from a patient with PCD and breast cancer was reacted with frozen sections of human cerebellum obtained at autopsy. There is no counterstain. The only cellular elements seen are those that react with the human IgG. **A**: A single cerebellar folium in which only Purkinje cells can be seen (×16). **B**: Detail of an area from A. Course granular staining of the cytoplasm of three Purkinje cells is evident. Nuclei and axons do not stain, although some dendrites do (×250). (From Furneaux, Posner, (1989). Paraneoplastic syndromes. In: B.H. Waksman (ed): Immunologic Mechanisms in Neurologic and Psychiatric Disease. New York: Raven Press, pp. 187–219. With permission.)

of PCD [24]. Reduced Purkinje cell numbers were especially pronounced in patients with ovarian cancer. In addition to diminished Purkinje cell numbers, reduced granular cell numbers were found in patients with ovarian cancer and lung cancer. These data suggest the occurrence of a clinically latent form of this disorder.

6. PATHOGENESIS

A number of theories have been proposed to explain the pathogenesis of PCD and its relationship to cancer. These include (1) a tumor-secreted substance toxic to Purkinje cells, (2) opportunistic infection, and (3) immunologic factors [25–27].

The first report of an antibody reactive with Purkinje cell antigens was by Trotter in a patient with Hodgkin's disease and PCD [28]. Using indirect immunofluorescence (IIF) staining techniques, an antibody at low titer (1:20) was found in the serum that reacted with the cytoplasm of normal human Purkinje cells. Reactivity above a 1:20 dilution of the patient's serum was not found; similar but less striking staining was found in patients with Hodgkin's disease without PCD, normal individuals, and a patient with multiple sclerosis. Subsequently, Stefansson [29] reported an antibody reactive with Purkinje and cortical neuron cytoplasm in another patient with Hodgkin's disease. Titers and further definition of the antibody by immunoblotting techniques were not reported.

In 1983, Greenlee and Brashear [30] and Jaeckle et al. [31] independently demonstrated a circulating antibody in patients with ovarian cancer and PCD that produced coarse, granular cytoplasmic staining of Purkinje cells (Figure 20-3). These antibodies were present at high titer. Greenlee and Brashear [30] reported a similar antibody in two neurologically normal patients with ovarian cancer; both patients died of their malignancies without developing neurologic symptoms. Postmortem examination of one of these patients, however, revealed moderate, diffuse Purkinje cell loss [32]. (We have not identified the anti-Yo antibody in any patient without PCD.)

In 1985, Jaeckle et al. [33] found an antibody with the same staining characteristics noted by Greenlee and Brashear in six women with PCD and ovarian or breast cancer (Figure 20-3). The antibody was polyclonal (IgA, IgG, IgM, and IgD). The antigen was stable to acetone, ethanol, ribonuclease, and deoxyribonuclease, and to 70°C. It was labile to formalin, trypsin, and proteinase, and partially labile to heating at 100°C or to pronase, suggesting it is a protein. The antigen is found in multiple species, including the guinea pig, rat, mouse, monkey, sheep, pig, cat, and human [34,35]. Using immunohistochemistry and electron microscopy, Rodriguez et al. localized the antigen to Purkinje cell ribosomes, rough endoplasmic reticulum, and the maturing face of the golgi apparatus [36].

Immunohistochemical staining is exclusive to cerebellar Purkinje cells. Other cells within the central nervous system and cells in systemic organs do

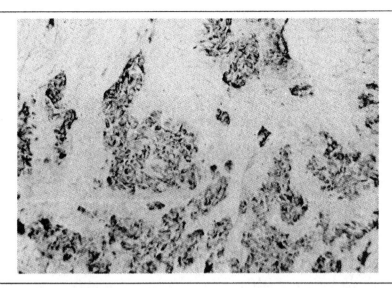

Figure 20-4. Frozen section of ovarian tumor reacted with biotinylated IgG from an anti-Yo-positive patient with PCD. The technique is the same as in Figure 20-1. There is no counterstain so that no staining is seen if IgG from normal serum is used. Most, but not all, of the tumor cells react with the patient's serum. Under higher power, the reaction is restricted to the cytoplasm and has a granular appearance. Elution of the antibody from the tumor gives a positive reaction with Purkinje cells, indicating that the antibody reactive with tumor and Purkinje cells is identical (×100). (From Furneaux, Posner, (1989). Paraneoplastic syndromes. In: B.H. Waksman (ed): Immunologic Mechanisms in Neurologic and Psychiatric Disease. New York: Raven Press, pp. 187–219. With permission.)

not stain. Ovarian and breast tumors of patients without PCD are, likewise, negative both by immunohistochemistry and by Western blot. However, the sera of patients with PCD react immunohistochemically and by Western blot with the ovarian and breast tumors from patients with PCD [37] (Figure 20-4).

On immunoblots of Purkinje cell extracts, the patients' serum and CSF identify two discrete groups of antigens with molecular weights of 62–64 kDa and 34–38 kDa (called CDR 62 and CDR 34) [38] (Figure 20-5). Antibodies with these characteristic histochemical and immunoblot findings have been termed *anti-Yo* (Yo are the first two letters of the last name of one patient reported by Jaeckle et al. [33]) to distinguish it from other anti-Purkinje cell antibodies associated with PCD. Reactivity with these antigens has not been demonstrated in PCD patients who are seronegative by immunohistochemistry for anti-Yo.

A cDNA clone coding for an epitope recognized by Anti-Yo positive sera has been isolated [39]. This gene, expressed exclusively in Purkinje cells,

Autoantibodies from Patients with PCD Recognize Two Purkinje Neuron Antigens

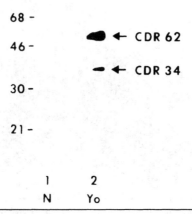

Figure 20-5. Western blot of serum against human Purkinje cell extract: Lane 1, normal serum (N); lane 2, anti-Y serum (Yo). Each patient's serum was diluted 1:500. The 62- (CDR 62) and 34- (CDR 34) kDa bands are clearly present in the Yo positive serum but are absent from the control patient. In our experience, these bands are not identified, except in patients with PCD usually associated with gynecologic cancer.

codes for the CDR 34 antigen. Both the CDR 34 and CDR 62 antigens are also expressed in tumor tissue from PCD patients, but not in normal tissues or in tumors from patients without PCD [37]. The gene coding for the CDR 34 antigen has been localized to the X chromosome; preliminary data suggests that the gene coding for the CDR 62 antigen resides on chromosome 16. The characteristics of anti-Yo antibody and the Yo antigen are listed in Table 20-3.

A number of anti-Purkinje cell antibodies (APCA) reactive with other Purkinje cell antigens have been described in PCD patients with systemic malignancies (Table 20-4). Using indirect immunofluorescence antibody techniques (IIF), Greenlee et al. [40] described an antibody reactive with the nucleus and cytoplasm of Purkinje and granule cells in the serum and CSF of a patient with PCD and oat-cell carcinoma of the lung. The antibody titers were relatively low, and further definition of the antigen using immunoblotting was not performed. Antigens with other staining characteristics and molecular weights have been identified in the serum from patients with PCD and small cell lung carcinoma, non-Hodgkin's lymphoma, adenocarcinoma of the lung, and breast cancer [2,21,41–44] (Table 20-4). This suggests that

Table 20-3. Characteristics of the Anti-Yo antibody and its antigens

ANTIBODY
Coarse granular cytoplasmic staining of Purkinje cell neurons and brain-stem nuclei using IIF and
  IIP techniques.
Polyclonal (IgA, IgG, IgM, and IgD); all have IgG
Present in CSF and serum
Evidence for synthesis within the CNS

ANTIGENS
62–64 kDa (CDR 62) and 34–38 kDa (CDR 34) molecular weight
Found in multiple species: human, guinea pig, rat, mouse, monkey sheep, pig, and cat.
Located in ribosomes, rough endoplasmic reticulum, and the maturing face of the Golgi
  apparatus
Stable to acetone, ethanol, ribonuclease, deoxyribonuclease, and heating to 70°C.
Labile to formalin, trypsin, and proteinase
Partially labile to pronase and heating to 100°C
Gene coding for CDR 34 located on X chromosome
Gene coding for CDR 62 located on chromosome 16
Both CDR 34 and CDR 62 antigens are expressed in tumor tissue from patients with PCD

many different antigenic epitopes may be involved in PCD. The antibody reactive with the 35- to 40-kDa antigen identified by Andersen et al. [2] in two patients with small-cell carcinoma of the lung and PCD is identical to the anti-neuronal antibody (anti-Hu) found in some patients with subacute sensory neuropathy (SSN) and encephalomyelitis [45,46]. In addition, Tsukamoto described a patient with small-cell lung cancer, PCD, and a serum antibody reactive with rat brain antigens of 38 and 40 kDa molecular weight [47]; this may be identical to the anti-Hu antibody.

Recently, Tsukamoto [47] reported five PCD patients with ovarian and uterine cancers and previously undescribed serum APCAs. Unlike the Anti-Yo antibody, the antibodies in these patients were found to react with neurons throughout the CNS, as well as with the cytoplasm of Purkinje cells. In addition, reactivity on immunoblots was confined to rat brain antigens of 52 and 58 kDa molecular weight; reactivity with antigens of 34 and 62 kDa (as seen in anti-Yo seropositive patients) was not reported. These findings are of interest in that they represent the first report of gynecologic cancer-associated PCD with an APCA that is not anti-Yo. Greenlee et al. [42] have also described antibody reactivity outside the cerebellum in PCD patients with gynecologic cancers.

Some investigators have discovered an APCA in patients with Hodgkin's disease [28,29,42]. The specific antigen to which these antibodies were directed was not defined with immunoblotting, however. Other investigators have not found antibody directed against Purkinje cell antigens in the serum or CSF of patients with Hodgkin's disease and PCD [2].

The uniting theory of pathogenesis is that a tumor antigen provokes a host immune response; the host antibodies then cross-react with antigenically

Table 20-4. Anti-Purkinje cell antibodies in PCD (Reported in the literature)

Name	No. patients	Cancer	Histochemistry	Immunoblot	Reference
1.	1	Hodgkin's disease	IIF—predominantly cytoplasmic; titers only to 1:20 dilutions		Trotter et al., 1976
2.	1	Hodgkin's disease	IIF—predominantly cytoplasmic		Stefansson et al., 1981
3. Anti-Yo	41	Ovarian cancer Uterine Fallopian tube Breast cancer ACA-unknown primary Lymphoma	IIF and IIP—coarse granular cytoplasmic staining	Reactive with 34–38 kDa and 62–64 kDa Purkinje cell antigen	Cunningham, 1987 Jaeckle, 1985 Greenlee, 1983, 1984, 1988 Andersen, 1988 Furneaux, 1989 Hammack, 1989 Wang, 1988 McLellan, 1988 Royal, 1987
4. Anti-Hu	3	Small cell CA lung	IIF and IIP—staining of nuclei throughout CNS, DRG, trigeminal ganglion	Reactive with 35–40 kDa nuclear Ag	Andersen, 1988 1988 Greenlee, 1986
5.	1	Small cell Ca lung	Staining of Purkinje cytoplasm and dendrites	Reactive with 85 kDa antigen	Brown, 1985
6.	1	NHL	Staining of Purkinje cytoplasm	Reactive with 250 and 110 kDa antigen	Tanaka, 1986
7.	1	Small cell Ca lung		Reactive with 98 and 68 kDa cytoplasmic antigens of cerebrum and cerebellum	Tanaka, 1987
8.	1	ACA lung	IIF and IIP—fine granular staining of Purkinje cytoplasm	Reactive with 56, 64, 68, and 80 kDa antigens	Anderson, 1988

#	Antibody	n	Cancer	Staining	Reactivity	Reference
9.	Anti-neuronal nuclear Ab (not Anti-Hu)	1	Breast cancer	IIF and IIP—staining of all neuronal nuclei	Reactive with 53–61 kDa and 79–84 kDa antigens	Anderson, 1988
10.	"Atypical APCA"	2	Non-small cell lung Colon cancer	IIF and IIP —staining of Purkinje cell cytoplasm	Negative	Anderson, 1988
11.		2	Benign monoclonal gammopathy	IIP—staining of Purkinje cell cytoplasm		Bourdette, 1987
12.		1	No cancer	IIP—diffuse cytoplasmic staining of Purkinje cells (looks like anti-Yo)	Negative	Anderson (unpublished)
13.	Anti-Nb	1	No cancer	IIP—staining of Purkinje cell neurons	Reactive with 170, 155, and 65 kDa antigens	Darnell, 1989
14.		4	Uterus Ovary	IIP—staining of Purkinje cell cytoplasm as well as cortical and deep cerebellar neurons	Reactive with 52 and 58 kDa antigens	Tsukamoto, 1989
15.		1	Small cell Ca lung	IIP—Staining nucleus and cytoplasm of neurons throughout the CNS	Reactive with 38 and 40 kDa antigens	Tsukamoto, 1989
16.		1	Breast	IIP—Staining nucleus and cytoplasm of neurons throughout the CNS	Reactive with 30 and 46 kDa antigens	Tsukamoto, 1989
17.		3	SCCL Breast	IIP—Cytoplasmic and nuclear staining of neurons throughout cerebrum, cerebellum, and brainstem	Not done	Greenlee, 1988
18.		1	Hodgkin's disease	IIP—Cytoplasmic staining of Purkinje and Golgi neurons	Not done	Greenlee, 1988
19.		1	Mesodermal sarcoma (ovary)	IIP—Cytoplasmic and nuclear staining of neurons and astrocytes throughout CNS	Not done	Greenlee, 1988

488   III. Etiopathogenesis of cerebellar disorders

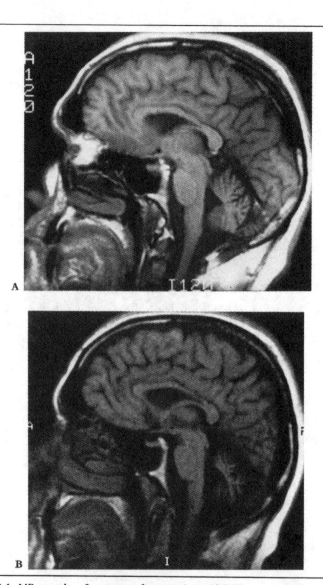

Figure 20-6. MR scans done 2 years apart from a patient with PCD associated with Hodgkin's disease. **A**: The first scan was done 3 months after the development of neurological symptoms and after the discovery of the underlying cancer. At that time, the patient was ambulatory but was disabled by dysarthria and oscillopsia. The MR scan shows an essentially normal cerebellum. **B**: Two years later, the patient had marked limb and truncal ataxia and was unable to walk unassisted. His lymphoma was in remission. The MR scan shows substantial atrophy of the cerebellum.

similar epitopes on Purkinje cells, causing Purkinje cell injury or destruction. Intrathecal production of anti-Yo has been demonstrated [48]. Cerebellar Purkinje cells have been reported to have the capacity to extract large molecules, including immunoglobulins, from the CSF [49]. This may explain how anti-Purkinje cell antibodies may reach intracytoplasmic antigens.

If immunity is involved in the pathogenesis of PCD, how can one account for those patients who lack detectable antibodies but otherwise have neurologic disease that is clinically and pathologically identical to that of their seropositive counterparts? It is unlikely that these patients have a completely different disorder with a different pathogenesis. Significantly, these seronegative patients occasionally have an additional paraneoplastic disorder, Lambert-Eaton myasthenic syndrome (LEMS). LEMS is a disease with a proved immune etiology [50,51], but without antibodies that can be demonstrated using conventional light microscopic IIF or immunoperoxidase (IIP) techniques. Perhaps the actual pathogenic factor in PCD, if not cell mediated, is undetectable by current methods because it is very small or sparingly located, as is the antigen (the voltage-gated calcium channel) in LEMS.

Even if not pathogenic, anti-Yo represents an immune response to an organ-restricted epitope expressed only in some tumors, and thus serves as a marker to identify specific tumors in a subset of PCD patients.

7. DIAGNOSTIC STUDIES

Computerized tomography (CT) and magnetic resonance imaging (MRI) of the head usually reveal no abnormalities early in the clinical course of PCD [13,18]. Greenberg [18] demonstrated that follow-up head CT performed 7–25 months later was more likely to demonstrate cerebellar atrophy. MRI abnormalities, including cerebellar and brainstem atrophy, as well as T2 signal abnormalities in cerebral and cerebellar white matter, have been reported in some PCD patients [13]. CT and MRI abnormalities did not correlate with the severity of neurologic disease. No radiographic differences between seronegative and seropositive patients have been found. Figure 20-6 demonstrates the progession of cerebellar atrophy (over 2 years) in a patient with PCD and Hodgkin's disease.

While not present in all patients, cerebrospinal fluid (CSF) abnormalities, consisting of elevated protein, elevated CSF IgG and IgG index, CSF oligoclonal bands, as well as CSF pleocytosis have been noted in both seropositive and seronegative patients. The CSF pleocytosis resolves with time. An abnormal CSF IgG index has been noted more frequently in seropositive patients [12,13]. As indicated above, comparisons of serum and CSF antibody levels generally show a relative increase in the amount of antibody in the CSF when serum and CSF antibody levels are compared with total IgG in their respective compartments. Such a finding strongly indicates central nervous system synthesis of the autoantibody, supporting an immune hypothesis for the disorder and suggesting that the antibody plays a direct role in pathogenesis. Interestingly, plasmapheresis, which can substantially lower the level of serum antibody, has a minimal effect on CSF antibody. This may partially explain the failure of plasmapheresis to affect the course of PCD [2,12,13].

Cerebral metabolic activity in 11 patients with PCD has been evaluated using 18-fluorodeoxyglucose (FDG)/positron emission tomography (PET) [17]. Significant reductions in metabolic activity were observed in most brain regions when compared with normal controls. While these findings are of uncertain significance, they suggest either that the neuronal damage is more widespread than clinical and pathological abnormalities suggest or that the loss of cerebellar cortical efferents may indirectly affect the function of extracerebellar neurons.

8. DIFFERENTIAL DIAGNOSIS

A number of other disease processes, related and unrelated to systemic malignancy, may cause acute or subacute cerebellar dysfunction in patients with cancer (Table 20-5). Cerebellar metastases, both intraparenchymal and meningeal, may present in this fashion, although a CT or MRI scan and CSF examination should effectively exclude these possibilities.

Epstein-Barr or varicella virus have been occasionally identified to cause a brain-stem encephalitis or a cerebellitis. This complication may be on a direct infectious or immune postinfectious basis. It is most commonly seen in children or young adults, and elevated serum and CSF antibodies to these viruses may be identified [52]. Opportunistic infection, such as progressive multifocal leukoencephalopathy (PML), *Listeria monocytogenes,* chronic tuberculous or fungal meningitis may present with predominantly cerebellar findings. This differential diagnosis is especially important in those patients with lymphoma. MRI, CSF examination, and clinical evolution should help to establish these diagnoses. Several chemotherapeutic agents, such as ara-C and 5-fluorouracil (5-FU), have been associated with cerebellar toxicity [53,54].

Disorders not directly associated with cancer, but that my cause a late-onset cerebellar disorder, include demyelinating disease, cerebellar infarction or hemorrhage, cerebellar abscess, and Creutzfeldt-Jacob disease. Metabolic disorders, including hypothyroidism, alcoholic cerebellar degeneration, mercury poisoning, and phenytoin toxicity, can be excluded on the basis of history, laboratory tests, and additional physical findings. Some forms of hereditary cerebellar degeneration may present in adult life. These disorders are usually more slowly progressive, and patients often have a family history of similar illness.

Most of the disorders listed in Table 20-5 have characteristic clinical or pathologic findings and make them easily separable from PCD. Few of these disorders cause an acute or subacute pancerebellar degeneration with a

Table 20-5. Differential diagnosis

Metastatic disease
   Parenchymal (cerebellar)
   Meningeal
Infections
   Bacterial
      Cerebellar abscess
      Listeria monocytogenes (rhombencephalitis)
   Viral
      Progressive multifocal leukoencephalopathy
      EBV
      Herpes simplex virus
      Varicella/zoster
      Postinfectious
   Mycobacterial
      TB meningitis
   Fungal
   Spongiform encephalopathy (Creutzfeldt-Jacob)
Metabolic/toxic
   Hypothyroidism
   Alcoholic cerebellar degeneration
   Mercurial toxicity
   Phenytoin toxicity
   ara-C toxicity
   5-flourouracil toxicity
Other
   Cerebellar infarct
   Cerebellar hemorrhage
   Demyelinating disease
   Heat stroke
   Inherited cerebellar degeneration

negative MR scan, CSF pleocytosis, and an elevated CSF IgG. In patients with such a history, it is likely that half or more suffer paraneoplastic cerebellar degeneration [11]. In those in whom a tumor is not found, even after a long period of careful follow-up, a definitive diagnosis is almost never made; in one such patient, an APCA different from anti-Yo (or any other reported autoantibody in PCD) has been noted [55]. This finding suggests that many cases of acute-onset, nonparaneoplastic cerebellar degeneration may, like PCD, have an autoimmune pathogenesis. In these cases, the inciting pathogenic factor is not a tumor. Figure 20-7 presents a clinical approach to pancerebellar syndromes occurring in adulthood.

9. PROGNOSIS AND TREATMENT

Most patients with PCD have relentless progression of their neurologic disease to near total disability; often the disorder stabilizes clinically only at a point when little cerebellar function is left to lose. This may occur despite cure of the systemic malignancy. There are case reports of early arrested

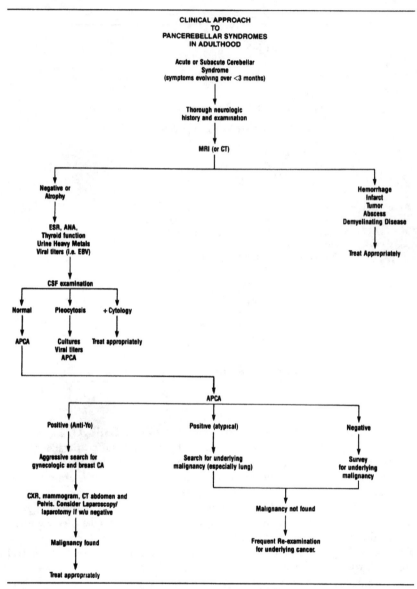

Figure 20-7. An alogrithm suggesting the clinical and laboratory approach to a patient presenting with the acute or subacute onset of cerebellar symptomatology. The upper portion of the figure demonstrates the clinical and routine laboratory evaluation. The lower portion of the figure shows the approach after measurement of anti-Purkinje cell antibodies (including Yo) in serum and/or CSF.

progression or complete clinical remission of the cerebellar syndrome following the treatment of the underlying cancer [13,15,56,57].

Working on the assumption that PCD is an autoimmune disorder, plasmapheresis and glucocorticosteroids have been used in attempts to remove or suppress the formation of presumed pathogenic antibodies. Most patients have derived no apparent benefit from these interventions, although a drop in APCA titer has resulted [13]. A few cases have been reported in whom plasmapheresis appeared to produce clinical improvement [58–60]; in one patient this was associated with a drop in antibody titer [59]. Our own experience has not been encouraging [2]. Occasional spontaneous remission of the cerebellar syndrome has been noted, unassociated with immunosuppressive therapy or treatment of the underlying cancer [61,62].

REFERENCES

1. Croft P.B., Wilkinson M. (1965). The incidence of carcinomatous neuromyopathy in patients with various types of carcinoma. Brain 88:427.
2. Anderson N.E., Rosenblum M.K., Posner J.B. (1988). Paraneoplastic cerebellar degeneration: Clinical-immunological correlations. Ann. Neurol. 24:559–567.
3. Brouwer B. (1919). Beitrag zur Kenntnis der chronischen diffusen Kleinhirnerkrankungen. Mendels neurologisches Zentralblatt 38:674–682.
4. Lhermitte J. (1922). L'astasie-abasie cerebelleuse par atrophie vermienne chez le vieillard. Revue Neurologique 38:313–316.
5. Casper J. (1929). Toxische kleinhirnatrophie bei brustkrebs. Zentralblatt fur die gesamte neurologie und psychiatrie 53:854–856.
6. Parker H.L., Kernohan J.W. (1933). Parenchymatous cortical cerebellar atrophy. Brain 56:191–192.
7. Greenfield J.G. (1934). Subacute spinocerebellar degeneration occurring in elderly patients. Brain 57:161–176.
8. Brouwer B., Biemond A. (1938). Les affections parenchymateuses du cervelet et leur signification du point de vue de l'anatomie et la physiologie de cet organe. Belge de Neurologie et Psychiatrie 38:691–757.
9. Zulch K.J. (1936). Uber die kleinhirnrindenatrophie. Zeitschrift fur die gesamte neurologie und psychiatrie. 156:493–573.
10. Kennard M.A. (1935). Clinical and histological observations on a case of primary cortical degeneration of the cerebellum. Koninklijke Nederlandsche Akademie van Wetenschappen. Series C: Biological and Medical Sciences. Proceedings 38:544–552.
11. Henson R.A., Urich H. (1982). Cancer and the nervous system. In: The Neurological Manifestations of Systemic Malignant Disease, 1st ed. Oxford: Blackwell.
12. Hammack J.E., Kimmel D.W., O'Neill B.P., Lennon V.A. (1988). Paraneoplastic cerebellar degeneration: A clinical comparison of patients with and without Purkinje cell cytoplasmic antigens. Neurology 38(Suppl. 1):3 (abstract).
13. Hammack J.E., Kimmel D., O'Neill B.P., Lennon V.A. (1990). Paraneoplastic cerebellar degeneration: A clinical comparison of patients with and without Purkinje cell cytoplasmic antigens. Mayo Clinic Proc. 65:1423–1431.
14. Horwich L., Buxton P.H., Ryan G.M.S. (1966). Cerebellar degeneration with Hodgkin's disease. J. Neurol. Neurosurg. Psychiat. 29:45–51.
15. Eekhof J.L.A. (1985). Remission of a paraneoplastic cerebellar syndrome. Clin. Neurol. Neurosurg. 87:133.
16. Brain W.R., Wilkinson M. (1965). Subacute cerebellar degeneration associated with neoplasms. Brain 88:465–478.
17. Anderson N.E., Posner J.B., Sidtis J.J., et al. (1988). The metabolic anatomy of paraneoplastic cerebellar degeneration. Ann. Neurol. 23:533–540.
18. Greenberg H.S. (1984). Paraneoplastic cerebellar degeneration. A clinical and CT study. J. Neuro-Oncol. 2:377.

19. Okita N., Ohara Y., Kobayashi K., et al. (1981). A case of lung cancer associated with subacute cerebellar degeneration and Eaton-Lambert syndrome. Igaku-no-Ayumi 118:156–166.
20. Satoyoshi E., Kowa H., Fukunaga N. (1973). Subacute cerebellar degeneration and Eaton-Lambert syndrome with bronchogenic carcinoma. Neurology 23:764–768.
21. Tanaka K., Tanaka M., Miyatake T., Yamamoto A., Kurahashi K., Matsunaga M. (1987). Antibodies to brain proteins in a patient with subacute cerebellar degeneration and Lambert-Eaton myasthenic syndrome. Tohoku J. Exp. Med. 153:161–167.
22. Patten J.P. (1971). Remittent peripheral neuropathy and cerebellar degeneration complicating lymphosarcoma. Neurology 21:189–194.
23. Vick N., Schulman S., Dau P. (1969). Carcinomatous cerebellar degeneration, encephalomyelitis, and sensory neuropathy (radiculitis). Neurology 19:425–441.
24. Schmid A.H., Riede U.N. (1974). A morphometric study of the cerebellar cortex from patients with carcinoma. A contribution on quantitative aspects in carcinotoxic cerebellar atrophy. Acta Neuropath. (Berlin) 28:343–352.
25. Gordon M.H. (1932). Rose research on lymphadenoma. Bristol, England: Wright and Sons.
26. Meyer J.S., Foley J.M. (1953). The encephalopathy produced by extracts of eosinophils and bone marrow. J. Neuropath. Exp. Neurol. 12:349–362.
27. Russel D.S. (1961). Encephalomyelitis and carcinomatous neuropathy. In: The Encephalitides, L. van Bogaert, J. Radermecker, J. Hozay, A. Lowenthal (eds): Amsterdam: Elsevier, pp. 131–135.
28. Trotter J.L., Hendin B.A., Osterland K. (1976). Cerebellar degeneration with Hodgkin disease. Arch. Neurol. 33:660–661.
29. Stefansson K., Antel J.P., Wollman R.I., Levin K.H., Larson R., Arnason B.G.W. (1981). Anti-neuronal antibodies in serum of a patient with Hodgkin's disease and cerebellar ataxia. Neurology 31(Suppl. 2):126.
30. Greenlee J.E., Brashear H.R. (1983). Antibodies to cerebellar Purkinje cells in patients with paraneoplastic cerebellar degeneration and ovarian carcinoma. Ann. Neurol. 14:609–613.
31. Jaeckle K.A., Houghton A.N., Nielsen S.L., Posner J.B. (1983). Demonstration of serum anti-Purkinje antibody in paraneoplastic cerebellar degeneration and preliminary antigenic characterization. Ann. Neurol. 14:111.
32. Brashear H.R., Greenlee J.E. (1985). Persistence of anticerebellar antibodies in neurologically normal patients with ovarian neoplasms. Neurology 35(Suppl. 1):174.
33. Jaeckle K.A., Graus F., Houghton A., Cardon-Cardo C., Nielsen S.L., Posner J.B. (1985). Autoimmune response of patients with paraneoplastic cerebellar degeneration to a Purkinje cell cytoplasmic protein antigen. Ann. Neurol. 18:592–600.
34. Greenlee J.E., Sun M. (1984). Variable immunofluorescent staining of nonhuman cerebellar tissue with sera from patients with paraneoplastic cerebellar degeneration: A cautionary tale. Neurology 34(Suppl. 1):88.
35. Smith J.L., Finley J.C., Lennon V.A. (1988). Autoantibodies in paraneoplastic cerebellar degeneration bind to cytoplasmic antigens of Purkinje cells in humans, rats and mice and are of multiple immunoglobulin classes. J. Neuroimmunol. 18:37–48.
36. Rodriguez M., Truh L., O'Neill B.P., Lennon V.A. (1986). Autoimmune cerebellar degeneration associated with cancer: Localization of Purkinje cytoplasmic antigen by immuno-electron microscopy. J. Neuropathol. Exp. Neurol. 45:322.
37. Furneaux H.M., Rosenblum M., Wong E., Woodruff P., Posner J.B. (1989). Selective expression of Purkinje neuron antigens in ovarian and breast tumors of patients with paraneoplastic cerebellar degeneration. Neurology 39(Suppl. 1):260.
38. Cunningham J., Graus F., Anderson N., Posner J.B. (1986). Partial characterization of the Purkinje cell antigens in paraneoplastic cerebellar degeneration. Neurology 36:1163–1168.
39. Furneaux H.M., Dropcho E.J., Barbut D., et al. (1989). Characterization of a cDNA encoding a 34-kDa Purkinje neuron protein recognized by sera from patients with paraneoplastic cerebellar degeneration. Proc. Natl. Acad. Sci. USA 86:2873–2877.
40. Greenlee J.E., Lipton H.L. (1983). Detection of anticerebellar antibodies in serum and cerebrospinal fluid of a patient with oat cell carcinoma of the lung and subacute cerebellar degeneration. Ann. Neurol. 14(1):142.
41. Brown R.H., Ronthal M., Come S., et al. (1985). Antibodies to 85,000 dalton protein in

paracarcinomatous cerebellar degeneration. Neurology 35(Suppl.):288.
42. Greenlee J.E., Brashear H.R., Herndon R.M. (1988). Immunoperoxidase labeling of rat brain sections with sera from patients with paraneoplastic cerebellar degeneration and systemic neoplasia. J. Neuropath. Exp. Neurol. 47(5):561–571.
43. Tanaka K., Yamazaki M., Sato S., Toyoshima I., Yamamoto A., Miyatake T. (1986). Antibodies to brain proteins in paraneoplastic cerebellar degeneration Neurology 36:1169–1172.
44. Anderson N.E., Budde-Steffen C., Wiley R.G., et al. (1988). A variant of the anti-Purkinje cell antibody in a patient with paraneoplastic cerebellar degeneration. Neurology 38:1018–1026.
45. Graus F., Cordon-Cardo C., Posner J.B. (1985). Neuronal antinuclear antibody in sensory neuronopathy from lung cancer. Neurology 35:538.
46. Graus F., Elkon K.B., Cordon-Cardo C., Posner J.B. (1986). Sensory neuronopathy and small cell lung cancer. Antineuronal antibody that also reacts with the tumor. Am.J. Med. 80:45.
47. Tsukamoto T., Yamamoto H., Iwasaki Y., Yoshie O., Terunuma H., Suzuki H. (1989). Antineural autoantibodies in patients with paraneoplastic cerebellar degeneration. Arch. Neurol. 46:1225–1229.
48. Posner J.B., Furneaux H., Rosa E. (1989). Central nervous system synthesis of autoantibodies in paraneoplastic syndromes. Neurology 39(Suppl. 1):244–245.
49. Borges L.F., Elliot P.J., Gill R., et al. (1985). Extraction of small and large molecules from the cerebrospinal fluid by Purkinje neurons. Science 228:346.
50. Lang B., Newsome-Davis J., Wray D., Vincent A., Murray N. (1981). Autoimmune aetiology for myasthenic (Eaton-Lambert) syndrome. Lancet 2:224.
51. Lennon V.A., Lambert E.H., Pollock H., DuPont B., Whittingham S. (1982). Lambert-Eaton myasthenic syndrome: An autoimmune disease. Muscle Nerve 5:s21.
52. Kramer D.S., Smitnik L.M., John K., Drake M.E. (1985). Acute cerebellar sydrome in infectious mononucleosis: Documentation of two cases with Epstein-Barr virus infection. J. Natl. Med. Assoc. 77(4):305–308.
53. Winkelman M.D., Hines J.D. (1983). Cerebellar degeneration caused by high-dose cytosine arabinoside: A clinicopathological study. Ann. Neurol. 14:520–527.
54. Riehl J.L., Brown N.J. (1964). Acute cerebellar syndrome secondary to 5-fluorouracil chemotherapy. Neurology 15:254.
55. Darnell R.B., Furneaux H., Posner J.B. (1989). Characterization of neural antigens recognized by autoantibodies in CSF and serum of a patient with cerebellar degeneration: Co-expression in Purkinje cells and tumor lines of neuroectodermal origin. Neurology 39(Suppl.):385.
56. Paone J.F., Jeyasingham K. (1980). Remission of cerebellar dysfunction after pneumonectomy for bronchogenic carcinoma. N. Engl. J. Med. 302:156–157.
57. Kearsley J.H., Johnson P., Halmagyi M. (1985). Paraneoplastic cerebellar disease. Remission with excision of the primary tumor. Arch. Neurol. 42:1208.
58. Cocconi G., Ceci G., Juvarra G., et al. (1985). Successful treatment of subacute cerebellar degeneration in ovarian carcinoma with plasmapheresis. Cancer 56:2318.
59. Royal W., Galasko D.R., McKhann G.M., Cunningham J.M., Dropcho E.J. (1987). Clinical course, immunologic, and biochemical features of a patient with paraneoplastic cerebellar dysfunction. Neurology 37(Suppl. 1):305–306.
60. Sapra R., Armentrout D., Margileth D., Opfell R. (1981). Subacute cerebellar degeneration (SCD) in Hodgkin's disease (HD)-Successful treatment with plasmapheresis. Proc. Am. Assoc. Cancer Res. 22:511.
61. Auth T.L., Chodoff P. (1957). Transient cerebellar syndrome from extracerebral cancer. Neurology 7:370–372.
62. Cairncross J.G., Posner J.B. (1980). Neurological complications of malignant lymphoma, In: P.J. Vinken, G.W. Bruyn (eds): Amsterdam: North-Holland, Handbook of Clinical Neurology, Vol. 39. Neurological Manifestations of Systemic Diseases, Part II. p. 27.
63. Bourdette D.N., Nilaver G. (1987). Cerebellar degeneration associated with anti-Purkinje cell antibodies and benign IgG monoclonal gammopathies. Neurology 37(Suppl. 1):291.
64. Greenlee J.E., Lipton H.L. (1986). Anticerebellar antibodies in serum and cerebrospinal fluid of a patient with oat cell carcinoma of the lung and paraneoplastic cerebellar

degeneration. Ann. Neurol. 19:82–85.
65. McLellan R., Currie J.L., Royal W., Rosenshein N.B. (1988). Ovarian carcinoma and paraneoplastic cerebellar degeneration. Obstet. Gynecol. 72:922–924.
66. Wang A.-M., Leibowich S., Ridker P.M., David W. (1988). Paraneoplastic cerebellar degeneration in a patient with ovarian carcinoma. AJNR 9:216–217.
67. Barraque-Bordas L., Ruiz-Lara R. (1952). Atrofia cerebelosa carcinotoxica: Observation anatomochemica. Revista Clin. Espanola 45:114–120.
68. Bocian J.J., Zealer D.S. (1957). Subacute cerebellar degeneration associated with bronchogenic carcinoma. Calif. Med. 87:37–39.
69. Alessi E. (1940). Lesioni parenchimatose del cervelletto da carcinoma uterino. Rivista di Patologia Nervosa e Mentale 55:148–174.
70. Brain W.R., Daniel P.M., Greenfield J.G. (1951). Subacute cerebellar degeneration and its relation to carcinoma. J. Neurol. Neurosurg. Psychiatry 14:59–75.
71. Brain W.R., Henson R.A. (1958). Neurological syndromes associated with carcinoma: The carcinomatous neuromyopathies. Lancet 2:971–975.
72. Brazis P.W., Biller J., Fine M., Palacious E., Pagano R.J. (1981). Cerebellar degeneration associated with Hodgkin's disease: Computed tomographic correlation and literature review. Arch. Neurol. 38:253–256.
73. Brouwer B., Schlesinger F.G. (1947). Carcinoma ovarii and cerebellar degeneration Koninklijke Nederlandsche Akademie van Wetenschappen. Series C: Biological and Medical Sciences. Proceedings 50:1329–1334.
74. Budde-Steffen C., Anderson N.E., Rosenblum M., Posner J.B. (1988). Expression of an antigen in small cell lung carcinoma lines detected by antibodies from patients with paraneoplastic dorsal root ganglionopathy. Cancer Res. 48:430.
75. Castaigne P., Buge A., Escourolle R., et al. (1964). L'atrophie cerebelleuse paraneoplastique. A propos d'une observation anatomo-clinique. Presse Medicale 72:2639–2644.
76. Castleman B., Towne V.W. (1989). Case records of the Massachusetts General Hospital, Case 40391. N. Engl. J. Med. 251(14):573–578.
77. Currie S., Henson R.A. (1971). Neurological syndromes in the reticuloses. Brain 94:307–320.
78. Dazzi P., Ferrari G. (1970). Sull'atrofia cerebellare paraneoplastica. A proposito di un caso gastrectomizzato per carcinome gastrico. Revisto Sperimentale di Freniatria 94:251–272.
79. Dropcho E.J., Chen Y.-T., Posner J.B., Old L.J. (1987). Cloning of a brain protein identified by autoantibodies from a patient with paraneoplastic cerebellar degeneration. Proc. Natl. Acad. Sci. USA 84:4552–4556.
80. Ellenberger C., Campa J.F., Netsky M.G. (1968). Opsoclonus and parenchymatous degeneration of the cerebellum. Neurology 18:1041–1046.
81. Garcin R., Lapresle J. (1956). Sur un cas d'atrophie cerebelleuse corticale subaigue en relation avec un epithelioma du larynx. Bulletin de la Societe Medicale des Hopitaux de Paris 72:761–769.
82. Greenlee J.E., Brashear H.R., Jaeckle K.A., Stroop W.G. (1986). Anticerebellar antibodies in sera of patients with paraneoplastic cerebellar degeneration studies of antibody specificity and response to plasmapheresis. Ann. Neurol. 17:82–85.
83. Anderson N.E., Rosenblum M.K., Graus F., et al. (1988). Autoantibodies in paraneoplastic syndromes associated with small-cell lung cancer. Neurology 38:1391–1398.
84. Hall D.J., Dyer M.L., Parker J.C. (1985). Ovarian cancer complicated by cerebellar degeneration: A paraneoplastic syndrome. Gyn. Oncol. 21:240–246.
85. Julien J., Vital C., Vallat J.M. (1972). Atrophie cerebelleuse paraneoplastique. Observation anatomoclinique. Bordeaux Medical 5:2461–2467.
86. McDonald W.I. (1961). Cortical cerebellar degeneration with ovarian carcinoma. Neurology 11:329–334.
87. Ang L.C., Zochodne D.W., Ebers G.C., et al. (1986). Severe cerebellar degeneration in a patient with T-cell lymphoma. Acta Neuropath. 69:171–175.
88. Malamud N. (1962). Atlas of Neuropathology. Berkeley, CA: University of California Press, p. 118.
89. Mancall E.L. (1975). Late (aquired) cortical cerebellar atrophy. In: Handbook of Clinical Neurology, Vol. 21. P.J. Vinken, G.W. Bruyn (eds): Amsterdam: North Holland Publishing, pp. 477–508.

90. Messert B., Blume W.G. (1969). Parenchymatous cerebellar degeneration associated with carcinoma of the lung. Wisconsin Med. J. 68:101–107.
91. Meyer M.A. (1984). Immunological similarities between T lymphocytes and Purkinje cells (letter). Ann. Neurol. 16:369.
92. Millefiorini M., Antonini G., Cortesani F., et al. (1980). Cerebellar paraneoplastic degeneration: Neuropathological and biological observations. Acta Neurol. (Napoli) 35:23–29.
93. Missen A.J.B. (1965). A case of carcinomatous encephalomyeloneuropathy. St. Bartholomew's Hospital Reports (London) 69(Suppl. 10):6–11.
94. Missen G.A.K. (1966). Intestinal malignant lymphoma and cerebellar cortical degeneration complicating idiopathic steatorrhoea. Guy's Hosp. Rep. (London) 115:359–385.
95. Monseu G., Vanderhaegen J.J., Stenuit J., et al. (1971). Etude clinique et anatomique de deux cas de degenerescence cerebelleuse subaigue. Acta Neurologica Belgica 71:324–334.
96. Morton D.L., Itabashi H.H., Grimes O.F. (1966). Nonmetastatic neurological complications of bronchogenic carcinoma: The carcinomatous neuromyopathies. J. Thor. Cardiovasc. Surg. 51:14–29.
97. Oelbaum M.H., Statham R. (1961). Carcinoma of the bronchus presenting as a cerebellar neuropathy. Postgrad. Med. J. 37:546–549.
98. Quadfasel F.A., Richardson E.P. (1954). Case records of the Massachusetts General Hospital-Case 40391. N. Engl. J. Med. 251:573–577.
99. Rewcastle N.B. (1963). Subacute cerebellar degeneration with Hodgkin's disease. Arch. Neurol. 9:407–413.
100. Sarbach B. (1977). L'atteinte paraneoplastique du cervelet. Etude anatomoclinique d'un cas porteur d'une atrophie corticale et de lesions inflammatoires des noyaux denteles. Acta Neurol. Belg. 77:363–372.
101. Spencer S.S., Moench J.C. (1980). Progressive and treatable cerebellar ataxia in macroglobulinemia. Neurology 30:536–538.
102. Steven M.M., Carnegie P.R., Mackay I.R., Bhathal P.S., Anderson R.McD. (1982). Cerebellar cortical degeneration with ovarian carcinoma. Postgrad. Med. J. 58:47–51.
103. Takagi H., Kato H., Okubo S., et al. (1968). A case of bronchial carcinoma and pulmonary silicosis associated with cerebellar degeneration. J. Jpn. Soc. Intern. Med. 57:1264–1269.
104. Tsapatsaris N., Wanger S.L., Steinberg D. (1979). Cerebellar degeneration and Hodgkin's disease. Arch. Intern. Med. 139:829–830.
105. Tsukamoto T., Yoshie O., Tada K., Iwasaki Y. (1987). Anti-Purkinje cell antibody producing B-cell lines from a patient with paraneoplastic cerebellar degeneration. Arch. Neurol. 44:833–837.
106. Victor M., Ferendelli J.A. (1970). The nutritional and metabolic diseases of the cerebellum. Clinical and pathological aspects. In: The Cerebellum in Health and Disease. W.S. Fields, W.D. Willis (eds): St. Louis, Mi Green, chap. 16.
107. Balla J.I.O. (1968). Cerebellar degeneration in association with carcinoma of the stomach. Br. Med. J. 1:34.
108. Zulch K.J. (1948). Uber die anatomische Stellung der Kleinhirnrindenatrophie und ihre Beziehnung zur Nonne-Marieschen Krankheit. Deutsche Zeitschrift fur Nervenheilkunde 159:501–513.

INDEX

Abetalipoproteinemia
  clinical features of, 190
  oculomotor deficits of, 291–292
Accessory olive, 169
Acetylcholine, 26, 90, 398
Acetylcholinesterase, 17, 26, 108, 148
3-Acetylpyridine, 6, 31, 65, 91–93, 370
Adenosine, 95, 106–107
  receptors of, 66, 106–107
Adrenergic receptors, 106, 148
Adrenoleukodystrophy, 191–192
Amino acid neurotransmitters, 3, 4, 89–121
AMPA/KA receptors, 64, 65, 96–97, 112, 114, 127, 134
  cloning of, 97
  subunits of, 97
Arachidonic acid, 94
Aspartate, 31, 65, 91, 124, 147
Ataxia telengiectasia
  B-cell differentiation and, 463
  B-cell receptor genes and, 463
  chromosomal translocations and, 464
  classification of, 189–190
  clinical features of, 189, 190, 461, 466
  cytogenetic abnormalities in, 461
  developmental neuroimmune defects in, 466

DNA repair in, 461–462
DNA synthesis in, 461
genetic recombination and, 469
genes for T-cell receptors and, 463–465
immunodeficiency in, 461–465
immunotherapy for, 465
immunoglobulin superfamily and, 466–468
lymphocyte defects in, 464–465
neoplasias and, 461–464
neuropathologic changes of, 466
Nijmegen breakage syndrome and, 461–462
oculomotor deficits in, 296
Purkinje cell loss in, 232
sinopulmonary infections in, 465
T-cell mediated immunity in, 462–463
T-cell receptors and, 463
thymic hypoplasia in, 465
Ataxias, *see* Cerebellar degenerations
Ataxic encephalopathies, treatable, 187–188

b-N-methyl-amino alanine, 98
Baclofen, 103, 112
Bechterew nucleus, 44
Bergmann glial cell, 38, 208

Bicuculline, 102–103
Bodian stain, 206–208

Ca^{2+}-activated ATPase, 109
Ca^{2+}-calmodulin, 100, 108
Ca^{2+}-conductance, 109
Calcineurin, 109
Calcium-binding protein, 109, 162–163
CaM kinase II, 101, 109, 112, 115
Canal-ocular reflex, 72
Caudate, 142, 148
Central gray matter, 142
Cerebellar
  afferent projections, 42–45, 91–96
  antigens, 108
  basket cells, 30, 35–37, 64, 65, 70, 71, 89, 109, 134
    in cerebellar degeneration, 206–224
  climbing fibers, 66, 69, 107–108, 132
    neurotransmitter of, 61, 65, 91–95
  development, 90, 91, 114, 115
  dorsal vermis and eye movements, 286–288
  flocculonodular lobe, 13, 44, 45, 69
  fusiform neurons, 64
  glomerulus, 21, 25, 110
  granular layer, 21, 130, 142
    neurotransmitters in, 26–29, 95, 96
  granule cells, 21, 61, 69, 89, 90, 96, 101, 103–107, 111
  lobules of, 308–331
  molecular layer, 21, 30, 35, 130
  mossy fibers, 64, 69, 95–96, 104, 108, 132
    neurotransmitter of, 26–29, 95, 96
  mutants, 103, 105–106, 130–135, 159–181
  ontogeny, 14
  parallel fibers, 61, 104–107, 132
    neurotransmitters of, 30–35
  paravermis, 284
  stellate cells, 64, 65, 70, 71, 89, 109, 134, 206–224
  vermis, 13, 72, 284
Cerebellar anatomy, 12–57
  afferent pathways, 43–45
  arbor vitae, 21
  archicerebellum, 14
  climbing fibers, 16, 20, 30–32, 40–44
  connectivity, 42–45
  corpus cerebelli, 13
  cortex, 20–38

  corticonuclear projections, 42–43, 71
  efferent projections, 43, 45
  folia, 21
  glial cells, 38
  glomerulus, 23–26, 28
  Golgi neurons, 23–36
  granular layer, 21–29
  hemispheres, 13–15
  longitudinal zonal organization, 14–18, 28, 29
  module, 17
  molecular layer, 29–38
  mossy fibers, 16, 23, 25, 40
  neocerebellum, 14
  nuclei, 38–42
  nucleocortical projections, 42–43
  olivocerebellar tract, 44
  parallel fibers, 16, 23, 33–35
  peduncles, 14, 16, 42–45
  pontocerebellar tract, 44
  pontocerebellum, 14
  posteriolateral fissure, 13
  primary fissure, 13
  Purkinje cell, 16, 22–24, 29–36, 39–40
  reticulocerebellar input, 44–45
  somatotopic organization, 19, 20
  stellate cells, 30, 35, 36
  synaptic organization, 20–45
  vermis, 13–15, 17, 41, 42
  vestibulocerebellum, 14
  white matter, 17, 21
    myeloarchitectural organization of, 17
Cerebellar ataxia
  abnormalities of force in, 266
  dysdiadochokinesia in, 266
  dysmetria in, 263–265
  dyssynergia in, 266–267
  electromyography in, 262–266
  hyporeflexia in, 272
  hypotonia in, 272
  motor programming disturbances in, 268–269
  movement termination disorders in, 263–265
  pathophysiology of, 261–279
  reaction times in, 262–263
  time perception disturbances in, 271–272
  tremor in, 249, 272–273
Cerebellar control of
  compound reactions, 75
  eye movements, 75, 76, 281
  learning, 81, 82

limb movement, 78, 79
locomotion, 76
mental function, 79
posture, 76
reaction, 79, 80
reflex, 79, 80
speech, 79
Cerebellar cortex
  anatomy of, 20–38
  electron microscopy of, 230–231
  histological stain of, 206
  in hereditary ataxias, 206–224, 230–231
  neuropathologic changes of, 206–224
Cerebellar degeneration
  cholinergic systems in, 362
  chronic alcoholism and, 273
  classification of, 185–204
  clinical features of, 305–365
  clinicopathologic correlations in, 357
  control of posture in, 273–276
  Dennmark kindred, 428
  diagnosis of, 198–199
  epidemiology of, 199
  genetic heterogeneity of, 392
  Houston kindred, 428, 433–435
  in Cuba, 199, 428, 435
  in Eastern Siberia, 199
  in Norway, 199
  Italian kindred, 428
  Japanese kindred, 435
  Louisiana kindred, 428
  Michingan kindred, 428, 433–435
  Minnesota, 428, 433–435
  Mississippi kindred, 428, 433–435
  mitochondrial abnormalities in, 391–401
  Nebraska kindred, 428, 435
  oculomotor abnormalities in, 281–303
  oxidative metabolism in, 396–398
  pathogenesis of, 231–233, 375–360
  peripheral neuropathy in, 345
  postural ataxia in, 273–276
  radiologic features of, 305–365
  recessive form, 343
Cerebellar physiology, 59–87
  cell membrane
    action potentials of, 60, 61
    Ca^{2+} channels of, 61
    K^+ channels of, 61
    Na^+ channels of, 61
    plateau potentials of, 61
    properties of, 60
  cellular processes, 60

signal processing, 68, 69
synaptic modulation, 65, 66
synaptic plasticity, 66–68
synaptic transmission, 64, 65
Cerebellin, 30
Cerebello-olivary atrophy
  autosomal dominant, 323
  autosomal recessive, 323
  classification of, 193
  clinical features of, 323–325
  hypogonadism and, 323
  neuroimaging abnormalities of, 325–327, 356
  neuropathologic findings of, 325
  oculomotor deficits of, 292, 323–324
Cerebro-cerebellar interactions, 77, 81
Cetacea cerebellum, 13
cGMP, 67, 93, 100–101
cGMP-activated protein kinase, 19, 30, 67, 94, 108, 162, 211
  in cerebellar degenerations, 218–219
Chloride channels, 65, 102
Chloride ions and glutamate binding, 124–132
Cholinacetyltransferase, 26, 113, 148
Citrate synthase, 394–397
Congenital cerebellar hypoplasia, 296–297
Cerebral cortex
  endorhinal, 142, 148
  primary olfactory, 142, 148
Cortical cerebellar degeneration, see cerebello-olivary atrophy
Corticotropin-releasing factor, 31, 91
Cuneocerebellar projections, 43
Cysteine sulfonic acid decarboxylase, 18, 108
Cytochrome C oxidase, 139, 142–143, 146–151
Cytotoxic lymphocytes, 468–469

Deafness
  in Refsum's disease, 190, 191
  in dégénérescénce systématisee optico-cochléo-dentelée, 191
  in olivopontocerebellar atrophy, 192, 197
Deiters nucleus, 30, 41, 42, 108
Dementia
  in hexosaminidase deficiency, 192
  in Gerstmann-Staüssler syndrome, 191
  in Jakob-Kreutzfeldt disease, 191
  in Marinesco-Sjogren syndrome, 192
Dendritic spikes, 109

Dentate nucleus, 38, 39, 42
  in cerebellar degenerations, 221–230
  in dyssynergia cerebellaris myoclonica, 224–225
  in Friedreich's ataxia, 225–229
  in olivopontocerebellar atrophy, 224–228
Desmosomes, 37
Dextromethorphane, 97
Dihydrolipoyl dehydrogenase, 139
Dihydrolipoyl transacetylase, 139
Dopaminergic neurons, 148
Dyssynergia cerebellaris myoclonica, 191

EDRF, 101
Electromyography in OPCA, 239–240
Enkephalin, 31, 91, 107
Excitatory amino acids
  in cerebellum, 91–115
  receptors of, 64, 92–94, 96–98, 104, 106
    in human cerebellum, 123–137
  sulfur-containing, 94
  toxic effects of, 3, 4, 113–115
  transduction systems of, 98
  trophic effects of, 3, 4, 112–115
Excitatory interneurons, 103–107
Excitotoxicity, 113–115, 356–360
Eye movements
  burst cells and, 286–287
  cerebellar ablation and, 281–288
  cerebellar flocculus and, 281–285
  cerebellar nodulus and, 285–286
  cerebellar paraflocculus and, 281–285
  dysmetric saccades, 288
  gaze holding abnormalities, 282
  integrator time constant and, 284
  ocular flutter, 286, 288
  opsoclonus, 286, 288
  optokinetic nystagmus, 282–283
  postsaccadic drift, 288
  pulse cells and, 286–287
  pulse innervation and, 286
  saccades, 282–297
  saccadomania, 288
  smooth pursuit, 282–297
  uvulla and, 285, 286
  vestibuloocular reflex, 283–285
Eye-blink reflex, 74, 75

Friedreich's ataxia
  age at onset, 311

DNA markers and, 319–320
interferon-b gene and, 319
auditory system in, 313
autonomic dysfunction in, 313
biochemical abnormalities in, 444–445
cardiac involvement in, 312
cerebellar changes in, 315–318
classification of, 189
neuroimaging in, 317–318, 356, 445
endocrine system in, 314–315
excitatory amino acids and, 448–449
GABA/benzodiazepine receptors and, 445
gene linkage in, 319–321
genetic defect of, 318–321
glucose hypermetabolism in, 446–449
local cerebral metabolic rate in, 446–449
medulla oblongata in, 317
muscarinic cholinergic receptors in, 445
neuroimaging findings of, 317–318
neurologic features of, 307, 311–312, 444
oculomotor abnormalities in, 291
pathologic findings of, 315–317, 444
positron emission tomography in, 444–449
sensory system in, 314
skeletal abnormalities in, 311–313
spinal cord atrophy in, 315–318
vermian atrophy in, 315–318
visual system in, 313–314
Fastigial pressor response, 244
Fumarase, 395–397
Foix-Alajouanine cerebellar atrophy, 293

G-protein, 65, 68, 98, 103, 107
GABA, 24, 25, 27–30, 36, 37, 39, 43, 89, 102–103, 107–112
  trophic effects of, 111–112
GABA/benzodiazepine receptors, 28, 36, 65, 102–103, 107–111
Genetic linkage
  GT-repeat markers and, 435–436
  in cerebellar degeneration, 433–438
  location score in, 433
  lod score in, 432
  map distance in centiMorgans and, 431–432
  methods of, 431
  multipoint analysis and, 433
  polymorphic markers and, 430–431
  rationale for, 429–431
  recombination hot spots and, 431–432

theta value and, 431–432
Gerstmann-Straussler syndrome, 191
Glial cell, 38
  in cerebellar degenerations, 221–224
Glial fibrillary protein, 38, 140, 143, 164–165
Glu R5 gene, 97
Glucose-6-phosphate dehydrogenase, 372
Glutamate binding
  in atrophic human cerebellum, 123–135
  in normal human cerebellum, 124–135
Glutamate dehydrogenase
  brain distribution of, 139–157
  characterization of, 379–381
  chromosomal mapping of, 385–387
  glutamatergic pathways and, 143
  immunocytochemistry of, 139–157, 358–360
  immunogold labeling of, 140–141
  isoforms of, 372–374, 379–380
  mitochondrial localization of, 140–142, 393
  purification of, 378–379
  -specific cDNAs, 387
  -specific mRNA, 387
Glutamate dehydrogenase deficiency
  brain biochemical changes in, 378–381
  clinical features of, 370–376
  genetics of, 384–387
  glutamate clearance in, 381
  glutamate metabolism in, 381
  neuropathology of, 376–378
  pathophysiology of, 381–383
Glutamate, 26, 27, 35, 39, 43, 64, 66, 89, 103–106
  receptors, see excitatory amino acid receptors
  neurotoxicity in cerebellar diseases, 99, 110, 151, 358–360, 369–370
Glutamic acid decarboxylase, 29, 39, 107, 109, 112, 113
  in hereditary ataxia, 210–211, 226–227
Glutamic acid pools, 143
Glutaminase, 113
Glutamine synthetase, 148
Glutathione reductase, 372
Glycine, 28, 106, 110–111
Golgi cells, 69, 77, 89, 90, 109–111, 134, 205–208
  in cerebellar degenerations, 206–224
Golgi impregnation, 208–210
Guinoxaline diones, 98

Harmaline, 93
Hereditary ataxias, see cerebellar degenerations
Herophilus, 1
Hexosaminidase deficiency, 191–192
  adult-onset form of, 405
  asymptomatic form of, 408
  cerebellar ataxia in, 406, 412–413
  clinical features of, 321–322, 403–410
  cytoplasmic inclusions in, 322
  biochemical abnormalities of, 410
  dystonia and, 407
  enzymatic abnormalities of, 410
  glycosaminoglycan storage in, 410–411
  G_{M1}-gangliosidosis in, 414
  G_{M2}-ganglioside storage in, 410–411, 414–416
  neuronal involvement in, 411, 412, 414–415
  in infantile Sandhoff disease, 409
  in infantile Tay-Sachs disease, 408–409
  infantile form of, 404
  juvenile form of, 405
  late-infantile form of, 405
  macular cherry-red spot in, 406–410, 412
  membranous cytoplasmic bodies in, 411
  motor neuron disease in, 407
  mRNA abnormalities in, 411
  neuroimaging changes in, 322, 356
  other types of, 409–410
  pathology of, 408–410
Hexosaminidase
  activator protein, 410–411
  cDNA, 411
  developmental regulation of, 416–417
  gene organization of, 411
Hippocampal formation, 143, 148
"Nervous" cerebellar mutant, 130–133
Histamine, 103
History of cerebellar research, 1, 2, 59, 60
Holmes cerebellar atrophy, see cerebello-olivary atrophy
Homocystate, 65, 124

Ibotenate, 124
Isocitrate dehydrogenase, 372
Interleukin-2, 468–469
  in Purkinje cell mutant, 168–169
Inhibitory interneurons, 109
Interleukin-1, 468–469
Ionotrophic glutamate receptors, 112

Inositol triphosphate (IP₃)

Jakob-Kreutzfeld disease, 191
Joubert's syndrome, 297

Kainic acid, 65
  binding of, 124, 127
Kearns-Sayre syndrome, 397
ketoglutarate dehydrogenase, 394

Lectin affinity cytochemistry, 210
Lipoamine dehydrogenase, 394, 399
Locus coeruleus, 148, 246
Long-latency responses, 274
Long-term depression, 66, 70, 71-75, 93, 107, 113
Long-term potentiation, 113
"Lurcher" cerebellar mutant, 166
Lugano cells, 215-217

Machado-Joseph disease
  linkage studies in, 435
  oculomotor deficits of, 293, 427
Marie's ataxia, 186
Malate enzyme, 393-394
Malate dehydrogenase, 395-397
Mitochondria abnormalities
  in cerebellar degeneration, 391-401
  in Friedreich's ataxia, 392-401
  in olivopontocerebellar atrophy, 231, 392-401
  pathophysiology of, 396-398
Marinesco-Sjogren syndrome, 192
Maturational stages of neurons, 114
MERFF, 394, 397
Metabotropic receptors, 65, 67, 98, 99
Methyl-4-phenyl-1,2,3,6-tetrahydropiridine, 6
Microtubule-associated protein, 211-214, 227
Migration of neurons, 114
MK-801, 97
Motilin, 30, 107
Multiple system atrophy, 151, 185, 246, 398, 449
Muscimol, 108
Myoclonus in

dyssynergia cerebellaris myoclonica, 191
JaKob-Kreutzfeldt disease, 191
olivopontocerebellar atrophy, 191, 249-250
dégénérescénce systématisee optico-cochléo-dentelée, 191

N-acetylaspartylglutamate, 91
N-CAM, 467
NMDA binding, 124, 127, 129, 133-135
NMDA receptor, 94-97, 100-101, 104, 106, 108-115, 143
  cloning of, 97
  purification of, 97
NADPH-diaphorase, 148
Nerve growth factor, 113, 468
Neurofilament proteins, 211, 215-218, 227-229
Neuron-specific enolase, 211, 214-215, 227
Nitric acid, 94, 101
Nociceptive signals, 70
Non-neuronal enolase, 211, 221-224
Noradrenaline, 45, 65, 89, 148
Nuclei
  dorsolateral pontine, 284
  external cuneate, 143
  hypoglossal, 147
  lateral vestibular, 148
  medial vestibular, 284
  motor of the trigeminal nerve, 148
  paramedian reticular, 45, 147
  perihypoglossal, 285
  pontine, 147
  sensory relay, 147
  septal, 143, 148
  solitary tracts, 143
  thalamic, 143
  vestibular, 285
5'-Nucleosidase, 17, 108
Nucleus
  ambiguus, 142
  basalis of Meynert, 142, 151
  dorsal motor of vagus, 142, 148
  emboliformis, 39, 42
  fastigius, 42, 44, 45
  globosus, 39, 42
  interposiyus, 39, 42
  lateral reticular, 142
  medialis, 39
  of the optic tract, 285
  prepositus hypoglossi, 72, 284-285

raphe, 45
rostal fastigial, 244–245
Nystagmus, 285
    caloric, 288
    centripetal, 288
    divergent, 288
    downbeat, 283
    optokinetic, 288
    periodic alternating, 285–286
    positional, 288
    of primary position, 288
    rebound, 283
    upbeat, 288

Olivo-cerebellar fibers, 66, 69, 91–95, 107–108, 132
Olivocerebellar tract, 44
Olivopontocerebellar atrophy, 31
    abnormal movements in, 249–251
    amyotrophy in, 340
    animal models of, 370
    autonomic neurophysiology of, 243
    ballistic movements in, 251
    baroreceptor neurons and, 248
    blindness in, 337–340
    brain stem evoked response in, 241
    cardiovascular reflex responses in, 245
    choline acetyltransferase in, 451
    classification of, 191, 193–198, 327, 426–429
    clinical features of, 341
    clinical neurophysiology of, 237–259
    Dejerine Thomas type, 327, 353–354
    dementia in, 340, 450
    dorsal motor nucleus of vagus in, 246
    dysautonomia in, 246–249, 347–349
    dystonia in, 250–251
    fastigial nucleus and, 248
    GABA/benzodiazepine receptors in, 451
    glutamate dehydrogenase deficiency in, 345, 356, 369–389
    glucose hypometabolism in, 454–456
    glutamate receptors in, 123–130, 383–384, 451
    glutamate toxicity in, 151, 357–360, 381–383
    inferior olives in, 248
    intermediolateral columns in, 248
    involuntary movements in, 251–252
    late-onset, *see* sporadic
    lateral reticular nucleus of medulla in, 248
    local cerebral glucose metabolism in, 454–456
    locus coeruleus in, 248
    myoclonus in, 340
    neuroimaging in, 332–356
    noradrenaline in, 451
    nucleus ambiguus in, 246
    oculomotor deficits in, 293–296, 333
    orthostatic hypotension in, 245–249, 347
    parasympathetic pathways in, 245–248
    Parkinsonism in, 353–355
    pathology of, 332, 335, 340, 348, 353, 354, 450
    Pattern-reversed visual evoked response in, 241–243
    plasma renin in, 245
    positron emission tomography in, 449–456
    pupillary changes in, 347–348
    recessive, 341–345, 356
    REM sleep in, 252–254
    substantia nigra in, 248
    seizures in, 340
    sleep apnea in, 252–254
    sleep disturbances in, 252–254
    sphincter dysfunction in, 246
    sympathetic pathways in, 245
    thyrotropin-releasing hormone and, 451
    tremor in, 251
    Valsava maneuver in, 245–248
Olivopontocerebellar atrophy sporadic
    classification of, 197
    clinical features of, 345–347, 352–353
    neuroimaging abnormalities of, 354–355
Olivopontocerebellar atrophy dominant
    classification of,
    clinical features of,
    gene locus of, 426–429
    Haymaker-Schut type, *see* Schut-Haymaker
    HLA-linked, 425
    linkage studies in, 429–438
    linkage to chromosome 6p, 425–441
    Menzel's type, 195, 327–333
    neuroimaging findings in
    ophthalmoplegia in, 340
    optic atrophy in, 353
    progressive bulbar palsy in, 340
    Schut-Haymaker type, 329, 332–333, 426–429
    Schut-Swier, *see* Schut-Haymaker
    with deafness, 197

with myoclonus, 251–252
with neuroretinal degeneration, 197, 337–341, 356
with peripheral neuropathy, 195–196, 239, 339–340
with slowed eye movements, 195–196, 339–340, 356
Olivopontocerebellar atrophy with dysautonomia
classification of,
clinical features of, 347–348
neuroimaging changes in, 349–352, 356
neuropathology of, 348–349
Optokinetic response, 74

Paraneoplastic cerebellar degeneration, 475–493
anti-Hu antibodies in, 485–487
anti-Purkinje cell antibodies in, 476, 482–489
anti-Yo in, 476–479, 482–489
basal ganglia degeneration in, 478
clinical manifestations of, 477–479
computerized tomography in, 489
CSF abnormalities in,
differential diagnosis of, 490
epidemiology of, 477
immunoblotting in, 483
immunohistochemical staining in, 482–483
inflammatory reaction in, 479
Lambert-Eaton myasthenic syndrome and, 489
magnetic resonance in, 489
pathogenesis of, 482–489
pathology of, 478–482
prognosis of, 491–492
Purkinje cell degeneration in, 478–482
Purkinje cell antigens in, 476
spinal cord degeneration in, 479
systemic malignancies and, 475–476
treatment of, 490–492
Parkinson's disease and GDH deficiency, 372
Pericellular basket, 36
Phospholipase C, 99
Physiology of the cerebellum, 59–87
Plasticity of neurons, 115, 123
"Reverse genetics," 5, 429
Polymerase chain reaction, 436
Polypeptide PEP-19, 162
Pontocerebellar tract, 444

Pontocerebellum, 14
Positron emission tomography
GABA/benzodiazepine receptors and, 455–456
in cerebellar degenerations, 443–459
in Friedreich's ataxia, 444–449
Protein kinase, 19
Protein phosphorylation, 99, 101, 102
Pyruvate clearance, 393
Pyruvate dehydrogenase, 392–394
Purkinje cell, 60, 61, 66, 69–77, 90, 91, 107–109, 130–134
bands of, 16–19, 30
degeneration mutants, 159–181
amino acid transmitters in, 175
cerebellar basket cells in, 167
cerebellar granule cells in, 166–167
cerebellar nuclei in, 173
cGMP-activated protein kinase, 177
GABA in, 174–175
glial fibrillary protein in, 164–166
glycolipids in, 178
inferior olive in, 168, 169
noradrenaline in, 169–171, 176
serotonin in, 171
trophic factors and, 179
dendrites of, 30–33, 61, 160
dendritic pathology of, 206–224
development of, 160
immunolabeling of, 143, 148
neuropathologic changes of, 206–233
neurotransmitters of, 30–32
-specific glycoprotein, 108
synaptogenesis of, 160
torpedoes,
Putamen, 142
Pyruvate dehydrogenase complex, 139–157

QNB binding, 126
Quisqualate
binding of, 124, 127, 132–134
receptor, 66, 96, 104, 114, 451

Refsum's disease, 190–191
Restform body, 43
Reticulocerebellar connections, 44
Retinitis pigmentosa
in abetalipoproteinemia, 190
in Refsum's disease, 190
in vitamine E deficiency, 190

Ricinus communis agglutinin I, 210, 224, 230
Rolando, Luigi, 1

Saccadic eye movements, 75
Septal nuclei, 143
Serotonin, 29, 45, 65, 89, 103
Shy-Drager syndrome, 246–249, 449
"Staggerer" cerebellar mutant, 109, 130–133
Smooth-pursuit eye movements, 77, 78
Somatosensory evoked response (SEP) in OPCA, 240–241
Somatostatin, 26, 107, 148
Spino-pontine atrophy, 197–198
Spinocerebellar degenerations, *see* cerebellar degenerations
Spinocerebellar pathways, 43
Spinocerebellum, 14
Substance P, 26
Succinate dehydrogenase, 395–397
Synaptic
  plasticity, 66–67
  potentials, 60, 64
  protein (synaptophysin), 211, 219–220, 229
  transmission, 64–66

Taurine, 18, 30, 91–93, 107–109
Thalamic nuclei, 143
Theory of cerebellar control, 79–82
Trans-synaptic degeneration, 160
Transcription factors, 115
Tremor classification, 240
Trigeminal nucleus, 19

Valine dehydrogenase, 395–397
Vermis agenesis in Joubert's syndrome, 297
Vestibular nuclei, 41, 45, 65
Vestibulo-ocular reflex, 42, 72–75, 79
Vestibulocerebellar pathways, 44
Vestibulocerebellum, 14, 41
Vestibulocolic reflex, 72
Visuopostural loop, 274
Voogd's lobules, 45

Weaver mutant mouse, 115

Xeroderma pigmentosum, 190

Zebrins, *see* cerebellar antigens

CPSIA information can be obtained
at www.ICGtesting.com
Printed in the USA
LVHW081255310522
720127LV00003B/8